Praise for *Anthracite Labor Wars*

Anthracite Labor Wars *is a remarkable book. In rich and engaging detail, it adds immeasurably to our understanding of the turbulent and often tragic history of labor relations in the dark and bloody ground of Pennsylvania's hard coal region. In bringing to the fore the critical role played by Italian Americans in the industry, in the coal mining communities of northeastern Pennsylvania, and in the struggle for workers' rights, the authors have made a major contribution to our understanding of Progressive Era labor history and beyond. Copiously illustrated and based on fresh examination of a wide range of primary sources,* Anthracite Labor Wars *also insightfully links worksite conflict to broad, international patterns of labor migration. It is an outstanding achievement.*

—Robert H. Zieger, University of Florida, author of *John L. Lewis* (1988) and *The CIO* (1995)

The authors aptly demonstrate the importance of thick description for deeper sociological understanding of the intricate, often hidden, relations between management and labor. In uncovering links between hard coal extraction practices, ethnicity, and organized crime, they provide a valuable template for looking at other organizations at other times and places. In the process, they also make clear the impact of local and political processes that temper and sometimes deny the legitimate aspirations of Italian and other workers. As a bonus, their well-wrought descriptions and analyses are visually enhanced by a diverse collection of images ranging from geological maps to historical photographs.

—Jerome Krase, CUNY–Brooklyn College, co-author of *The Status of Interpretation in Italian American Studies* (2012)

Anthracite Labor Wars *is a major contribution to twentieth-century U.S. labor and immigration history. Wolensky and Hastie have illuminated a forgotten chapter in coalmining labor wars and reminded historians of the radicalness of rank-and-file miners' struggles with owners, their own union, and organized crime.*

—Thomas Dublin, Binghamton University–SUNY, author of *When the Mines Closed* (1998) and co-author of *The Face of Decline* (2005)

This meticulously researched study of Pennsylvania anthracite history is an impressive achievement. By making excellent use of the best primary and secondary sources available, the authors break new ground in the fields of labor, corporate, ethnic, and Pennsylvania history. Before this book, there was no comprehensive examination of the history of violence in anthracite during the first half of the 20th century. The volume can take its place among the half dozen or so major works on northeastern Pennsylvania's story. It is highly recommended for all readers interested in the history of one of America's most important industrial regions.

—Walter T. Howard, Bloomsburg University of Pennsylvania, author of *Anthracite Reds* (2004) and *Anthracite Radicals* (2005)

ANTHRACITE LABOR WARS

Tenancy, Italians, and Organized Crime in the Northern Coalfield of Northeastern Pennsylvania, 1895–1959

Robert P. Wolensky 1-22-16

Robert P. Wolensky

and

William A. Hastie Sr.

Published by Canal History and Technology Press
National Canal Museum
2750 Hugh Moore Park Road
Easton, PA 18042

www.canals.org

in association with the Smithsonian Institution

ISBN 978–0–930973–42–1

Library of Congress Control Number: 2012956541

Cover images:

Front: Pennsylvania Coal Company's No. 6 Colliery (background);
labor leader Rinaldo Cappellini; feature story from *The Scranton Republican;*
official logo of the Erie Railroad
Back: *Il Quarto Stato* (*The Fourth Estate*) by Giuseppe Pellizza da Volpedo, 1901

Canal History and Technology Press

National Canal Museum
Easton, PA 18042

DEDICATION

To the tens of thousands of anthracite mineworkers
who fought the battles for economic and social fairness,

especially
those who sacrificed their lives.

And also to

those who continue to keep anthracite history alive,

and to

Molly Hagan Wolensky
for 42 years of love, happiness, and support.

And to

the late Betty Jane Hastie and Megan Hastie
for their enduring love and support.

We are working for democracy, for humanity, for the future, for the day that will come too late for us to see it or know it or receive its benefits, but which will come, and will remember our struggles, our triumphs, our defeats, and the words which we speak.

— Clarence Darrow, 1902

Thirty-six per cent, or more than one-third, of all the workers engaged in Number Six Colliery come from Italy, almost entirely from Sicily. To them the [sub]contract system is a symbol of their old bondage, of the years when they were forced to work in the mine for wages and under conditions against which they were helpless to protest.

— Ben Selekman, "Miners and Murders," 1928

The anthracite miner is a peculiar creature. As an individual he is unknown. Only collectively does he make his presence felt.

—George Korson, *Minstrels of the Mine Patch*, 1938

Subcontracting practices readily undermined both stints [worker production quotas] and the mutualistic ethic [among workers] . . . and they tended to flood many trades with trained, or semi-trained, workers who undercut wages and work standards.

—David Montgomery, *Workers' Control in America*, 1979

History is a house of many mansions and its narratives change over time.

—Raphael Samuels, "Heroes below the Hooves of History," 1989

This, after all, is another function of oral history: to bring into the vision of history aspects of experience that have been ignored and left out, and at the same time to challenge and stimulate the historical self-awareness of the people we interview.

—Allesandro Portelli, "What Makes Oral History Different," 2009.

ACKNOWLEDGMENTS

A FIFTEEN-YEAR RESEARCH PROJECT manages to accrue countless debts of gratitude. No doubt we will have missed some persons or institutions without whose assistance the research would have suffered; however, with that caveat in mind, the authors would like to offer the following thanks.

For ideas, consultation and inspiration we would like to acknowledge Mike Auktakalnis, Ed Ackerman, Roger Beatty, Eric Bella, Bill Best, Fred Brutko, Audrey and Tom Calvey, Tom Dublin, Erica Funke, Ben Jones, John McKeown, John "Fiddler" Mikulski, Phil Mosley, Sandra and John Panzitta, Ed Philbin, Connie Richards, the late Ellis W. Roberts, Josephine Slusser, Thomas Supey Jr., John "Lefty" Usefara, Phil Voystock, the late Carol Wolensky, Jack Wolensky, Ken Wolensky, and Mary Ann Yockers.

For assistance with research materials, we would like to thank the late Harold Aurand, Eric Bella, Evelyn and Kate Gibbons, the late William Graham, the late Joe Keating, Barry Kernfield, Michael Knies, Michael Korb, Chester Kulesa, Michael Kuhns, Charles Kumpas, Joseph Panzitta, Zach Petroski, David Philbin, Ed Philbin, Mary Portelli, Barbara Powell, Jim Quigel, Margaret Reese, and Robert Sable.

We are especially indebted to the many individuals listed in the References as having contributed oral histories. Much of the information included in this volume could have been obtained only through oral memoirs. Louis Casterline, Joe Costa, and Joe Stella were especially helpful in providing "inside" information about the operation of the Erie Coal Companies and their tenancy systems.

For most welcomed criticisms and suggestions on the entire manuscript, or on single chapter drafts, we express our deep gratitude to Megan Hastie, Ann Larsen, Jack Larsen, George Turner, James Guthrie, Frank Adams, Steven Colatrella, Victor Greene, Edward Miller, Ronald Slusser, Patrick Lynch, Allan Murray, and Robert Enright.

The study would not have prospered without the documents and/or images provided by the National Canal Museum Archives, the Pennsylvania State Archives, the Pennsylvania Historical and Museum Commission, Indiana University of Pennsylvania Archives, Historical Collection and Labor Archives at Penn State University, the Anthracite Heritage Museum, the Lackawanna Historical Society, the Luzerne County Historical Society, the Osterhout Memorial Library, the Corgan Memorial Library at King's College, the Weinberg Memorial Library at the University of Scranton, the Pittston Memorial Library, the British Library, Pagnotti Enterprises, the Office of Surface Mining in Wilkes-Barre, the Albright Public Library, the United Mine Workers of America, the Historical Society of Wisconsin, the Library of Congress, and four northeastern Pennsylvania newspapers—the *Citizens' Voice*, *Scranton Times*, *Pittston Dispatch*, and *Times Leader*.

A special thanks to Allyssa Burch, who spent countless hours improving the quality of most of the images in the book. Walter Howard deserves special thanks for allowing us to use his extensive northeastern Pennsylvania newspaper files that date from the 1920s and 1930s. Many thanks also to John Dziak for loaning microfilm copies of the *Pittston*

Gazette, and to Patrick Lynch for loaning research items related to his study of the Industrial Workers of the World.

For assistance with securing images we are very grateful to Ann Bartholomew, Bill Best, Tony Brooks, Jim Burke, Joe Butkiewicz, John Dziak, Patricia Eide, Andy Ewonishon, Pauline Gallagher, Megan Hastie, Sam Lucchino, Sal Mecca, Lance Metz, Tom Mooney, Mary Ann Moran-Savakinus, Carl Orechovsky, F. Charles Petrillo, Zach Petroski, Mary Portelli, Betty Gugger Rose, Richard Stanislaus, Harrison Wick, and Chuck Yungkurth.

George Turner, the late Fr. Robert Cornell, the late Harold Aurand, and Victor Greene provided generous consultation on various aspects of anthracite labor history. James Guthrie and Frank Adams supplied vitally important information and knowledge about the history of the Erie Railroad, PaCC, and HC&I. The late federal Judge Max Rosenn offered consultation on legal questions in mining law, as did Attorney Hopkin T. Rolands. We benefitted greatly from the excellent work of Frank Adams and his website www.northernfieldinfo.com concerning the northern field's many coal companies and mines.

The University of Wisconsin-Stevens Point provided various types of support that facilitated the research and writing. Here we would like to thank the Department of Sociology, the Center for the Small City, the College of Letters and Science, the University Personnel Development Committee, the Learning Resources Center, and Interlibrary Loan; as well as several persons at the university: Colleen Angel, Lin Vogel, Allyssa Burch, Beth Bradley, Zachary Hudson, Shannon Duenow, Samantha Schmid, Rachel Whilemi, Justin Nooker, Matt Oldenburg, Nicole Pechinski, Lauren Nischan, Ashley Flogel, and Rachel Gaetz.

At Kings College, Fr. Thomas O'Hara, Fr. Patrick Sullivan, and Nick Holidick provided a home away from home for the first author as well as opportunities to present parts of our research to local audiences. The Institute for Research in the Humanities at the University of Wisconsin-Madison provided a fellowship that helped shape the direction of the study in its early stages; here we would like to acknowledge the late Paul Boyer, David Lindberg, and Loretta Freiling. The University of Wisconsin-Milwaukee's Center for 21st Century Studies offered three fellowships where the first author conducted research and drafted certain chapters; at that institute we would like to acknowledge Dan Sherman, Merry Wiesner-Hanks, Kate Kramer, and John Blum. The British Library, London, provided space and research materials where the first author completed a full draft of the manuscript.

The volume would certainly not have been possible without the Canal History and Technology Press and three persons associated with it. We are immensely grateful to Ann Bartholomew for her exceptional editorial, graphic, and layout work. A true professional, Ann has provided over 30 years of outstanding service to the Press. Lance Metz, retired National Canal Museum Historian, offered not only a forum where we presented our ideas at an annual symposium, but strong encouragement and direction regarding publications. Tom Stoneback, executive director of the National Canal Museum and the Press, supported and endorsed this volume from the beginning, even during some difficult economic times.

Our families deserve a special thanks. Molly Wolensky endured regular periods of "research widow-hood"—both within and outside of Stevens Point—and for her patience and understanding, Bob will be forever grateful. A native of the Wyoming Valley, her ancestors mined coal in northern England and worked on the railroads that transported anthracite in northeastern Pennsylvania so this is, in part, her story as well. Both authors are especially grateful to Megan Hastie who assisted with archival and genealogical research, read the manuscript at various stages, provided important editorial suggestions, and secured and improved upon some of the images. Her efforts were vital to the project.

Robert P. Wolensky
William Hastie Sr.

PREFACE

Those who have wished to emphasize the sober constitutional ancestry of the working-class movement have sometimes minimized its more robust and rowdy features.

—E.P. Thompson, *The Making of the British Working Class*, 1963

IN *THE MAKING OF THE BRITISH WORKING CLASS*, E.P. Thompson accentuated not only the "robust and rowdy features" of working-class movements, but also his dissatisfaction with versions of history where "[t]he blind alleys, the lost causes, and the losers themselves are forgotten."[1] He wanted to include accounts of ostensibly less-important events and personages that are, in fact, packed with instruction. Thompson, like American labor historian Herbert Gutman, believed that working people can make their own history, even when efforts fail.[2]

We based *Anthracite Labor Wars* on the same premises. Although hard coal's labor history has received greater consideration in recent years, many untold stories remain. The present volume tells the story of a "thirty-year labor war" (from approximately 1905 to 1935) and its long-term consequences (up to 1960) for the workers and the industry. It was an evolving conflict not only between labor and management, but also between labor and labor, and labor and organized crime. Much of the fighting occurred between and among employees of the Pennsylvania Coal Company (PaCC) and the Hillside Coal & Iron Company (HC&I)—both owned by the Erie Railroad and, therefore, called the Erie Coal Companies—in the northern anthracite field around Wilkes-Barre and Scranton. The book details the determined efforts by workers to organize against various workplace grievances, especially two forms of tenancy—the subcontracting system and the leasing system—initiated by management to achieve greater workplace control, productivity, and, ultimately, profits. The protest movement against tenancy that began at the Erie Coal Companies in the 1910s spread throughout the northern field in the late 1920s and early 1930s. The protestors broadened the list of demands and grievances to include equalization of work time, wage-rate violations, unemployment, and corruption. One consequential result was the formation of a new labor union in 1933, the United Anthracite Miners of Pennsylvania, formed to challenge President John L. Lewis and his previously dominant United Mine Workers of America (UMWA).

Authors have written about the distinction between mining history and mineworkers' history.[3] Mining history focuses on industry-related concerns such as markets, investments, transportation, competition, geology, and technology. Mineworkers' history, on the other hand, centers on the worker at the coalface, on the surface, and in the broader work environment and community. It looks at him as an employee, a union member, a labor activist, a family member, and a community citizen. Our approach was decidedly in the mineworkers' camp. We wanted to study the work culture, labor-management relationships, workplace grievances, and labor militancy, including participation in, or opposition to, union movements—all as related to subcontracting and leasing.

We also wanted to explore the role of ethnicity in the labor-management dynamic of the northern anthracite field. In the course of our research we discovered that Italian immigrant workers constituted over one-third of the Erie Coal Companies' employees (the largest single group) and played a prominent role among the so-called *insurgents* who led the drive against tenancy and on behalf of unionism and economic fairness. Moreover, Sicilian immigrants constituted the largest group within the Italians—as many as a quarter of the companies' work forces. The majority had been miners and laborers in their homeland's sulfur industry where many had been radicalized in a leftist direction by a broad-based Socialist movement during the late nineteenth century.[4]

Upon arriving in the U.S. between 1890 and 1920, large numbers of sulfur miners settled in Pittston, Dunmore, and Old Forge. For reasons that are not entirely clear, many secured work at Erie collieries. Although they were strong supporters of the general strikes of 1900 and 1902, led by the UMWA and its president, John Mitchell, they soon began to leave the union out of dissatisfaction with its conservative leadership and policies. By 1905, they were collaborating with militant workers from other ethnic groups in a series of wildcat strikes. In 1907, the Industrial Workers of the World (IWW) began organizing in the northern anthracite field and the Italians at PaCC and HC&I were among the most receptive to the union's philosophy. It marked the beginning of a decades-long labor-management and labor-labor conflict.

Our concern for workplace dynamics further required that we research another Sicilian immigrant group with a significant presence at the Erie Coal Companies—organized crime. While various researchers, journalists, and government investigators have studied the criminal element in northeastern Pennsylvania, this volume presents the first academic study of organized criminals' involvement as workers, subcontractors, leaseholders, and labor leaders in the anthracite industry.[5]

In recognizing that mineworkers' history is influenced by mining history, we also delved into matters such as industrial competitiveness, geology, technology, and corporate culture. The period under investigation witnessed a fast-changing coal industry beset by innumerable economic, managerial, organizational, and technological problems. The situation had direct and indirect repercussions on the workers, the workplace, and the community. The Erie Coal Companies formed the initial focus of our study but we soon realized the need to broaden the inquiry to include workers at other companies during the late 1920s and early 1930s.

We relied on primary sources such as coal-company records, government reports, newspapers, U.S. Census documents, passenger and naturalization records, as well as secondary sources including studies in labor, economic, and ethnic history by journalists and academic researchers. Yet, several important subjects had left little or no documentary or research trail and, in these cases, we relied on oral-history accounts. Oral historians have found the method especially useful in uncovering what historian Luisa Del Giudice referred to as "hidden histories," or those narratives previously omitted from scholarly or community discourse due to ignorance, neglect, omission, or self-imposed silence.[6] Oral history can also function to supplement the written record and we have employed it in that capacity as well. The References section of the book contains the names of each person who contributed an oral history. Their recordings and transcripts have been included in the Northeastern Pennsylvania Oral and Life History Project.

The book has eight chapters. Chapter One provides an introduction to the anthracite industry and the concept of tenancy in the form of subcontracting and leasing. Chapter Two presents an historical sketch of the Erie Railroad and its subsidiaries, PaCC and HC&I, which were the foremost purveyors of tenancy systems among northern coal field companies. Chapter Three turns to the labor protests over tenancy and other grievances at the Erie Coal Companies following the strike of 1902, including the strike of 1916 called by the IWW. Chapter Four examines the growing insurgent movement that culminated in major strikes at PaCC and HC&I in 1920 and 1924. Chapter Five details one of the most violent work suspensions ever to hit an anthracite company—the confrontation at PaCC's No. 6 Colliery involving the insurgents, the UMWA, the company, and organized crime. Chapter Six dissects a more expansive and dynamic form of tenancy, the leasing system. Erie's most prolific leaseholders—organized crime-affiliated firms among them—are catalogued and discussed. Chapter Seven delves into the role of tenancy in the decline of northern field anthracite in the 1930s, 1940s, and 1950s. Chapter Eight takes up the ethnic theme and addresses the heretofore untold story of Italian mineworkers' militancy on behalf of collective bargaining and workers' rights. In considering why this account has remained one of the hidden histories of anthracite, we survey four "narratives" related to the Italian-American experience in the U.S.: The Unfit Immigrant; Hard Work and Assimilation; Organized Crime; and Workers' Rights and Economic Justice. The analysis presented in the chapter is squarely in the last narrative. Two appendices follow: Appendix I provides a glossary of terms and Appendix II offers biographical sketches of most of the persons discussed herein.

On a personal note, we were introduced to the PaCC and HC&I tenancy schemes through our separate studies of the Knox Mine Disaster, the northern field's last major mining catastrophe, which occurred on January 22, 1959, at the Knox Coal Company's River Slope mine in Port Griffith, Jenkins Township, Luzerne County. Knox became a PaCC leaseholder in 1943 and grew into one of the field's larger producers. Bill Hastie worked for the company during the 1950s and was on the scene, but not in the mine, on the day of the disaster. A student of hard coal's history, he began speaking to community groups about the infamous company and its catastrophe in the late 1980s. Around the same time, Bob Wolensky began research for an academic book on the calamity. He read various government inquiries, newspaper accounts, and court cases while conducting over 50 oral history interviews. Bill Hastie provided one of the interviews in July 1989. We continued meeting for several other recorded

Authors Bob Wolensky (left) and Bill Hastie Sr. speaking at the Huber Breaker Preservation Society, 2008.

and unrecorded interviews (see References) and countless informal discussions. During more than 25 years of dialogue, research, public speaking, and writing, we have gone well beyond the Knox tragedy to analyze the labor, ethnic, and economic history of workers at PaCC, HC&I, and other hard coal companies.[7]

Along another personal dimension, in many ways this story intertwines with our own stories. Bill Hastie's family tree includes scores of anthracite mineworkers on both his maternal and paternal sides going back three generations in Pennsylvania, and many scores of lead and coal miners who lived in Scotland and Wales. His grandfathers and a number of their relatives participated in the 1902 strike, while his father was forced to leave school at age 12 in order to supplement the family's income during the strike. Hastie was immersed in the Pittston area's anthracite industry and culture from his youth in the 1920s to the industry's decline in the late 1950s. Although Bob Wolensky was never employed in or around a coal mine, he has visited many of them and, more importantly, his family history shows hard coal workers on both the maternal and paternal sides going back two generations in Swoyersville, Pennsylvania, and many more in the sulfur mines of Sicily on the maternal side. His grandfathers and uncles participated in the general strikes of 1922 and 1925–26, and many localized walkouts.

Consequently, in line with the work of authors such as Ronald Berger and Richard Quinney, our biographies are closely related to the contents of this volume.[8] As such, we are not entirely objective observers. On the other hand, our knowledge of, and experience within, the anthracite regions have allowed us to gather valuable information and rely upon persons, resources, and insights to which researchers with a more distant orientation may not have had access.

Finally, let us add that, like many sagas within the anthracite regions, the chronicle contained in these pages has been difficult to tell, not simply because several incidents and characters proved elusive, but because of the disheartening violence, anguish, and corruption that we have discovered. Anthracite's exploitative and brutal past has produced long-lasting social and cultural "wounds" that can still jolt the public conscience. There should be little surprise that the entrenched patterns of dishonesty and conflict—forged by companies, governments, unions, courts, and individuals between the 1830s and the 1960s—have endured to the present.[9] Deep-seated cultural patterns die hard.

This having been said, however, another enduring regional trait has been a citizen and working-class activism that has regularly mobilized on behalf of workers' rights as well as numerous social, economic, and environmental alarms.[10] At root, we have tried to record and analyze the broader dynamics surrounding a little-known but, in our view, remarkable example of the area's militant tradition: a series of labor movements undertaken against powerful forces by workers who were unwaveringly set on pursuing their own vision of the just workplace and community.

Bob Wolensky, Stevens Point, Wisconsin

Bill Hastie, West Pittston, Pennsylvania

October 29, 2012

TABLE OF CONTENTS

List of Figures

CHAPTER THREE:

CHAPTER FOUR:

CHAPTER SEVEN:

CHAPTER EIGHT:

ABBREVIATIONS

ABC—Anthracite Board of Conciliation

AF of L—American Federation of Labor

AMUP—Anthracite Miners Union of Pennsylvania

CIO—Congress of Industrial Organizations

CPUSA—Communist Party of the USA

D&H—Delaware & Hudson Company

DL&W—Delaware, Lackawanna, & Western Railroad

HC&I—Hillside Coal & Iron Company

HCLA—Historical Collections and Labor Archives, Penn State University

ILD—International Labor Defense

IWW—Industrial Workers of the World

KKK—Ku Klux Klan

LC&N—Lehigh Coal & Navigation Company

L.U.—Local Union

LVCC—Lehigh Valley Coal Company

L&WBCC—Lehigh & Wilkes-Barre Coal Company

MG—Manuscript Group

NCMA—National Canal Museum Archives

NMU—National Miners Union

NPOLHP—Northeastern Pennsylvania Oral and Life History Project

NYS&WRR—New York, Susquehanna & Western Railroad

NYS&WCC—New York, Susquehanna & Western Coal Company

PaCC—Pennsylvania Coal Company

PHMC—Pennsylvania Historical and Museum Commission

PSA—Pennsylvania State Archives

PSU—Penn State University

TPC—The Pittston Company

UAMP—United Anthracite Miners of Pennsylvania

UMWA—United Mine Workers of America

USGPO—United States Government Printing Office

WBA—Workingmen's Benevolent Association

Anthracite Regions of Pennsylvania

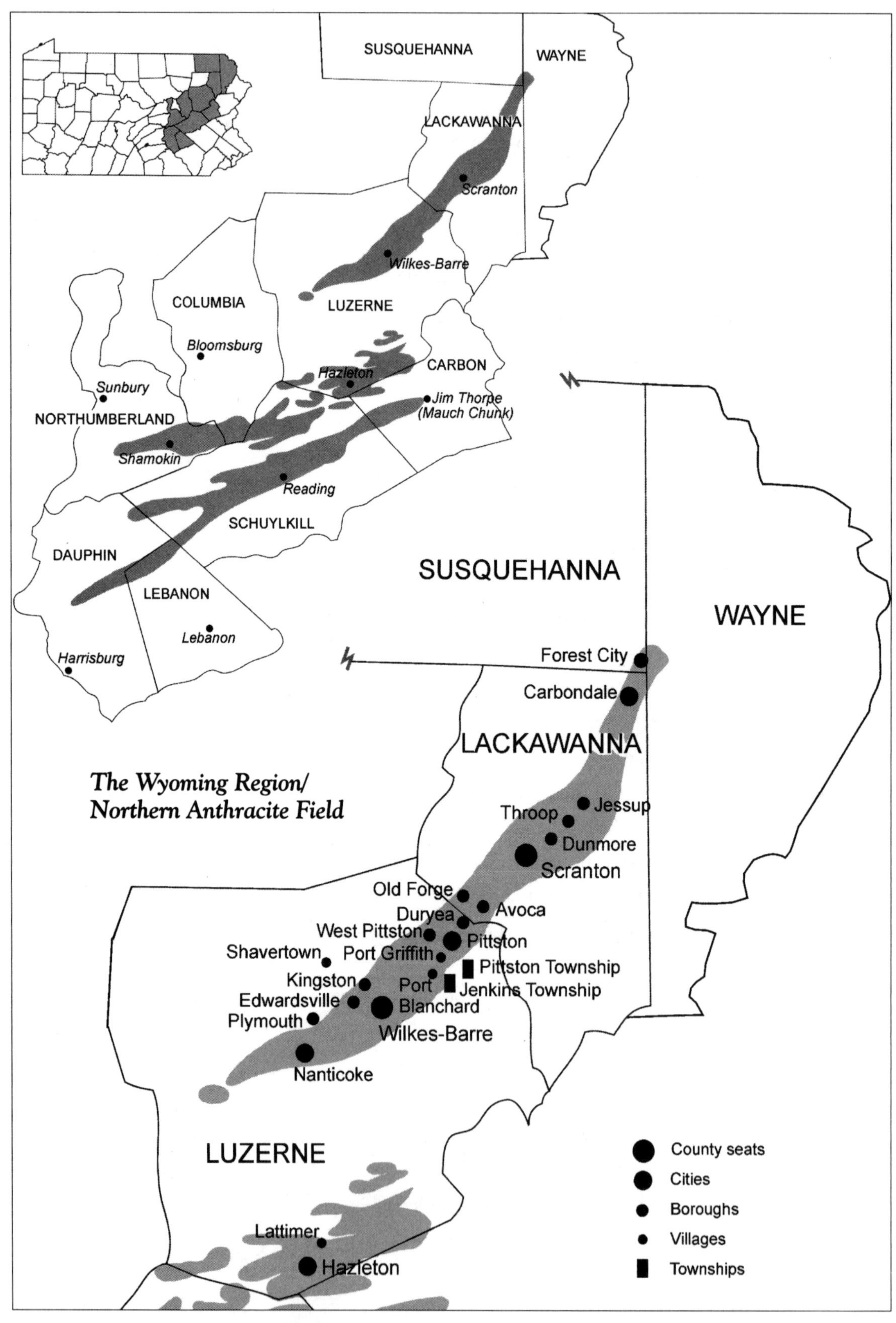

PART I

The Subcontracting System at the Erie Coal Companies

John Mitchell Memorial, Scranton, Pennsylvania

CHAPTER ONE

Control, Competitiveness, and Conflict: An Introduction to the Subcontracting System

Commissioner Watkins: Would your organization permit a miner to engage three laborers if he could employ them with safety and load ten cars, say?
Mr. Mitchell: No. The organization is opposed to miners employing more than one laborer. The anthracite miners find that the other system [subcontracting system] introduced in this field has done them a great deal of harm.

— UMWA President John Mitchell, in testimony before the Anthracite Strike Commission, 1902

Even though the [sub]contractors make tremendous wages, the advantages to the company are many.

— Powers Hapgood and Mary Donovan, 1928

Subcontracting practices readily undermined both stints [worker-induced production quotas] and the mutualistic ethic [among workers]... and they tended to flood many trades with trained, or semi-trained, workers who undercut wages and work standards.

— David Montgomery, 1979

The Early Anthracite Industry

By 1900, the United States had become the world's leading coal producer and Pennsylvania ranked first in output among the states, a position it held for 100 years between the 1820s and the 1920s. The Keystone State held two types of coal deposit: anthracite (hard coal) in the northeast, and bituminous (soft coal) in the central and western areas.[1] Anthracite proved vital to the nation's industrial development in the early nineteenth century due to the coal fields' proximity to manufacturing plants in Eastern states.[2] The fuel remained the nation's leading source of industrial energy from the 1830s until the post–Civil War years when the more plentiful bituminous took the lead.[3]

Unlike bituminous, which is found in 32 states, mineable quantities of anthracite are found only in a ten-county territory in northeastern Pennsylvania. At the start of the mining era during the second quarter of the nineteenth century, the section contained up to 75 percent of the world's anthracite reserves.[4] The coal measures lay in four basins or *fields*, which, for marketing and transportation reasons, were divided into three *regions*: the Wyoming Region, encompassing the northern field (the focus of this study); the Lehigh Region, covering the eastern-middle and eastern tip of the southern field; and the

Figure 1. Panorama of Wilkes-Barre, Pennsylvania, 1889.

Schuylkill Region, comprising most of the southern and all of the western-middle fields. Locational advantages and a well-developed canal system enabled Schuylkill to lead in output and employment between the 1830s and 1850s.[5] Following the Civil War, the northern field surged ahead and became the most productive, urbanized, and populous of the regions. The field's two foremost municipalities, Wilkes-Barre and Scranton, developed into the largest and most affluent of anthracite's cities.[6]

Because the northern field's remoteness and geology required significant capital investments, the state legislature allowed the large, well-financed railroads to own and operate coal mines, a privilege denied the southern field's operators until 1869. The railroad-owned or "Line" coal companies became the driving force behind the north's ascendancy. By the end of the century, J. Pierpont Morgan and his financial empire controlled anthracite's six main rail lines which, in turn, owned the main anthracite companies in each of the four fields. The so-called "Coal Combination" developed an elaborate system of interlocking directorates with key Morgan officials dominating the boards of directors of the participating railroads and coal companies.[7] By 1900, anthracite had become a virtual monopoly directed from New York.[8]

The Subcontracting System in the Northern Field

Despite their dominance, the Line-owned coal companies had great difficulty controlling one factor of production: labor. Human toil constituted up to 75 percent of operating costs and the workers—mainly British and German between 1840 and 1880, and Eastern and Southern European by 1900—wanted a fair share of the proceeds.[9] They also sought protection from arbitrary firings and safety lapses, company stores and physicians, and excessive dockage and high blasting-powder prices. They furthermore desired the "free-

Figure 2. Panorama of Scranton, Pennsylvania, 1890.

dom," as economist Carter Goodrich termed it, to work without overbearing supervision by bosses.[10] Perhaps most importantly, they wanted to abandon the individually signed contract and secure a collectively bargained agreement negotiated by a labor union.[11] The divergences between labor and management over collective bargaining, wages, working conditions, and workplace control resulted in some of the most acrimonious conflicts in American labor and business history.

To help deal with "the labor problem," some northern-field companies implemented a *putting-out* (or *tenancy*) plan during the 1890s based on the *individual contract* and its derivative, the *subcontracting system.*[12] Putting-out had been used in various nineteenth-century industries such as wool, cotton, garments, stone quarrying, and iron mining. Large manufacturers used it to reduce costs by contracting with small producers who had lower overhead, cheaper labor, and other advantages. Workers generally opposed the structure because it generally functioned to de-skill trades, reduce wages, and speed up production.[13]

The individual contract began in anthracite when, based upon the British practice, the companies required a miner to sign an agreement obligating his work for a specific period of time.[14] By the end of the nineteenth century, it had evolved from a general contract for mine work to an agreement for a specific project. Such agreements, hereafter referred to as *subcontracts*, were issued for tasks such as *robbing the pillars*[15] of coal left behind from a first mining; conducting *development work*, which involved driving new entryways and haulage roads often through virgin coal; and quarrying entrances to new shafts and slopes.

Company Logos | *Figure 3. Erie Railroad Logo.*
Figure 4. Delaware & Hudson Railroad Logo.
Figure 5. Delaware, Lackawanna, and Western Railroad Logo.

Four northern corporations—Pennsylvania Coal Company (PaCC); Hillside Coal & Iron Company (HC&I); Delaware & Hudson Coal Company (D&H); and Delaware, Lackawanna, and Western Railroad's coal division (DL&W)—began using subcontracts at some of their collieries in the 1890s. As detailed later, PaCC and HC&I—known as the Erie Coal Companies because they were owned by the Erie Railroad—emerged as the leaders in offering subcontracts, while D&H and DL&W employed the scheme to a more limited extent. The subcontract precipitated profound worker unrest as it became enmeshed in broader labor-management conflicts over workplace control and industrial competitiveness.[16]

Who exactly were the subcontractors? To understand their status, a classification of mining positions must be considered. The term *mineworker* applied to all employees in or around a colliery regardless of occupation. They included subterranean men and boys such as miners, subcontractors, laborers, haulers, fire bosses, nippers (boy door tenders), driver boys (young mule drivers), and stablemen (mule tenders); as well as above-ground workers such as breaker men, engineers, and weighmen.[17]

In the underground world of shafts, slopes, gangways, and coal faces, *miners* were the highest-skilled employees. They traditionally attained the position through years of apprenticeship; however, a state law enacted in 1889 required a miner to pass an examination and secure state certification or "mining papers."[18] In 1900 there were three categories of certified miners in anthracite: contract miners, company miners, and subcontractors. *Contract miners* constituted the largest group. The term originated in eighteenth-century British mining, as a result of the requirement that each miner sign an individual contract obligating his work for a specific period of time. As skilled craftsmen, contract miners worked with little supervision and enjoyed wide autonomy in harvesting coal at the face. They hired, paid, and mined alongside one or two laborers in a single chamber. Contract miners were paid by the piece (car of coal) or by weight (tonnage).[19] By law, only they could set and execute the charges for black powder and, later, dynamite, to blast coal from the face. They developed a strong sense of independence, or so-called "freedom," and took great pride in their craft. As Priscilla Long observed: "It should come as no surprise, therefore, that most coal mines, even quite large ones, represented a proliferation of small underground workshops."[20]

Company miners constituted the second group of certified miners. They toiled under the direct control of management and were paid by the day for sundry projects such as clearing cave-ins, conducting development work, securing gangways, making repairs, and facilitating underground transportation. Because their tasks often involved the mining of coal, they were required to hold a miner's certificate.

Subcontractors comprised the third and smallest group of certified miners. As in the British industry, where subcontracting practices existed in the eighteenth and nineteenth centuries, subcontractors were typically entrepreneurial contract miners who secured an agreement with a company to take coal in a specific vein or undertake some special pro-

Figure 6. Anthracite mineworkers and mules, circa 1900. (Courtesy of Carl Orechovsky)

Figure 7. Miners and laborers, circa 1900. (John Mitchell, Organized Labor, *1903)*

ject.[21] As a relatively independent agent, the subcontractor was remunerated by the overall job and enjoyed a larger income than the ordinary contract miner.[22] He was not a petty operator because he was not incorporated, nor did he own any of the means of production; yet his independence, mineral-rights access, and comparatively large work gang allowed him to function as such. He not only hired and paid a larger labor crew than the contract miner, but often hired other certified miners to assist in harvesting the coal. Taking a subcontract disqualified a miner from membership in a labor union such as the United Mine Workers of America (UMWA) because he was viewed as an ally of management. According to sociologist and clergyman Peter Roberts, who studied the anthracite industry around the turn of the century, "[contract miners] often have two and three laborers to work for them, but when a miner takes a [sub]contract which enables him to hire miners and laborers he is disqualified as a member of the union, although he pays the standard wage in the colliery to the men he hires." [23]

The subcontractor and subcontracting system reorganized access to mineral rights as it also brought a nascent form of mass production to mining. Carter Goodrich described the system as it functioned in the 1890s:

> A more systematic step in the same direction was taken by a number of operators in the nineties who divided the work between groups of "shooters" [subcontractors] who were to do the timbering and the shooting [blasting] and all the skilled work, and "loaders" [laborers] who were merely to shovel the coal [into coal cars] ...[24]

Economist Frank J. Warne characterized it in 1903:

> Under this plan a certain section of a seam of coal, comprising a number of breasts, or working-places, was let out under contract to a skilled miner by the foreman; that is, the miner agreed to take out a certain part of a seam for a stated sum of money. He would then employ miners and laborers to do the work, bidding them against each other for the places. The lower the wage for which these men worked the greater the earnings of the [sub]contract miner.

Warne highlighted one of the benefits to the company: "This [sub]contract system aided capital materially in reducing wages throughout the region. ..."[25]

Unskilled *laborers* constituted the largest category of mining employees. During the late nineteenth century, up to 80 percent of the northern field's laborers were employed by contract miners or subcontractors, while the other 20 percent worked directly for the company in various jobs above and below ground. The laborer(s) received about one-third of the miner's gross earnings, a percentage that the subcontractor usually, but not always, honored.[26] The subcontractor employed labor teams of up to a half-dozen workers in the 1890s, which expanded to as many as 60 laborers by the mid-1920s.[27]

According to historian Clifton Yearley, during the nineteenth century approximately one-third of all adult anthracite workers were contract miners or subcontractors, seven percent were company miners, and the balance were inside or outside laborers.[28] The certified miners—contract miners, company miners, and subcontractors—were typically of Welsh, Scottish, or English background, while German and Irish immigrants were usually the laborers. By 1900, the latter two groups had moved up to the skilled and supervisory positions and the laborers were most often Slavs, Italians, Lithuanians, Hungarians, or other recent immigrants.[29] The ethnic-occupational divisions reflected the region's demographic history, where the first mine-working immigrants from mainland Britain were followed by Germans and Irish after 1840, and then by Eastern and Southern Europeans after 1875. Poles, Ukrainians, Russians, Slovaks, Czechs, Slovenians and other Slavs constituted only two percent of the anthracite regions' population in 1880, but 46 percent by 1900. Italians, who figure prominently in this volume, constituted less than one percent of the population in 1880, and about two percent in 1900. By 1910, Eastern and Southern Europeans formed over one-half of the anthracite labor force.[30] Over two dozen nationality groups called the anthracite fields home during the 1920s.

Labor-Management Conflict over the Subcontracting System

The subcontract became a labor-management issue just as the UMWA arrived in the anthracite fields during the mid-1890s to organize a workforce that had grown to over 140,000.[31] After a slow start, the union was accepted as the vehicle to help the workers share in the prosperity of a growing industry with a capitalization approaching $200 million ($5.36 billion in 2010 dollars).[32] They also saw the union as a bulwark against unsafe mining, poor living conditions, and arbitrary treatment. Founded by bituminous men in Columbus, Ohio, in 1890, the UMWA established a presence in anthracite following a series of strikes between 1894 and 1899.[33] It became hard coal's first enduringly successful union and represented the workers in three district offices: District 1 in the northern field, District 7 in the middle fields, and District 9 in the southern field. UMWA members from anthracite and bituminous elected the conservative and charismatic John Mitchell as president in 1898. A former breaker boy and soft-coal miner from Illinois, Mitchell became a revered leader who displayed great skill in unifying the diverse labor force while skillfully negotiating with company executives.[34]

Mineworkers at the affected northern-field companies wanted the UMWA to address three complaints regarding the subcontract. The first concerned favoritism: employees argued that the agreements were issued to favored (e.g. anti-union) or otherwise preferred (e.g. through ethnicity, religion, or politics) mineworkers, and/or to those willing to pay kickbacks. The preferential treatment included the subcontractor's access to "easy coal"

Figure 8. Pennsylvania Coal Company miners at No. 6 shaft, Pittston, PA. (Library of Congress, Lewis W. Hine Collection)

Figure 9. Breaker boys, Pennsylvania Coal Company, Pittston, PA, circa 1910. (Library of Congress, Lewis W. Hine Collection)

Figure 10. Boy door tender. (Courtesy of Carl Orechovsky)

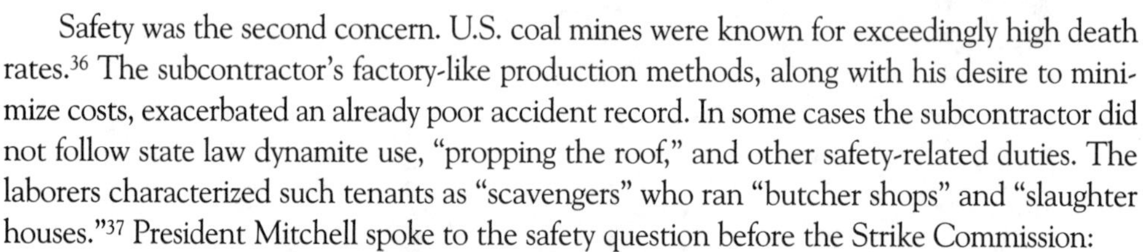

Figure 11. John Mitchell, President, UMWA, 1902. (Library of Congress)

while the ordinary contract miner was assigned to more difficult veins. The critics also protested the subcontractor's unfair advantage in securing empty coal cars, which resulted in their ability to load more coal in a shorter time. In reference to one local union's resolution against the subcontract in 1901, President Mitchell addressed the favoritism issue before the Anthracite Strike Commission, which settled the major strike of 1902 (see below):

> The resolution was simply intended to prevent some men who were more favored than others by the companies from getting all the coal they could load and employing extra labor to do it for them, while some other man not so fortunate would remain in the mines all day without getting his share of coal.[35]

Safety was the second concern. U.S. coal mines were known for exceedingly high death rates.[36] The subcontractor's factory-like production methods, along with his desire to minimize costs, exacerbated an already poor accident record. In some cases the subcontractor did not follow state law dynamite use, "propping the roof," and other safety-related duties. The laborers characterized such tenants as "scavengers" who ran "butcher shops" and "slaughter houses."[37] President Mitchell spoke to the safety question before the Strike Commission:

> [Elimination of subcontracting] is simply a regulation that the miners of the anthracite field find is necessary to protect themselves against ... one man going out and hiring a lot of laborers and bringing them in and having them work for him, his being unable to protect them properly against the dangers of mining.[38]

Potentially drastic changes in wages and work norms constituted the third objection. Many subcontractors wanted to pay as little as possible, regardless of the union having negotiated a wage schedule, or rate sheet, in 1900. As historian Douglas Monroe observed, "Not a particularly scrupulous brand of entrepreneurs, these [sub]contractors were also not above attempting to cheat the miners out of their rightful wages."[39] When it came to mining, the subcontractor demanded discipline and productivity: "When a miner displeased a [sub]contractor or failed to deliver a superhuman effort, he was quickly replaced by a more compliant worker."[40] Just as important, the contract miners realized that, if the plan were to spread, the industry would need fewer certified miners, forcing many to become laborers in order to make a living.

For their part, the coal operators endorsed the subcontract for several self-interested reasons. The first reflected Frederick W. Taylor's contemporary writings on Scientific Management, which held that managers, not labor, should control the workers and the work process.[41] While the anthracite managers were not "Tayloristic" to the same degree as their counterparts at Bethlehem Steel Corporation, who hired Taylor to implement his ideas at their mills, the hard-coal executives clearly wanted greater discipline and control.[42] Another motive had Social-Darwinist underpinnings: management held that some men possessed superior entrepreneurial, physical, and mental abilities, and should be encouraged to take as much coal as possible.[43]

Figure 12. A patch town at Shamokin, PA, circa 1920.
(Courtesy of James Burke)

Figure 13. A miner at the face in a Delaware Lackawanna & Western Coal Company mine.
(Courtesy of National Canal Museum Archives)

Figure 14. Hazleton, PA, coal miners, 1905.
(Library of Congress, B.L. Singley Collection)

A third argument involved the worker-sanctioned production limits or "stints," which management abhorred. As labor historian David Montgomery pointed out, "Subcontracting practices readily undermined both stints and the mutualistic ethic [among workers] … and they tended to flood many trades with trained, or semi-trained, workers who undercut wages and work standards."[44] Subcontractors in garment manufacturing, iron molding, stone quarrying, and railroad-locomotive manufacturing were known to have surmounted production quotas. William Graham, the last president of the Penn Anthracite Collieries Company in Scranton, recalled the benefits to his company as late as the 1940s and 1950s:

> These people [subcontractors and lessees] could do what we couldn't do. … They could do things that the company wouldn't, couldn't do … such as get the extra car of coal from the men. …[45]

The fourth argument stated that the industry could not remain profitable and competitive with an "expensive" work force. In 1901, Frank J. Warne reported that the cost of mining increased 31 cents per ton between 1880 and 1890, while the value per ton increased by only 11 cents. High transportation charges allowed the mine-owning railroads to earn an extra 40 cents for every ton sold (an amount not figured into Warne's calculations); however, government regulations and public opinion endangered this source of income.[46] Eben B. Thomas, chairman of the PaCC board of directors, and president of the Erie Railroad, articulated the operators' concern in 1902: "Every advance in the cost of production of anthracite tends to benefit its competitors in the bituminous field."[47] Labor activists Powers Hapgood and Mary Donovan summarized the benefits of subcontracting from the companies' perspective:

> Even though the [sub]contractors make tremendous wages, the advantages to the company are many. Under the agreement between the coal companies and the United Mine Workers, the companies themselves cannot reduce wages. If they let their mines out on special contracts, however, the [sub]contractors can employ their fellow-union men to work for them, pay them the regular day wages as laborers, and force them to produce from 50 to 100 per cent more coal than the men would have to produce for the same wages if working directly for the company. This system makes it possible for the companies to under-cut the scale agreement without repudiating their [side of the] agreement.[48]

A summary of President Mitchell's testimony before the Strike Commission of 1902 encapsulated the workers' view of the matter:

> Commissioner Watkins asked if there was any rule of the union forbidding [sub]-contract work. Mr. Mitchell said he did not know of any written rule, but he knew the union looked with disfavor on a miner having more than two laborers, favored one laborer for a miner, and preferred, if possible, to have men work as partners. If a man is allowed to have a large number of laborers, Mr. Mitchell explained, he becomes a small operator. He lives off the work of other mine workers. It is objectionable, also, on the ground that one miner cannot properly protect a large number of laborers. If the thing were not checked, there is danger of the system extending to the chambers and the number of miners reduced.[49]

In a more critical tone, Elizabeth Gurley Flynn of the Industrial Workers of the World (IWW) referred to subcontractors as "'cockroach' small shop capitalists who exploit workingmen and their families in these forlorn places. ..."[50]

Figure 15. After a mass rally in Scranton, PA, Anthracite Strike of 1902. (Catholic University of America, Mother Jones Collection)

Figure 16. Anthracite Strike Commission, 1902–03.

(Library of Congress, William H. Rau Collection)

Figure 17. Coal railroad presidents during the Strike of 1902.

Top row, left to right: E.B. Thomas, Erie RR; Geo. F. Baer, Phila. & Reading; Wm. H. Truesdale, DL&W RR.

Bottom row: Thos. P. Fowler, New York, Ontario & Western RR; R. M. Olyphant, Del. & Hudson RR; Alfred Walter, Lehigh Valley RR.

In a very real sense, a clash between two ideologies of work had emerged. One insisted on the mineworkers' freedom while the other emphasized management control; one upheld craft principles, while the other emphasized productive efficiency. The differences would undergird four decades of labor-management conflict in the northern field.

The Strike of 1902

Even though the bituminous executives in the Central Competitive Field of Western Pennsylvania, Ohio, and Illinois had negotiated a labor contract with the UMWA in 1897, their anthracite counterparts would allow no such arrangement. Their intransigency incensed the workers who wanted a collectively bargained agreement, pay raises, and the elimination of the individual contract, among other changes. The demands were proffered by the UMWA in the labor-management talks preceding the historic strikes of 1900 and 1902.[51] While pay gains were attained in both strikes, management would not negotiate the individual contract in either strike.[52]

The 164-day Strike of 1902, which began on May 12, was especially noteworthy because it changed the *status quo* of labor-management relations. Not only was it the first anthracite strike to gain wide public support, but the union also showed that workers from diverse ethnic backgrounds could maintain solidarity over an extended period of time.[53] Just as important, President Theodore Roosevelt decided to intervene when the shutdown entered October and threatened a fuel shortage for the winter months. The president initiated meetings with union and company executives in search of a settlement. He was as exasperated with the executives' rigidity and arrogance as he was impressed by Mitchell's demeanor and sincerity. After much negotiation, which eventually included financier and banker J.P. Morgan, the executives consented to arbitration through a presidentially appointed, seven-member Anthracite Strike Commission. The commission conducted hearings over a three-month period in Scranton and Philadelphia at which 558 witnesses, including mineworkers, union officials, company representatives, and others testified.[54]

The commission's *Final Report*, issued March 21, 1903, established the terms of settlement that continued for three years. Among other outcomes, the ruling granted an average ten percent wage increase, established the Anthracite Board of Conciliation to adjudicate grievances, allowed workers to hire and pay a check-docking boss, and reduced shift hours from ten to nine, in most cases.[55] The UMWA had achieved a great victory.

Besides federal involvement, the strike brought another new resource to bear on the side of the workers. Many Episcopal, Methodist, Catholic and other clergy supported the workers. Reformed Episcopal Bishop Samuel Fallows of Chicago, Methodist minister James Moore of Avoca, and Catholic priest James V. Hussie of Hazleton criticized the reports of worker violence and other attempts to portray the strikers in negative light. Mitchell wrote that Catholic priests supported the strike "almost universally."[56]

By the turn of the century, the majority of the anthracite workers were Roman or Greek Uniate Catholics. Pope Leo XIII had issued an encyclical (a teaching "letter") termed *Rerum Novarum* (*Of New Things*), on May 15, 1891.[57] The Pope titled the document "Rights and Duties of Capital and Labour," and gave it the subtitle, "On the Conditions of Labour." The encyclical focused on the worker in industrial society because Leo had concluded that "the condition of the working classes is the pressing question of the hour."[58] He emphasized economic fairness in labor-management relations: "If

Figure 18. Left to right: Bishop Michel J. Hoban, President Theodore Roosevelt, and Rev. John J. Curran, Scranton, PA, 1912. (Courtesy of F. Charles Petrillo)

through necessity or fear of a worse evil the workman accepts harder conditions because an employer or contractor will afford him no better, he is made the victim of force and injustice."[59] He spoke of "safeguarding the interests of the wage-earners ... "[60] and, to this end, advocated workers' associations of which "the most important of all are workingmen's unions."[61] Indeed, he concluded, "it were greatly to be desired that they should become more numerous and more efficient."[62] The prejudice against Catholics in anthracite and elsewhere notwithstanding, the Pope clearly supported the concept of unionization and collective bargaining.[63]

In the Diocese of Scranton, Rev. Michael J. Hoban, whose immigrant father had worked for PaCC, was appointed bishop in 1899.[64] He endorsed Leo XIII's encyclical such that he allowed Fr. John J. Curran, the pastor of Wilkes-Barre's Holy Saviour Church, to participate in the negotiations as a workers' advocate. Curran became sensitized to the hardships of the mineworkers from his youth as a PaCC breaker boy and mule driver.[65] He attended meetings with Mitchell and Roosevelt, joined union rallies, and testified before the Strike Commission. The Catholic influence in the strike was seen not only in Bishop Hoban's support and Fr. Curran's exertions, but also in President Roosevelt's appointment of Bishop John Lancaster Spalding of Peoria, Illinois, as a member of the Anthracite Strike Commission, and in Mitchell's philosophical and moral approach to the dispute.[66] Therefore, in addition to the federal government's involvement, the workers enjoyed two other new institutional advantages: the union and the churches.[67]

The strike and its outcome marked an industrial relations watershed for another reason. If previously the operators could improve their balance sheets by lowering wages, enlarging coal cars, increasing dockage, raising company-store prices, and issuing subcontracts, the new balance of power meant that the old methods no longer applied.[68] Company executives realized that in order to remain competitive and continue the pattern of high returns to stockholders, they needed to implement new technological, managerial, and organizational efficiencies. Frank Warne defined the problem from the operators' perspective: "If there is any one thing certain as to the future of the anthracite coal industry it is that the cost of mining must be reduced more and more."[69] Tenancy in the form of subcontracting emerged as a key cost-cutting strategy at the aforementioned four coal companies even before the strikes of 1900 and 1902.[70]

Early Skirmishes over the Subcontracting System

The workers' hostility toward the subcontract could be seen in the labor-management clashes at the DL&W, D&H, PaCC, and HC&I. DL&W engaged in subcontracting at only one colliery, the Woodward in Edwardsville, near Wilkes-Barre. In 1901, the UMWA District 1 president, Thomas D. Nicholls, enthusiastically stated: "The [sub]contract system, while it is still in vogue at some of the collieries in our District, is, in the main, on the decline" although, he noted, the plan still flourished at the Woodward.

Nicholls contacted the Woodward Colliery's top man, Superintendent Loomis, to discuss the matter but was rebuffed. He tried a second time but Loomis "still held aloof from meeting us, and said that while he would not take away the [sub]contracts from the two men who had them, he would at once issue an order to have the cars equally divided among all the miners." Nicholls persisted until Loomis finally assented to a conference:

> When I called upon him he received me courteously, and had no objection to talking business. After some argument about doing away with the [sub]contract system ... (which he said had been in vogue for about five years), he agreed to consider the question of giving only one place to one miner instead of two or three, as then was the case. A few days later he issued a notice stating that only one place would be given to [a] miner ... but that those who now had [sub]contracts would be allowed to keep their places.[71]

At the D&H, much of the strife involved the "heading men," whose subcontracts for development work usually involved removing large quantities of rock and coal. In November 1901, the UMWA local union at the company's Jermyn No. 3 Colliery in Manville, Lackawanna County, accused the heading men of overproduction. Union officials ordered the mule driver boys to give them only a "fair share" of empty cars, rather than the larger number they demanded. When the driver boys limited the cars, the company dismissed them, whereupon the entire workforce marched out. The boys were soon rehired and mining resumed, but not before the local union passed a motion declaring: "Unless all heading men cease loading more than their share of cars you will be expelled from the union." To enforce the directive, some militant workers entered the mine and destroyed tools belonging to Michael McHale, Harry Gilbert, and Henry Richards, three heading men who had disregarded the notice.[72] Their surnames indicate that D&H's first subcontractors were of British (including Irish and Welsh) background, a pattern no doubt resulting from these groups' dominance of the certified-miner ranks.

Figure 19. Woodward Colliery, Delaware, Lackawanna & Western Railroad, Edwardsville, PA, 1907. (Courtesy of Plymouth Historical Society)

Figure 20. Erie Colliery, Hillside Coal & Iron Company, Carbondale, PA, 1889. (Courtesy of National Canal Museum Archives)

The driver boys recommenced delivering empties, but several laborers refused to load the subcontractors' coal. Convinced that the union lay behind the boycott, management closed the colliery on December 31, 1901. They reopened it two weeks later after receiving pledges that no further "molestation" of the heading men would occur. In the meantime, the local union insisted that the subcontractors adhere to the established production stint of two cars per man per shift. When heading man John Sobey refused to comply, saboteurs wrecked his home with dynamite.[73]

Because the tenancy systems at PaCC and HC&I are examined in later chapters, for now suffice it to say that the companies were the foremost proponents of subcontracting, while the employees were among the most vigorous protestors. The Erie Companies' workers passed a resolution against the plan, which was read at the UMWA Tri-District convention in March 1902, shortly before the strike: "We the employees of the Erie Co. ask this convention to stand by us in whatever stand we may take to abolish this [sub]contract system."[74] The larger body of delegates at the Tri-District convention heard the complaint and approved a supportive resolution:

> Whereas, it is a well-known fact that in many sections of the anthracite region there is a system of contract in vogue which is, and has been very obnoxious and vicious in its fulfilment. Inasmuch that in many cases one man employs from four to twenty laborers, and in some cases the [sub]contractor seldom enters or comes near the work for one or two weeks at a time, and this in itself is not the cause of principle advocated by the U.M.W. of A.; therefore, be it resolved that any member of the organization who shall contract for such work as well as necessitate the employment of more than two laborers, excepting such contracts as shaft sinking, slope sinking or tunnel driving, shall be expelled from the U.M.W. of A.; and we, the members of the U.M.W. of A., absolutely refuse to work with any man so expelled from the union; and be it further resolved, that we condemn the employment of laborers [for this purpose].[75]

A handful of large companies were not the only ones issuing subcontracts. Smaller firms also dispensed the agreements. For example, the independent (i.e., not affiliated with a railroad) Forest Coal Company in Archbald, near Scranton, issued subcontracts in the 1890s. A strike against the company in 1896, led by a contingent of Italian immigrant workers who walked out in a wage disagreement, is detailed in Chapter Eight. Furthermore, certified miners were not the only ones securing the subcontracts. As President Mitchell told the Strike Commission, "[T]here are instances of saloon keepers taking pillar-robbing contracts and turning them over to practical [contract] miners. This means that all the men working on the [sub]contract must go to the saloon to be paid, and, of course, leave some of their money on the bar."[76]

The Subcontract and the Stint

If the companies envisaged stints as economically "irrational" norms that hindered efficiency, the workers saw them as necessary and integral to the work culture. David Montgomery concluded that dedication to "rationally restricted output" reflected "unselfish brotherhood," "personal dignity," and "cultivation of the mind." He added that stints supported mutuality and cooperation among workers and lessened the possibility of "hoggish behavior," "anarchic competition," and other forms of individualism.[77]

The benefits particular to mining included the reinforcement of safety procedures, which often deteriorated in the subcontractor's frenzied work environment, and protection from overproduction, which inevitably led to a market glut causing lower prices and wages. The stint also provided a defense against unemployment because work could be paced and prolonged. Finally, the quotas allowed some workers—those fortunate enough to toil in "easy coal"—to meet their output rather quickly and go home. They could then pursue personal and community interests related to family matters, ethnic associations, brass bands, choirs, church groups, learning circles, civic organizations, and the myriad other activities available in mining communities. Priscilla Long described the fundamental difference between labor and management in this regard: "The miners' concern with the quality of life clashed with the operators' concern with profits."[78]

In testimonies before the 1902 Strike Commission, miners stated that an output norm dating to at least 1895, and probably much earlier, called for two or three cars per worker per shift, depending on the size of the car.[79] The practice remained a cornerstone of the work culture well into the twentieth century, as Italian immigrant miner and, later, PaCC superintendent Lewis Casterline recalled about the 1920s:

> Whether they were Poles, Slovaks, Lithuanians, Italians, or whatever, they'd always do like that [puts up two fingers] and I was always wondering why they were doing that. [So his father told him:] We couldn't talk Irish, we couldn't talk this or that, but when you saw them two fingers [that meant] two cars a man, don't load no more than two cars a man.[80]

Figure 21. Clarence Darrow, 1902. (Library of Congress)

Clarence Darrow, the UMWA's chief attorney during the Strike Commission hearings, accused the companies of hypocrisy when he showed that coal managers had often limited the product going to market in order to raise prices.[81] He also denied that production lagged because of the stint. The real problem, he told the commissioners, had been an insufficient number of coal cars in the mines, as indicated by an article published in a local newspaper, from which he read:

> Today nearly all the collieries in the Pennsylvania Coal Company in Pittston and vicinity are idle because of the scarcity of cars, and yet the company's representatives are in Philadelphia telling the Anthracite Commission that the output has been restricted by the union. At several collieries in this city the miners have not been able to make a full shift in three weeks because of the lack of cars. The local unions should pass resolutions denouncing the stories of the corporations.[82]

Moreover, he added, even when the union became involved in directing the distribution of cars, it was a matter of fairness:

> I do not deny [that] miners have sought to regulate the crusts that have been thrown to them so that one man should not have a loaf while the other has nothing but they have not done it because there are too many [empty coal] cars. If those men were furnished the work and the cars according to their own statements, there will be no restrictions anywhere in this region. Let them take their own written records furnished to this Commission and see the story they tell. I do contend ... that where there is not work enough to go round, common justice and common human-

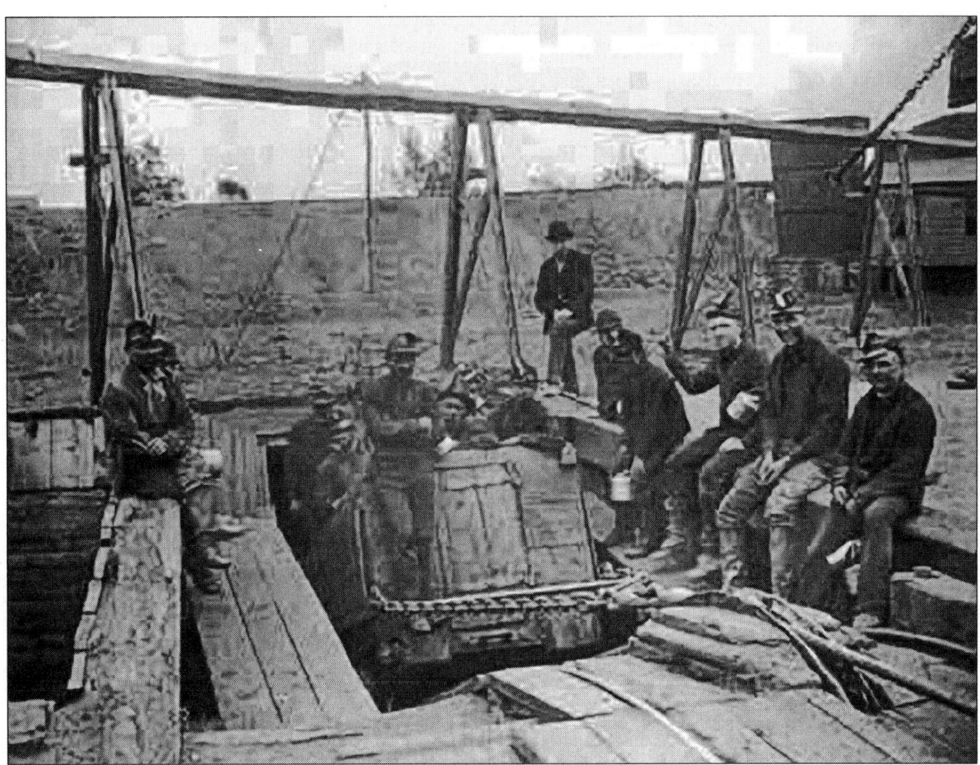

Figure 22. Shift change at a slope mine.
(Courtesy of Pennsylvania Historical and Museum Commission)

> ity would say that one man should not jump in and eat all there is and leave the rest hungry, and that is all the miners have attempted to do in this case.[83]

In opposition to such arguments, the operators regarded the quotas as defiant and uneconomic acts that challenged their authority and profitability. One company official criticized the miners' "independence" and, by implication, the stint: "He comes out when he pleases," he said. "We are not about to control him."[84] Just as important, the operators saw the UMWA as an obstacle to higher yields. D&H foreman Thomas R. Thomas emphatically stated that he did not want "the Union to run the company"; rather, he "wanted to control the men" in order to insure greater output.[85] Predictably, therefore, the operators refused to recognize the UMWA even though they unofficially agreed to bargain with the union's leaders.[86]

The Persistence of the Subcontracting System

Warne's assessment in 1903 that the anthracite subcontracting system "is gradually being abolished through the efforts of the United Mineworkers of America" missed the mark, for the union only temporarily slowed its advance.[87] Carter Goodrich's declaration in 1925 that subcontracting "was checked by the increase of the union's strength" after 1900 reflected the same error.[88] More recently David Montgomery, in 1979, mistakenly concluded that subcontracting remained a quarrelsome matter "until large numbers of Italians, enrolled in the IWW, fought for the suppression of the [sub]contract system in 1916."[89] The UMWA's official historian, Maier Fox, drew a similar conclusion in 1990: "Through the grievance process [following the labor-management agreement of 1916] the miners then won another long-standing goal, the elimination of a [sub]contracting system under which low-paid laborers were hired to do most of the work on a day-wage scale."[90]

Subcontracting, in fact, endured far past the Strike of 1902 into the 1950s, despite the UMWA's persistent demands for its elimination. The UMWA called for an end to the individual contract in every labor-management negotiation between 1902 and 1923, and again in the 1930s, 1940s, and 1950s. The demand as stated in 1906 was typical: "That no contract miner shall have more than one working place at the same time," nor employ "more than two laborers at the same time." The operators continued to resist the idea:

> We have no such legal or moral right to establish such arbitrary rules. Men must be free to act for themselves. Where the conditions are such as to safely permit men to work we cannot consent to limiting the ability and ambition of industrious men by arbitrarily agreeing to restrict their opportunities to earn increased remuneration.[91]

Even though the UMWA regularly presented the demand, the union hierarchy invariably set it aside in favor of "business unionism" priorities related to "bread and butter" concerns such as wages and benefits.

The grassroots opposition nevertheless continued. At a UMWA meeting prior to the labor-management negotiations of 1916, PaCC and HC&I delegates remonstrated that one subcontractor and his large labor crew were excavating as many as 30 chambers concurrently. The critics demanded union action against the scheme. When the negotiations reached a settlement and the coal operators for the first time granted the UMWA partial recognition, International union President John B. White mistakenly concluded that "this form of recognition ... destroyed the subcontractor with his *Padrone* system."[92] White's confidence proved unfounded, however, for it soon became apparent that the plan would continue as long as the labor-management agreement did not specifically prohibit it.

White's *Padrone* (Italian for "Boss") reference related to the contract labor system common in Sicilian sulfur mines and elsewhere in European and American industry and agriculture during the nineteenth century.[93] Padronism was originally structured around a labor contractor (the *Padrone* or boss) who gained control over bonded or indentured workers often through nefarious means. He usually paid their way to America or some other country and hired them out to employers in various enterprises. The *Padrone* received payment for his minions' toil and, in return, provided shelter, food, and simple services. In a later stage, the workers received direct payment for their labor, but were required to pay a portion of their wage to the *Padrone*. In both systems the *Padrone* was well known to dispense exploitation and abuse to his charges.[94]

Reports of Padronism in anthracite dated to at least the 1890s. Terrence V. Powderly, former head of the Knights of Labor and mayor of Scranton, addressed the matter before the Immigration Investigation Commission in 1895: "The 'Padrone' system does ex-

Figure 23. John B. White, UMWA President, 1911–1918. (Courtesy of United Mine Workers of America)

Figure 24. Italian Children Rescued from the Padrone Ancarola, 1888. (New York Society for the Prevention of Cruelty to Children, The Italian Padrone Case: U.S. Against Antonio Giovanni Ancarola, *New York: Styles & Cash, 1880)*

ist in every large and small center, and particularly in the mining regions."[95] Instances of labor agents importing Slavic workers were reported in the 1880s.[96] Historian Michael La Sorte researched the Italian Boss System and found that it thrived even after the U.S. Congress passed the Foran Law in 1885 to prevent the immigration of persons who had signed work contracts before entering the U.S. La Sorte reported that many of the emigrants on a ship bound for America in 1902 were laborers who had been secured by two *Padrones* representing "the Lackawanna mines." These same agents had recruited 300 men to work in hard coal one year earlier. La Sorte maintained that the Boss System endured in the U.S. until the end of World War I.[97]

Beginning in the 1910s, PaCC and HC&I employed a small number of Sicilian subcontractors who hired fellow countrymen as laborers in a manner reminiscent of Padronism. As discussed in later chapters, the companies' large

Figure 25. Andrew Chippa, breaker boy, Markle Coal Company, 1902. (Courtesy of Historical Society of Wisconsin)

Figure 26. No. 5 Colliery, Pennsylvania Coal Company, Jenkins Twp., PA, built 1884. (Courtesy of National Canal Museum Archives)

contingent of Italian, especially Sicilian, workers fought against the subcontract in part because of its *Padrone* parallels, as well as the involvement of alleged organized criminals as mining subcontractors.

Although the IWW led a major strike against subcontracting in 1916 (Chapter Three), the system actually continued at PaCC and HC&I for over three more decades. It precipitated numerous strikes, including those conducted under the banners of the United Mine Workers of America in 1920, 1924, and 1928 (Chapters Four and Five), and the United Anthracite Miners of Pennsylvania in the early 1930s (Chapter Six). The system was finally abolished in the northern field in 1952 when District 1 officials bargained for its elimination (Chapter Seven). Yet the goal that occupied three generations of hard-coal workers came as a hollow victory. With a new PaCC subsidiary, The Pittston Company (TPC), leading the way, a more expansive and complex form of tenancy called the *leasing system* surpassed subcontracting during the 1930s. PaCC and TPC began issuing leases for large sections of mines and even entire collieries to a new breed of independent operator, many of whom had been subcontractors. Some of the former subcontractors had ties to organized crime (Chapter Six). As with the fight against subcontracting, Italian immigrant mineworkers were among the most militant participants in the movement against leasing (Chapter Six and Chapter Eight).

Because this volume focuses on the tenancy systems at PaCC and HC&I, the next chapter examines the founding, expansion, and corporate cultures of these Erie Railroad subsidiaries.

Miners Memorial in Jessup, Pennsylvania

CHAPTER TWO

The Erie Coal Companies and the Subcontracting System

That American standard of living to which our friends on the other side refer, covers something besides wages. It embraces the right to sell one's labor without let or hindrance or intimidation or abuse.

— Major Everett Warren, PaCC spokesman, to the Anthracite Coal Strike Commission, 1902

And be it further Resolved, That we the employees of the Erie Co., ask this convention to stand by us in whatever stand we may take to abolish this [sub]contract system.

— Resolution, UMWA representatives from the Erie Coal Companies, 1902

The Founding of the Pennsylvania Coal Company

On April 16, 1838, the Pennsylvania General Assembly passed a law simultaneously chartering two corporations "for the purpose of mining coal, and for transacting the usual business of companies engaged in mining, transporting to market, and selling of coal, and the other products of coal mines."[1] One charter went to the Washington Coal Company, which was founded by a group of entrepreneurs from Carbondale, Lackawanna County. The second went to the Pennsylvania Coal Company (PaCC) whose shareholders came primarily from New York City. The firms had some common investors but distinct boards of directors.[2]

In line with the period's strict business regulations, PaCC's charter permitted a capitalization of $200,000 and ownership of 100 acres in Pittston Township, Luzerne County. The company shipped most of its coal through the North Branch Canal to Harrisburg and other towns along the Susquehanna River to the south. To connect its collieries with the New York and New England markets, the company constructed a gravity railroad that opened in 1850. The line transported coal from the mines in Pittston and Dunmore to the Delaware and Hudson (D&H) Canal at Hawley, Wayne County, 47 miles away.[3]

In 1847, another investor group composed largely of D&H stockholders formed the Wyoming Coal Association. They wanted to buy coal lands, develop mining operations, and send the final product over the D&H Canal. The association purchased the Washington Coal Company and then worked with PaCC directors to combine the companies. The state's General Assembly approved the transactions in 1849 and permitted the firms to merge into one entity called the Pennsylvania Coal Company. The enterprise also received permission to buy 12,000 acres of coal lands in Dunmore, Lackawanna County,

Figures 1 and 2. Two views of the Pennsylvania Coal Company's gravity railroad. (Courtesy of Sal Mecca)

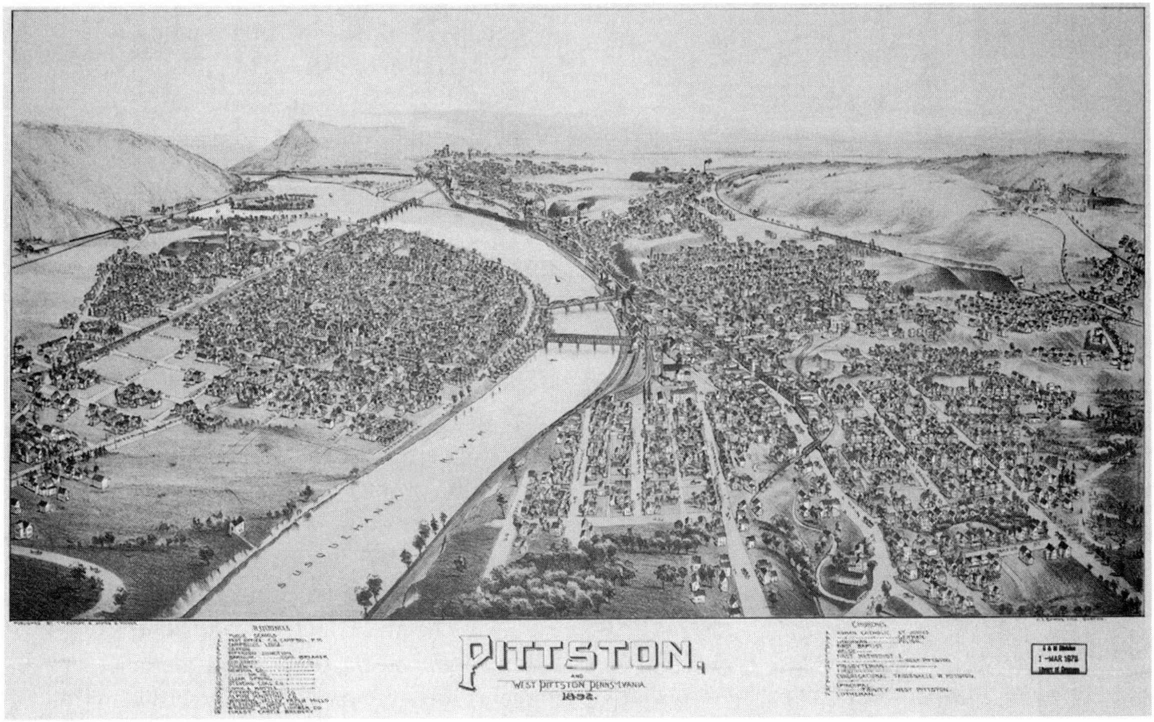

Figure 3. Panorama of Pittston, Pennsylvania, 1892.

and in Pittston Township, Luzerne County. Although publicly traded, the majority of shares were held by a small number of local and New York families who also managed the company.[4] PaCC established the town of Pittston in 1853, making it "both a coal town and a company town right from its founding."[5]

Growth and Expansion of the Pennsylvania Coal Company

In 1850, PaCC marketed 111,014 tons of anthracite and more than tripled output to 361,017 tons the following year. Through the remainder of the decade, production totaled more than 500,000 tons annually. Between 1850 and 1865, the company sent 9.3 million tons to market, for an average of 581,775 tons per year.[6] By 1867, PaCC was operating 22 mines that yielded over 4,000 daily tons, or 1.2 million tons per annum. Unlike other Line companies that arranged purchase contracts with independent operators, PaCC enjoyed the distinction of having "mined all the coal sent by their road—purchasing none from other parties."[7]

At mid-century PaCC ranked as the second-largest producer in the northern field, after D&H.[8] By the mid-1860s, it had developed a large labor force. In the Pittston area, 1,600 colliers worked underground with another 78 on the surface. In the Dunmore area, 320 men toiled in the pits and were sup-

Figure 4. John B. Smith, superintendent of Pennsylvania Coal Company's gravity railroad, 1850–1886, and president of the Erie & Wyoming Valley Railroad, 1887–1895. (Courtesy of Lackawanna Historical Society)

Figure 5. Main office of the Pennsylvania Coal Company, Dunmore, PA, 1870. (Courtesy of Sal Mecca)

ported by 437 others in surface positions, including rail operations. The gravity road employed 63 people in Hawley and 279 along the line, while 164 others handled maintenance.[9]

Early company presidents included Charles T. Pierson (1838–1848), William R. Griffith (1848–1850), Ira Hawley (1850–1851), and John Ewen (1851–1877).[10] John B. Smith served as the superintendent of "The Gravity" from its founding in 1850 until the year after it closed in 1885. He then became general manager of the company's Erie & Wyoming Valley Railroad in 1886 and the line's president from 1887 until 1895.[11]

New York and local investors dominated the board of directors, and Scottish immigrants constituted most of the labor force. A wave of Scots originally settled in Carbondale during the second quarter of the nineteenth century to work on the

Figure 6. Central Colliery, Pennsylvania Coal Company, Duryea, PA, built 1874. (Courtesy of National Canal Museum Archives)

Figure 7. Pennsylvania Coal Company employees, 1882. (Courtesy of Carl Orechovsky)

D&H canal. Many relocated to Hawley around mid-century to earn a living on the gravity road. In a second migration, several moved to Pittston to gain employment in PaCC's burgeoning mining business.[12] Migrants were often related through blood or marriage and most belonged to Pittston's First Presbyterian Church.[13] Among the most prominent were Andrew Bryden and William Law, colliery superintendents, and James McMillan, a high-ranking executive. John B. Smith was another prominent Scot. The Scottish worker and managerial presence at PaCC lasted well into the twentieth century.[14]

Early labor relations seemed fairly harmonious, perhaps because of a relatively homogeneous workforce and a pattern of steady employment and growth. In 1863, PaCC signed one of the industry's first labor agreements with mineworkers in Pittston.[15] The contract likely figured into the workers' reluctance to participate in anthracite's first industry-wide strike in May 1869, called by the Workingmen's Benevolent Association. The PaCC men hesitantly joined 35,000 other strikers in June, but they were among the most willing to end the shutdown in late August.[16] During the national railroad strike of 1877, when arsonists set fire to a section of the gravity railroad, most employees condemned the act and expressed a preference for work.[17]

In 1862, toll disputes with D&H prompted PaCC to contract with the recently reorganized Erie Railway to build a broad-gauge line between the gravity road terminus in Hawley and the Erie's hub at Lackawaxen, Pennsylvania, from where the coal was transported to New York City.[18] PaCC and Erie transacted numerous joint construction and operation agreements over the next three decades.[19] With a rail connection to New York and other points along Erie's routes, PaCC's output grew to nearly 1.5 million tons per year by the mid-1870s. In 1875, the company operated 13 collieries.[20] At the turn of the century, the harvest surpassed two million tons a year, making PaCC anthracite's largest independent operator (i.e., not owned by a railroad), yet the smallest of the six main northern field companies.[21]

Figure 8. Postcard of the Butler Colliery, Pennsylvania Coal Company, Pittston Twp., PA, circa 1910.

Figure 9. Main office expansion of the Pennsylvania Coal Company, Dunmore, PA, 1896. (Courtesy of Sal Mecca)

While a full history of the Erie Railroad is beyond the scope of this study, it is noteworthy that after a series of bankruptcies and reorganizations during the last quarter of the century, J.P. Morgan interests gained control of the line in 1901.[22] Because debt remained a recurring problem, Erie President Eben B. Thomas sought to expand in the highly profitable coal business during the latter half of the 1890s. Accordingly, Erie resumed control of the Hillside Coal & Iron Company (HC&I) in 1895 at a cost of $1 million. The House of Morgan financed the deal.[23] An earlier Erie incarnation had taken over HC&I in 1871 at the behest of former investor and board member Jay Gould, and the company remained a subsidiary through various bankruptcies and reorganizations, including those in 1893.

Established in 1867, HC&I had been a highly profitable enterprise that helped bolster Erie's lagging revenues. The coal company reincorporated in 1873, at which time several small coal entrepreneurs joined the fold.[24] HC&I operated four anthracite collieries in 1895: the Butler and Consolidated in Luzerne County, the Erie in Lackawanna County, and the Forest City in Lackawanna, Susquehanna, and Wayne counties.[25] HCI's output of 934,306 tons in 1899 made it one of the smallest Line companies.[26]

In 1898, Erie gained control of a competing railroad called the New York, Susquehanna, & Western (NYS&W), which, in turn, owned an anthracite coal contracting subsidiary, the New York, Susquehanna, & Western Railroad Coal Company (NYS&WCC). Morgan again provided the financing. The firm produced no coal but contracted for the output of numerous independents, whose coal was sold under the NYS&WCC label. Erie placed HC&I's management in charge of NYS&WCC's coal operations.[27] As important as these holdings were to Erie's coal strategy, however, the big prize was yet to be had.

Erie's Takeover of the Pennsylvania Coal Company

As PaCC reincorporated under a new state law in 1895, it began to fight against the freight increases of its former ally, Erie.[28] To gain independence from the now J.P. Morgan-controlled railway, PaCC joined with a group of other independent operators and railroads in 1900 to build a new rail line. PaCC would own 51 percent of the stock.[29]

Erie and the other Morgan-dominated railroads did not welcome the competition. Before construction began, President George F. Baer of the Philadelphia & Reading Railroad maneuvered with Morgan's Guaranty Trust Company to purchase the Simpson & Watkins Coal Company, which controlled one of the railroads that wanted to invest in the new venture. With one of the key shareholders gone, the project collapsed.[30]

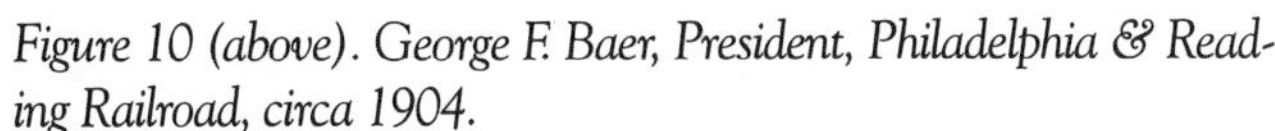

Figure 10 (above). George F. Baer, President, Philadelphia & Reading Railroad, circa 1904.

Figure 11 (right). J.P. Morgan, financier. (Library of Congress)

PaCC's managers remained undeterred and, in 1901, joined with new partners to build the competing line. Because Erie shipped most of PaCC's product and would suffer the greatest losses, company president Eben B. Thomas convinced Morgan and his banks to finance the purchase of almost all PaCC stock. The hostile takeover cost $32 million ($847 million in 2010 dollars).[31] It included one PaCC-owned railroad, the Erie & Wyoming Valley, and the right of way for the planned railway. At an average price of $552 per share, the deal earned J.P. Morgan & Company a $5 million commission. Erie then bought PaCC from the Morgan syndicate at a price described by the U.S. Industrial Commission as the highest ever paid for this type of property.[32] PaCC became the last major independent company to fall to what contemporary critics called "Morganization."[33]

The investors undoubtedly recognized PaCC's financial potential, for "The Pennsylvania [Coal Company] has long been a prosperous company and has earned large dividends."[34] Coal shipments of over 2.9 million tons in 1902 remained on the Erie and, in addition to lucrative transport charges, fresh cash flowed into the railroad's coffers from new coal sales. When PaCC's production was added to the output from HC&I and the coal contracts held by the NYS&WCC, the Erie companies accounted for nearly six million tons in 1902.[35]

Erie officials turned PaCC over to HC&I's management, and the two firms (along with the smaller NYS&WCC) became known as the Erie Coal Companies. Although the firms were under the same management, they had different work rules and pay systems, reflecting separate histories and corporate cultures. Frederick D. Underwood had been president of the Erie Railroad and became the first president of the Erie Coal Companies in 1901, a position he held until 1926. Captain William A. May, an HC&I official, was ap-

Figure 12. Breaker boys at the No. 9 Colliery, Pennsylvania Coal Company, Hughestown, PA, 1913. (Library of Congress, National Child Labor Committee Collection, Lewis W. Hine)

Figure 13. Breaker boys at the Ewen Colliery, Pennsylvania Coal Company, Jenkins Twp., PA, circa 1913. (Library of Congress, National Child Labor Committee Collection, Lewis W. Hine)

Figure 14. New main office of the Pennsylvania Coal Company, Dunmore, PA, circa 1920. (Courtesy of National Canal Museum Archives)

Figure 15. Pennsylvania Coal Company locomotives, circa 1910. (Courtesy of the Lackawanna Historical Society)

pointed as the first manager of the coal subsidiaries. He directed the companies from corporate headquarters in Dunmore, near Scranton.[36]

While the profit margins of the PaCC and HC&I were not disclosed early in the twentieth century, their earnings between 1913 and 1922 were reported as exceedingly high. PaCC stockholders received an average annual return of just under 30 percent during the period, making it the second-most-profitable of the Line-owned producers. The smaller HC&I stood out as the most profitable of any anthracite company, with an average annual return of nearly 50 percent.[37]

Figure 16. Frederick D. Underwood, President of the Erie Coal Companies, 1901–1926.

For the workers who invested their labor in producing coal, however, the return was quite the inverse of those who invested money. The Anthracite Strike Commission of 1902 found that, among the Line companies, PaCC paid the lowest average annual wages and the lowest rates per day. HC&I paid slightly below average on both counts.[38] Clearly, low wages were the other side of the high-profits coin. Because of Erie's perennially precarious debt and profit situations, it seems clear that management continued to expect, and need, high returns from the coal division. Managers no doubt also sought to reduce costs especially in wages, which, as mentioned, constituted up to 75 percent of operating expenditures.

As the twentieth century began, therefore, the Erie Coal Companies' corporate cultures were accustomed to a demand for the highest returns at the lowest costs. Needless to say, the situation was ripe for labor-management conflict. From a company with comparatively harmonious labor relations in the 1860s, PaCC became known for acrimony and discord by the early 1900s. HC&I had a similar history. The combined workforces became perhaps the most strike-prone within an industry well known for industrial strife. With this background in mind, the Companies' position as the leader in fostering the cost-cutting and labor-disciplining subcontracting system and, later, the leasing system, becomes discernible.

The Individual Contract and the Subcontracting System

On the eve of the anthracite strike of 1902, PaCC and HC&I had gained reputations for profitability as well as labor conflict. The companies' vehement anti-union stance was apparent during the strike and in the testimonies by corporate officials before the Anthracite Strike Commission. According to Major Everett Warren, who represented the companies before the Commission:

> This particular organization [UMWA] is an anomaly. It seeks to put every miner on an equality [*sic*] with his neighbor. You cannot do this. Given equal opportunities, the results are unequal. Two factors have heretofore entered into the whole matter of wages—the ability and disposition of the worker is the greatest factor, the place and surroundings the other. The United Mine Workers tend to destroy the first. It is "labor's war on labor."[39]

The companies' general manager and future president, Captain William A. May, criticized the UMWA in his testimony:

> Q: Now, what is the effect as you have observed it, upon the work, the earning capacity, the opportunity to labor of your men since the entrance of the union there?
>
> WM: Why, there is less efficiency and there is a lack of discipline. There is interference with authority and there is tyranny of other workmen.
>
> Q: What effect has it on the ability of the honest worker to earn money?
>
> WM: Why, it limits him.[40]

In an officially prepared statement, the companies described the UMWA as a calamity for the industry: "Until its advent in the anthracite field in 1899, peace and contentment had reigned for a quarter of a century in the mines of this company. Unrest, agitation, turmoil and financial loss have followed its appearance."[41]

When it came to output, the companies joined with other operators in condemning the worker-imposed stints and other aspects of the work culture. They repudiated the demand for a shorter workday and charged that too many men were ignoring the existing nine-hour rule by loading the standard two cars per man and going home early. PaCC superintendent Henry McMillan decried his employees' reluctance to produce:

> ... the witness told that, in 1901, one of the shafts at the Barnum Colliery was thrown idle. He asked the men at the other shaft to load a little extra coal to help out the company. They refused to do so, saying they had an agreement to send out not more than "eight hours" of coal. The witness explained that this means coal enough to supply the breaker for eight hours.[42]

Figure 17. A group of miners.
(Rosamond D. Rhone, "Anthracite Coal Mines and Mining," American Monthly Review of Reviews 26 *[1902], 54–63)*

Figure 18. Judge George Gray, Chairman, Anthracite Strike Commission.
(Library of Congress)

Figure 19. Breaker boys at the No. 9 Colliery, Pennsylvania Coal Company, 1916.

Officials further demeaned the "spirit of insubordination" among the workers, particularly since the strike of 1900.[43] Victor L. Peterson, general superintendent at HC&I, testified that one miner, Richard Holland, "loaded 10 cars [per shift] prior to 1900, and who now cannot be induced to load more than six." When asked what reason men like Holland gave for the reduction, Peterson answered: "They simply say, 'That's enough.'" When Strike Commission Chairman Judge George Grey asked: "Can they load more?" Peterson replied: "They have loaded more."[44] Management blamed the UMWA for condoning the stint and, therefore, obstructing the drive for greater productivity.

As the chief proponents of the subcontracting system, the Erie companies experienced the most severe criticisms from the workers. At the UMWA Tri-District (anthracite Districts 1, 7 & 9) convention in March 1902, shortly before the historic strike, Erie delegates offered a strongly worded resolution against the tenancy plan:

> Whereas, The [sub]contract system which some of our men are subject to at present and which an effort is being made by the companies to put in force in several of their collieries under a large scale, which is detrimental to our organization.
>
> And Whereas, Those working for said [sub]contractors are violating the laws of our Union in every respect ...
>
> And Whereas, This system of work has caused more men to become traitors to our organization than any other system of work ever introduced into our coal fields, and subject our men who remain faithful to us by being cut short in their cars to supply [sub]contractors on days when the collieries are idle.
>
> And Whereas, The Erie Coal Compan[ies] *are the foremost in having this system put in force* as through this system they can have their coal mined thirty per cent less by giving it out to [sub]contractors who will make at least thirty percent on each man who works for him. ...
>
> *Resolved:* That this convention takes into consideration the danger of this [sub]-contract system *especially under the Erie Coal Co.*, and force said company to give up said system of work and have our men work the same, as [they] worked before the [sub]contract system was introduced.
>
> And be it further *Resolved,* That we the employees of the Erie Co. ask this convention to stand by us in whatever stand we may take to abolish this [sub]contract system.[45]

Some local unions imposed strict penalties on members who took the agreements, as in the case of William Zarn, a miner from Dunmore who said he lost money when he was forced to surrender a subcontract:

> He took a [sub]contract to rob pillars and employed thirteen men. The union protested and summoned him to come to a meeting of the local. He told the local he had lost money on the [sub]contract and asked to be given a little more time, that he might recoup some of his losses. The local voted to let him take out ten cars a day for two months. Three days later, the driver boy gave him only six [empty] cars, and when asked why he was not getting ten, [he was] told that the president of the local instructed [the driver] to deliver him only six cars.[46]

Figure 20. Santo Volpe, subcontractor of the Pennsylvania Coal Company, 1939.
(Courtesy of William A. Hastie)

PaCC spokesman Wayne MacVeagh Esq. condemned the union's opposition to subcontracting and held fast to the company's legal right to issue individual contracts: "We have only two radical and irreconcilable differences [with the union]. The limitation of the hours of labor upon the men desiring to work longer, and the right to employ such people as the owners of property think fit ...."[47]

A review of the companies' individual contracts between 1890 and 1902 indicated that the number of subcontractors remained small, perhaps no more than a dozen.[48] The figure was in line with economist Frank J. Warne's report in 1901, which found that nearly 90 percent of the certified miners in the northern field worked as contract miners. The remainder were employed either as company miners or subcontractors.[49]

After greatly reducing the number of subcontracts following the strike of 1902, the Erie companies resumed the program at some collieries in 1909.[50] In that year John J. Boland took a subcontract to mine virgin coal in Dunmore. Over the next 21 years, Boland secured ten additional subcontracts and leases.[51] In 1911, the Yost Mining Company signed a lease that permitted access to second minings and virgin takings in Duryea, Avoca, Hughestown, and Pittston Township.[52]

In 1913, the general superintendent, Joseph P. Jennings, moved to expand subcontracting at PaCC's No. 6 Colliery in Jenkins Township, near Pittston. He offered agreements to Frank McHale and Thomas Mitchell, who declined the solicitation because they feared resentment from fellow workers who, no doubt, recalled the conflict surrounding subcontracting in the 1890s and early 1900s. However, two other mineworkers accepted the proposition: Santo Volpe, the alleged regional crime boss, and Charles Consagra, an alleged criminal associate of Volpe.[53] Both were immigrants from Sicily, and Volpe (and probably Consagra) had experience working in the sulfur-mining industry of their native land. The covenants marked the beginning of organized crime's involvement with the Erie Coal Companies' tenancy business. Over the next three decades, under presidents William A. May, Frederick D. Underwood, C.S. Goldsborough, and Michael Gallagher, the companies remained at the forefront in issuing subcontracts and leases, many of them to criminally affiliated tenants.[54]

Other examples of tenancy dealings in the 1910s included the Nay Aug Coal Company, which secured a lease to conduct second and third minings in Scranton in 1916.[55] The Carney & Brown Coal Company obtained a similar agreement in 1917, to take "certain coal in the Clark Vein, Dunmore Borough." This tenant subsequently obtained rights to additional seams in the same area in 1919 and 1920.[56]

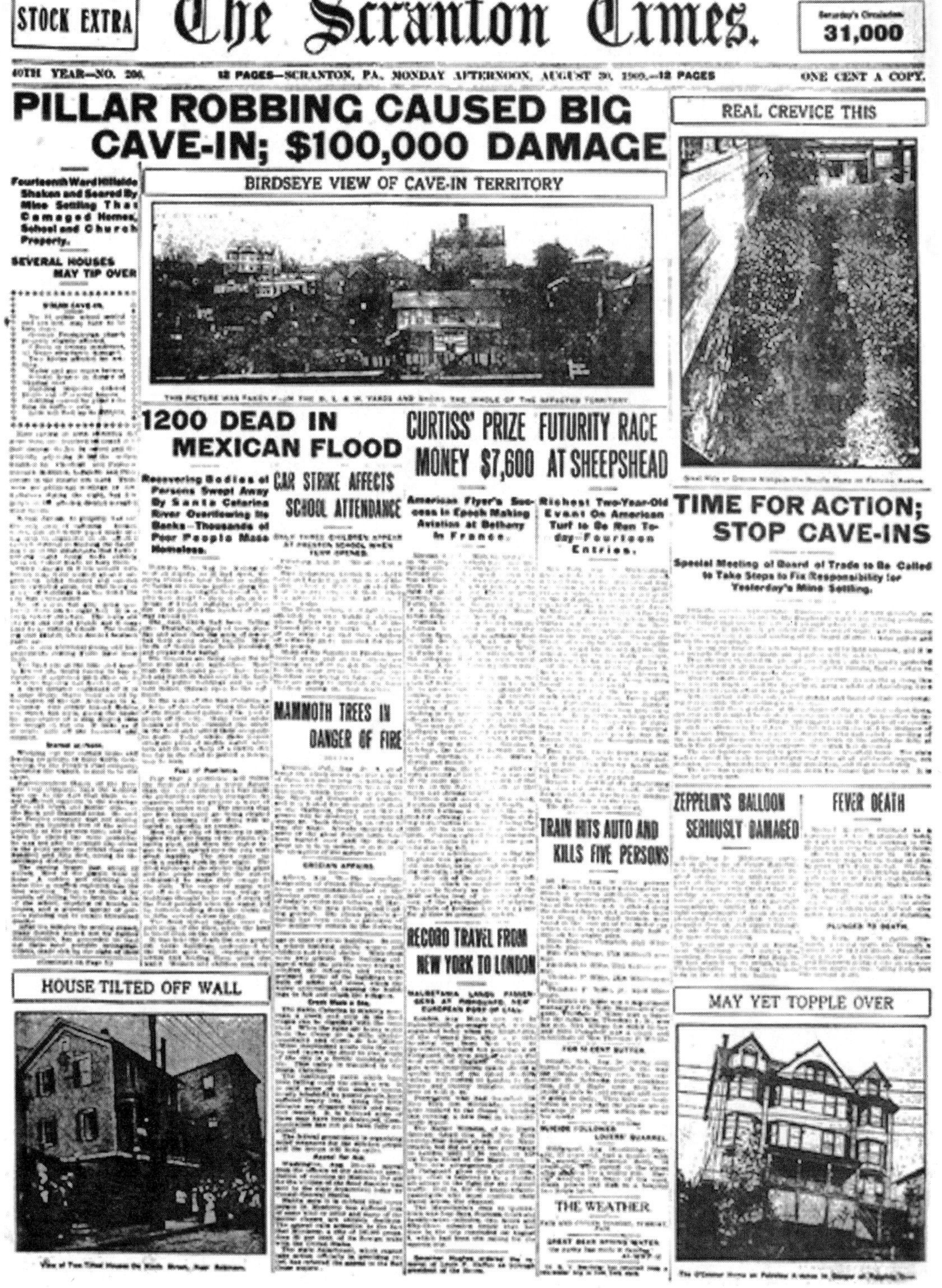

STOCK EXTRA

The Scranton Times.

Saturday's Circulation 31,000

40TH YEAR—NO. 206. 12 PAGES—SCRANTON, PA., MONDAY AFTERNOON, AUGUST 30, 1909.—12 PAGES ONE CENT A COPY.

PILLAR ROBBING CAUSED BIG CAVE-IN; $100,000 DAMAGE

Fourteenth Ward Hillside Shaken and Seared By Mine Settling That Damaged Homes, School and Church Property.

SEVERAL HOUSES MAY TIP OVER

BIRDSEYE VIEW OF CAVE-IN TERRITORY

REAL CREVICE THIS

1200 DEAD IN MEXICAN FLOOD

Recovering Bodies of Persons Swept Away By Santa Catarina River Overflowing Its Banks—Thousands of Poor People Made Homeless.

CAR STRIKE AFFECTS SCHOOL ATTENDANCE

CURTISS' PRIZE MONEY $7,600

American Flyer's Success in Epoch Making Aviation at Bethany in France.

FUTURITY RACE AT SHEEPSHEAD

Richest Two-Year-Old Event On American Turf to Be Run To-day—Fourteen Entries.

TIME FOR ACTION; STOP CAVE-INS

Special Meeting of Board of Trade to Be Called to Take Steps to Fix Responsibility for Yesterday's Mine Settling.

MAMMOTH TREES IN DANGER OF FIRE

ZEPPELIN'S BALLOON SERIOUSLY DAMAGED

FEVER DEATH

TRAIN HITS AUTO AND KILLS FIVE PERSONS

RECORD TRAVEL FROM NEW YORK TO LONDON

HOUSE TILTED OFF WALL

MAY YET TOPPLE OVER

THE WEATHER

Figure 21. Scranton cave-in problem, August 30, 1909. (Courtesy of *Scranton Times*)

By 1920, PaCC and HC&I employed some 30 subcontractors who, in turn, hired up to 1,000 miners and laborers out of a total workforce of 12,000.[57] Despite their small numbers, the subcontractors exerted a good deal of influence over various aspects of mine work. For example, they enforced harsh disciplinary and output demands not seen among the regular contract miners and laborers. Moreover, tenants with organized-crime affiliations introduced a level of fear and intimidation that the workers clearly recognized. The gangsters buttressed a corporate culture bent on managerial control, high output, greater profit margins, and little tolerance for independence. As detailed in Chapter Six, after the Van Sweringen brothers gained control of Erie in 1924, they expanded subcontracting and leasing through a new subsidiary called The Pittston Company and alleged organized criminals began to secure a large number of the agreements during the 1930s and beyond.

Surface Subsidence and Subcontracting

Another indicator of the Erie companies' corporate culture can be seen in management's approach to mine subsidence, including the frequency with which mining practices led to surface caving. While surface landfalls had long been an environmental problem in anthracite, a rash of incidents hit Scranton and other towns in 1910. The crisis prompted Scranton city officials and school board members to seek some remedy.[58] Mayor E.B. Jermyn later appointed an engineer to an ad hoc group that became known as the

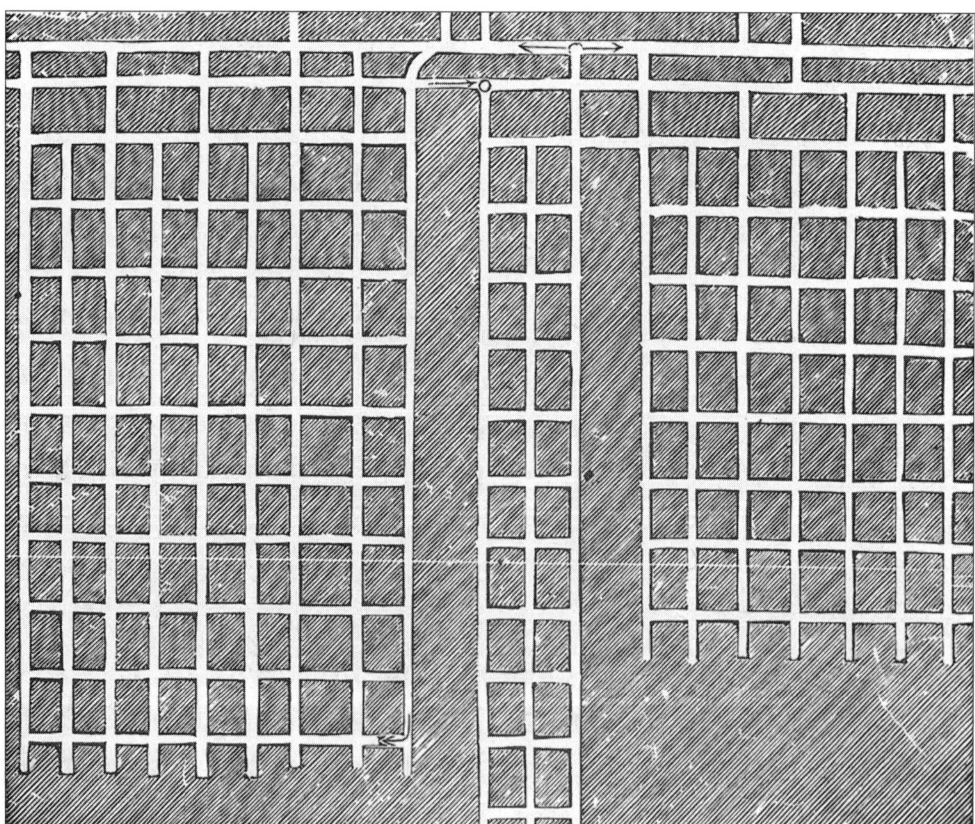

Figure 22. Mined-out section of a coal vein showing rooms, pillars, and gangways. (Andrew Roy, A History of the Coal Miners of the United States, *Columbus, OH: Trauger, 1903, 140)*

Scranton Mine Cave Bureau.[59] The body brought public attention to the problem but could do little more than discuss and protest because the law held the companies blameless.

Random caving incidents persisted over the next several years until 1918, when an alarming increase in ground falls motivated prominent citizens and elected officials to undertake a more comprehensive plan. They joined with three coal companies—PaCC, the Scranton Coal Company, and the Delaware, Lackawanna, & Western Railroad's Coal Division—in establishing the three-person Scranton Mine Cave Commission. The panel was authorized to evaluate subsidence cases and place responsibility when it could not otherwise be determined. The commission did nothing to prevent collapses but facilitated reconstruction where it could establish blame. The responsible companies made repairs of up to $5,000 per incident, and the commission raised an additional $100,000 to help pay the overages by assessing the coal operators on a tonnage basis.[60]

In 1928, after litigating two decades of subsidence cases, Scranton attorney P.V. Mattes wrote about anthracite's "scavenger operators." He described them as "small corporations composed of a group of favored individuals to whom one of the large operators would lease upon royalty a limited section of pillar coal, which the landlord did not care to be responsible for mining."[61] Subcontractors and leaseholders from PaCC and HC&I caused much of the damage by "robbing the pillars" under Scranton, Pittston, and other cities.

Subsidence remained a hotly contested regional issue over the next four decades.[62] Pittston formed a Mine Cave Bureau in the 1920s to bring political pressure and public awareness to what had become an epidemic of land falls that resulted in damage and injury. To seek redress, concerned Pittstonians formed the Citizens' Mine Cave Association under the leadership of merchant and labor activist James A. Joyce. In 1921, the group

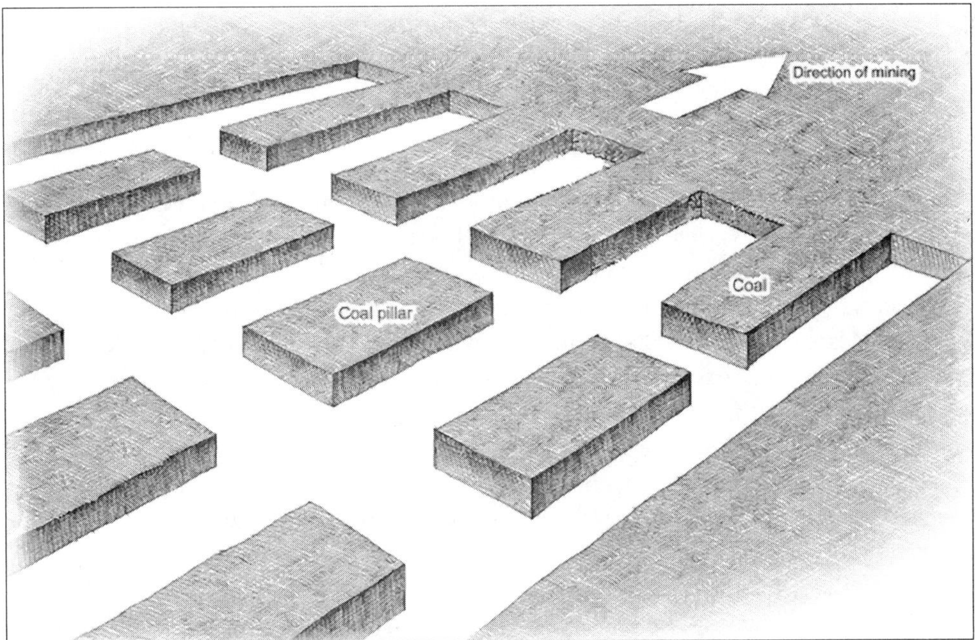

Figure 23. "Roofless" mine showing pillars of remaining coal following a first mining. (Based on George Korson, Black Land, *Evanston, IL: Row & Petersen, 1941)*

brought legal action against PaCC for causing damage to a street in upper Pittston, a case the company fought in court even as its tenants continued to rob pillars.[63] The Erie Companies considered subsidence an unfortunate but unavoidable consequence of their legal right to take all the coal from a particular vein. Their indifference to the human and environmental consequences pointed to a business culture that placed output and income above all else.[64]

In response to the crisis, the Pennsylvania Legislature passed the state's first anti-subsidence law, the Davis Act, in 1913. The anthracite companies immediately claimed the statute was unconstitutional and, after a series of trials, the courts agreed. Citizen pressure did not let up and, eventually, lawmakers passed another measure, the Kohler Act, in 1921. The law made it illegal to mine coal "so as to cause the caving-in, collapse, or subsidence" of public buildings, streets, railroad tracks, homes, cemeteries, and other places.[65]

PaCC took the lead in overturning the measure. On May 27, 1921, the same day the Kohler Act took effect, the company notified home owners Attorney H.J. and Mrs. M.C. Mahon of imminent mining under their property. PaCC had deeded a parcel of land to the Mahons in 1878, which granted the owners surface rights but retained the mineral rights for the company. The deed stated that the Mahons waived all claims to property damage that might result from mining. The company, therefore, argued that the new law was unconstitutional because it abrogated its right to remove all of the coal from the mine, including the pillars. The Mahons sued in the Court of Common Pleas to enjoin PaCC from taking the coal.

The court agreed with the company and ruled the law unconstitutional. Upon appeal, the Pennsylvania Supreme Court decided in favor of the Mahons and upheld the Kohler Act as "a legitimate exercise of the police power" of government.[66] PaCC then appealed to the U.S. Supreme Court. On December 11, 1922, in what became known as "the Penn Coal Case," the high court declared the law unconstitutional by a 5–4 margin. Justice Oliver Wendell Holmes wrote for the majority and Justice Louis Brandeis for the minority. The landmark decision placed clear limits on government regulation of mining companies and, by implication, other private businesses. It clearly established the companies' right to mine coal without regard to surface repercussions.[67]

To promote good will, a few coal operators repaired damaged buildings or compensated owners when cave-ins occurred, but they had no legal obligation to do so. For PaCC and other firms, the successful challenge to the law meant that pillar removal by subcontractors and leaseholders could expand unabated.[68] The laissez-faire outcome from the Penn Coal Case remained in effect until 1987 when the U.S. Supreme Court ruled in *Keystone Bituminous Coal Association, et al. v. Nicholas DeBenedictis* that companies were responsible for damage from cave-ins.[69]

The Erie companies' tenancy policies caused extensive community conflict. As Pittston and Scranton-area residents protested the personal losses and environmental degradations caused by subsidence, the Erie workers grew more resentful towards the subcontractors and leaseholders who caused the problems. The antagonisms were fueled by knowledge that the homes and businesses of prominent families—such as the so-called Anthracite Aristocracy in Wilkes-Barre who owned coal, banking, retail, and other enterprises—had not been undermined.[70]

More important for this study, the companies' tenancy and other workplace policies led to recurring labor-management strife. Even though thousands of PaCC and HC&I colliers began leaving the UMWA after 1903 out of dissatisfaction with the union and its leaders, as the next chapter demonstrates, they displayed a remarkable degree of solidarity in staging wildcat strikes without any union affiliation in 1905 and 1910, and again in 1916 under the banner of the Industrial Workers of the World.

PART II

Labor-Management Conflict at the Erie Coal Companies

Miners Memorial, Forest City, Pennsylvania

CHAPTER THREE

Grassroots Solidarity: Italians, Wildcatters, and the Industrial Workers of the World, 1903–1916

The men have lost confidence in the promises of the company's officials and don't want to end the strike without a written agreement from Captain May.

— Joseph Manos, UMWA organizer, June 1, 1910

If the Pennsylvania Coal Co. officials do not take all possible measures to adjust the grievances of the 12,000 miners who have been on strike for weeks but who are returning to work this morning, the blame will fall largely upon the shoulders of Chevalier Fortunato Tiscar, Italian Consul, Scranton.

— *Wilkes-Barre Record*, June 8, 1910

I stand rigid and deep-rooted with reference to the fight against the existence or perpetuity of such an organization [the IWW] in our community.

— Fr. John J. Curran, Pastor, Holy Saviour Church, Wilkes-Barre, 1916

Labor-Management Conflict Following the Strike of 1902

The United Mine Workers of America (UMWA) did not call a general strike in the years immediately following the strike of 1902, but that did not prevent dissatisfied workers at several coal companies from engaging in wildcat actions. UMWA president John Mitchell and his immediate successors tried to negotiate the disputes but had little success. The companies owned by the Erie Railroad—Pennsylvania Coal Company (PaCC) and the Hillside Coal & Iron Company (HC&I)—experienced several unauthorized walkouts between 1903 and 1916. The actions were among the many labor conflicts in the most strike-prone industry in America.[1]

One shutdown occurred in 1904 when the men and boys at two PaCC collieries, the Barnum in Duryea and the Old Forge in Old Forge, protested against excessive dockage and short-weighing. The workers censured the weighmen at both collieries who automatically deducted a relatively large percentage for waste no matter what the actual contents of a coal car. They cited evidence uncovered by their own check-weighman that the company scales were off as much as 1,000 pounds on a three-ton car, amounting to a wage decrement of $2.10 per worker each day. The company vouched for the weighmen and the scales, and refused to address the complaints. The workers looked to the UMWA for support but found little. They filed grievances with the Anthracite Board of Conciliation (ABC), which had been the arbiter of disputes since the strike settlement in 1903. The body made no decision

Figure 1. Barnum Colliery, Pennsylvania Coal Company, Duryea, PA, circa 1920. (Courtesy of National Canal Museum Archives)

and, instead, referred the matter to Judge George Gray of Wilmington, Delaware, who had served as chairman of the Anthracite Strike Commission.[2]

In 1905, a year when the industry achieved a record output of 61,410,201 tons, workers at PaCC and HC&I walked off the job over the same issues. As part of an effort to keep wavering men within the union, while not condoning an unauthorized strike, Mitchell maligned the company as one of the most anti-worker and anti-union operations in anthracite. The president's words did nothing to end the stoppage. The colliers grew even more disillusioned with the union's inability to remedy their problems.[3]

The deep-seated grievances provoked large numbers at PaCC and HC&I to leave the UMWA.[4] They were not alone. Total membership in the anthracite regions dropped from 85,000 in early 1903 to fewer than 43,000 by the end of 1904. In District 1 the numbers fell from 39,000 in July 1903 to 23,000 in January 1905.[5]

Mitchell tried to hold the union together. He realized the organization's weakened position during the 1906 contract negotiations and argued against a general strike, but colliers throughout the hard-coal fields felt otherwise. The president gave in and authorized an industry-wide work suspension that lasted 36 days. Among UMWA's demands were better wages, a detailed "rate sheet" or wage schedule for over 150 different jobs, specific rules for selecting check-weighmen and check-docking bosses, and payment by weight rather than by the car.[6] When the strike ended the workers saw no wage increase; however, they did achieve modest gains in some of the other areas.

Figure 2. No. 14 Colliery, Pennsylvania Coal Company, Jenkins Twp., PA, circa 1900. (Courtesy of National Canal Museum Archives)

Figure 3. Ewen Colliery of the Pennsylvania Coal Company, Jenkins Twp., PA, built 1915. (Courtesy of National Canal Museum Archives)

The Wildcat Strike of 1910

Motivated partly by dissatisfaction with the terms of the labor-management agreement of 1906, the PaCC and HC&I men called wildcat strikes in 1907 and 1908. When the companies did not address their complaints in either instance, growing tensions boiled over into a general shutdown on May 18, 1910. Nearly 12,000 employees walked off the job, shuttering all ten PaCC collieries and three of four HC&I collieries.[7] The strikers presented management with a list of ten objections that included excessive dockage and low pay. The display of solidarity appeared all the more remarkable when considering that fewer than 100 belonged to the UMWA.

The feeble commitment to the UMWA disappointed the new District 1 leadership, including President Benjamin McEnaney, who characterized the two Erie-owned companies as "practically speaking unorganized."[8] Yet, ironically, after failing to secure a settlement with the protesters, company managers asked District 1 officials to intervene and negotiate a resolution. The strikers argued that, since the great majority belonged to no union and were not bound by any labor agreement, they would ignore the UMWA and deal only with company officials. The ABC, which included President McEnaney as a member, reviewed the case and ordered a work resumption. The strikers ignored the ruling.[9]

On May 24, 1910, the strike turned violent. A pitched battle between picketers and state troopers broke out at the No. 14 Colliery at Port Blanchard, near Pittston. The combat began when a "band of Italian women" gathered around the mine to help enforce the closing. When a strikebreaker thrust one of the women aside, fists, batons, and even bullets began to fly. Both sides inflicted blows.[10] Officer Jaspar Offendafer sustained injury when Pietro Zurla, a "burly Italian," beat him about the head after he had fallen from his horse. Police arrested and jailed Zurla. Three protestors suffered wounds when state troopers and company officials fired indiscriminately into the crowd. Colliery foreman Thomas Muir was accused of shooting an Italian striker who was hospitalized, but survived.[11]

A second disturbance occurred the same day at the nearby Ewen Colliery in Jenkins Township. Before Sergeant Hennig's State Police arrived on the scene, a docking boss named King had been injured, "the Italians first felling him with a storm of stones after which he was kicked and beaten."[12] News reports described the tumult as an "open conflict" and indicated that Italian mineworkers, who constituted about one-quarter of the companies' workforces, were among the most active participants.

Another outburst took place two days later at the PaCC's Central Colliery in Duryea, near Pittston. Luzerne County Sheriff Frederick Rodda led deputies against 300 strikers who had gathered around the mine. In near-anarchical conditions, Rodda's troops staged an assault on the protesters. They prevailed and the sheriff arrested and jailed the group's "ringleader," Angelo Campamassi. It turned out that several of the picketers were not Ewen employees but were from PaCC's Old Forge Colliery in Lackawanna County. Old Forge had become a stronghold of the Industrial Workers of the World (IWW), the radical union that began organizing the anthracite fields in 1907. The UMWA claimed that the Wobblies (as the IWW members were also known) had stirred

Figure 4. Luzerne County Sheriff Frederick Rodda. (Samuel Hudson, Pennsylvania and its Public Men, *Philadelphia, 1909)*

Figure 5. Breaker boys, Ewen Colliery, 1913.
(Library of Congress, National Child Labor Committee Collection, Lewis W. Hine)

up the Central workers and caused the disturbance. However, the evidence indicated otherwise, nor did the IWW claim any credit for the uprising.[13]

The press referred to the boycotters as a "surging mob," a term with dual meaning since Italian immigrants were commonly associated with organized crime. The *Wilkes-Barre Record* reinforced the criminal image by editorializing against the "Black Hand terror" that was evident in certain towns. Though perhaps not intentionally, the editorial associated the striking Erie mineworkers with the Black Hand (*Mano Nero*) gangs. Immigrant Black Handers from Sicily and Calabria had indeed been tied to violence and murder, including a brutal slaying in Pittston that received wide coverage. Yet the inference hurt the efforts of the Italian mineworkers and their colleagues who believed they had legitimate grievances.[14] In fairness to the Italian community, the *Record* also chronicled a group called the White Hand Society, formed by law-abiding sons of Italy who were determined to overcome the stereotypes perpetuated by the criminal element.[15]

The UMWA wavered in responding to the turmoil. District 1 officials debated the wisdom of expending resources on non-dues-paying colliers who worked for perhaps the most anti-union company in the field. The 1902 strike settlement established the quasi-legitimate status of the union within the three anthracite regions, yet PaCC and HC&I management had effectively ignored the organization until situations such as a wildcat strike, which violated the labor-management accord, arose. As historian Joseph Gowaskie observed: "Obviously, the operators had the best of both worlds. They held the UMWA officials responsible for enforcing the settlement, yet they refused to recognize them as union officials whenever it was convenient not to do so."[16]

President McEnaney sought counsel from UMWA International President T. L. Lewis, who immediately sent in two Italian-speaking organizers. Despite a few weeks of rallies and personal meetings, the organizers failed to secure a settlement or boost membership.[17]

The violence continued as the strike carried into June. Six men accosted and shot William Zeto, a watchman at the Ewen Colliery. The attack was said to have occurred because Zeto remained on the job during the strike. A bullet wound to the stomach caused him to lapse into critical condition. An official investigation into the case concluded that the shooting occurred not because of the strike but because Zeto had been attempting to extract money from his underlings. Two Italian miners were arrested for attempted murder, adding to the broad sense of lawlessness in the northern field.[18]

Public fears were exacerbated when two Italian miners from Old Forge were bruised and beaten, presumably by fellow countrymen, when they tried to enter a PaCC mine for the morning shift. Hoping to avoid picketers, they had arrived early but were caught and forced to suffer the consequence.[19] In another instance, an "irate Italian" striker shot at but missed Edwin Chaplain, an assistant foreman at the No. 1 Colliery in Dunmore, as he walked to the pit. The perpetrator belonged to a group of mainly Italian strikers determined to prevent the boss from entering the colliery grounds. The police failed to apprehend the assailant, who escaped into the crowd. Civil authorities called in Lackawanna County Sheriff Connor to establish order in Dunmore.[20]

Figure 6. Captain William A. May, General Manager, Pennsylvania Coal Company. (Courtesy of Lackawanna Historical Society)

In spite of the involvement of protestors from various nationalities, the newspapers continued to attribute the unrest mainly to Italians. As ethnic historians have recently argued, Italian immigrants were, in fact, among the most dedicated social and political activists during the first two decades of the twentieth century.[21] Their militancy derived in large part from the exploitative and repressive economic and political conditions they had experienced in Italy. Once in the U.S., where they were free of many old-world constraints, they stood at the forefront in organizing Italian and other immigrants in demanding better pay and working conditions. As examined further in Chapter Eight, this is a story that has been not only forgotten but overwhelmed by other Italian-American "narratives," including that of the organized criminal. The actions at PaCC and HC&I pointed to a different story, grounded in labor activism and the pursuit of workers' rights and economic fairness.

To help end the stalemate, the strikers elected five representatives to meet with the Erie companies' general manager, Captain William A. May.[22] The group consisted of three Irishmen, one Lithuanian, and one Italian.[23] In line with his position during the Anthracite Strike Commission hearings, May conveyed a willingness to meet only after work had resumed.[24] With the general manager and the demonstrators holding seemingly unmovable positions, a resolution seemed far off.

Figure 7. T. L. Lewis, President of the United Mine Workers of America, 1908–1910. (Courtesy of United Mine Workers of America)

The Erie Railroad's board of directors summoned Captain May to New York for consultation. Board members were troubled by the stagnant mining operations and the consequent loss of revenue. They wanted to know why the strike was called and what could be done to restart the mines. Rumors circulated about a shake-up in Erie's coal division.[25]

Within the UMWA, National President T. L. Lewis and District 1 President McEnaney redoubled their efforts to end the strike. The union dispatched two new organizers to the area. Like their predecessors, they had no success in brokering a compromise. One of their rallies ended in a confrontation when a large group of Italians marched to the meeting hall and, finding the door locked, broke it down and took over the proceedings. They pushed the organizers aside and insisted that the strike continue until the company changed policies.[26]

In a referendum vote on the closings, the men at the Barnum, Butler, Old Forge, and Central collieries voted to restart operations on June 1, 1910. However, the employees at the other idle mines remained adamant and voted to hold the strike. To maintain solidarity, all of PaCC's and HC&I's operations except one stayed shut. The outlier was HC&I's Consolidated Colliery in Avoca where, for reasons that remain unclear, the employees had ignored the strike from the beginning. However, their recalcitrance changed on June 2 when a group of Italian militants from Old Forge intercepted a large contingent of work-bound Consolidated men and convinced them to stand out. The Consolidated employees subsequently supported the strike.[27]

Why were the boycotters so determined? The strike represented the culmination of years of frustration and mistrust. In the 1904 and 1905 walkouts, as well as in a three-day shutdown in 1907, and another protest in 1908, Captain May did not live up to his pledges to redress various grievances. According to UMWA organizer Joseph Manos of Wilkes-Barre: "The men have lost confidence in the promises of the company's officials and don't want to end the strike without a written agreement from Captain May."[28]

To break the impasse, a group of company, union, and community leaders appealed to the Italian Consul, Chevalier Fortunato Tiscar, based in Scranton. Although he had no formal authority in the case, he agreed to use "moral suasion" and talk to both sides. On June 6 Tiscar met with 30 strikers' representatives. To their dismay, he urged a work resumption followed by negotiations. The idea of accepting a plan that essentially echoed Captain May's position was out of the question. The strikers would return only after the company took the first move. That same day the employees at PaCC's ten collieries conducted another ballot on the strike and, in an outcome closer than many expected, six produced affirmative majorities, so the shutdown continued.[29]

When first asked to participate in the talks, Captain May refused to meet with Tiscar and the representatives. However, he eventually agreed to confer with a smaller group. The Italian Consul then formed a committee composed of six workers' delegates, two Italian-born priests (Rev. John Babolo and Rev. William Gislon), and two prominent Pittston

Figure 8. Chevalier Fortunato Tiscar, Italian Consul, Scranton (fourth from left, standing), circa 1940s. (Stephanie Longo, Italians of Northeastern Pennsylvania, *Chicago, IL: Arcadia, 2004.)*

Italians (Joseph Feraini and Vito Branco). The initial negotiating session showed no movement.[30] Tiscar then decided to sustain the dialogue on a personal level, whereupon he achieved some results. As he told an interviewer: "I am greatly encouraged with the progress made and have hopes that the men will return to work pending a settlement."[31]

On June 7, thousands of strikers attended mass meetings in Dunmore, Old Forge, and Pittston. UMWA officials addressed the crowds with an offer to secure a written agreement from management stating a willingness to address grievances within 48 hours of a work resumption. The reaction was predictable: no guarantee, no work. On the same day, Tiscar spoke to a large group of No. 6 Colliery strikers at St. Aloysius Hall in Pittston. He received a standing ovation when introduced, but the cheers turned to jeers when he read a letter from Captain May reiterating the company's position.

The Consul nevertheless persevered and urged negotiations and compromise. His equanimity and sense of fairness, coupled with the economic strains facing the workers and their families, finally had an effect. With much trepidation, the strikers accepted a Tiscar-brokered truce. It essentially upheld Captain May's position requiring a work continuation followed by company actions to redress the grievances. The workers voted to end the strike on June 8.[32]

In the final analysis, mineworkers from numerous ethnic groups trusted the Italian Consul and his deal, as the *Wilkes-Barre Record* editorialized:

Figure 9. An Italian quarter or patch in an anthracite colliery village. (Jay Hambidge, The Century Magazine, *April 1898)*

> If the Pennsylvania Coal Co. officials do not take all possible measures to adjust the grievances of the 12,000 miners who have been on strike for weeks but who are returning to work this morning, the blame will fall largely upon the shoulders of the Italian Consul, Fortunato Tiscar of Scranton.[33]

Yet the *Record* put the onus for the strike on the workers because they had ignored the union:

> The strike of employees at the Pennsylvania Coal Company collieries has been prolonged because of the unwillingness of many miners, not members of the union, to abide by the decision of the Strike Commission [of 1902–03] or to follow the course laid down by it. The miner who refuses to affiliate himself with the union and take an interest in it, helps to build up trouble for himself and for the whole community.[34]

The *Scranton Times*, on the other hand, blamed management and characterized the episode as "a three week strike against corporate wrongs." Had managers investigated the complaints earlier, the *Times* editorialized, the strike could have been averted. The company's dockage practices were the most unacceptable aspect: "Of all damnable and outrageous grafting and stealing, the docking of miners for coal is the worst without the coal being seen or weighed...."[35]

The Erie companies' approach to the matter had apparently not changed since Captain May told the Strike Commission that miners consistently produced coal with up to 21 percent waste.[36] Although the Commission's award permitted the workers to hire and pay a check-weighman as well as a check-docking boss, PaCC and HC&I bosses resisted the idea because, they claimed, it was inconvenient to collect funds from each man's wages, keep records, and pay the checkers. The workers asserted that the opposition derived not from inconvenience but from a company's desire to retain control of an important and "profitable" aspect of the trade—namely docking excessively, short-weighing cars, and cheating the employees.[37] Even when the men at the No. 6 Colliery elected and agreed to directly pay labor activist Alex Campbell as check-weighman, the company refused to allow him on the property on more than one occasion.[38]

However, both the *Record* and the *Times* agreed with Tiscar's final charge to the workers: "Join the union. It is a great organization. Hold your meetings twice a month. Get together and study the award."[39] About one-half of the companies' personnel followed the advice, but within a year membership began to slide.[40] Still, the strike of 1910 demonstrated the workers' determination and unity. As Peter Roberts observed in 1901: "Strikes, above all else, promote the sense of the solidarity of labor."[41]

In the aftermath of the protest, a false rumor circulated that PaCC would no longer hire Italians at the newly rebuilt No. 9 Colliery because of their militancy. More important than the rumor, however, was the new company policy stating that all cars coming out of the mines would be weighed. The workers appreciated the change but protested that the cars were now being run past the scale so quickly that the "stealing" continued. They insisted that the weighman stop the cars on the scale to insure an accurate reading.[42] On a related front, the strike prompted District 1 officials to consider a universal policy on paying check-weighmen and check-docking bosses.[43]

Within two weeks of the armistice, the workers sent a list of seven grievances to Captain May. The items included excessive dockage, payment for "dead work," furnishing supplies such as tools, and abolishment of the "card of recommendation."[44] The company responded to some of the complaints but ignored others, causing broad discontent among the workers.[45]

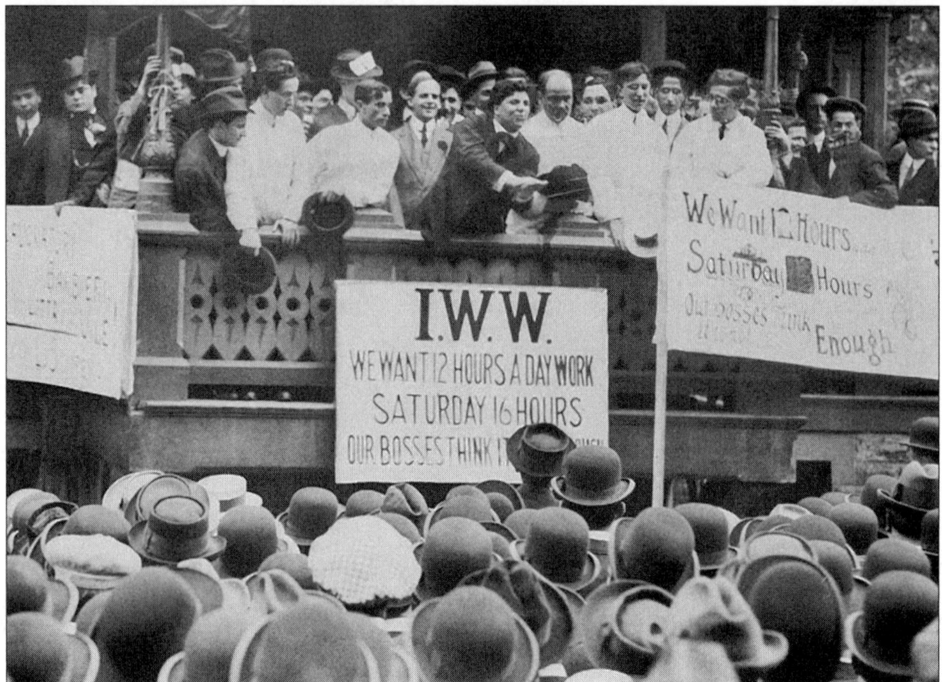

Figure 10. Joseph Ettor of the Industrial Workers of the World addressing striking Brooklyn barbers in Union Square, New York, May 17, 1913. (Library of Congress)

The Industrial Workers of the World and the Strike of 1916

A dedicated group of Anarchists, Socialists, Communists, and other leftists founded the Industrial Workers of the World in Chicago in June 1905. The organization opposed the conservative, craft-oriented policies of the American Federation of Labor, abhorred the wage-labor system, and highlighted the irreconcilable differences between labor and management. Members wanted to establish "One Big Union" for all workers. The group's motto, "An injury to one is an injury to all," grew out of an idealistic, class-based commitment to fairness, equality, and solidarity. It was one of the first American unions to recruit blacks, women, and workers of different skills, while also promoting unity across ethnic and religious lines.[46]

The wave of wildcat actions following the 1902 strike undoubtedly encouraged the Wobblies to enter the anthracite fields in 1907. When the union's organizers arrived, the industry employed nearly 169,000 men and boys whose annual production topped 90 million tons. The new organization gained support in many quarters but experienced greatest acceptance in the northern field among employees at PaCC and HC&I collieries, as well as the Jermyn No. 1 Colliery and the Greenwood Colliery of the Delaware & Hudson Coal Company (D&H).

In late 1908, the union dispatched Joseph Ettor to the area. Ettor was an executive board member and top organizer who was born in Brooklyn, New York, of Italian immigrant parentage. Using his language skills (fluent English, Italian, and Polish; understanding of German, Hungarian, and Yid-

Figure 11. Industrial Workers of the World Logo.

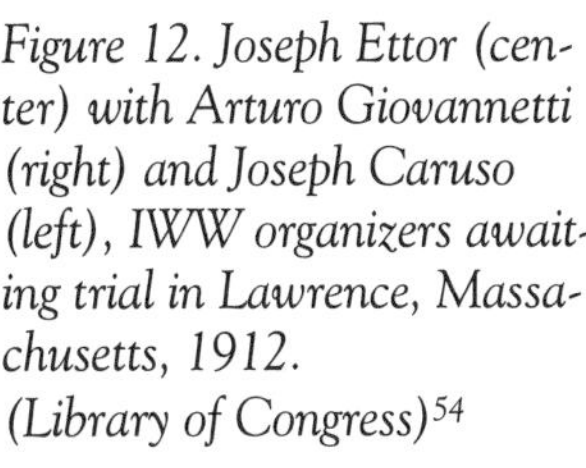

Figure 12. Joseph Ettor (center) with Arturo Giovannetti (right) and Joseph Caruso (left), IWW organizers awaiting trial in Lawrence, Massachusetts, 1912. (Library of Congress)[54]

dish), he worked to build one of the first IWW locals at PaCC's Old Forge Colliery.[47] He focused on immigrant mineworkers because they constituted the majority, were the most aggrieved, and were very accepting of the union's ideology. As historian Walter T. Howard wrote: "The Italians, Lithuanians, ... and other Eastern Europeans who joined, or otherwise supported the Wobblies, found the revolutionary rhetoric of class warfare attractive and empowering."[48]

Just as Italian immigrants were instrumental in building the IWW nationally, the Italians at the Erie companies were the most supportive of the union's message and methods.[49] The organization directed resources to each of PaCC's ten and HC&I's four collieries, which together employed nearly 12,000 mineworkers.[50] As the Wobblies' northern field membership grew to between 5,000 and 10,000, Erie's Italian employees comprised the largest ethnic contingent. They constituted one-third of the companies' combined labor forces by 1916, with Sicilians comprising about one-quarter of the total. The Sicilians were perhaps the most receptive to the union because of their opposition to the subcontracting system, which became the signature issue in 1915 and 1916.[51] As historian Patrick M. Lynch concluded: "Many Italians followed the IWW because the organization opposed the [sub]contract miner system which kept wages low. ..."[52]

However, wages were not the only issue. There were the related matters of control, fairness, *Padronism*, and organized crime. As discussed in Chapter Two, PaCC began issuing subcontracts in 1913 to immigrants from Sicily who had worked in the island's sulfur mines. Indeed, experienced sulfur miners composed 56 percent of the male Sicilian immigrants to the Pittston area in 1910, while mine laborers constituted 32 percent.[53] Dunmore and Old Forge were other towns with large numbers of Sicilian sulfur men, most of them employed at PaCC and HC&I. It was not clear why so many Sicilians gained employment at the Erie coal companies. However, it may have been no coincidence that the companies were the premiere architects of the subcontracting system, for Sicily's mining industry had long been structured around petty subcontractors who hired relatively large work crews, often with the assistance of *Padrone* labor contractors.

For reasons that very likely related to their prior experiences with subcontracting, PaCC and HC&I hired a number of Sicilian miners as subcontractors. They, in turn, took on fellow countrymen as laborers and drove them with "pushers," "hustlers," and "enforcers " in a manner similar to the old-world pattern. The Sicilian workers had seen the sys-

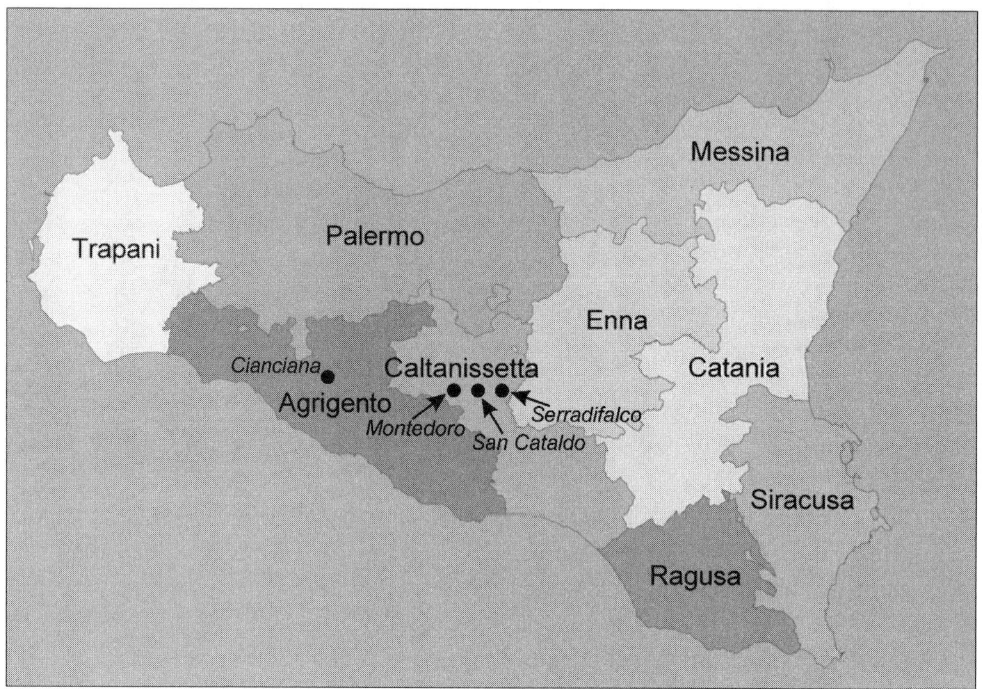

Figure 13. Map of Sicily showing the sulfur provinces of Agrigento, Caltanisetta and Enna, and the towns of Cianciana, Montedoro, Serradifalco, and San Cataldo, from which many sulfur miners emigrated to northeastern Pennsylvania.

Figure 14. Sulfur miners' memorial, Cianciana, Sicily.

tem's harmful consequences in the mines around their former homes in Montedoro, San Cataldo, Serradifalco, Cianciana, and other sulfur towns in south-central Sicily.[55] They knew that organized-crime-affiliated miners were prominent among the subcontractors and they were determined to keep the system, as well as its criminal foundations, out of their American workplaces.[56]

Grassroots support for the IWW grew in 1914 following the murder of Charles Attardo, an Italian immigrant living in Pittston, who was a subcontracting critic and Wobbly supporter. The region's organized-crime element allegedly ordered the assassination because the victim had vociferously criticized gang members and their subcontracting practices.[57]

The companies and the UMWA moved against the Wobblies, reinforced by disapproval from city and county officials, newspaper men, and religious, ethnic, and business leaders. Erie's management even backed the UMWA's button strikes, where all workers had to wear a UMWA lapel button or else a mine would close.[58] Luzerne County District Attorney F.A. Slattery joined with Sheriff George Buss in starting deportation actions against all "rabblerousing" Wobblies, even if it involved false accusations and civil rights violations.[59] Newspapers carried articles and editorials unfavorable to the union. A group of Italian clergymen in Luzerne County arranged a meeting with Sheriff Buss and Italian Consul Tiscar where they denounced the IWW as a radical and un-American organization and affirmed the need for law and order. Presumably reflecting the views of his church, the locally well-known Catholic cleric Fr. John J. Curran of Wilkes-Barre, who had remained a supporter of the workers since the strike of 1902, extolled the UMWA and condemned the IWW. "I stand rigid and deep-rooted with reference to the fight against the existence or perpetuity of such an organization in our community," he emphatically stated.[60]

Figure 15. Luigi Galleani, anarchist and Industrial Workers of the World organizer.

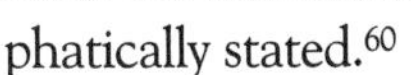

If the IWW received no community institutional backing, the UMWA did not garner much approval from the workers. Indeed, the Wobblies prospered largely because of the wide discontent with the UMWA's leaders and policies. As the IWW's influence grew, the UMWA's membership virtually disappeared at the Erie companies. One PaCC miner, a UMWA member who served as a delegate to the union's Tri-District convention in 1916, decried the Wobblies' broad following: "In speaking of the UMWA in District No. 1, ... we are weaker up there today than we were four years ago. In the Pennsylvania [Coal Company] collieries we have absolutely no [United] Mine Workers you can speak of, nothing but IWWs that governs; the same at [other collieries in] the City of Pittston, Duryea, Avoca, Old Forge and Dunmore ..."[61]

Figure 16. Luigi Galleani on the cover of Cronaca Sovversiva.[69]

Figure 17. Greenwood Colliery, Delaware & Hudson Coal Company, Moosic, PA, 1915. (Courtesy of Carl Orechovsky)

Figure 18. Hudson Coal Company Logo.

The workers' antagonism toward the UMWA had a number of sources. They knew that wages remained frozen between 1902 and 1912, and though a 5.9 percent increase was achieved in 1912, incomes still lagged.[62] Management demanded ever more output while simultaneously violating wage standards and docking excessively. Moreover, promised redresses were few and far between. And, through it all, the workers wanted to know, "Where was the established union?" The UMWA's membership drives after 1912 were intended to boost numbers in anticipation of the 1916 labor-management talks (the existing contract expired on April 1, 1916), but also to prevent the rival IWW from gaining ground.[63]

The situation climaxed in 1916. The Wobblies had established a dozen local unions at collieries in Pittston, Old Forge, Moosic, Dupont, Dunmore, Wilkes-Barre, Scranton, and other towns.[64] They joined in convention on February 6, and set up a district organization with standardized dues and initiation fees. Joe Ettor returned to the Pittston area and, by mid-March, had formed a local with 500 members.[65] When Ettor departed, Wobbly strategists sent in Luigi Galleani, a renowned leader and organizer. Well educated and multi-lingual, Galleani recruited large numbers to the cause.[66] Under his leadership the organization created a list of demands that included an eight-hour day, half-day on Saturday, higher pay rates, abolition of subcontracting, union-appointed mine inspectors, full shift pay when an accident closed a mine, and company-supplied equipment and supplies.[67]

A series of strikes began early in 1916 and continued over the next several months. In January, workers at the D&H's Greenwood Colliery walked out in a wage dispute that lasted four weeks.[68] In April, IWW members set up picket lines at HC&I's Butler Colliery and held

Figure 19. Central Colliery Washery, Pennsylvania Coal Company, Duryea, PA, circa 1915. (Courtesy of National Canal Museum Archives)

Figure 20. Old Forge Colliery, Pennsylvania Coal Company, Old Forge, PA, circa 1915. (Courtesy of Carl Orechovsky)

them for over three weeks until the facility finally closed. In one altercation at the pit, a policeman's billy club sent picketer Tony Policca to the hospital with scalp lacerations.[70]

The protests carried into May when over 1,000 IWW members, accompanied by 200 cheering women, took over the Jermyn Coal Company's No. 1 and No. 2 collieries. The strike occurred because UMWA car runners refused to supply Wobbly-affiliated miners with empties.[71] The boycotts endured throughout the summer into August. Late in the month, over 500 IWW men surrounded PaCC's Central Colliery and forced it to close. Twelve men were arrested and chose jail over fines.[72] Also in late August, 2,500 Wobblies staged a walkout at PaCC and Jermyn Coal Company mines in Old Forge.[73] When Martin Memolo, the Burgess of Old Forge who was also a UMWA organizer, refused to order the State Police out of the borough, he received well-publicized threats, presumably sent by IWW backers. One intimidating letter threatened his life and that of his family:

> Dear Mr. Memolo,
>
> Because we like you, we want to advise you at the earliest possible moment. If you, on Saturday, September 8, do not have the police out of Old Forge, your life will be in our hands, also the lives of your family. If you don't want to die with your family and your house sent up in the air, you must have the state police sent out of Old Forge. So you must do your duty and chase the state police from Old Forge as soon as you can.
>
> Signed, THE IRON HEAD [74]

Memolo relocated his family to a secret destination in Scranton.

In September, the Old Forge and No. 6 collieries were scenes of violent clashes between the competing unions. The union-friendly *Scranton Times* editorialized against the IWW's "arrogance" and "frightfulness," which had "thrown Old Forge into a state of terror."[75] On September 14, the Lackawanna County police force, under the direction of Sheriff Ben S. Phillips, joined with troopers from the State Police, led by Lieutenant Price, in raiding an IWW meeting in Old Forge where they arrested some 267 Jermyn Coal Company employees. Those detained had convened an IWW gathering in spite of the Burgess having denied the required permit. The Wobblies were charged with conducting a clandestine meeting and inciting a riot. The leaders called the gathering to discuss the ongoing labor conflict at the Jermyn mines, where the workforce was evenly split between the IWW and the UMWA. One of the arrested men possessed a list of IWW demands, which the police confiscated and released to the press:

> [An] eight hour day; half holiday on Saturday; $4 a day for miners; $3.50 a day for laborers; $4 for carpenters, engineers, and motor runners; $2.50 for mule leaders; $2.75 for headmen; $2.50 for footmen; $2 for breaker boys; $4 for track layers and $3.50 for helpers; full day when colliery closes down because of accidents; [and] release of all Industrial Workers [who were] prisoners in Minnesota jails.[76]

Joe Nozzi was the IWW leader in Old Forge. Nearly all of the persons arrested were of Italian background, prompting Nozzi to contact the Italian Consul Tiscar, who applied for their habeas corpus release. News reports indicated that, prior to being discharged, 200 of the men repudiated the IWW and agreed to join the UMWA, perhaps because John R. Edwards, legal counsel to Sheriff Phillips, along with Assistant District Attorney Frank Benjamin, started deportation proceedings against non-citizens.[77]

During the course of the nine-month rebellion, picketers shut down all PaCC and HC&I operations, as well as D&H's Greenwood Colliery and Jermyn No. 1 and No. 2 collieries. As many as 12 separate mines were on strike at the same time.[78] While both the UMWA and the IWW perpetrated the violence, the Wobblies received most of the blame. Police arrested scores of them for inciting riots, unlawful assemblies, vagrancies, and other charges. The courts showed little mercy in dispensing penalties. Seventy-five percent of those detained were foreign-born, with Italians constituting the largest single group. After posting bail for some of his members, IWW organizer Arturo Calvini blamed the strikes on tenancy: "the [sub]contract system of the Pennsylvania Coal Company caused the unrest."[79]

Each shutdown brought another loss for the protesters. The combination of police and court biases, stiff jail sentences, threatened and real deportations, unfavorable press stories, and negligible community support shattered the union's efforts. Its seemingly unpatriotic strikes in the midst of the World War I-induced fuel shortage did not help the cause. The same can be said about the arrest of local Wobbly leader Joseph Graber of Scranton for alleged espionage as a German spy.[80] The union also received negative national and local press over a major strike on the Masabi Iron Range in Minnesota, which led to the imprisonment of former northern-field organizer Joseph Ettor, as well as national organizer Carlo Tresca.[81]

In 1917, the federal government arrested and indicted 166 IWW members in a nationwide crackdown. Several deportations followed. Even though 11 alleged members were detained in the northern field as late as 1919 for "spreading anarchistic ideas among the miners in the Nanticoke and Mocanaqua regions," the IWW's days in hard coal effectively ended in 1916 with the failed strikes in the Pittston–Old Forge–Dunmore areas.[82] Despite having been defeated, the Wobblies underscored the severe contentions surrounding the subcontracting system and other workplace grievances, thereby providing a foundation for similar insurgencies in the 1920s and 1930s.[83]

As the leaders of the only viable union in anthracite, District 1's UMWA officials attempted to capitalize on the Wobbly threat by reminding Erie company executives that a business-unionism approach, which emphasized good-faith bargaining, cooperation, and practical issues, was preferable to the IWW's radical vision. The tactic proved unsuccessful and companies remained thoroughly anti-union. The demand for coal during World War I could have facilitated unionization but the outcome proved otherwise.[84]

As part of the industry's record harvest of over 100 million tons in 1917, the Erie companies enjoyed a bumper yield. Yet for the employees, union affiliation and a successful answer to their complaints appeared remote. As the 1920s began, with John L. Lewis as the newly elected UMWA president, PaCC and HC&I still employed the only non-union workforces among the large, railroad-owned coal corporations. It was a situation that would change following a long and bitter strike in 1920, led by union organizers Alex Campbell and Rinaldo Cappellini.

Figure 21. John L. Lewis, President of the United Mine Workers of America, 1920. (Courtesy of Historical Society of Wisconsin)

Miners Monument, Olyphant, Pennsylvania

CHAPTER FOUR

The Labor Rebellions of 1920 and 1924–1925: The United Mineworkers of America and the Erie Coal Companies

Finally in 1920 the men revolted. Ten thousand strong, they came out under the leadership of Campbell and Cappellini. … For seven months they held firm.

— Ben Selekman, *Survey Graphic*, 1928

Rather than have 10,000 men out of work because, as they say, we were still at the Butler Colliery, we now stand ready and willing and do hereby resign entirely as [sub]contractors.

— Subcontractors' resignation letter, Butler Colliery, 1920

You men of Pittston made me and you men of Pittston can break me. You made me District President and Cappellini is big enough and honest enough to acknowledge that. … Cappellini does not want to stay where he is not wanted. If you men of Pittston are tired of me, I will give you the opportunity now to get rid of me.

— Rinaldo Cappellini, UMWA District 1 President, 1925

More Trouble Brewing in the Northern Anthracite Field

The defeat of the Industrial Workers of the World did not end the colliers' desire to overthrow the subcontracting system. The Erie Companies' managers, however, savored the victory and moved to expand the scheme. They were led by a management team consisting of President William A. May (pictured as general manager in Chapter Three, p. 52), Major William W. Inglis, the new general manager, and Joseph P. Jennings, the long-standing general superintendent.

At the same time, the anti-subcontracting forces continued to organize at the Pennsylvania Coal Company (PaCC) and Hillside Coal & Iron Company (HC&I). They also sought to expand their so-called *insurgent movement* to other workers within District 1. Dissidents from numerous local unions joined to convene an unprecedented district-wide "rump" convention for January 29, 1919. In doing so, they circumvented the official District 1 UMWA administration, in which they had lost faith, in order to provide a forum to air complaints and seek remedies. They invited John L. Lewis, then International vice president of the United Mine Workers of America (UMWA), to attend. Lewis declined the invitation but he could not help but realize that trouble was brewing among workers in the northern anthracite field.[1]

The Strike of 1920

UMWA membership at PaCC and HC&I collieries in 1920 had not changed since 1910, when no more than 100 belonged.[2] In addition to company opposition to the union, organizer and former PaCC employee Rinaldo Cappellini observed that the UMWA had earned a bad reputation among Erie workers because by this time subcontractors held most of the elective offices and thereby dominated the local unions.[3] The larger body of employees deplored the apparent collusion among the subcontractors, locals, and managers in a strategy to bolster the subcontracting system and otherwise control the workplace. They viewed the UMWA as essentially a "company union" with corrupt leaders who had no real concern for the ordinary mineworker.

Figure 1. William W. Inglis, General Manager, Erie Coal Companies. (Courtesy of Lackawanna Historical Society)

The workers' approach to the problem changed in the early summer of 1920. Recognizing that the UMWA was the only viable labor organization in the region, the men at PaCC's No. 6 Colliery decided to take over L.U. 1703. Its membership quickly expanded to nearly 2,000 men. During an election of new officers, the members rebuffed a group of subcontractors who ran as incumbents and instead chose Alex Campbell as president and Joseph Yannis as vice president. Campbell was a union stalwart while Yannis had served as interim District 1 president between 1918 and 1919. The membership also elected Campbell as check-weighman, a post to which he had been chosen several times but was usually denied access to the colliery because of company opposition.[4]

Figure 2. William P. Jennings, General Superintendent, Erie Coal Companies. (Courtesy of Lackawanna Historical Society)

Despite a deepening national recession in 1920, the growing use of the subcontract along with other grievances prompted the newly established General Grievance Committee, representing all Erie collieries, to call a general strike on July 16, 1920.[5] The members at L.U. 1703 fully endorsed the idea and saw it as the first step in disposing of the hated tenancy system. The Grievance Committee characterized Erie as the "worst offender in the operation of the [sub]contract system," and resolved that "the principal fight now is for the abolishment of the system."[6] The entire body of 10,000 PaCC workers, along with 2,000 others from HC&I, suspended operations for 200 days. One-third of the strikers were Italian immigrants, with the majority from Sicily.[7]

In addition to the subcontract, the protestors made two other demands. The first concerned dockage. To remedy unfair deductions for rock and other waste, they reaffirmed their right to elect a check-docking boss and check-weighman, and insisted that the company stop obstructing the checkers' access to the collieries. The second dealt with

Figure 3. Alex Campbell and brothers. Left to right: James, William, John, Alex, David, and Robert. (Courtesy of Betty Gugger Rose)

payment for all sizes of coal. The companies had not been paying for the so-called "fines" or small sizes because there had been no demand. However, markets had developed for this formerly useless grade and the employees wanted a fair share of the proceeds.[8]

Well before the strike commenced, Campbell recognized the speaking and organizing abilities of PaCC mineworker Rinaldo Cappellini, a charismatic Italian immigrant who was particularly influential with his countrymen. Campbell was successful in convincing the District 1 office to hire the young man as an organizer.[9] Campbell and Cappellini led the strike movement with assistance from seasoned militants such as Yannis.[10]

RINALDO CAPPELLINI,
President District No. 1, U. M. W. of A.

Figure 4. Rinaldo Cappellini, woodcut circa 1923–24. (Courtesy of Anthracite Heritage Museum)

Campbell began working for PaCC as a boy in the late 1880s. A devoted unionist, he had been involved in the industrial actions at the turn of the century and during the 1910s, and was a prime mover in re-invigorating L.U. 1703 and establishing the General Grievance Committee in 1920.[11] Cappellini was born in 1895 in the central Italian province of Perugia. He emigrated to the U.S. with his

Figure 5. Joseph Yannis, Interim President, UMWA District 1, 1918–19. (Courtesy of Special Collections, Indiana University of Pennsylvania)

Figure 6. No. 6 Colliery, Pennsylvania Coal Company, Jenkins Twp., PA, built 1898. (Courtesy of National Canal Museum Archives)

mother and siblings in 1902. The family settled in Pittston where the eight-year-old lad found work as a PaCC breaker boy, hand picking rock waste out of freshly mined coal.[12] Within a few years he entered the mines as a mule driver. At age 15, a "trip" of coal cars ran over his right arm and left hand. He lost the arm as well as parts of three fingers. A financial settlement brought him nearly $10,000 and the promise of a job for life.

Figure 7. Angelo Ross, breaker boy at No. 9 Colliery, Pennsylvania Coal Company, Hughestown, PA, 1911. (Library of Congress, National Child Labor Committee Collection, Lewis W. Hine)

In 1919, at the age of 24, the compelling Cappellini—described as "young and boyish in look[s]," with "luminous dark eyes" and "a shock of black hair waving above a well-shaped brow"[13]—decided to run for a municipal elective office. Opposed by a company-sponsored candidate, he lost in the primary amid charges of ballot fraud but won the general election as a write-in candidate. According to his wife, Marie, the victory cost him his job. The ill treatment by the company turned him to labor organizing.[14] The next year he joined Campbell as a UMWA organizer in the drive to unionize PaCC and HC&I and, in the process, defeat the subcontract.[15] The movement rested upon what historian David Montgomery called the employees' "sense of moral work relations" because they saw tenancy as an essentially immoral structure populated by thugs acting in conspiracy with the company.[16]

When a national journalist visited Pittston in 1928 to write about another subcontract-related strike (discussed in Chapter Five), he learned that the bitter confrontation of 1920 still affected the mineworkers:

> Even after other operators had accepted the union, the Pennsylvania Coal Company continued to fight bitterly every attempt to organize its employees. The old [sub]contracting system proved an effective weapon in its hands. By picking off immigrant leaders and allotting contracts to them, the company succeeded in erecting barriers in the way of the union organizers. But for a time only. The [sub]contractors, untutored men from the immigrant ranks themselves, were not slow to exploit their unwonted power for their own benefit. There was talk of graft between them and company officials. Repeating the history of contractors elsewhere, as in the sweatshops of New York and Chicago, they ground the men down, paid them low wages, and drove them hard. Finally in 1920 the men revolted. Ten thousand strong, they came out under the leadership of Campbell and Cappellini. ... For seven months they held firm. ...[17]

Strike participant Lewis Casterline admired Alex Campbell's honesty and dedication: "Alex Campbell had eight [*sic*, seven] children. He was as honest as they made them. The company offered him anything that he wanted [i.e., good work places, money, security] and he never wanted it. He fought with them." Casterline also respected Cappellini's organizing abilities as well as his determination to "get rid of the [sub]contractors."[18]

The Murder of Detective Salvatore Lucchino

On July 21, 1920, in the midst of the strike, assassins killed detective and former PaCC mineworker Salvatore "Samuel" Lucchino, age 33, as he walked near his home on East Railroad Street in Pittston. Described as "a tower of strength to the Pittston police force," especially in "ferreting out crimes that have been committed by foreign-speaking people," the well-known policeman had found corruption after investigating the Erie companies' subcontracting system.[19] Police inquiries suggested what the mineworkers already knew: hired gunmen ambushed Lucchino "presumably because of his success in fighting Italian outlaws to justice."[20]

Moments before he died, Lucchino told investigators: "A stranger shot me, but it came through the hands of Charles Consagra." A powerful and affluent subcontractor at the No. 6 Colliery, Consagra was an alleged leader in the region's criminal gang (see Chapter Two). He was suspected of having engineered the slaying to halt Lucchino's inquiry into subcontracting and because the detective would not intervene to break the strike. Police arrested Consagra and found a key witness against him; however, the witness changed his story, purportedly in response to threats, and the charges were dropped.[21]

Figure 8. Salvatore "Samuel" Lucchino, Pittston detective and former Secret Service agent. (Courtesy of Samuel J. Lucchino)

Salvatore Lucchino's story began in the sulfur-mining town of Montedoro, Sicily, where he was born on July 15, 1885. He arrived in the U.S. in August 1903 as part of a family that had organized-crime dealings in the old country. He soon became active in the Pittston area's Black Hand gang. In 1907, he was charged, with Charles Consagra, Santo Volpe, Stefano LaTorre, and seven others, for conspiracy to extort protection money from mineworkers. Nine of the eleven men were convicted, including Lucchino, and each spent one year in solitary confinement in the Luzerne County jail.[22]

Soon after his release, for deeply personal reasons perhaps associated with his prison time as well as his family's criminal past,[23] Lucchino rejected the underworld and became a law officer. He left Pittston to work for the U.S. Secret Service on federal probes, including the well-known Lupo-Morello case in New York City.[24] That investigation dealt with a band of notorious counterfeiters and murderers led by Ignazio Saietta, called Lupo the Wolf, and his brother-in-law, Giuseppe Morello, known as the Gray Fox. Officials called them "the most villainous Italian gang that ever supplied cases for the Homicide Bureau."[25] Lucchino went undercover for the Secret Service and gained the gang's trust, including that of one member whose territory included Pittston. The hoodlum provided Lucchino with a cache of counterfeit bills for distribution in the city, the bogus money later serving as key evidence in the government's case against the gang members.

Despite the obvious dangers, Lucchino served as the main witness in a trial that began on January 26, 1910. All of the Lupo-Morello men were convicted and given prison sentences. The two leaders received 25 and 30 years, respectively, in the Atlanta federal penitentiary. Lucchino confided to his boss in the case, agent William J. Flynn, that he agreed to go undercover because "he had some ancient wrong to right."[26] Following the trial, Lucchino returned home and joined the Pittston police force, where he often cooperated with Luzerne County detectives on organized crime and other cases.[27]

About a year later, on February 7, 1911, an assailant shot Lucchino in the back of the neck as he walked up Railroad Street alongside none other than Charles Consagra. While at first Consagra was not a suspect because he rendered first-aid, further investigation led to a charge of attempted murder. Following Lucchino's astonishing recovery, a trial against Consagra ended in acquittal because of insufficient evidence.[28] After a period of convalescence, Lucchino returned to police work. He survived three other attempts on his life over the next several years.[29]

Over 6,000 people marched in Lucchino's funeral procession, the largest in Pittston's history:

WS SERVICE OF THE UNITED PRESS—TODAY'S NEWS

THREE CENT

CIRCULATI
OPEN TO AD

on Streets and at

THURSDAY, JULY 22, 1920

DAILY EST. BY THEO. HART, 1882. WEEKLY ESTABLISHED 1850

DETECTIVE LUCCHINO ASSASSINATED AT THE DOOR OF RESIDENCE

City Detective Samuel Lucchino, who for a number of years has been a tower of strength to the Pittston police force, especially in ferreting out crimes that have been committed by foreign-speaking people in this city and in capturing criminals from other sections who have come to this part of the coal region, lies cold in death—the victim of an assassin's bullets. Lucchino, returning at the close of what had been an uneventful and peaceful day to his pleasant home, 29 East Railroad street, where his wife and little daughter were awaiting his coming, was shot down at 10:30 o'clock last night just as he turned from the walk to ascend the steps to his home. Four shots were heard to ring out in rapid succession. Mrs. Lucchino, seated in her home, heard the shots and ran outside the door only to find her husband lying on the walk, with blood streaming from his wounds. He was already so weak that he could not speak to her. Mrs. Lucchino's cries quickly brought her neighbors to the scene, and as soon as possible the mortally wound-

[illegible] was removed to Pitts[illegible] [illegible], where he passed away [illegible] after being admitted.

There is universal mourning mingled with intense indignation among the people of the city today over the tragic death of a loyal officer who had, by reason of his devoted service to the cause of law and order, gained the respect and admiration of the law-abiding people of the community. Not only was the dead detective held in high respect, both by the people of his own race and others, on account of his skill as a detector of criminals, but his private life was such as to commend him to all who knew him intimately. Gentlemanly in his manner, attractive in his bearing, he was most courteous to all who had dealings with him. His home life was ideal, and he enjoyed nothing better than with his little family in their comfortable home. A social service worker who knew him intimately told this morning of numerous instances where he had been of great practical help to her in her efforts to improve social conditions and look after the welfare of the foreign-speaking people of the community. His interest in her work, she said, was shown by his voluntary contributions from time to time in aid of unfortunate people of his own race. That he fell a victim to his devotion and loyalty to the cause of law and order was the very general belief expressed on the streets today. His activity and success in ferreting out criminals, and the prominent part that he had taken in the prosecution of criminal cases, evidently made him a mark for some man who took his life in a most cowardly and dastardly manner, when he was unable to defend himself and had not the slightest chance to save himself. One suggestion heard today was that the detective was put out of the way because he knew too much about the ill deeds of certain men.

Shot Down at Door

Detective Lucchino was on duty yesterday, from noon on, and at 10:20 o'clock left City Hall to go to his home. In view of the numerous attempts which had been made previously to end his life, one or more of his friends, in most cases his brother Peter, always accompanied him to his home, but his brother being out of the city, and no other friends being about, he walked the entire distance to his home alone. Realizing that he was a hunted man, Lucchino always carried his loaded automatic pistol in his coat pocket, prepared to defend himself, if attacked. He had walked up East Railroad street to a point opposite his home, where he crossed the street. As he started to ascend the stone steps leading to his residence, the first of the assassin's bullets rang out, wounding the officer in the back, near the spine.

Lucchino wheeled about when a realization of his plight came to him and he seized his revolver, but a second bullet struck him beneath the heart, and he was felled to the ground. It was then alleged that two men sprung from ambush a short distance away, and two other shots were heard. The two men then slunk away. Mrs. Lucchino, wife of the mortally wounded man, hurried from the home and discovered her husband in the throes of death. Officers Hession and McManus, who were on duty in the lower end of the city, were soon on the scene and directed the removal of their fellow officer to Pittston Hospital, where he expired at 11:45 o'clock.

Chief of Police Leo Tierney and County Detective Thomas Allardyce were on the scene in quick time and every effort was made to apprehend the two men implicated in the shooting. Chief Tierney at once notified police officers of neighboring municipalities to be on the lookout for the two men, a description of whom had been secured from a neighbor of

Figure 9. Assassination of Pittston detective Salvatore "Samuel" Lucchino, July 21, 1920.
(Pittston Gazette, *July 22, 1920*)

> ... the homage paid the dead detective this morning was by far the greatest ever given a citizen of Pittston. What seemed to be the entire Italian population of Pittston and surrounding towns were present. County and city officials joined with the working people in showing their respect for the man who during the past several years did much in preserving law and order and in running down criminals among the foreign speaking element.[30]

Burial took place at St. John's Cemetery in the city, where the UMWA erected a headstone acknowledging Lucchino's contribution to the mineworkers' cause. He left a widow and seven children.[31]

The nature of the slaying, coupled with Consagra's iniquitous reputation, indicated that sinister forces lay behind the plot. Lucchino had been warning fellow officers that "the Consagro [*sic*] crowd were the most vicious element in the city."[32] Workers and citizens alike knew that leading subcontractors such as Consagra, Stefano LaTorre, and the alleged top boss, Santo Volpe, wielded strong influence over the subcontracting system. Indeed, the workers saw them as agents of the company on behalf of a plan that resembled labor and business racketeering.[33]

The employees also knew it was highly unlikely that the murder could have occurred without the approval of the main crime leaders. The extraordinarily long lines at Lucchino's wake along with the immensity of his funeral cortege—composed mainly, but not only, of Italians—were not merely intended to honor the slain lawman. They were also demonstrations of resistance against the instigators, who were well known in the community.[34]

Former mineworker and co-author William A. Hastie of West Pittston obtained reliable information on organized crime's involvement in Lucchino's removal:

> Miss Emily Johnson, from West Pittston, was the daughter of a well-to-do family. She was well-educated. She used to write for the *Saturday Evening Post* under a pen name. She was Luzerne County Judge Fuller's secretary for thirty-five years, and that gave her a window to certain information. ... She also befriended many immigrant families in Pittston, poor Sicilian families, Lithuanian, Polish families. She would help them and buy confirmation dresses for the children and things like that.
>
> Some of those Sicilian families confided in her and told her things about the Mafia that they weren't allowed to tell anybody. She told me about such things as apropos to the subcontracting system. Many of the subcontractors were Sicilian. She mentioned a term that I never heard anyone else use, "Mafia bonded slaves," like indentured servants, that were brought over here from Sicily by these Mafioso, the subcontractors, and forced to work for a dollar a day and that sort of thing. That gave those subcontractors a great advantage.
>
> Also, the night that Sam Lucchino, the Pittston city detective, was murdered on Welsh Hill in Pittston by Volpe [and his gang], Emily Johnson, a spinster, a single woman, was walking around, by herself, trying to head it off. She knew it was coming.[35]

After an extensive investigation, police arrested Peter Erico and Antonio Puntario, two hired assassins from Trenton, New Jersey, and charged them with first degree homicide. Court proceedings revealed that they received a few thousand dollars for the job. Arthur James, the Luzerne County District Attorney and a future governor of Pennsylvania, prosecuted the accused killers in separate trials. Juries convicted them of first degree

Figure 10. Arthur H. James, District Attorney of Luzerne County 1920–1926, and Governor of Pennsylvania 1939–1943. Gov. James worked as a breaker boy and mule driver in his youth.
(Courtesy of Luzerne County Historical Society)

murder.[36] They were sentenced to death by electrocution, the punishment being carried out at Rockview State Penitentiary in Centre County on September 25, 1922. Puntario became the first person to die by electrocution for a crime committed in Luzerne County.[37] The condemned men never revealed who hired them. Consagra was the prime suspect in part because of Lucchino's last words but also because he spent a great deal of money on their defense. He died shortly thereafter in a manner similar to Lucchino: a gangland "hit," in Buffalo.[38]

Seeking a Strike Settlement

Lucchino's demise hardened the strikers. Volpe addressed 2,000 of them at a mass rally in August 1920, urging a return to work and even volunteering to relinquish his subcontracts if they complied. The men snubbed the offer and insisted on the complete elimination of the system.[39] A few days later, at another large gathering where Pittston police maintained order, Volpe derided the strikers for being duped by their union leaders, whom he referred to as "agitators." When one of them, Joseph Yannis, declared, "The Pennsylvania Coal Company is a highway robber," Volpe snapped, "Prove it." A number of men converged on the crime boss causing the police to intervene and terminate the meeting.[40]

Born in Montedoro, Sicily, in 1880, Volpe was a sulfur miner who emigrated to Pittston in 1904. By 1908, he had joined his brother-in-law and fellow Montedoran, Stefano LaTorre, in establishing one of the earliest criminal gangs in the U.S., known as the "Men of Montedoro."[41] According to historian Stephen Fox:

> The coal mines of northeastern Pennsylvania brought a Sicilian colony to Pittston and other areas nearby. About a hundred families from one Sicilian town, Montedoro, lived in the Brandy Patch section of Pittston [Township]. There Stefano LaTorre and Santo Volpe, both immigrants, formed "The Men of Montedoro," a second early Mafia family in America. Starting among the coal miners, their influence spread to the United Mine Workers and even the mine owners. Still based in the Brandy Patch, they eventually took over the UMW in their part of the state.[42]

State government authorities acknowledged Volpe as "the first boss of the northeastern Pennsylvania Cosa Nostra family ... [who] began as an impoverished coal miner but, through manipulation, obtained an interest in a number of coal mines."[43] Others knew him as the "King of the Night" for his dark and violent propensities.[44] He purportedly remained the region's crime chief between 1908 and 1933. The U.S. Senate's hearings on organized crime in 1958, chaired by Senator John McClellan, recognized his role in American mob circles.[45] Like other criminal organizations, Volpe's gang operated by "feeding off the common laborer's honest toil and claiming to serve as a means of easing adjustment to American society."[46]

ERICO AND PUNTARIRO ARE ELECTROCUTED TODAY FOR MURDER OF LUCCHINO

Slayers Are Comparatively Calm as They Are Lead to the Chair—No Statements Are Issued—Bodies Are Unclaimed and Are Buried in Prison Yard of Western Penitentiary at Rockview — Final Chapter of the Slaying of Pittston Detective

EXECUTION IS WITNESSED BY TIMES-LEADER REPORTER

Erico is First Led to Chair, His Lips Twitching Nervously as He Glances About—Both Men Are Accompanied to Death Chamber by Rev. B. A. O'Hanlon, of State College—Prisoners Spent Sleepless Night

PAY THE DEATH PENALTY

PETER ERICO.

TONY PUNTARIRO.

By F. X. Welsh, Times-Leader Staff Correspodent.

Western Penitentiary, Rockview, Pa., Sept. 25.—Antonio Puntariro and Peter Erico, convicted gunmen of Trenton, today paid the penalty of death for the murder of Sam Lucchino, Pittston detective, for which crime they were convicted in this county in September, 1920.

From early morning, after a sleepless night, both prisoners sat in their cells with Rev. B. A. O'Hanlon, of State College, for whom they had asked, listening attentively and with noticeable nervousness to the confessors words of encouragement.

At 7:13 o'clock, Peter Erico was led from his cell to the death room, built to the right of the main prison on the third floor, a small bare room, with six windows, the shades tightly drawn and containing a death chair and a circular bench for the witnesses and newspapermen. Erico was guarded by two of the death watch, one of the most cold-blooded crimes ever committed in Luzerne County. The Pittston detective was shot down within thirty yards of his home in full view of his wife who sat in a swing on the porch awaiting his homecoming. When Tony Puntariro, a Trenton gunman, stepped out of ambush in front of the detective he lev-

Figure 11. Peter Erico and Antonio Puntario, convicted murderers of detective Salvatore "Samuel" Lucchino, receive the death penalty.
(Times Leader, *September 25, 1922*)

A few weeks after Lucchino's murder, Cappellini led a group of strike representatives to New York City for a conference with the Erie companies' board of directors. The strikers gave board members a packet containing affidavits that Lucchino had collected, documenting numerous subcontracting-related corruptions. The offenses included wage and safety violations, kickbacks and payroll paddings, and threats and violence.[47] Despite the evidence, the board would not debate the issue, which they defined as a private business matter. Yet they knew that for each day the strike persisted the market missed 20,000 tons of coal and Erie lost tens of thousands of dollars.[48]

The strike continued into August, the same month when the industry-wide labor-management agreement was set to expire. Negotiations between UMWA leaders and operators' representatives quickly reached gridlock, prompting President Woodrow Wilson to appoint another Anthracite Coal Commission. The commission's final decision supported few of the workers' demands, and prompted thousands to threaten a strike if the President effected the recommendations. Wilson rebuffed the protestors and reaffirmed the commission's work, whereupon virtually all anthracite men walked off the job on an impromptu "vacation" that began on August 12, 1920. Though not officially a strike, the wildcat action halted operations across the coal fields. When the president threatened federal action to restart the mines, the workers returned to their jobs after 18 days.[49]

Because the UMWA's first demand again called for the elimination of individual contracts, subcontracting figured prominently in the commission's investigations. While the commissioners recognized a company's constitutional right to issue contracts and, therefore, did not prohibit the subcontract, they did attempt to impose limits:

> The Commission recognizes, however, that abuses of the [sub]contract system of mining are possible and has listened with sympathy to a recital of some of these alleged abuses. In order to remove so far as possible all such abuses the commission hereby directs that upon the complaint of any employee affected, the Board of Conciliation shall review the practices in existing or proposed individual agreements and contracts ... The Board of Conciliation, by way of appeal, shall promptly give consideration to the complaint and render decision in the case.[50]

The commission further directed the Anthracite Board of Conciliation (ABC) to "consider and decide the question of the terms of the contract involving rates of pay and other conditions in such a way as to *protect and conserve the rights of all the employees in the colliery affected*."[51] The commissioners were influenced by the views of Neal J. Ferry, labor's representative to the body, who convinced members to allow for individual contracts only in "isolated cases" such as those caused by "abnormal [mining] conditions," or for tasks such as "development work."[52]

Therefore, the Erie strikers believed that President Wilson's commission had ruled against the subcontract for ordinary mining, which management nevertheless continued to allow and even encourage. The problem was that, while a subcontractor might have secured an agreement for an allowable special condition, such as development work, his crew often ended up mining coal after they had completed their task of opening a new vein. It was up to members and officers in the local union to detect the unapproved mining but, by then, it was usually too late.

President Wilson's threats notwithstanding, the PaCC and HC&I men remained undeterred. They did not have to join the "Vacation Strike" because they were already boycotting their workplaces. More importantly, they defied the president's return-to-work

Figure 12 (left). Santo Volpe, subcontractor at the Pennsylvania Coal Company and reputed organized crime leader. (Courtesy of National Canal Museum Archives)

Figure 13 (right). Stefano LaTorre, subcontractor at the Pennsylvania Coal Company and reputed organized crime leader.

order and remained on strike.[53] They persevered despite entreaties and warnings from the Anthracite Board of Conciliation, Pittston city officials, and District 1 and International UMWA officers. On the other hand, community leaders such as Rev. R.D. Jordan, a Catholic priest, backed the strikers and legitimized their cause. At an outdoor rally of some 5,000 men, Rev. Jordan pleased the throng when he pronounced: "The false impression has gone out that this meeting has been called for the purpose of ending the strike." The stoppage would not end, he continued, until the company addressed the outstanding grievances. The crowd reacted jubilantly, as one report indicated: "From thousands of throats there came yells of 'No, no, the strike is to go on!' Strikers then started cheering and for ten minutes the meeting was in a veritable uproar."[54]

The protestors enjoyed support from various religious groups, newspapers, and other local institutions. However, because some used violent tactics, endorsements began to subside among certain constituencies. Although not a sanctioned strategy, powder and dynamite charges presumably set by strikers became regular occurrences. Militants were blamed for the blasts that hit a number of subcontractors' homes in August and September 1920, including that of Stefano LaTorre. While the subcontractors' agents often retaliated and perpetrated other violent acts, the strikers received most of the blame. *The New York Times* characterized the bitter confrontations between insurgents-versus-company and insurgents-versus-subcontractors as "the keystone in the situation surrounding labor unrest in the northern field."[55]

As the shutdown entered the tenth week, community leaders solicited the intercession of Judge William Tracy of the Pennsylvania Bureau of Mediation and Arbitration. Tracy initiated a series of meetings that included President May, Mayor Walsh of Pittston, workers' representatives, and local businessmen. He stressed the companies' legal right to issue contracts, but the workers insisted on the termination of the overall system as a precondition to further negotiations. The *Scranton Times* supported the strikers in an editorial that acknowledged the "deep seated grievances" leading to the strike, including PaCC's "notorious" hostility to the UMWA and its use of intimidation to enforce subcontracting and frustrate unionization.[56] At the same time, however, in spite of the ruling by the 1920 coal commission regarding the need to establish "special conditions" for subcontracting, the workers did not trust the ABC to judge in their favor, so no official grievances against the subcontract had been filed. The workers would deal only with management.

Growing public pressure, coupled with management's promises to rectify the most pressing complaints, finally brought an armistice. The strike ended during the last week of September, 1920. Over 6,000 workers paraded through downtown Pittston to celebrate the presumed victory over the company and the subcontractors. Coal production commenced on September 27. Within days of the settlement, however, complaints began to mount. The subcontractors returned to a business-as-usual mode and the company seemed in no hurry to implement the promised reforms. Little had apparently changed and, in disapproval, the workers began staying home. After only one week, the General Grievance Committee, under the direction of Cappellini and Campbell, voted for another wildcat strike.

James Joyce, a Pittston shopkeeper who had become a leading negotiator for the men, addressed a group of strike leaders and explained the stakes:

> James A. Joyce, militant merchant and now militant labor leader, aroused 200 members of the executive committee in charge of the strike. ... Joyce said that the officials of the company told him that the Pennsylvania Coal Company was fighting the battle of the Lehigh Valley [Coal] Company and other companies, and that they would have the union smashed to smithereens within a few months. He credited the superintendent with outlining how they would proceed in the work of disrupting the organization.[57]

Apparently, the activism among L.U. 1703 members, coupled with the company-wide strike, provoked the region's coal operators to such a degree that they wanted to use the Erie companies as a vehicle to break the UMWA. The idea paralleled the national anti-union crusades within numerous industries during the 1920s. The workers' fight against anti-unionism at PaCC and HC&I prompted social critic J. Louis Engdahl to extol "the spirit that has made the miners' union [at the Erie Companies] the hope of the American labor movement, capable ... of breaking the nationwide open-shop drive against all organized labor."[58]

As the suspension persisted and revenues continued to fall, the board of directors initiated another meeting with a strikers' committee that included Cappellini, Joyce, and Mayor Walsh. After a brief discussion, and to the representatives' immense surprise, the board agreed to eliminate subcontracting at all Erie coal mines. Astonished and jubilant, the negotiators traveled to company headquarters in Dunmore for a meeting with President May. As promised, the chief executive consented to remove all subcontractors and their crews after the protestors returned to work.[59] The resolute and united mineworkers had achieved a momentous triumph.

On October 5, 1920, seven subcontractors at the Butler Colliery issued a public letter of resignation:

> We the undersigned, comprising the entire [sub]contracting force of the Butler Colliery, ... desire to make an explanation of our side of the case.
>
> During the past several weeks we have been reading daily that some eight or ten thousand men were out of work because of the [sub]contract system. It was a surprise to us that the men were dissatisfied with their employment, as we never had a

word of complaint that was not straightened out with them. We were led to believe that they were perfectly satisfied with conditions. ...

Rather than have ten thousand men out of work because, as they say, we were still at the Butler Colliery, we now stand ready and willing and do hereby resign entirely as [sub]contractors for the above-mentioned company and reserve only the right to ask for a separate chamber for ourselves in order that we may earn a living. We trust this will be sufficient to bring about an immediate end to the strike that has been a menace to the entire community.

Respectfully yours,

Pasquale Adonizio, Joe Adonizio, James D. Adams, Mike Naples, Enrico Wassinci, Charles Adonizio, Joe Ferretti.[60]

The New York Times reported that 23 additional subcontractors resigned at other collieries, meaning that PaCC and HC&I had employed at least 30 such tenants.[61] The action reinforced sociologist Charles Tilly's observation regarding collective action: some grass roots campaigns persevere so forcefully that "authorities have made concessions."[62]

On October 7, 1920, thousands of elated mineworkers returned to their workplaces with the belief that the subcontracting system was, at last, gone. The cumulative effect of industrial shutdowns, Lucchino's death, worker solidarity, dogged negotiations, the commission's rulings, and community support had led to an apparently historic outcome. The company returned all chambers to contract miners who worked in the traditional manner at one coal face with one or two laborers. Production stayed at the culturally sanctioned two or three cars per man. With the detested structure removed and the company weakened, 12,000 PaCC and HC&I employees joined the UMWA en masse. All of the northern field's major coal producers now had the union shop.

However, the triumph proved short-lived. After only three months, in January 1921, over 1,000 employees at HC&I's Butler Colliery walked out when the company reinstated 13 former subcontractors despite claims that they would not hold subcontracts or supervise laborers in the usual manner.[63] The men did not trust the company or the subcontractors and took the rehirings to mean that the dreaded system would soon return. Violence again flared. Numerous fights erupted on the picket lines. James Joyce's home as well as his business establishment suffered extensive bomb damage, apparently at the hands of strikers who thought he was too willing to compromise.[64] The wildcat closing was eventually settled but the men realized more than ever that, despite statements to the contrary, the company was not about to surrender the subcontracting system without a prolonged fight.[65]

Figure 14. Pasquale Adonizio (aka Tony Rose), subcontractor at the Pennsylvania Coal Company. (Collection of Robert P. Wolensky)

Figure 15. Butler Colliery, breaker and washery, Hillside Coal & Iron Company, Pittston Twp., PA, circa 1920.
(Courtesy of National Canal Museum Archives)

Figure 16. Scales at the Ewen Colliery, Pennsylvania Coal Company, Jenkins Twp., PA, circa 1920.
(Courtesy of National Canal Museum Archives)

District 1 Politics and Subcontracting: The Cappellini Presidency

Between 1919 and 1923, PaCC's and HC&I's mines worked no more than four consecutive months without a strike. In one instance, in May 1921, over 1,000 men at the No. 6 Colliery walked out when Santo Volpe, who had taken a rock subcontract to quarry a tunnel between coal veins, was found mining coal. The workers argued that the takings violated the 1920 commission's ruling and, in a show of solidarity, they shuttered the operation. In July 1921, when Volpe arrived at the Ewen Colliery's No. 4 Shaft after having secured another rock tunnel subcontract, grievance committee members met him and his crew and warned: "The men do not want you here." The episode precipitated a 12-week strike.[66] Numerous other walkouts occurred involving subcontracts as well as wage rates, work rules, and dockage.[67]

Other anthracite companies also experienced labor protests over these and similar matters. The general worker dissatisfaction carried over into the District 1 election in 1921, when union activists William J. Brennan (a PaCC employee) and Enoch Williams (a Vacation Strike leader and Hudson Coal Company employee) were elected president and secretary-treasurer, respectively. The vote signified the district-wide displeasure with work and labor relations, including the growing unemployment from the national economic downturn of the early 1920s. The new administration promised to stabilize the job market and confront objectionable practices.

As one of the leading union organizers, Cappellini spoke at the District 1 convention in July 1921. He reproved the subcontractors not only for violating the labor-management agreement, but also for their moral degradations:

Figure 17 and Figure 18. Rinaldo Cappellini addressing mineworkers, 1922. (John Jennings photographs, courtesy of F. Charles Petrillo)

Figure 19. William J. Brennan, President, UMWA District 1, 1921–23. (Courtesy of Special Collections, Indiana University of Pennsylvania)

> [Subcontractors] were not satisfied to put you in the mines and make you load six and seven cars a day for the forty and fifty dollars in two weeks ... no, they even wanted your dear wife and daughter to play with while you were working nights. They wanted your very flesh and blood and they wanted the honor of the poor innocent daughter and wife, which has been proven in different hotels in Lackawanna and Luzerne counties where they slept.
>
> [I]n the Pittston District you had no organization. Your organization was only in existence to this extent that the president of the local would be an assistant foreman; the secretary of the local would be a [sub]contractor; the vice president of the local was a driver boss and the committee [were] all politicians.

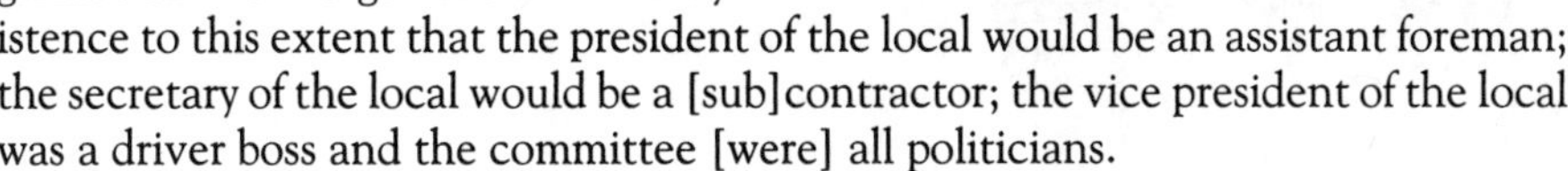

> [A] man would have to go to the foreman and say, "Mr. Man please give me a job, please," after giving fifteen or twenty years of his life to the company; or go to the [sub]contractor and "please, have you got a job for me." And their earnings, forty-five or fifty [dollars] for some, but others sixty or twice as much because he stood in with the boss.[68]

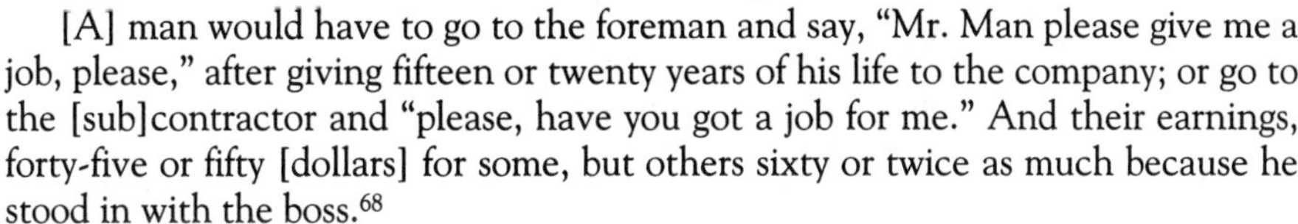

Although Cappellini supported Brennan's candidacy, they soon split over a policy issue that led to Cappellini's dismissal.[69] He then ran against Brennan in 1923 and won the presidency.[70] Elimination of the subcontracting system stood out as Cappellini's main campaign promise. Lewis Casterline recalled the election: "He was against the [sub]contractors. Get 'em out. That's how he won."[71]

Many thought that UMWA President John L. Lewis would not allow the results because of Cappellini's militant reputation. However, Brennan had also been a rebel who was not afraid to challenge Lewis. To complicate the situation, Brennan and his supporters believed that Cappellini won through ballot fraud. They wanted to overturn the election at the upcoming District 1 convention. However, to the great surprise of the delegates, Cappellini and Brennan decided to put their differences aside early in the convention:

> The one-armed leader rose and went to the platform. "You wonder what has been patched up," he exclaimed. "You might just as well remove that wonder from your minds. You showed confidence in me when you elected me President. If you haven't the same confidence now, I don't want the office." Thereupon he changed the subject and talked about corruption in the district. As he started to leave the platform Brennan came forward. They shook hands and Brennan raised his voice to the audience: "If for no other reason than that we are shaking hands, you ought to be satisfied. This is the first time I have spoken to brother Cappellini in a year. I want his administration to be a success." The delegates sat in mute astonishment. Then they began to cheer.[72]

Figure 20. Rinaldo Cappellini organizing near the Harry E Colliery, Swoyersville, PA, 1922. (John Jennings photograph, courtesy of F. Charles Petrillo)

No doubt Lewis played a key role in brokering the truce after deciding that he could work with Cappellini. As author Cecil Carnes concluded: "It became just another 'Lewis convention.' Young Cappellini went to the president and said he was 'for the administration' and 'wanted to go along.' "[73]

The other successful candidates on the Cappellini ticket were George Isaacs, vice president; Enoch Williams, secretary-treasurer; and Alex Campbell, International board member. The new administration promised to expedite grievance filings and encourage cooperative industrial relations. The president said that he anticipated fewer wildcat strikes when the tenancy issue had been solved. Within one month of the election, he convinced the PaCC General Grievance Committee to call off a subcontracting strike so that he could pursue a negotiated settlement.

Figure 21. UMWA President John L. Lewis, 1922. (Library of Congress)

Figure 22. White House Coal Delivery, Washington, D.C., October 1922. (Library of Congress)

The Strike of 1924–1925

Reflecting the wide discontent over wages, unemployment, and working conditions, the UMWA International office authorized industry-wide anthracite strikes in 1922 and 1925–26.[74] Elimination of the subcontract was again among the union's demands in 1922. After the parties reached an impasse, President Warren G. Harding appointed yet another Coal Commission that issued a temporary, 12-month settlement, which merely extended the existing contract. The commission gave the subcontracting issue detailed consideration and reinforced a corporation's constitutional right to engage in contracting. By citing cases such as Locher v. New York (1905), where the U.S. Supreme Court struck down a law limiting the workday to ten hours on grounds that it infringed upon a company's freedom to contract within the meaning of the due process clause of the fourteenth amendment, the body seemed to come down more on the side of the companies than had the Commission of 1920.[75]

The stopgap agreement expired one year later, on August 31, 1923, whereupon company and union negotiators began a new round of deliberations. For the first time in two decades, the UMWA did not demand the elimination of the individual contract, but adopted a new strategy by insisting that all subcontractors conform to the negotiated wage rates and work rules. The operators supported the idea and both sides agreed that "No contracts shall be made with individual employees at less than the prescribed scale rates or not in keeping with customary practices."[76] The overall compact, which brought no wage increase or other major benefits, was set to run for two years. Lewis called the settlement a victory because arbitration had been avoided and union members suffered no wage losses.

Figure 23. Marie and Rinaldo Cappellini, circa 1940. (Courtesy of Kim Cappellini)

Because of the 1923 armistice, the UMWA did not mention individual contracting when the agreement expired in 1925. Nevertheless, the negotiators' inability to resolve various differences led to anthracite's longest strike to date—170 days. The coal operators saw the sanctioned closing, which carried into 1926, as the last straw. Samuel Warriner, president of the Lehigh Coal & Navigation Company and the operators' chief spokesman, called it "a finish fight" through which he hoped to break the UMWA and expel it from anthracite.[77] The walkout garnered little public support and helped erode numerous markets. The final settlement brought no real gains for the workers and merely extended the existing agreement for five years. Lewis could claim only that the pay cuts demanded by the operators were successfully resisted.[78]

Meanwhile, beginning in 1924, the Erie Coal Companies had a new owner—the Van Sweringen brothers of Cleveland, who were building one of the largest railroad empires in the nation (see Chapter Six). Despite the change, PaCC and HC&I continued to have the most quarrelsome labor relations in the industry. When the Butler Colliery's grievance committee shut the operation in January 1923, and promised to keep it closed until every worker joined the UMWA, the company threatened to abandon the facility

Figure 24. Samuel D. Warriner, President, Lehigh Coal & Navigation Co. (Courtesy of Special Collections, Indiana University of Pennsylvania)

Figure 25 (top). James A. Musto, Member, Pennsylvania House of Representatives, circa 1954. (The Pennsylvania Manual, 1954–1955, *Harrisburg, PA: State Bureau of Publications)*

Figure 26 (below). Daniel Hart, Mayor, City of Wilkes-Barre, 1919–1933. (Courtesy of Claire Hart Cummings and Luzerne County Historical Society)

and shift production elsewhere. Most colliers favored the action but some were becoming much less enamored with constant strikes. In 1925 local union gatherings became highly contentious affairs requiring police to maintain order. Pittston Mayor P. R. Brown blamed Cappellini for the turmoil and banned him from the city. When a local union announced a meeting, Brown ordered police to stop all cars at the municipal line to search for the District president. Cappellini defied the mayor on one occasion and slipped through a sentry post. He secretly entered an assembly hall whereupon he was arrested and jailed.[79]

Marie Cappellini recalled that she, her husband, and their two sons faced constant danger: "We had to get a trained police dog to protect our children because we got threats."[80] Cappellini's followers also suffered recriminations. Blacklisted by PaCC and threatened by hoodlums, James A. Musto of Pittston could find no work in mining and became a barber and grocer, until his election much later as a Representative in the Pennsylvania Legislature.[81]

Organizing activities by Communist groups such as the Workers' Party and the Progressive Miners' Union added to the district's upheaval and polarization.[82] When a group of deputized American Legionnaires broke up a Workers' Party gathering in Wilkes-Barre in early 1924, Mayor Daniel L. Hart applauded the move. He subsequently prohibited "Anarchists or Communists" from meeting in the city. The *Wilkes-Barre Record* agreed with the decision:

> We have in our midst a large number of impressionistic foreigners who come into this country influenced by the fearful oppression they experienced abroad: their minds [are] filled with resentment against all existing forms and order of law. They are to be pitied rather than condemned. But their wiser leaders are in for trouble and they should be curbed in their efforts to sow the seeds of discord.[83]

Several other companies experienced labor strife in 1924. In February, President Cappellini, who had become more conservative under John L. Lewis's influence, warned the 22,000-person workforce at the Glen Alden Coal Company, as well as the 14,000 colliers employed by the Lehigh & Wilkes-Barre Coal Company, that he would "take drastic action against the union leaders" if unauthorized shutdowns occurred. The "drastic actions" would involve the loss of local union charters and banishment from the UMWA. Nevertheless, over 400 men at Glen Alden's Peach Orchard Colliery staged a wildcat strike

STATEMENT IN BEHALF OF THE PENNSYLVANIA COAL CO.

The grievances appearing in the public press and published by a committee, terming itself the "Pennsylvania [Coal Company] General Grievance Committee," are skillfully incomplete and are not representative of the facts. Inferentially they allege a disposition on the part of the Pennsylvania Coal Company to ignore rates and reduce wages through various devices wherever possible.

The Pennsylvania Coal Company denies any attempt to reduce established rates or to depart in any manner whatsoever from the provisions of the existing wage agreement, and it disclaims knowledge of a single instance wherein a grievance has not been heard and decided with promptness.

But decisions which do not fully grant the demands which are made, regardless of the facts involved, are not regarded by the General Grievance Committee as an adjustment of those demands.

The General Grievance Committee has repeatedly instigated strikes to enforce such demands, ignoring the agreement between the operators and the mine workers and the methods therein provided, through the Board of Conciliation and an Umpire, for the determination of all grievances and demands.

The result of the present strike, so instigated, is that eight collieries involving about 11,000 employees, have been continuously idle since November 24th, and the loss in wages to the men has amounted to approximately $1,685,000.00.

PENNSYLVANIA COAL COMPANY

By A.K. Morris

Vice-President and
General Manager

December 20, 1924

Figure 27. Strike statement published in newspapers by the Pennsylvania Coal Company. (The Evening News, *December 24, 1924)*

over deficient payments for the removal of rock in advancing the coalface (so-called *yardage*). Another unsanctioned walkout occurred among 1,800 workers at the company's Woodward Colliery who protested the hiring of non-union laborers.[84] The Lehigh Valley Coal Company experienced a similar cessation of work at one pit.

Cappellini blamed the disorder on Bolshevists, Communists, Industrial Workers of the World, the Workers' Party, and other radicals. He specifically attributed the strikes at the Lehigh Valley's Westmoreland Colliery and the Glen Alden's Woodward Colliery to "Communistic agents."[85] The strikes, he exclaimed, were clear examples of the professed Communist strategy of "boring from within." He added that extremist elements were agitating not on behalf of genuine grievances, but to overturn and undermine industrial standards and established institutions.[86]

Figure 28. Underwood Colliery, Pennsylvania Coal Company, Olyphant, PA. (Courtesy of National Canal Museum Archives)

The dissidents vociferously asserted that Communism was not the problem. At PaCC and HC&I, the great majority favored less radical means and defined themselves as UMWA insurgents who could fight their own battles even if the "official" UMWA did not back them. The main issue remained what it had been for well over a decade—subcontracting—supplemented by complaints over excessive dockage, unfair weights, and low pay. When the colliers discovered that management had recently hired several subcontractors at certain collieries, the PaCC General Grievance Committee did not seek permission from the international or district office to call a general strike; they did it on their own. Over 12,000 mineworkers in ten PaCC and HC&I local unions followed the order and, in late November 1924, another stoppage ensued.

The General Grievance Committee asked Cappellini to call a special district convention to address several district-wide problems. The General Grievance Committee at the Lehigh & Wilkes-Barre Coal Company seconded the idea.[87] However, Cappellini ignored the requests and ordered all protestors to take their grievances to the ABC. It was becoming clear that, since he had assumed the district's leadership position, Cappellini had altered his views on militancy. Despite his own use of unauthorized boycotts in earlier days, he now saw all wildcatters as embarrassments to his administration, violators of the negotiated contract, and threats to his relationship with John L. Lewis. No doubt feeling pressure from the union president as well as company executives, Cappellini insisted that workers honor the labor agreement clause specifying that only the UMWA could authorize a strike. He charged the PaCC and HC&I men with fomenting a dual union and demanded a return to work. As so often happened in the past, the members rebuffed the command.[88]

Figure 29. Powder House at No. 14 Colliery, Pennsylvania Coal Company, Hughestown, PA. (Courtesy of National Canal Museum Archives)

The episode turned violent on the morning of December 24, 1924, when saboteurs ignited a ton of dynamite at PaCC's Underwood Colliery in Olyphant, near Scranton. The initial charge hit the main powder house and set off the entire cache of explosives. The concussion demolished colliery buildings, phone lines, vehicles, and rail cars. It tossed Throop residents from their beds and blew out every window in town. People heard and felt the blast 80 miles away in Port Jervis, New York. Scores were injured but, incredibly, no deaths occurred. Police were quick to blame the act on radicalized strikers. Lewis demanded an end to the boycott but his exhortations went unheeded.[89]

To consider solutions to the standoff, the strikers convened a conference of their ten local unions in Wilkes-Barre. They analyzed the main grievances and reduced them to two. They agreed that if the company addressed both issues, the work suspension could end:

> First, that 62 miners at the Underwood Colliery be placed in the [sub]contractors' section of the mine where they previously worked, but independent of the [sub]-contractors; the company and the grievance committee to arrive at a rate of wages immediately upon resumption of work.
>
> Second, that 50 of the men at the Ewen Colliery, who were idle before the strike, be taken back within three weeks and that the remainder of the idle men at this colliery be placed three-handed [three-person work crews] and to load six cars of coal where previously two men loaded four cars. In the event the men do not get six cars to load they [are] to remain in their places for eight hours and be paid at the rate of $8.40 per miner and $7.70 per laborer. All other grievances to be taken up at once by the company and the grievance committee after work is resumed.[90]

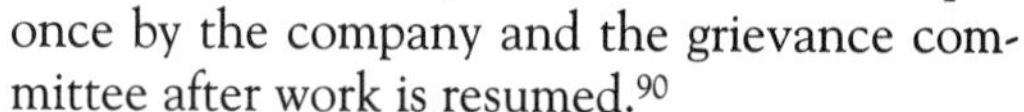

Rev. John J. Curran of Wilkes-Barre attended the gathering. A PaCC breaker boy and mule driver in his youth, he had remained a mine-workers' advocate since he assisted President Roosevelt and John Mitchell with the 1902 strike settlement. Curran agreed to serve as an envoy and submit the proposal to PaCC Vice President A.K. Morris. A meeting was arranged and, despite the clergyman's negotiating skills, Morris dismissed the tender and presented no counter offer.[91]

Figure 30. Rev. John J. Curran, 1919. (Courtesy of F. Charles Petrillo)

Lewis and Cappellini retaliated against the rump gathering by revoking the charters of the ten striking locals. Still, the disorder troubled Cappellini and he tried to find a compromise. At a meeting of 3,000 boycotters on his home territory of Pittston, he offered to resign if the men did not trust his leadership:

> You men of Pittston made me and you men of Pittston can break me. You made me District President and Cappellini is big enough and honest enough to acknowledge that. ... Cappellini does not want to stay where he is not wanted. If you men of Pittston are tired of me I will give you the opportunity now to get rid of me. Some one of you get up and make a motion that Rinaldo Cappellini resign January 7, 1925 and put it to a vote. If a majority wants Cappellini to quit, Cappellini will step down and out.[92]

With the Italians in the lead, the audience rejected the offer. However, although they reaffirmed their former colleague and local union leader, the clear majority stood resolute on holding the strike. Cappellini listened to their complaints but disregarded the pleas for a special convention. As industrial relations at PaCC, HC&I, and other northern-field companies deteriorated, some workers began asking Cappellini to step down.[93]

On January 6, 1925, assassins brutalized Samuel Spachia, age 31, with 11 bullets. A miner at PaCC, he had served as a member of the General Grievance Committee and vice president of the Ewen Colliery's L.U. 1487. A reserved and noncontroversial person, he had recently attended a special meeting at the Underwood Colliery to consider ways to end the strike. The police held two suspects but filed no charges.[94] The motives for the slaying were vague and some thought that Spachia's presence at the meeting, during which he was purported to have said nothing, led to the slaying by workers who wanted to keep the strike going.[95] However, the culprits and their motives remained unknown. Hundreds attended Spachia's wake and funeral.[96]

To broaden the strike, leaders implored District 1's 60,000 mineworkers to close down the entire northern field. Nearly all of the field's more than 150 local unions conducted ballots on the proposal, with 90 voting against the idea, much to Cappellini's satisfaction. The vote also pleased Neal J. Ferry, chairman of a special commission appointed by Lewis to investigate the Erie companies' labor problems.[97]

The strike finally ended on January 23, 1925, after seven weeks. The company offered no concessions. The workers lost nearly $4 million in pay as well as their union charters. Lewis and the UMWA applauded the outcome because an unendorsed work stoppage had been thwarted. The *Scranton Times* editorialized in favor of the result and added that the shutdown should never have occurred.[98] Ferry's commission met with Cappellini to discuss the conflict. As they had no authority to address the fundamental cause—subcontracting—they merely considered restoring the charters.[99]

The UMWA could have used the Ferry Commission to address subcontracting and other areas of discontent. However, Lewis and the international leadership did not understand the gravity of the complaint, nor did they see any solution. Although Lewis later disparaged another form of tenancy, the leasing system, which expanded in the late 1930s (see Chapter Six), the UMWA missed an opportunity to confront the cause of what would become the most violent episode to date—the "Feud at the No. 6 Colliery" in 1928.[100]

Miners monument, Pittston, Pennsylvania

CHAPTER FIVE

The Subcontracting Rebellion at the No. 6 Colliery, 1928

The conditions that prevail in Pittston now might be looked upon as a volcano; it is not ejecting lava or smoke at present ... but the situation continues, the fire is not extinguished, there is bitterness, there is hatred existing there to a greater extent than most people realize.

— Pittston Mayor William H. Gillespie, March 1928

The old abuses which were stamped out in 1920 are again in force. Mine [sub]contractors get sections of mines allotted to them. They are favored by having cars provided for them while the regular miner must wait until the company gets ready to provide cars. Every favoritism is shown to the mine [sub]contractor as against ignoring of common courtesy to the miner.

— William J. Brennan, former president, UMWA District 1, April 1928

To the men the [sub]contract system is an unmitigated evil.

— Ben Selekman, *Miners and Murder*, May 1928

My heart and soul has always been with the miner[s] and will always be with them.

— Rinaldo Cappellini, president, UMWA District 1, July 1928

The 1920s in Anthracite

The stock market reached new highs when economic conditions improved during the second half of the 1920s. As the power of business surged, however, management sought to overturn the gains made by unions and workers during the Progressive Era. With feeble labor laws and an unconcerned Coolidge administration, numerous industries moved against labor in pursuit of lower wages and greater workplace control.[1]

Because of a militant workforce, the Erie Coal Companies continued with difficult labor relations despite anthracite's grim circumstances. The hard-coal industry experienced precipitous declines in employment, demand, production, and revenue, especially following the 1925–26 strike. Employment rose from 161,926 in 1921, to 168,734 in 1926, but fell to 151,171 by 1930.[2] In response to weak demand, production dropped 12.5 percent over the same time span. Company revenues fell more than 17 percent between 1924 and 1928.[3] Although coal still provided 67.4 percent of the nation's energy supply in 1928—with oil and natural gas together contributing only 18.2 percent—anthracite's falling demand contrasted sharply with the surge in bituminous usage.[4] As industrial applications waned, hard coal became more confined to the home-heating market.

United Mine Workers of America (UMWA) President John L. Lewis criticized the naysayers and tried to present an optimistic view of anthracite in 1927:

> The mine workers have not and do not now join the great chorus of voices whose public lament for years past was that the industry was decadent and was rapidly traveling to its goal of ultimate disintegration and dissolution. A peculiar psychology has enveloped the industry like a malarial fog and its throttling effects have been more than manifest to any student who sought to analyze its industrial modalities.[5]

Still, in protest over the workers' deteriorating position, the UWMA called general strikes in 1922 and 1925–26 (see Appendix I: Glossary). Lewis termed both final settlements "victories" because they maintained wages and avoided arbitration. However, the detractors were not as sanguine. Workers realized few gains and many agreed with social critic Anna Rochester who characterized the 1925–26 strike compact as having "marked a definite further stage in placing the union at the service of the operators."[6]

As they fought the union and its demands, the operators simultaneously launched various promotional strategies to strengthen the bottom line. In 1928, nearly two dozen firms hired a New York advertising firm for a three-year, $500,000 publicity campaign.[7] They also formed an Anthracite Boosters Club in Philadelphia to promote hard coal in Pennsylvania's largest city and beyond. They joined with UMWA officials and community leaders to create the Anthracite Cooperative Association to develop public relations and educational programs.[8]

As part of the renewal strategy, several companies introduced unique marketing tactics. The Hudson Coal Company labeled its anthracite "Lackawanna Coal" and "Sterling Coal." It was advertised as "Cone-cleaned Guaranteed," a reference to the new Menzies Cone technology, and a phrase that appeared on each of 2,000 orange paper "scatter discs" deposited in each truckload. The Payne Coal Company dispensed similar orange discs.[9] The Glen Burn Company seeded its coal with red baseball-like rounds that contained the legend "Glen Burn, Ball of Fire." Pagnotti collieries used gold-colored discs with "Gold Nugget Anthracite" printed on them. The Glen Alden Coal Company chose a different strategy: it sprayed the final product with blue dye and called it Blue Coal.[10] Reading Anthracite not only used red speckled discs but covered its anthracite with red dye. Some local dairies used industry-sponsored milk bottles embossed with the phrase, "Buy Burn Boost Pennsylvania Anthracite."[11]

These and other advertising ploys yielded only marginal gains, for even the best commercials could not overcome the industry's deep-rooted problems. Many critics agreed with Bernard Heller's assessment in *The Nation* magazine in early 1929 when he called anthracite "a critically sick industry," and described the "pall of gloom [that] hangs over the sections of Pennsyl-

Figure 1. Hudson Coal Company and Pagnotti Coal Company scatter discs.

Figure 2. Payne Coal Company scatter disc.

Figure 3. Reading Anthracite dealer card and scatter disc.

vania which are dependent on the mining of coal."[12] Needless to say, the Wall Street crash of October 1929 only worsened an already weak local economy. Living and working conditions in the four coal fields dropped to levels unseen in 30 years. With mining accounting for 40 percent of male employment in Luzerne and Lackawanna counties, and up to 49 percent of total unemployment, the northern field's relief rolls were among the highest in the country.[13] The crisis led to one of the largest urban out-migrations during the first half of the twentieth century. Over 150,000 and possibly as many as 200,000 anthracite citizens relocated to Philadelphia alone.[14] Analysts predicted that the men and boys of hard coal would never again see the employment levels of the past.[15]

Figure 4. Hudson Coal Company, Sterling Coal dealer card.

Figure 5. "Blue Coal" logo of the Glen Alden Coal Company.

UMWA affiliation reflected the decline. There were 70,878 union members in 1921 in a workforce of nearly 162,000. During the strike of 1925–26, the numbers rose to 119,895, but by 1929 dipped to 74,755.[16] The membership fall-off precipitated a severe revenue decline for the organization.

As the economic emergency deepened, the companies moved to slash expenditures. A main strategy involved running only the most profitable mines and collieries. Between 1929 and 1932, the first wave of what social scientists in the 1980s called *de-industrialization* hit the anthracite fields. The number of northern field mines dropped from 163 to 113. Management also turned to what businesses in the 1990s described as *downsizing* by eliminating thousands of jobs.[17] The policy of laying off senior and higher-paid men only added to labor's dissatisfaction.[18]

For their part, the workers saw *equalization* as one answer to the problem. The concept involved the sharing of available work at each pit so that all employees had some income. From the colliers' communalistic perspective, part-time labor for everyone was better than full-time work for a few.[19] The Erie companies, like other northern operators, resisted equalization. On January 5, 1928, workers at the Hillside Coal & Iron Company's (HC&I) Butler Colliery in Pittston staged a one-week strike over the issue. They were incensed that most other Erie operations worked six, seven, and eight days over a two-week period while they put in just two days.[20] Although the three anthracite districts and their presidents were on record as supporting equalization, President Lewis did not endorse the idea because it challenged management's established right to determine work crews and schedules.[21]

Consequently, only the determined colliers in the Panther Valley area of the southern field were successful in implementing a relatively long-lived job sharing program.[22] Yet, even within a corporation known for rancorous labor relations, the Erie Coal Companies established a temporary equalization plan in July 1928.[23] It included all but one of the corporation's 14 collieries—the Pennsylvania Coal Company's No. 6, which, because of incessant strikes, worked a total of only 30 days during all of 1928.

Protracted Strikes at the No. 6 Colliery

The No. 6 Colliery was described as "the principal [mining] operation in the Pittston area" and one of the largest and most important in the Erie system.[24] In August 1927, Erie's coal operations had a new president, Michael Gallagher, appointed by the Van Swerigen brothers of Cleveland who now owned Erie and its subsidiaries (see Chapter Six). Gallagher convened an emergency meeting with the District 1 president, Rinaldo Cappellini, to discuss a problem at the No. 6. Gallagher declared that one of the colliery's four mines—the No. 6 Shaft—had accrued an $80,000 deficit over the previous six months.[25] To rectify the shortfall he wanted to reduce wages, lay off workers, install mechanical loaders, and hire subcontractors.[26]

Figure 6. Michael Gallagher, president, the Erie Coal Companies. (Courtesy of Pauline Gallagher and Patricia Eide)

Despite resistance from a number of L.U. 1703 members, a joint union-management committee studied the matter. After a few months, the committee decided that two of the No. 6 Shaft's four separate mine workings could reopen if changes were made to wages and operations (meaning the use of machinery and subcontractors), but the other two would have to close, resulting in the elimination of 200 or more jobs.[27] The workers rejected the proposal as a ploy to expand subcontracting. They instead demanded the removal of all subcontractors from each of the colliery's mines. To emphasize the balance-sheet distress, Gallagher locked out the No. 6 Shaft employess on December 31, 1927.[28]

He then proceeded to restart the No. 6 Shaft and hire six subcontractors. When the resulting position losses were announced in January, the shaft's 400-person workforce, many of whom had participated in the 1916, 1920, and 1924 anti-subcontracting strikes, walked off the job.[29] Cappellini, president of District 1 since 1923, had worked at the No. 6 Colliery and belonged to L.U. 1703, but his background and

Figure 7. Rinaldo Cappellini, 1928. (J.E. Russell photograph in Ben Selekman, "Miners and Murder," Survey Graphic, *60, 1 May 1928, 154)*

office did nothing to settle the dispute between the workers on one side and Gallagher and Superintendent Glover on the other. Described by *Time* magazine as one of anthracite's most formidable labor leaders, Cappellini rose to power in the early 1920s on a tide of wildcat strikes, as discussed in Chapter Four.[30] However, since attaining high office, he became much less sympathetic to unauthorized walkouts.

The men at the No. 6 Colliery's three other pits—1,300 in total—joined the shutdown. About 36 percent of them were immigrants from Italy, mainly Sicily. To strengthen their position, the dissenters petitioned L.U. 1703 members to support closure of the three pits. When the local union's sitting officers refused to consider the matter, the mutineers convened an unofficial meeting of the local. They claimed that the officers were part of the problem for they were virtually all subcontractors and other "company men" who had gained office through controlling jobs and receiving favors from the bosses. The insurgent faction that had taken over the local in 1920 (see Chapter Four) had lost control to a new element that enjoyed a cozy relationship with management. In a move to retake L.U. 1703, the insurgents called an unofficial meeting of the local on January 11, 1928, and ousted the incumbents, including President (and subcontractor) Patsy Paglioca.[31] They selected new representatives who promised to remove the subcontractors and their machines and move forward on a strike of the entire No. 6 Colliery workforce.

Cappellini investigated the unsanctioned election with the assistance of August J. Lippi, the district's International Board member. In consultation with Lewis, they negated the results and demanded reinstatement of the original officers. Erie's managers applauded the decision and said they would not recognize the rump ballot. The insurgents proclaimed the legitimacy of the outcome because their votes constituted the majority of local union members.[32]

The Feud at No. 6 Turns Violent

As the strike continued through January and into February, "The Feud at No. 6," as it became known, turned brutally violent. Four former or current L.U. 1703 officers were murdered, one was severely wounded, and numerous others sustained injuries. The *Scranton Times*' description of the crisis as a "reign of industrial war and terror" was echoed by *The New York Times*, which wrote about "Pittston in Terror from Mine Killings."[33] The four deaths were among the 30 homicides in the Pittston area since 1916, many of them subcontracting-related.[34]

Thomas Lillis, age 43, a former grievance committee member, L.U. 1703 treasurer, and check-weighman, was the first casualty. On January 19, 1928, gunmen ambushed the Pittstonian as he walked home from a heated union meeting that lasted until midnight. After administering a beating, the assassins riddled Lillis's body with shots to the arms, shoulders, neck, and head.[35] He had left the presence of union leader and fellow insurgent Alex Campbell only minutes earlier.

Figure 8. Engine and fan house, No. 6 Colliery, Pennsylvania Coal Company, Jenkins Twp., PA. (Courtesy of National Canal Museum Archives)

Figure 9. Mineworkers at the No. 6 Shaft of the No. 6 Colliery, Pennsylvania Coal Company, 1911. The young boy in front is Joe Puma of Pittston, who worked as a nipper. (Library of Congress, National Child Labor Committee Collection, Lewis W. Hine)

The next day, Thomas Lewis of Pittston, age 40, was struck in the back with six bullets. Passersby found his body lying along the Erie Railroad tracks. Police pursued two suspects who fled by rail but made no arrests. Lewis was an outspoken critic of subcontracting who had served as president of L.U. 1703 in the early 1920s.[36]

To quell the violence, Cappellini proposed a second election at L.U. 1703. The members agreed and the district office assigned two administration men, Lippi and Frank Agati, to supervise the poll. In a hotly contested vote on January 25, 1928, with local police maintaining order, members again selected the anti-subcontracting slate. Sam Bonita was chosen as president, Joseph Victor as vice president, and Peter Reilly as recording secretary.[37] Despite the favorable outcome, the protestors declined Cappellini's plea to end the strike at the No. 6 Shaft. They also spurned the mediating appeals of the Anthracite Board of Conciliation (ABC).

In early February, under the new officers' direction, the workforce voted to shut down the No. 6 Colliery.[38] About 36 percent of the workers were immigrants from Italy, mainly Sicily, but as was typical of the hard-coal regions, Slavic, Scottish, Welsh, Irish, English, German, Lithuanian, Hungarian, and workers from other ethnic groups participated in the strike. As discussed in Chapter Three, the Sicilians were particularly opposed to subcontracting because they had seen its adverse consequences in the old country's sulfur pits, including the involvement of organized crime.[39]

The No. 6 strikers took hope when, soon after they closed the colliery, the company agreed to withdraw all subcontractors from the nearby Butler Colliery, which had been on strike since early January over the use of mechanical loading machines in subcontracting crews.[40] However, during negotiations in Scranton involving both Butler and No. 6 men, company officials informed the No. 6 workers that the colliery would not open until the profitability of the No. 6 Shaft was guaranteed.[41]

In search of broader support, the boycotters asked workers at the ten other PaCC and four HC&I collieries to join the mutiny.[42] The nine local unions that represented Erie's 12,000 employees met in several stormy sessions to consider a company-wide shutdown. Eight of them voted to remain on the job. The L.U. 1703 members had little choice but to continue on their own. The newly elected officers held a series of failed talks with Erie management, which remained firm on the need to reorganize the No. 6 Shaft.[43]

On February 16, Frank Agati, a former subcontractor and president of L.U. 1703, became the third fatality.[44] He died in a gun battle at District 1 headquarters in the Miner's National Bank building in downtown Wilkes-Barre.[45] Sam Bonita, the recently elected president of L.U. 1703, was charged in the shooting along with accomplices Steven Mendola and Adam Moleski, two other new officers. The three had traveled to the administrative center to discuss the strike with Agati and other union officials. Bonita claimed self-defense: "I killed him, but Agati fired the first shot," he declared.[46] Despite the evidence to support Bonita's argument, the three insurgents were charged with the shooting on February 25.[47] Juries convicted them of voluntary manslaughter.[48] Judge William S. McLean gave Bonita the maximum six to 12 years sentence, while Mendola and Moleski received the maximum sentences of four to eight years each.[49] Bonita wrote a letter from prison in which he blamed the violence on the subcontracting system and the Cappellini administration:

Dear Comrades and Brothers of the United Mine Workers:

I am behind the bars now and I cannot be with you, brothers, at the present time, so do not let this discourage you, but let it make you fight all the harder to overthrow and expose the Cappellini [sub]contractor system. In 1920 we United Mine Workers of America fought and eliminated the [sub]contractor system. In 1925 the [sub]contractor reappeared after Cappellini revoked the charters from 10 different locals in the Pittston district. These said [sub]contractors influenced Cappellini to violate the laws of our constitutions. Several men have lost their lives and we others have taken chances of losing our lives and our freedom in our activity in opposing this [sub]contractor system and these tactics of Cappellini and his associations. Our constitution states that any one of its members violating its laws must be thrown out of the organization. Now brothers, this is our only chance to stop all this, give our ranks a thorough cleaning out and save our union. Mere words on paper cannot express my great thanks for the brotherly help and support of the organization.

Fraternally yours,

Sam Bonita.[50]

Agati, the father of seven children, had resigned as L.U. 1703's president two years earlier and joined the Cappellini administration where he served as an organizer, bodyguard to Cappellini, and assistant to Lippi. Ironically, a few weeks before his demise, he tendered a resignation citing a desire to spend more time on private business; however, the crisis at No. 6 delayed the departure.[51] Agati's interment motorcade of 121 cars was "possibly the most impressive funeral procession ever held in the Pittston district." Eleven of the vehicles carried "floral tokens" valued at over $3,000 ($39,400 in 2011 dollars) and were described as "the most beautiful ever seen in this section."[52] A 41-piece band played dirges at the head of a cortege that flowed from the Agati residence to nearby St. Rocco's Catholic Church for a Requiem Mass.[53] Agati was alleged to have been an organized crime agent within District 1 officialdom.[54]

On February 19, "Big Sam" Grecio, a 35-year-old member of L.U. 1703's grievance committee, became the fourth target. A devotee of the insurgent movement, he was waylaid as he and his wife walked along a Pittston street at 9:00 p.m. Two culprits lurched from the dark and grabbed Grecio while a third fired three rounds to the head. The attack purportedly occurred in retaliation for Agati's death. Grecio miraculously survived the assault, although he lost sight in both eyes and incurred other disabilities. As he drifted in and out of consciousness, he summoned Alex Campbell to his hospital bed. Grecio warned the unofficial leader of the No. 6 Colliery's workers that he would be the next quarry.[55]

Fear spread throughout the Pittston area. Previously unarmed workers began carrying pistols and knives for protection. When L.U. 1703 convoked a special meeting at St. Aloysius Hall, Pittston, following Grecio's attack, police frisked each entering person. Even though an arms search had been well publicized, six men, all Italian, were arrested for carrying concealed weapons.[56] A meeting four weeks later led to the arrest of three workers, again all Italian, on the same charge.[57]

Grecio's premonition proved correct. Ten days later, in the late afternoon of February 28, insurgent leader Campbell—called "the idol of Pittston"—perished alongside his friend and colleague, the recently elected L.U. 1703 secretary, Peter Reilly.[58] The murders

Figure 10 (above left). Frank Agati, murder victim, February 16, 1928.

Figure 11 (left). Miners National Bank building, Wilkes-Barre, postcard view.

Figure 12 (above right). Frank Agati murder scene, Miner's National Bank, Wilkes-Barre. (Courtesy of Times Leader)

occurred 100 yards from Campbell's home on Railroad Street as the duo were returning from a fundraising event for Bonita, Mendola, and Moleski. Four hired gunmen in a stolen Peerless sedan sprayed them with a fusillade of bullets as they rode up the street in Reilly's Hudson automobile. The assassins used 12-gauge Remington automatic shotguns to attack the insurgent leaders with some 18 rounds each. The weapons had been modified so that one pull of the trigger discharged six rounds.[59] *Time* magazine described the homicides:

> In Pittston, Pa., where hard coal comes out of the earth, a Hudson closed car turned into hard-boiled Railroad Street, closely followed by a Peerless sedan. Crowding the Hudson to the gutter, the Peerless paused to belch a noisy blast of powder and lead slugs from several pump [*sic*] guns. Then it vanished toward the neighboring hamlet of Moosic, where it was abandoned, the occupants slipping away into a dense forest. In the shattered Hudson on Railroad Street lay Alexander Campbell, labor leader, and his friend Peter Reilly, both of them horribly dead.[60]

SAM BONITA

Set Free

ADAM MOLESKI

Set Free

STEVE MENDOLA

BONITA - MOLESKI - MENDOLA

Man who shot Agatti And ran.

Boy who gave up 3 hours later

Man who had two guns on him when captured on Hampton St

The Facts:--

On January 19, 1928, Thomas Lillis, an insurgent miner against the Cappellini machine, was killed. No arrests have been made.

On February 20, Sam Grecio, another insurgent, was shot and lay at the point of death for several weeks. He will probably remain an invalid for life. No arrests were made.

On February 28, Peter Reilly, a fighter in the union against the Cappellini regime, returning home from a visit to his brothers in jail, was killed. No arrests have been made.

On the same day Alexander Campbell, leader of the rank and file miners fighting against the contractor system and the Cappellini machine, was brutally murdered. Though the automobile from which the killing of Campbell and Reilly was done was found, and in spite of the fact that the murder was committed in broad daylight, no arrests have been made.

On February 16, Frank Agati, contractor and Cappellini supporter, was killed. Three arrests were immediately made.

Bonita, Moleski and Mendola, in opposition to the Cappellini outfit, are being held in connection with the death of Agati.

The Needs:--

All efforts must be made to make clear to the workers of this country, and to the coal miners in particular, what the flow of blood at Pittston really means.

We must make clear why it is that three insurgent miners are arrested for the death of one contractor while no one is taken into custody in connection with the murder and attempted murder of our four brothers, Lillis, Grecio, Reilly and Campbell.

We must hold meetings for this purpose and get wide publicity in all the labor press.

At the same time we must put up a strong legal defense for our brothers and employ the best possible services for this end.

To do all this, thousands of dollars will be necessary, as we know from the experience of workers in other cases, what can be expected.

The flow of blood which has occurred shall not choke our voices of protest!

The headline scares in the capitalist press about "deporting foreigners" shall not blurr our vision!

We know what the forces at work are. We know why arrests were made in connection with the death of one Cappellini supporter, and none in connection with the killing and attempt to kill off four opposed to the Cappellini machine.

We must fight on in the spirit of our martyred dead.

Bonita, Moleski and Mendola must be set free!

Help us to do it. Send your contribution at once and make it as large as possible.

NATIONAL BONITA—MOLESKI—MENDOLA DEFENSE COMMITTEE.

(Stanley Dziengielewski) Secretary. (Charles Licata.) Treasurer.

X Mine worker from Scranton 1703 Pittston

NAME	ADDRESS	AMOUNT

Figure 13. Poster for the defense of Bonita, Mendola, Moleski. (President/District Correspondence, 1920–1933, GST/2/06.02, Box 1, Folder 11, UMWA Papers, PSU)

A father of seven children, the 45-year-old Campbell had received several death threats over the years. He and his family had narrowly escaped injury when dynamite struck their home on March 13, 1924. Shortly following his most recent election as check-weighman on January 15, 1928, Campbell received "a Black Hand letter threatening death" that he turned over to police.[61] A few weeks before the assassination, he told his wife about the thugs who might want to kill him.[62] Campbell's labor leadership and militancy were the evident motives for the slaying.[63]

Reilly, a 28-year-old Lithuanian immigrant bachelor who had changed his name from Saudargis, had likewise received death warnings. Bombers damaged the residence where he lived with his mother and sister on the same day in 1924 when Campbell's dwelling sustained ruin. Reilly was widely known as a dedicated union man who had participated in anti-subcontracting strikes over the years. His colleagues elected him as secretary of L.U. 1703 following the strike of 1920, and he had served in that capacity until 1925 when he lost to a subcontractor-backed opponent. He admitted driving Bonita, Moleski, and Mendola to the District office in Wilkes-Barre on the day of Agati's killing, but convinced authorities that he was not part of a murder conspiracy.[64]

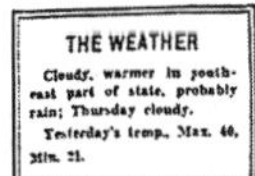
THE WEATHER
Cloudy, warmer in southeast part of state, probably rain; Thursday cloudy.
Yesterday's temp., Max. 40, Min. 21.

THE LARGEST MORNING DAILY CIRCULATION IN PENNSYLVANIA OUTSIDE OF PHILADELPHIA AND PITTSBURGH

The Scranton Republican

Boost Buy Burn Anthracite

ESTABLISHED 1867 VOL. 142, NO. 51 MEMBER OF THE ASSOCIATED PRESS SCRANTON, PA., WEDNESDAY, FEBRUARY 29, 1928 TWENTY PAGES • • PRICE TWO CENTS

2 UNION LEADERS SLAIN BY GUNMEN IN PITTSTON 'WAR'

BREAKUP OF SOLID SOUTH THREATENED

Dry Leaders Will Urge Democrats There to Bolt Party If Smith Wins Nomination For President—Demand Clear-Cut Prohibition Planks By Both Parties—Drafting of McAdoo and Third Party Is Discussed

WASHINGTON, Feb. 28 (A.P.).—Two movements one designed to put a damper on the presidential aspirations of Governor Smith of New York, the other to force both the Republican and Democratic parties to toe the mark on prohibition in the coming campaign—got underway here today under the direction of a group of dry leaders.

The offensive directed at the two political parties was launched at a conference in which representatives of thirty or more national temperance organizations participated. Resolutions were adopted demanding clear-cut prohibition planks and standard bearers genuinely dry.

Call Southern Conference.

While the meeting was in progress plans were disclosed for an Anti-Saloon League conference next month in St. Petersburg, Fla., in which friends of prohibition in nine southeastern states will be invited to have a hand. At that time, those in charge of arrangements predicted a counter-offensive against Gov-

BISHOP TALBOT FUNERAL TO BE HELDTHURSDAY IN BETHLEHEM

Body of Revered Protestant Episcopal Clergyman Will Be Laid at Rest

REPUBLICAN'S PEN ARTIST SKETCHES DOUBLE KILLING

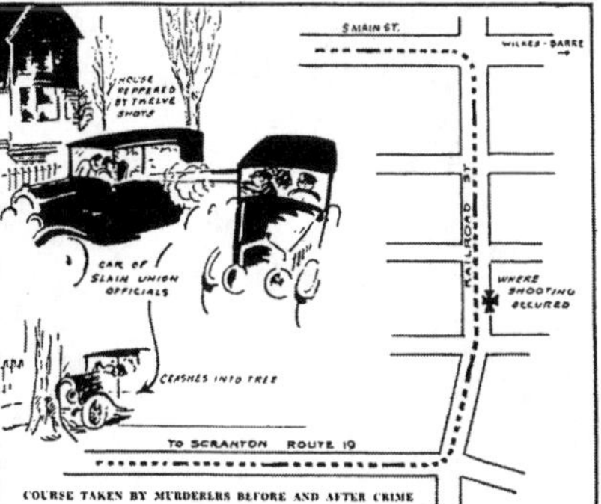

FRENCH FLIERS

MAYOR WALKER

VICTIMS RIDDLED BY BULLETS FROM AMBUSHED AUTO

Alex Campbell, Leader of Insurgent Miners in Pittston, and Peter Reilly, Checkweighman at No. 6 Colliery, Sprayed With Lead From Shotguns and Revolver As They Proceed Home in Automobile After Visiting Benito in Jail—Car Used By the Three Assassins Is Found Abandoned in Moosic—Crime Latest in Feud Over Agati Murder

Taking bloody toll for the murder of Frank Agati, mine union organizer, two Remington pump guns and a 38 caliber Smith and Wesson rained bullets into a Hudson automobile on East Railroad street, Pittston, at 5:05 o'clock last night, riddling the heads of Alexander Campbell and Peter Saudargas (Reilly). The killing was done from a Peerless seven-passenger sedan, which pulled up beside the other car opposite No. 86 on East Railroad street, by three men.

The Peerless sedan was found abandoned on Scott street, in Moosic, at 5:40 o'clock. Three men, muffled to the chin in overcoats, were seen to jump out of the car and board a D. & H. freight train. Scranton police are cooperating in the hunt for the killers, which is under personal direction of Captain William Clark of the State police.

VICTIM

ALEX CAMPBELL

Figure 14. "2 Pittston Leaders Slain by Gunmen in Pittston War."
(Scranton Republican, *February 29, 1928)*

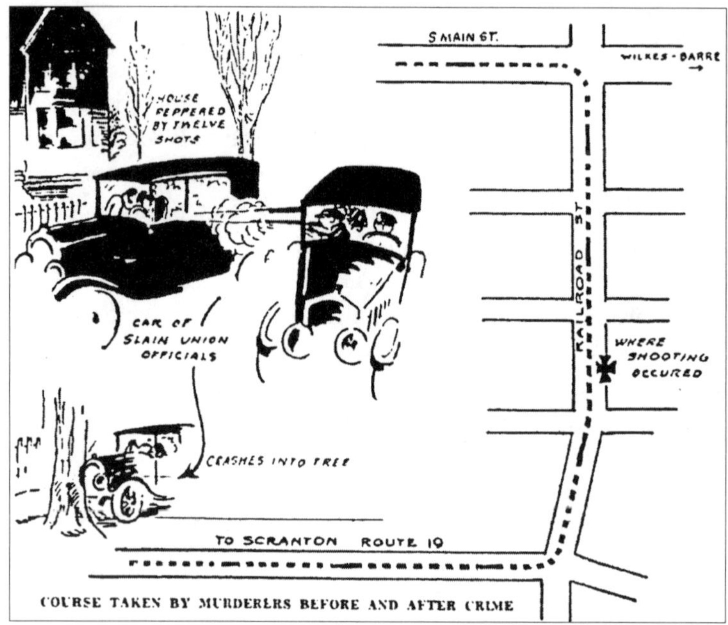

Figure 15. Magnified sketch of Campbell and Reilly murder scene.
(Scranton Republican, *February 29, 1928)*

Photograph by J. E. Russell

Alec Campbell, leader of the insurgents, on a picnic with his family before he was murdered on February 28

Figure 16. Mr. and Mrs. Alex Campbell with six of their seven children. (J.E. Russell photograph in Ben Selekman, "Miners and Murder," Survey Graphic, *60, 1 May 1928, 155)*

Figure 17 (above left). Campbell family home, Railroad Street, Pittston, following the murders. (Courtesy of Times Leader*)*

Figure 18 (above right). "Bullets from a Veritable Arsenal on Wheels." The Campbell home was struck with at least 18 errant bullets. (Courtesy of Times Leader*)*

The mourners at the funeral services for Lillis, Lewis, Campbell, and Reilly numbered in the tens of thousands. They memorialized the fallen men as heroes and martyrs. Local unions passed resolutions similar to the one approved for Campbell and Reilly by L.U. 1407 at a Delaware, Lackawanna, & Western company mine in Wilkes-Barre:

> Whereas, the brutal murders of Brothers Campbell and Reilly is a continuation of the campaign of the [sub]contractors and certain coal operations to exterminate every union man who dares oppose corruption in our union, and the robbery of miners by [sub]contractors and operators,
>
> Therefore, be it resolved that local union No. 1407, United Mine Workers of America, Wilkes-Barre, extends its sympathy to the families of our murdered brothers, that we pledge ourselves to support the heroic struggle of the Pittston miners.[65]

The violence shook Cappellini. He lamented that: "the whole affair is regrettable and I pray to God Almighty for the restoration of peace and harmony in the Pittston field."[66]

Figure 19. Alex Campbell's funeral at the First Presbyterian Church, Pittston. (J.E. Russell photograph in Ben Selekman, "Miners and Murder," Survey Graphic, *60, 1 May 1928, 155)*

COMPLETE TELEGRAPHIC SERVICE OF THE UNITED PRESS MEMBER AUDIT BUREAU OF CIRCULATION

Pittston Gazette.

THE HOME NEWSPAPER

CIRCULATION BOOK OPEN TO ADVERTISERS

78th YEAR — PITTSTON, PA., WEDNESDAY, FEBRUARY 29, 1928 — TEN PAGES

NO TRACE OF MURDERERS OF CAMPBELL AND REILLY

TWO MINE LEADERS CRUELLY SLAIN BY GUNMEN LAST NIGHT

Deadly warfare between desperate factions of the Miners' Union was resumed in Pittston last night. Alexander Campbell, of Railroad street, prominent as a miners' leader in this district for many years, and his friend and supporter, Peter Reilly (Saudargis), of Inkerman, were both cruelly slain by a band of gunmen.

The horrifying tragedy occurred on East Railroad street, above Vine street, about 100 yards distant from the residence of Mr. Campbell. Reilly was driving Campbell home in a Hudson closed car at 5:30 o'clock, when a Peerless sedan occupied by the assassins followed the Hudson up Railroad street, finally crowded the Hudson to the curb, and then literally riddled the car and its two occupants with numberless bullets from pump guns. Both Campbell and Reilly were instantly killed.

The assassins quickly sped away from the scene of the crime in their car, passing up Railroad street, eastward by way of Tedrick road to the Scranton highway, passing through Dupont and Avoca, and abandoning their car and weapons in Moosic borough.

Every available police service in Northeastern Pennsylvania, including the State Police, was pressed into service in the hope of capturing the men guilty of the double murder. Numerous trails are being followed up, but at three o'clock this afternoon no arrests had been made, and it was feared that the murderers have made good their escape through the dense forest east of Moosic.

Residents of Pittston and of the entire valley were in a state of intense excitement throughout the night, and general indignation over the cold-blooded killings had not diminished today. On every lip is heard denunciation of the deadly warfare that is being carried on by the miners' factions, last night's tragedy being generally recognized as an outgrowth of the murder of Frank Agati, organizer of the Miners' Union, in Wilkesbarre, last week. Reilly, who was slain last night, was the driver of the car in which the three men accused of killing Agati went to Wilkesbarre just prior to the murder of Agati.

This morning Mayor Gillespie sent a telegram to John L. Lewis, International President of the United Mine Workers, urging him to come to Pittston and use his influence and power in an effort to put an end to the factional feud among the miners to which last night's murders and several crimes that preceded them are generally attributed.

ALEXANDER CAMPBELL

Mayor Urges Lewis To Come And Try To End Reign of Terror

Mayor William H. Gillespie this morning sent the following telegram to John L. Lewis, International President of the United Mine Workers of America, at Indianapolis, Indiana:

"In the name of the tens of thousands of terrorized people in this city and vicinity, I appeal to you to come here at once to use the great influence and power of your office as head of United Mine Workers of America to restore peace and order in the ranks of the mine workers of this community. The hostile factions in the local organization have created a reign of terror by their lawlessness. Dynamitings, murders and attempts to murder are of frequent occurrence. Two prominent leaders of the mine workers were murdered in cold blood in the heart of this city last evening. Our city is in a state of terror and turmoil. The cause of this bloody feud or vendetta is known to every intelligent person in the anthracite coal regions. This disgraceful and tragic situation is attributed directly to the bitter hostility that exists between the mine officials, mine contractors, union labor leaders and insurgent labor leaders connected with the Pennsylvania Coal Company. The public believes that yourself, District President Cappellini and the head of the Pennsylvania Coal Company can end these hostilities and bring this campaign of crime to a close if you meet together at once and make an honest effort to settle this deadly dispute. The good name of your organization, the good name of our city and the future safety of our lives and property depend largely upon your prompt response to this appeal."

WILLIAM H. GILLESPIE,
Mayor of the City of Pittston.

CAPPELLINI TO CALL MEETING OF DISTRICT BOARD

5 ARE INDICTED IN CHICAGO MAIL TRAIN ROBBERY

TRYING TO TRACE LICENSE PLATES ON THE DEATH CAR

LINDY DISREGARDS DAMAGE TO PLANE AND LANDS SAFELY

AIRSHIP ARRIVES AT CUBA

THOUSANDS DIE IN CHINA WHEN DYKES GIVE WAY

DYNAMITING IN ELIZABETH, N. J., CREATES HAVOC

BULLETINS

4 MARINES ARE SLAIN BY REBELS

36 AIRCRAFTS ARE SEARCHING FOR MISSING PLANE

Figure 20. "No Trace of Murderers of Campbell and Reilly."
(Pittston Gazette, *February 29, 1928*)

Confronting the Murders

The region struggled to make sense of the mayhem. Why was it happening? Who was behind it? How could it be stopped? A *Times Leader* editorial described the crisis simply as a civil war between UMWA factions:

> No other period of lawlessness in the history of Luzerne County has had the chilling, stunning effect on a community as has the bloody feud now raging between factions of the United Mine Workers in the Pittston district. Not even in the days of the feared Molly Maguires … has human life been held as cheaply. Here is a vendetta, whose death-dealing influences find their inspiration in deep burning hatred and their expression in open warfare.[67]

One news account said it was "alleged to be with IWW ramifications," a contention that mineworkers denied and for which there was no evidence.[68] In a featured article in *Survey Graphic*, journalist Ben Selekman described Pittston as a scene of riots, violence, and hand-to-hand fighting. He interviewed numerous workers, company men, and community leaders and concluded that the problem resulted from the subcontracting system.[69] The *Wilkes-Barre Record* also tied the murders to subcontracting and recalled the assassination of Pittston detective Samuel Lucchino in 1920:

> Yesterday's tragedy was the most horrifying in the crime annals of northeastern Pennsylvania, and it is said [that it] can be traced back to the fight against the [sub]contract mining system of Pennsylvania Coal Co., ... when Sam Luchino [*sic*], Pittston city detective, was murdered near his home on Railroad Street, a short distance from the scene of yesterday's double murder.[70]

The region's Communist directors, Stanley Dziengelewski and George Paucun of Wilkes-Barre, said the slayings were part of a conspiracy perpetuated by "the agents of the operators and [sub]contractors in the union supported by the Lewis-Cappellini machine."[71] Pittston police also associated the killings with the subcontracting system, as did *The Coal Digger*, a newspaper with Communist affiliation published for workers in the hard and soft coal fields.[72]

Organized crime remained an unspoken element in the bedlam. None of the authorities, journalists, or commentators directly mentioned the role of mob leaders or "soldiers."[73] However, the insurgents had few doubts about their participation.[74] In the simplest terms, the aggressions pitted the subcontracting system's supporters against its opponents. The strongest backers were, of course, the subcontractors and company managers. The opponents were the dissident colliers who saw the arrangement as the most unsatisfactory aspect of the workplace. While Erie preferred to occupy a neutral middle ground in an apparent battle between union factions, the No. 6 employees argued that company bosses had schemed with the subcontractors, including alleged organized criminals who, in effect, acted as agents on behalf of the plan. According to William A. Hastie:

> Alex Campbell, a fiery red-headed Scot, was the leader of the local union at No. 6 Colliery. He could not be intimidated nor bribed by organized crime boss Santo Volpe. He and Peter Reilly, who was the secretary of the local, were gunned to death while they were sitting in Campbell's car. Three New Jersey gangsters were tried and convicted of these murders. They were sentenced to life imprisonment. One of them, "Jimmy the Dressmaker," died in prison. But not one of them violated the oath of *Omerta*, the Mafia oath of silence, so Santo Volpe went unscathed. ...
>
> Most of these murders were over competition for the subcontracts, although Volpe had Campbell and Reilly killed because they were also standing by the men. Alex was also elected as the union check-weighman. And there is where he could have been bribed by Volpe, to cheat the men. See, you have the weighman, he's the company man inspecting coal cars coming out of the mine, weighing them and inspecting them for rock, slate, and fine coal. And that way they'd rob the miner. ... They'd take the best part of it and the miner would be docked. The union got the clause where at each operation there would be a union check-weighman. Alex Campbell was the check-weighman at No. 6, and they knew he wouldn't go along with Volpe. Alex's youngest daughter, Margaret Campbell Robertson, told us many times that there was no question in the world who had them murdered.[75]

The paroxysm disabled the whole of Pittston. Mayor William H. Gillespie took command of public safety and issued a series of orders and communiqués. He forced the cancellation of a union meeting at the No. 9 Colliery for fear of violence, and he forbade a mass gathering at the Pittston Armory organized by the Communist- affiliated Save-the-

Figure 21. Santo Volpe, subcontractor of the Pennsylvania Coal Company.

Union Committee. "I will permit no labor mass meetings to be held in this city until the people are composed again and the present crime wave has receded," he decreed.[76] The organizers of the latter gathering were forced to relocate the event to the Italian Hall in the Italian section of Swoyersville, near the Harry E Colliery where Sicilian immigrant Charles Siracuse served as president of L.U. 452.[77]

Mayor Gillespie arranged a "peace conference" that included No. 6 representatives, company executives, and ABC members.[78] He had

Figure 22. Harry E Colliery, Temple Coal Company, Swoyersville, PA, circa 1920. (Courtesy of Eric Bella)

run for office a few years earlier promising to bring peace to the city (his campaign slogan was "Gillespie or Guns"). When the current crisis erupted, he criticized the perpetrators and appealed for calm:

> As mayor of this city I deem it my duty to call your attention to the deplorable condition of affairs that exists in this city. I speak for the peaceable law-abiding men, women and children who are terrified by the reign of lawlessness that now exists in this community.

The mayor lamented that "Dynamitings, murder and attempted murder are of frequent occurrence," and said it must end. He saw the cause of the conflict much like the editors of the *Times Leader*, that is, as a feud between factions of "the foreign element":

> A bloody feud, or vendetta, has been started among the so-called foreign element, and when it will end, or where it will end God Himself only knows. It is enough to know that this community may become a second Herrin overnight, and that lynchings and murder may easily follow in the wake of a storm of mob rule, riots, and lawlessness. This disgraceful and tragic situation is laid at the door of the labor troubles

> in this city and vicinity, and is attributed directly to the hostility that exists between the mine officials, mine [sub]contractors, union labor leaders, and insurgent labor leaders, connected with the Pennsylvania Coal Company.[79]

He censured the main actors but made no reference to organized crime.

Figure 23. "Around the town rise a half-dozen gaunt shafts, breakers, and tipples; and in the distance gigantic banks of culm, dull and gray" (original caption). (Herbert Pullinger drawing in Ben Selekman, "Miners and Murders," Survey Graphic, *60, 1 May 1928, 152)*

Figure 24. "The whole town has a disheveled appearance. It is a typical mining community, rows of frame houses, gutted streets, muddy alleys" (original caption). (Herbert Pullinger drawing in Ben Selekman, "Miners and Murders," Survey Graphic, *60, 1 May 1928, 153)*

Gillespie made an urgent plea to John L. Lewis. "In the name of the tens of thousands of terrorized people in this city and vicinity," he wrote in a publicly released telegram, "I appeal to you to come here at once to use the great influence and power of your office … to restore peace and order in the ranks of the mineworkers of this community."[80] Lewis procrastinated in responding, such that the *Pittston Gazette* chastised him for disregarding the mayor's plea.[81] When he did reply, the union president said he would not visit Pittston nor would he comment on the killings because to do so would "create difficulties." Writing later in the *United Mine Workers Journal*, he blamed the chaos on Communists and other radicals:

> I express the hope that every member of our organization will remain calm throughout this trying period and that the activities of the agents of the Communist Party who are now in Luzerne County seeking to cause further confusion and disorder may not be able to influence our people to depart from the orderly, constructive policies of the United Mine Workers of America.[82]

In the name of the tens of thousands of terrorized people in this city and vicinity, I appeal to you to come here at once to use the great influence and power of your office as head of United Mine Workers of America to restore peace and order in the ranks of the mine workers of this community. The hostile factions in the local organization have created a reign of terror by their lawlessness. Dynamitings, murders, and attempts to murder are of frequent occurrence. Two prominent leaders of the mine workers were murdered in cold blood in the heart of this city last evening. Our city is in a state of terror and turmoil. The cause of this bloody feud or vendetta is known to every intelligent person in the anthracite coal regions. This disgraceful and tragic situation is attributed directly to the bitter hostility that exists between the mine officials, mine contractors, union labor leaders and insurgent labor leaders connected with the Pennsylvania Coal Company. The public believes that yourself, District President Cappellini and the head of the Pennsylvania Coal Company can end these hostilities and bring this campaign of crime to a close if you meet together at once and make an honest effort to settle this deadly dispute. The good name of your organization, the good name of our city and the future safety of our lives and property depend largely upon your prompt response to this appeal.

WILLIAM H. GILLESPIE
Mayor of the City of Pittston.

Figure 25. Telegram, Mayor William H. Gillespie to John L. Lewis, February 28, 1928. (President/District Correspondence, 1920–1933, Box 6, Folder 30, GST/2/06.02, UMWA Papers, PSU)

Lewis also penned a letter to District members reproving the insurgents and calling for loyalty among true UMWA men.[83] His response to the Pittston tumult mirrored his usual approach to dissidents, as historian Robert H. Zieger observed: "In the 1920s, Lewis had been among the foremost red-baiters in the labor movement … [and] lost few opportunities to discredit his critics by calling them Communists or Communist dupes …"[84] The widely publicized arrest of radical activists Powers Hapgood and Mary Donovan on March 5—for

Figure 26. John L. Lewis greeting Senator Robert La Follette Jr. of Wisconsin, 1927. (Library of Congress)

"inciting a riot" by leading a march through Pittston after Gillespie had refused a permit—fanned suspicions about Communist-inspired insurgents while also deflecting attention from the substantive concerns underlying the strike.[85] As for Lewis's hesitant reply, some workers thought that a strike by over 85,000 bituminous men in western Pennsylvania, Ohio, and West Virginia distracted his attention from anthracite.[86]

Cappellini also held subversives responsible. He issued a statement urging "all loyal, good thinking union men not to tolerate any interference from the Communist element now in our District, whose only means of existence appears to be in creating dissension and turmoil."[87] In point of fact, Communists were endeavoring to organize the anthracite fields. The Communist Party of the United States of America (CPUSA) had gained a small number of hard-coal members from various ethnic groups.[88] They hoped that the No. 6 upheaval, along with the growing dissatisfaction with Cappellini and Lewis, would force more Erie hands into the fold.[89] A statement published in the *Daily Worker* by the Save-the-Union Committee, which had been challenging the UMWA in both the anthracite and bituminous fields, called for an end to the Cappellini administration:

> The time is now to abolish the infamous individual contract system. The first step in doing this is to eliminate the Cappelini [*sic*] gang from control of our union. They are agents of the operators in maintaining the [sub]contractor system. Our slogan shall be: "Cappelini [*sic*] Must Go!"[90]

However, the protestors at No. 6 wanted nothing to do with the Communists or any other outside group.[91] "We want it made known," they wrote, "that we are not Communists or IWW's, and do not know of any members of the local union that are in such an organization." They resented the radical innuendoes:

> We are satisfied to be termed insurgents as we were known back in 1920 when he [Cappellini] was on the payroll of our insurgent General Grievance Committee prior to the elevation to the office of District president. Newspaper reports have such organizations as the Communists and I.W.W. in this territory but so far we have not had any dealings with them, nor do we intend to have any. We won our conflict back in 1920 without them with Campbell as our pilot and we believe that we can do the same now.[92]

As the community braced for more violence, Cappellini seemed marked as the next victim, perhaps in retaliation for Campbell and Reilly. *The New York Times* reported that participants in a lottery on the murders at a Pittston speakeasy agreed that it was Cappellini's turn: "It was significant that the man who drew the slip with Cappellini's name was congratulated as being the certain winner."[93] Although he did receive several threats, there was never an attempt on the District president's life.

The dynamiting of homes, public buildings, and churches, allegedly by gangsters as well as radical unionists, added to the public's sense of lawless disorder.[94] Insurgent and UMWA leaders, workers who crossed picket lines, and picketers who taunted scabs knew they risked having their houses attacked in what had become an epidemic of bombings. The terrorism extended beyond anthracite to other local industries, as when John W. Hopkins and his brother, Julian, were indicted and pleaded guilty to dynamiting the McLane Silk Company factory in Scranton.[95]

The Sicilian participation in organized crime presented another disturbing facet. Although newspapers continued to disregard the local mob and its leaders, the mineworkers and community members were constantly reminded of the criminal element's role in coal and other aspects of community life—gambling, alcohol, vice, and political corruption.

Figure 27. Frank Sobeck family home, one of many wrecked by dynamite in Pittston.
(Pittston Gazette, *May 29, 1928*)

Figure 28. Hoisting engines at the No. 6 Colliery, Pennsylvania Coal Company. (Courtesy of National Canal Museum Archives)

The workers furthermore witnessed the gang's soldiers around the mines, people such as Volpe's enforcer, Billy "The Bird" Tutsellino, who roamed the pits handling "trouble makers."[96] The Sicilian workers were especially concerned that they were experiencing the same culture of corrupt mining practices they had left behind in Europe.

Amid the pandemonium, the General Grievance Committee at the Glen Alden Coal Company adopted a resolution calling for Cappellini's resignation: "We feel that for the peace and harmony of District 1, members and their dependents, you should tender your resignation at once." The president replied that union members would have a chance to vote on his retention in the upcoming election.[97] At the same time, Erie management offered to reopen the No. 6 to end the bloodshed. The strikers spurned the gesture and voted unanimously to maintain the boycott until all subcontractors and their deputies were removed, a decision made in defiance of President Lewis who had ordered a work resumption.[98] The strikers took comfort in a unanimous vote against the subcontracting system by the General Grievance Committee at PaCC.[99]

Meanwhile, workers at the Butler Colliery decided, in late February 1928, to "clean house" at L.U. 265 when the company began rehiring subcontractors who used loading machines to speed production. The local union members threw out the sitting officers and elected Barney Guzior as president along with a group of other insurgents.[100] Guzior was a former official of the No. 6 Colliery's L.U. 1703 before securing a job at Butler. When the new officers called a strike of the colliery's 1,200-person workforce over subcontractors and mining machines, many worried that the Butler would become another No. 6.[101] The colliery remained idle until the end of March, when the workers and the company came to terms.[102]

A similar housecleaning occurred at PaCC's No. 9 Colliery in late April 1928, when the insurgent faction overthrew the "Ernest Orlando regime" that had been running L.U. 1495

Figure 29. Mule barn, Butler Colliery, Pennsylvania Coal Company, Pittston Twp, PA. (Courtesy of National Canal Museum Archives)

for some time. The dissidents elected Jacob Collier as president. He and his fellow officers promised to oppose the Cappellini establishment and rectify the workers' grievances against subcontracting and other matters. However, Orlando and his colleagues refused to step aside and the workforce remained split. Collier confiscated the local's charter, seal, check book, and equipment, whereupon Orlando took him to court for larceny.[103]

Back at the No. 6, the ABC arranged the peace conference for March 3. It included Erie managers, District 1 officials, L.U. 1703 representatives, and ABC members. The participants proposed an armistice that called for reopening the colliery for 30 days, and continued talks on mutually acceptable wage rates. Should the negotiations fail, the strike would resume.[104] However, two days later the No. 6 workers rejected the plan because it did not include the complete purging of the subcontractors.[105]

When the District 1 executive board convened in early March 1928, members voted to banish the individual and special contract from the northern field. "It has always been our policy to oppose the '[sub]contractor system' of mining," said Secretary Enoch Williams. The board authorized Cappellini to take the matter to the ABC in a special appeal. Cappellini reminded them that he had already filed a grievance against the Hudson Coal Company's use of subcontractors in machine mining.[106]

In mid-March, Cappellini called a special meeting on the No. 6 strike that included UMWA Vice President Thomas Kennedy, three representatives of L.U. 1703, and the presidents of the two other anthracite districts, Andrew Mattey of District 7 and Christopher Golden of District 9. Kennedy, as well as the two presidents, wanted to learn more about the structure and operation of District 1's subcontracting system and why it was causing such unrest at the Erie Companies. After Cappellini provided a brief overview and stated his opposition to subcontracting, Frank McGarry, the recently elected president of L.U. 1703 following Bonita's imprisonment, took the floor. He emphasized that the UMWA was simply ineffectual at the No. 6 and other Erie collieries because the largest subcontractors were ignoring the union:

> I want to say that the No. 6 men are not fighting among themselves. There are a few men at that colliery who are [sub]contractors. The biggest [sub]contractor in No. 6

Figure 30. Mechanical loader, Hudson Coal Company mine, circa 1920s. (Courtesy of Lackawanna Historical Society)

> Colliery doesn't belong to the No. 6 local, that is Sandy Volpe. He is not a member of our local.

When asked by Andrew Mattey if Volpe belonged to any local union, McGarry responded, "Not that I know of. …" Mattey was dumbfounded. "He would not work ten minutes in my district," he retorted. "[The men would say:] 'Either I quit or you quit.'" McGarry replied: "The No. 6 men already expressed the opinion that they won't work with the subcontractor"; however, he added, the state of affairs in District 1 were not the same as in District 7.[107]

James Lamarca, another L.U. 1703 representative at the meeting, addressed one of the central problems:

> I think you don't know of the situation we have in Pittston. These [sub]contractors in Pittston … if they know anything that you talk against a [sub]contractor, you are not going to live long. If they say they are going to get you, they don't fool around. That means sure death, which we have proof today that they have done that already.[108]

Without providing specific names, Lamarca in effect stressed that the subcontracting system at PaCC and HC&I had been infiltrated by an element not seen at other anthracite companies in District 1 or Districts 7 or 9: organized crime.

Another subcontracting-related death gripped Pittston in mid-March, 1928. An 18-year-old named Samuel Alfili defended his father against a gun assault by two intruders to their home. The teenager used a shotgun to kill Dominick Aellio, age 33, and seri-

Figure 31. Twin tunnels, Hudson Coal Company mine, built 1928.
(Courtesy of Lackawanna Historical Society)

Figure 32. A Susquehanna Collieries Company breaker, Nanticoke, PA.
(Courtesy of National Canal Museum Archives)

ously wound the other assailant, Frank Fordutto. The senior Alfili, age 56, received a minor injury from a pistol fired by one of the trespassers. All three adults worked at the No. 6 Colliery where they had fought over the subcontracting system.[109]

The hopefulness surrounding the ABC's investigation into PaCC's subcontracting policies faded when the body offered no prospects for amelioration. Indeed, it affirmed the company's right to issue subcontracts in special circumstances, which remained open to interpretation.[110] The ABC reissued the appeal for a 30-day return-to-work "test period" at No. 6, which the workers snubbed.[111] Around the same time, Charles F. Neill, the ABC umpire, delivered a blow to the strikers when he ruled in favor of the Hudson Coal Company's right to issue subcontracts for machine mining. Neill concluded that he had no authority to disallow the agreements without violating the 1916 labor–management agreement.[112] As the strike carried on, the No. 6 Colliery had worked only two days in January 1928, no days in February, and the prospects for the remainder of March looked just as bleak.

Fomenting a District-Wide Revolt

PaCC was not the only company experiencing labor-management disarray during the first half of 1928. Almost all of the northern field's producers faced discontent. Layoffs, under-employment, declining wages, forced unpaid overtime, and other problems fostered a siege mentality among the workers.[113] A sample of disputes within Luzerne County alone could include:

- The Susquehanna Collieries Company in Nanticoke: on February 23, about 3,000 workers set up picket lines at two mines after the company hired several "outsiders" as laborers and "forced [them] on miners" in violation of seniority rules.[114]
- The West End Coal Company in Mocanaqua: during early March, management locked out 900 men in a confrontation over the height of the topping on mine cars.[115]
- The Exeter Colliery of the Lehigh Valley Coal Company in Exeter: in April, 700 workers went on strike after management declined to reinstate miners who had allegedly violated rules about new controversial check-in and check-out procedures.[116]
- The Loree Colliery of the Hudson Coal Company in Larksville: in May, some 3,000 employees staged a wildcat strike when the General Grievance Committee could not resolve a work-rule dispute; the company's total labor force of 22,000 considered a sympathy strike.[117]
- The Maxwell No. 20 Colliery of the Lehigh & Wilkes-Barre Coal Company in Ashley: a work-rule controversy in July forced the operation to close.[118]

None of the stoppages was sanctioned by the UMWA. As with the wildcat actions following the Strike of 1902, the District and International offices seemed incapable of preventing the actions.

The unsatisfactory arrangement prompted several local unions, including those at the Hudson Coal Company and the Glen Alden Coal Company, to adopt resolutions demanding a special District convention.[119] Cappellini understood the displeasure, yet declined to convene the meeting, arguing instead for negotiations and established grievance procedures.[120] In a maneuver against the District president on March 18, forty-six of the District's 130 local unions sent delegates to an unofficial assembly in

Wilkes-Barre sponsored by the General Grievance Committee of the Lehigh & Wilkes-Barre Coal Company.[121]

The representatives discussed the many issues facing the District and produced a six-part resolution that demanded, among other things, a full investigation of the Pittston killings, elimination of the subcontracting system, and equalization of work.[122] They passed another resolution asking President Lewis to convene the special convention since Cappellini would not.[123] They referred to Section 2, Article 7, of the UMWA constitution, which stated that a special convention can be called upon the request of five or more local unions. A letter to Lewis summarized the appeal:

> Dear Sir and Brother:
>
> Our understanding is that many local unions in District No. 1 have requested a special convention as provided by Section 2, Article 7, of our constitution. From public statements and circular letters sent to locals by President Cappellini and Enoch Williams, secretary-treasurer, we understand our request for a special convention is denied.
>
> We, therefore, appeal to you as President of the United Mine Workers of America to use whatever power that is vested in your office to force the responsible officers of this district to grant said convention to investigate the underlying causes in the Pittston district and the alleged violations of the contract in this district.
>
> Signed by John F. Cavanaugh (L.U. 996), Frank Sobers (L.U. 699), John Orr (L.U. 898), James Gallagher (L.U. 2533), Edward Hogan (L.U. 1616).[124]

The delegates elected a temporary committee to plan the next steps in the growing insurgent movement. They chose former District 1 President William J. Brennan as chairman.

Lewis replied to the communiqué in a characteristically procedural manner:

> Dear Sir and Brother:
>
> I acknowledge receipt of your night letter, dated March 24, signed by yourself and four others. This office is advised that the question of a special convention has never been acted upon in any official and authoritative manner by the executive board of that district. The matter, therefore, is not within the jurisdiction of the International union. I am further advised that the executive board of District 1 will, at its earliest convenience, take up and pass upon this matter and authoritatively advise the members of District 1 of the action taken.
>
> Very truly yours,
>
> John L. Lewis[125]

Cappellini issued a disparaging statement describing the rump gathering as nothing more than "petty politics" and threatened to expel the participants for dual unionism.[126]

A modest recovery of anthracite markets during the spring of 1928 did not diminish the workers' militancy.[127] Again ignoring official admonitions, 150 insurgent delegates from 50 locals met in Wilkes-Barre on April 22, and passed resolutions to continue planning a special conclave.[128] They took a conciliatory tone by reiterating their opposition to dual unionism and refusing to censure Lewis or Cappellini, both of whom continued to ignore their pleas. It was evident, however, that they saw the convention as the first step in replacing Cappellini and his allies. As proof of their chief's remoteness, delegates from 14

Figure 33. Exeter Colliery, Lehigh Valley Coal Company, Exeter, PA. (Courtesy of William A. Hastie)

of the assembled locals said they had sent special convention requests to Cappellini but received no replies.

The interminable strike at No. 6 loomed over the gathering. A spokesman declared that one purpose of the special convention would be "to delve into the affairs of District 1, and more especially into the situation in the Pittston section where four murders [had] been committed as a result of the [sub]contract system feud."[129] In the day's most important decision, the delegates set May 21 as the date for the special convention in Scranton.

Cappellini saw the rearguard action as "positively a dual movement and nothing else." He said it would be dealt with "in the same way that other dual organizations are handled."[130] Lewis wanted to know more about the threat and summoned Cappellini and the District 1 executive board to Washington, D.C.[131] Officially, the International president adopted a hands-off policy and said Cappellini was in charge but, in fact, Lewis held a great deal of control.

Lewis sent another letter to L.U. 1703 members directing them to "return to work and comply with the terms of the existing agreement and take up the grievances through regularly constituted union channels."[132] The missive arrived on the desk of L.U. 1703 President Frank McGarry who, a few years earlier, had been fired for labor activism but gained reinstatement after an appeal to the ABC.[133] McGarry, described by one newspaper as "the stormy petrel of the District," had been a Campbell protégé who, upon his mentor's assassination, not only replaced him as the Erie companies' most prominent labor leader, but also assumed his former position as the No. 6 checkweighman.[134] He read the letter to a large gathering of strikers, and further readings followed in Italian, Lithuanian, Polish, and Slovak. McGarry then retook the floor and ridiculed Lewis' dispatch:

> The principal grievance among the miners at No. 6 Colliery has been the [sub]contract system and thus far we have failed to find where the District officers made a move of any kind to abolish the system in the collieries of the Pennsylvania Coal Company. In reading the letter of President Lewis one would be led to believe that the District president and the District board have been doing everything within their power to eliminate the [sub]contract system, while as a matter of fact nothing was done in this matter until Alex Campbell and a few others started the movement some months ago.[135]

The strikers took the same stance their counterparts had assumed as far back as the 1910s: mining would recommence only when the company had met their demands.

The No. 6 Strike Continues

By late March, the economic state of the No. 6 protestors had become so difficult that they accepted food and other assistance from the Workers' International Relief Society of New York City.[136] Nevertheless, in early April they unanimously rejected another instruction from Lewis to resume mining and begin negotiating.[137] Federal mediator Thomas Davis of Wilkes-Barre attempted to intervene, but to no avail.[138]

In pursuit of a breakthrough, Cappellini invited McGarry and other L.U. 1703 officers to yet another meeting.[139] An air of optimism pervaded the session because of the recent surge in hard-coal orders. Indeed, PaCC and HC&I had increased production at all operations except, of course, the No. 6. The talks were cordial but the men remained unmoved. One result was a request by McGarry to have Cappellini send a telegram to President Gallagher appealing for a new round of negotiations. Gallagher received the telegram and turned down the offer.[140]

Even though Cappellini tried to broker a compromise, McGarry unrelentingly condemned the District president for his unwillingness to convene a special convention. McGarry also threatened a larger industrial action over the subcontract: "A general strike of the Pennsylvania Coal Company would not eliminate the subcontractors, but an industry-wide shutdown of 81 local unions [the number believed to support the insurgents] will get rid of [them] for all time."[141] Selekman concluded that the legacy of Erie's regressive labor policies undergirded the conflict:

> As for the company, the trouble at Pittston is in a sense a harvest from seeds sown long ago—the hiring of cheap foreign labor in large numbers and the extensive use of the [sub]contract system in the fight against the miners' union. … Given the history of Colliery No. 6, the re-introduction of the [sub]contract system was bound to lead to trouble. Perhaps the management thought of this eventuality and took a chance, feeling as it does that above all it must make its operations pay. … But the tragedies in Pittston—where the real grievance underlying them was not brought out into open discussion until after it had driven the men to revolt—point to the need for some further measure that may operate continually not as much to adjust difficulties as to remove the causes that create them.[142]

Former District 1 President William J. Brennan explained the insurgents' position in an open letter:

> The old abuses which were stamped out in 1920 are again in force. Mine [sub]contractors get sections of mines allotted to them. They hire men at a daily wage and force production. They are favored by having cars provided for them while the regular miner must wait until the company gets ready to provide cars. Every favoritism is shown to the mine [sub]contractor as against ignoring of common courtesy to the miner. And the mine [sub]contractor is often not even a miner, but an individual who is the head of a monopoly. He rules like a king in his domain and pockets enormous profit, not for work performed, but for work which he forces others to perform.[143]

A break in the standoff came in late April. With over three months on the picket lines having depleted their resources, the No. 6 men consented to a 30-day truce. In return, District and company leaders promised to conduct an investigation. However, difficulties immediately arose concerning the company's refusal to let Frank McGarry assume the position of check-weighman and James Lamarca the post of check-docking boss. The workers held an emergency meeting and voted to stay on strike until McGarry and Lamarca were allowed to begin their duties. After some hesitation, Superintendent Benjamin Milton consented. The most rebellious mine in anthracite's largest District restarted on April 30, 1928, under a temporary ceasefire.[144]

The resumption did not quiet the demands for a special convention. The Save-the-Union Committee, under the guidance of Stanley Dziengelewski and George Papcun, fully backed the idea. Papcun envisioned the gathering as a chance to depose "the Cappellini gang." He told supporters: "We call upon all mine workers to clean out Cappellini and to support the special district convention in spite of the terrorizing tactics of the police who are trying to keep Cappellini and Lewis in power."[145] With Communist and other radical boosting, the Save-the-Union group had run an unsuccessful campaign against Lewis during the International union election of 1926. Their candidates—Central Pennsylvania bituminous leader John Brophy for president, and northern anthracite's (and No. 6 employee) William J. Brennan as vice-president—lost by a wide margin.[146] The loss, along with Lewis's denunciation of the rebels for having a "pernicious influence" on the UMWA, did not stop them from organizing a national convention in Pittsburgh in April 1928. They also ordered a strike that some 10,000 bituminous workers honored (but none in anthracite). At least 29 persons from District 1 attended the Pittsburgh convention.[147] Communist organizer Steve Nelson reported that the CPUSA had about 50 members across the four anthracite fields at the end of the 1920s.[148] The small number who belonged to L.U. 1703 at the No. 6 Colliery tried but failed to convince President McGarry to break with the UMWA and join their cause.[149]

Figure 34. John Brophy.
(From John Brophy, A Miner's Life, *Madison, WI: University of Wisconsin Press, 1964)*

The Insurgents' Special Convention

As the May 21 special convention date approached, Cappellini and the District board sent a letter to all local union members with admonitions against attending. For their part, the dissenters mounted a drive to bolster the turnout. Mrs. Alex Campbell promoted the assemblage in a letter to the editors of local newspapers:

> In reading accounts, I find that my late husband's friends are carrying on the fight for the things that he died for. I hope that his friends will be successful in removing those terrible evils to which he was opposed and for which he was murdered.
>
> I also hope that members of the great United Mine Workers of District No. 1 will stand by Alex's friends and Frank McGarry, president of Local Union No. 1703, and his associate officers and give assistance, both moral and financial, to free Benito [*sic*], Moleski, and Mendola from jail and return them to their wives and children as free men.
>
> Also remove the [sub]contractor who has caused so much sorrow to myself and others and which has brought about hardship to miners of District No. 1. I pray that the men be determined in their demand for a special convention, as Alex often said it may be the means of bringing peace and happiness to the miners and their families in District No. 1.[150]

Cappellini responded with another warning: "To participate in the business affairs of the dual organization opens the way for severe punishment such as handed to John Brophy, Powers Hapgood and others termed leaders in dual, outlaw organizations."[151] The local men knew that Brophy, Hapgood, and others were banned from the organization and, therefore, from employment in union mines.[152]

Shortly before the convention, the insurgents again asked Lewis to participate, this time as chairman. If he could not attend, they asked that he send a representative. The invitation helped dispel the dual-union accusations that were sure to follow. As anticipated, no response came from Lewis or anyone else at the UMWA's main office in Indianapolis.[153] The organizers also invited James J. Davis, secretary of the Coolidge Administration's Department of Labor, as well as William Green, president of the American Federal of Labor, neither of whom replied.[154]

The subcontractors at Erie operations were not taking the gathering lightly. According to insurgent spokesman Brennan:

> In the Pittston district we have found that the [sub]contractors who have dragged our organization into the mud [are] pooling from $250 to $1,000 each to interfere with the convention. They are using all the money and power they have gained through the laxity of district officials to defeat the progressive movement of real union men.[155]

Perhaps not by coincidence, as the convention approached, all Erie collieries were fully engaged with work, for the first time in 1928. Even the No. 6 was working because the men agreed to the aforementioned 30-day truce.[156]

On the appointed date, only 57 of the District's 156 locals sent delegates to the Regal Hall convention center in Scranton. Mrs. Alex Campbell greeted the delegates with a letter, read by convention secretary Edward McCrone, urging participants to "straighten out the existing evils which confront the UMWA of District 1," while criticizing "certain groups who have no regard for the rank and file of the UMWA and who have their own self in mind." The delegates praised her husband and the other L.U.

1703 murder victims.[157] The first session saw the approval of several resolutions, including one supporting the innocence of Bonita, Mendola, and Moleski. Contrary to rumored concerns about Communist agitators or Cappellini spies, the convention was a peaceable event. However, Communists were not allowed to participate and the sergeant-at-arms ejected George Papcun on the first day.[158]

The initial turnout proved insufficient, however, because the District had 139 locals in good standing, meaning that 13 more were needed for a majority. Since the proceedings would last a fortnight, organizers decided to send recruiters to the unrepresented locals. News reports indicated that some companies used intimidation to prevent employees from partaking.[159] Some locals stayed away because they were split between the dissident and the pro-administration factions. At the Ewen Colliery, a heated argument over sending delegates erupted in a fist fight.[160] By the fifth day, 63 locals were present.[161] By May 26, the number had grown to 81, which surpassed the needed majority. The joyous rumpers celebrated by staging a noon-time parade through downtown Scranton.[162]

In the main order of business on June 1, the delegates upheld their insurrectionary reputation by impeaching Cappellini and the other officers.[163] The vote was 170 to 12.[164] They elected new leaders including Frank McGarry as president, George Isaacs as vice president, Walter Harris as secretary-treasurer, and Ray Delaney as International Board member.[165] The new officials agreed to serve temporarily until the regularly scheduled District elections in June 1929, when Cappellini and his colleagues were expected to run again.

McGarry's experience and charisma made him the logical choice for the presidency. Not only had he been chosen head of L.U. 1703, chairman of the No. 6 colliery's general grievance committee, and colliery check-weighman, but like his predecessors, Campbell and Cappellini, he displayed strong leadership skills.[166] In addressing the insurgent convention, McGarry focused on the problems facing the ordinary mineworker, including the individual contract. "If I could be sure that the [sub]contract system would be eliminated today," he professed, "I'd be satisfied to be planted in Shanty Hill tomorrow."[167]

Soon after the convention, the insurgents established a headquarters in the Miller Building in downtown Scranton.

VICTORY PARADE AS CONVENTION GETS A MAJORITY

The special mine convention of District No. 1, United Mine Workers of America, "went over the top" today in the drive to obtain representation of a majority of the 139 locals in good standing in the district. Eleven new locals were represented at the morning session of the special convention today, boosting the number of locals represented to seventy-four. Wild scenes of enthusiasm greeted the announcement that considerably more than the number required for a "legitimate and legal" majority, had sent representatives to the special meeting.

Delegate Dougher, of Archbald, one of the most active of the insurgent leaders, asked that a parade of victory be staged at noon in Scranton so as to celebrate the accomplishment of obtaining a majority representation. "Don't forget to go up Spruce street and pass the Miller Building," shouted Ray Delaney, of West Avoca, who is representing a Lehigh Valley Coal Company colliery. The suggestions were acted upon.

It is the plan of the convention to call President Cappellini and other officials upon the floor of the convention to defend charges which have been made against them. In the event that they do not appear to defend themselves, the charges will be regarded as well founded and dismissal of the officers will follow.

Delegates to the special convention explained that, after the district officials are ousted, no attention will be paid to President Cappellini and his corps of workers. Dues will be paid to officers of the new organization, which is bound to cause a split in the [illegible] forces, serious enough to require the intervention of International President John L. Lewis or a capable representative.

The following additional locals were represented at the convention today:

No. 844, Coalbrook, a big Hudson Coal Company colliery. Delegates John Branick and John Leasch.

No. 605, known as the "Tony Rose" colliery, Pennsylvania Coal Company, Yatesville, John Butler, delegate.

No. 1132, No. 5 Loree, Plymouth, Hudson Coal Company. Delegates Frank Zelensky, Edward Kondrak, John Lucas and George Nice.

No. 452, of Brodericks, Pa., Luzerne County, Temple Coal Company. Delegates, Anthony Maniskus and John Snyder.

No. 1024, Erie local in Mayfield, where John Kneisch was slain a couple of years ago. Delegates Thomas Daley and A. G. Neary.

No. 1025, Jermyn colliery, Hudson Company, delegates Paul Kregesky and John Hodgson.

No. 917, Miles Slope, Olyphant, Hudson Company, delegates John Barron and John Zernosky.

No. 1877, Johnson colliery, Dickson City, Scranton Coal Company, delegate, Charles Parchinski.

No. 1004, Price-Pancoast, Throop, delegates, George E. Bezek and Wil-

Figure 35. "Victory Parade as Convention Gets Majority." (Pittston Gazette, *May 28, 1928)*

Figure 36. Insurgent officers McGarry, Isaacs, Delaney and Harris. (Wilkes-Barre Record, *June 28, 1928)*

The rump officers wrote to Lewis regarding the "new" administration and requested that he dismiss the "former" principals.[168] Unsurprisingly, Lewis characterized the congress as "an ill-conceived dream" directed by "a designing but unwise leadership" who acted in violation of the UMWA Constitution.[169] Neither he nor the International Executive Board would recognize the outcome. He underscored the legitimacy of the Cappellini administration and demanded union loyalty. And yet, he could not deny the widening rift between the District leadership and the rank and file. It was a fight that clearly threatened the legitimacy of the UMWA in the northern field and all of anthracite.

With his "fiery temperament, blunt speech, and great activity," Cappellini fought back, undoubtedly with Lewis's consent.[170] He distributed another letter to the District membership disparaging the convention as an attempt by persons "interested only in destroying the efficiency and stability of our organization."[171] In public comments he censured the gathering as "unconstitutional, illegal and without authority under our laws." Taking a page from Lewis, Cappellini said that the attendees had been duped by Communists affiliated with the Save-the-Union group.[172] He also insisted that he had made progress on subcontracting, wage violations, equalization, and the growing controversy over machine mining. To show that he meant business, Cappellini began proceedings to expel the insurgents from the UMWA.[173]

In the meantime, the 30-day temporary work agreement at the No. 6 was set to expire in late May. PaCC officials insisted that the No. 6 Shaft's unprofitability remained the primary concern. The workers continued to demand fair wages and an end to subcontracting. PaCC Vice President A.K. Morris threatened to close the facility if the parties could not come to terms. A joint union–company committee studied possible solutions and submitted another wage plan to keep the colliery running. McGarry urged his charges to consider the rate offer as another temporary compromise.[174] On April 19, they accepted his advice and voted to extend the truce for another 30 days.[175] The negotiations proceeded. At the same time, the No. 6 men rejected a permanent settlement proffered by Lewis.[176]

Erie's civil war carried on through June as the local unions at PaCC and HC&I began to fracture along insurgent and administration lines.[177] The meeting of L.U. 1703 on June

27 was a raucous affair where a smaller pro-Cappellini faction clashed with the larger pro-McGarry forces. Although he was still a member of the local, McGarry was prohibited from speaking because he was now the insurgent president. When it became clear that the meeting would make no progress, the police consulted with Mayor Gillespie and cleared the hall for fear of a fight. McGarry proceeded to lead a march to a nearby field where he delivered a speech on the District's problems.[178]

In early July, with the truce period having expired, the picket lines went back up at No. 6. However, a significant turn of events came two weeks later, on the eve of the official District 1 biennial convention. Over 1,200 No. 6 men gathered at St. Aloysius Hall and, with McGarry presiding, voted to end the strike.[179] What seemed like a meaningful resolution grew out of private negotiations that began several weeks earlier when McGarry sent General Grievance Committee member James Kearney to secret conferences with President Gallagher and Vice President Morris.

The parties reached a tentative agreement that included favorable pay rates for miners at the No. 6 Shaft's Red Ash Vein and for laborers who worked with mechanical loaders in the Clark Vein. The latter negotiation proved especially difficult since some of the work crews were led by subcontractors.[180] Lewis welcomed the constructive step and thought the dialogue "may well result in the restoration of normal conditions when negotiations are continued for a permanent understanding between the company and the men."[181] A guard of 22 state troopers accompanied 1,600 miners and laborers as they returned to the pits in late July, 1928.[182] Many workers remained dissatisfied because the subcontractors were still underground.

It proved to be a brief respite. A few weeks later, despite opposition mainly from "American" or English-speaking (British, Irish, and German) workers, and in disregard of "the considerable suffering [that] is prevalent among the families of the miners at No. 6 Colliery," the insurgents instigated yet another walkout.[183] Disputations over the usual grievances provoked the action. The PaCC General Grievance Committee called what became one of the most turbulent meetings yet held to consider a general-strike motion. Pittston police were enlisted to maintain order. Committee members rejected the company-wide strike idea after learning that most of those present wanted to remain on the job. However, as a gesture to the No. 6 men, they approved a resolution calling for another rump convention to consider a District-wide shutdown.[184]

Disorder plagued other PaCC and HC&I mines during the long summer of 1928. Management sought to divide the workers by idling all collieries except the Ewen and the No. 14, which continued on a full-time basis. The Ewen was the only Erie colliery (out of 14) that did not send delegates to the rump convention. The tactic infuriated many Ewen workers who proceeded to unseat George Moleski, president of L.U. 1487, and elect insurgent Anthony Mancini in his place.[185] The Butler Colliery closed in June and July over a wage disagreement marred by the destruction of the Fernwood Slope's power station by dynamite.[186] At No. 14, a split between the insurgent camp, which supported McGarry, and the Cappellini faction, under Frank Cardoni, resulted in a lockout of 1,700 men in mid-July.[187] The No. 9 Colliery closed for two weeks in late July when 1,600 workers protested seniority-rights violations.[188] Continuing rows at the Ewen caused unpredictable work schedules into August. Police broke up a meeting of the colliery's local union when factional infighting erupted.[189]

Figure 37. Ewen Colliery, Pennsylvania Coal Company, Jenkins Twp., PA.
(Frank Hizny photograph)

Cappellini meanwhile remained determined to purge the insurgents from local union offices. He declared July 1 as the "expulsion date" when all locals had to remove any officials associated with the rump convention.[190]

Trouble for Rinaldo Cappellini

Again, the Erie Coal Companies were not alone in experiencing labor unrest. Workers at the Maxwell No. 20 Colliery of the Lehigh & Wilkes-Barre Coal Company walked out over a wage disagreement.[191] The No. 5 and No. 7 Collieries of the Susquehanna Collieries Company in Nanticoke went idle when a mine foreman assaulted a motor runner.[192] The Florence Colliery of the Lehigh & Wyoming Valley Coal Mining Company closed in July and again in August over wage-related quarrels.[193]

Exhortations mounted for Cappellini's removal. A few locals filed charges against him for malfeasance in office.[194] Rumors circulated about an imminent resignation, but the once-popular District president denied having any plans to step down. He characterized a published report in the *Scranton Times* on July 18, which forecast his exit within 48 hours, as "nothing but a false alarm." Indeed, he said the uprising that began at No. 6, spread to other PaCC operations, and now extended throughout the District "was about on its last legs and was no cause of worriment to the District officers."[195] The rumors nevertheless persisted. One suggested that Cappellini had become disillusioned with the backbiting and would leave to encourage fence mending.

The gossip mill proved correct. The three-term president of 60,000 mineworkers tendered his resignation on July 10, 1928, six weeks after having been "impeached" by the

Figure 38. Engine and fan house, No. 9 Shaft of No. 9 Colliery, Pennsylvania Coal Company, Hughestown, PA. (Courtesy of National Canal Museum Archives)

rump conventioneers. Earlier in the week, he had traveled to Indianapolis to confer with Lewis and withdrew upon returning home. Cappellini said he bowed out for health reasons but he also admitted that he had lost control of District affairs.[196] He regretted vacating his duties: "My heart and soul has always been with the miner[s] and will always be with them."[197]

Lewis schemed and bent union by-laws to have Scrantonian John J. Boylan appointed as chief executive.[198] A stalwart member of the executive board, Boylan had been sending private messages to Lewis throughout the recent protests.[199] Contrary to those who thought that Cappellini's downfall and Boylan's rise would calm the waters, the disputations did not subside. Boylan was rightly seen as a Lewis lieutenant. McGarry remained one of the loudest oppositionists. He welcomed Cappellini's resignation but also condemned the "unseen forces" (i.e. Lewis) who were controlling District 1 affairs. While he did not hold Cappellini responsible for every problem, he lamented that the former president had presided over an era of "loss of work, discrimination, and intimidation."[200] Former District President William J. Brennan blamed Lewis for the current problems: "No individual is bigger than the

Figure 39. John Boylan, President, UMWA District 1, 1928. (Courtesy of Special Collections, Indiana University of Pennsylvania)

Figure 40. Maxwell Colliery, Glen Alden Coal Company, Ashley, PA, circa 1938. The Huber Breaker is being constructed in the background. (Courtesy of Luzerne County Historical Society)

rank-and-fine, and John L. Lewis is no super-man. Right and justice will prevail, regardless of the arbitrary actions of the men who are living off the fat of the land through our efforts."[201]

On the corporate side, the upheaval prompted the Erie companies to transfer four superintendents and five foremen. Among them was James C. Johnson, superintendent of the No. 6 Colliery, who was transferred to the same post at the Ewen Colliery. In turn, Superintendent Benjamin Milton left the Ewen for the No. 6. Similarly, John Wynne departed the No. 9 Colliery for the Central Colliery, and Thomas Ridgeley was transferred from the Central Colliery to the No. 9.[202] The company also accepted the resignation of General Superintendent Joseph P. Jennings, a long-tenured official who had been in charge of Gallagher's plan to expand tenancy.[203] Although the dissenters took some satisfaction in the personnel changes, they grew ever more distressed by the company's obdurate positions on subcontracting, machine mining, equalization, pay rates, and other matters. Many joined the growing ranks calling for a separate anthracite union.[204]

The labor wars at the No. 6 Colliery and elsewhere in District 1 continued over the next several months. Lewis's machinations for Cappellini's resignation and Boylan's appointment continued to offend. As McGarry quipped: "If John L. Lewis thinks that Cappellini's resignation will cause harmony in the District, he is only kidding himself."[205] Insurgent Secretary-Treasurer Walter Harris doubted the new president's sincerity on subcontracting: "If John Boylan is so anxious to eliminate the [sub]contract why is it that there are more [sub]contractors in his inspection district [in Scranton] than in any other?"[206] McGarry reiterated his belief that a general strike was the "only way to clean up the [sub]contractors."[207]

Amidst the crisis of legitimacy and authority, the insurgents convened another convention in August.[208] In preparation for the affair, they issued a public statement listing several loathsome aspects of the anthracite workplace under Cappellini, Boylan, and Lewis, including

> ... discrimination on the part of the coal companies against the mineworkers, especially those fighting for rights in the insurgent movement, retention of and favoritism toward [the] mine [sub]contract system by companies with abusives [*sic*] that come with it, and numerous other points of complaint along with evident cooperation of officials of the [UMWA] District.

The disorders, they maintained, "force us to believe that the companies are looking for trouble." Yet they wanted to emphasize that

> ... we don't want trouble. The thing needed now is peace. We know the general public does not want trouble. But coal operators are not acting with pacific intentions, and actions with the so-called District officers' cooperation make us believe they are trying to force us to the wall. ... We have been kidded long enough and have been restrained in our stand in the hope that the facts before this would have convinced the operators and so-called [District] administration forces of the justice of our claims. But it has been no use. Instead we are being forced into a step which cause[s] sufferings for the District.

Still, they declared, they would not back down: "Our men will fight for their rights as true union men fight. They prefer the just way out. They will battle to the end to obtain what they know is due them."[209]

The second convention of the so-called "Committee of 81" insurgent locals who participated in the first confab picked up where they had left off in May. The delegates discussed numerous issues and problems and passed several resolutions. After threatening a general strike, however, they decided to keep working but they agreed to withhold dues from the UMWA and send the monies to the dissident union's main office. As a consequence, within weeks 48 locals were sending the "per capita tax" from dues to Boylan, 46 to McGarry, and 36 to neither.[210] Of course, criticisms and warnings flowed from Boylan regarding the rump gathering, and PaCC officials said they would not meet or negotiate with any insurgent leaders.[211]

Despite some internal dissent and a full attack by the UMWA, the protests endured through the second half of 1928. The insurgents, the District 1 administration, and the Communists continued to vie for the workers' loyalties. Unauthorized strikes persisted at Erie and other companies. The No. 6 Colliery miners experienced nonstop tensions. During one walkout at No. 6, the Communist-led National Miners Union sent one of its organizers, Tony Minerich, to take control of the strike. The workers ignored Minerich and did not complain when the State Police jailed him on a spurious charge of planting dynamite. The strike soon failed.[212] In another instance, the colliery experienced a two-week strike when Superintendent Brown refused to negotiate with the recently elected insurgent officers of L.U. 1703.[213] The Ewen and No. 9 witnessed regular disturbances.[214] Dynamite destroyed the engine house at the Butler later in the fall. Insurgents were the likely culprits; no one was arrested.[215] Another shooting death occurred in Pittston when a young PaCC workman refused to honor a picket line and fired into the strikers when they pursued him.[216]

The *Wilkes-Barre Record* editorialized against the chronic distress in Pittston and elsewhere:

> As long as the division in the ranks excites such fierce animosities as have appeared for months, the whole anthracite industry is in danger of permanent damage. ... Peace in the anthracite region is necessary to assure outside consumers that they are not in danger of being left again without an adequate supply.[217]

Boylan proved a capable leader who mounted a vigorous counter-offensive and debated dissenters whenever possible. He wrote a cutting letter to the District membership condemning each of the "bogus" insurgent officers.[218] He devoted greater attention to subcontracting: "I feel the [sub]contract system should be eliminated," he confirmed, "as it has been the cause of all the turmoil in the District."[219] In an appearance before the ABC, he argued for the complete elimination of the structure.[220] Under the direction of a local sheriff, he joined Frank McGarry in a peace conference.[221]

As 1928 drew to an end, Boylan had succeeded in tightening the UMWA's hold on District 1 affairs. To leave little doubt that deviation from UMWA policies and procedures would not be tolerated, he moved to destroy the dissident faction by expelling 45

Figure 41. Underwood Colliery, Pennsylvania Coal Company, Olyphant, PA, built in 1914; this photograph taken in 1936. (Courtesy of James Guthrie)

"renegade" members for making plans to establish an alternative organization called the Anthracite Mine Workers Union (AMWU).[222] Although the dual organization never advanced much beyond the planning stage, among those banished were rump officials Frank McGarry, Walter Harris, Henry Schuster, and Joseph Dougher.[223] More than a few fearful insurgents began rejoining the UMWA.

As one of the most tumultuous years in anthracite mining history ended, there was one significant breakthrough on subcontracting. In December, William Cooney, chair of the Underwood Colliery's grievance committee, announced that "Rigo Madtucci has been removed as head of the [sub]contract system at the Underwood and beginning today that system will be a thing of the past. ..." The company had apparently decided to take the proactive step in order to quell the growing dissatisfaction at one of its most compliant mines. According to the *United Mine Workers Journal*, the Underwood workers had "always remained in the administration ranks and always loyal to President Boylan, not once even dickering with the insurgent cause."[224] Some observers saw the move as a victory for the militants, but others thought the Underwood workers had been coopted. The nearly 100 men who had toiled in the colliery's subcontracting system were re-employed as traditional contract miners and laborers working in one chamber at a time. Although the company and the union may have been trying to show the rebels that progress lay in moderate, not activist, methods, the program did not last nor did it diminish the larger uprising.

Political Realignment: An Outcome of the Labor Wars

As difficult as the labor-management conflict was for the mineworkers and their families, the effects appeared to have extended well beyond the workplace and the family into the larger community. Lewis Casterline, who began as a miner and was eventually promoted to a PaCC superintendent position, maintained that the homicides during the winter of 1928, as well as the continuing insurrection at Erie and elsewhere, did more than take lives, cause strikes, and shock the public. The killings had far-reaching consequences for the social and political life of the community.

As one case in point, Casterline described how tenancy and murder mobilized the Italian community and increased its resolve against organized crime. He pointed out that the killings at No. 6 triggered a political realignment during the rule of Luzerne County judge and political boss, John S. Fine:

> When Fine took control of the county [in the 1920s] he did that by [doing] something that he should have never done. He's starting to play [cooperate with] certain elements. Murder Inc. who were coming in here killing labor leaders who didn't want to go along. ... We had a lot of this in the anthracite region. ... Frank [*sic*] Lillis, Alex Campbell, John [*sic*] Reilly—now they were fighting for the rank-and-file. So they were cut [down]. ... When the Italian people began to see that, they said: "He's [Fine] supposed to be a friend of ours?" Now this town that was a hundred percent registered Republican begins to be ninety percent registered Democrat and ten [percent] Republican. So Fine knows that I've been his opposition ever since I can remember. He calls me in one day and he says, "Lewis, why is it that Yatesville and Keystone and Hilldale [Pittston suburbs], the Italian people [there], are now ninety percent registered Democrats?" I said, "Why don't you go and ask them?" But I knew what was working, see.

> They didn't like his connection [to organized crime] and you know how they were.[225]

Angelo Siracuse, an American-born mineworker whose parents emigrated from the sulfur-mining region of Sicily during the early 1900s, worked at the Harry E Colliery in Swoyersville along with his father and brothers. He told a similar story of political reorganization in that small mining town during the early 1930s. Swoyersville's predominantly Italian and Slavic workers rebelled against Fine and the Republicans, whose most prominent community leader was Lorenzo Ferraro, the alleged local crime boss:

> Well, the Ferraros, they had control here of the Italian people, you know. He [Lorenzo Ferraro] was a big politician in his own way. It was nothing for him to talk to an Italian guy and say we want you to vote this way in politics. You'd say "no" to them, they'd slap you in the face; gangster, you know. He was the guy that used to boss everybody. He used to try to control my old man and everybody. [For example] he didn't want my brother Charlie to run [for city council as a Democrat]. And when he said it, it was like he was the boss, you know. [My father said] "No way. If Charlie wants to run, he's running." And he ran. And he knocked the ears off them. Our Charlie got elected and that's when the Democrats took the Republicans over [surpassed them]. We've been Democrats [in Swoyersville] ever since. Ever since the thirties, thirty-three.[226]

Along with local political changes, the northern field's labor upheavals may have influenced the presidential election of 1928 when, for the first time since before the Civil War, the majority of Luzerne County and Lackawanna County voters chose Democratic Party candidates. Democratic presidential nominee Al Smith of New York carried both counties, which had been Republican strongholds reinforced by the coal companies' control of politics.[227] The two northern-field counties were among only three Pennsylvania counties to prefer Smith over Hoover; the other was Elk County.[228] The election demonstrated that citizens began to question "the wisdom of continuing the established economic policies" and, from the 1928 election onward, "many working class people began to consider themselves Democrats, at least on the national level."[229] Continuing the trend, majorities in Luzerne and Lackawanna counties voted for Franklin D. Roosevelt in 1932. The counties have remained predominantly Democratic in registration and national election results ever since.[230]

Prologue to another Rebellion

The highly politicized atmosphere of 1928—in both the workplace and the community—was fed by, and contributed to, the insurgent challenge. Despite a dual-union failure, insurgent expulsions, editorialists' columns, and hungry families, the movement endured. The bitter strikes at the No. 6 Colliery, as well as the deteriorating working conditions throughout the District, served as the prologue to another, more successful, insurgency motivated by tenancy, equalization, unemployment, and other grievances.

As controversial as it was, however, subcontracting would not remain the main source of consternation among Erie workers. After the Van Sweringen brothers gained control of the Erie Railroad in 1924, they decided to take their anthracite subsidiaries beyond subcontracting into another form of tenancy, the *leasing system*. They established a new subsidiary within PaCC, The Pittston Company (TPC), whose main purpose was to issue and oversee coal leases. TPC and the leasing system would contribute to another round of violently disruptive labor wars that spread throughout District 1. The new uprising led to the most significant dual union anthracite had ever seen. The movement would be led by none other than Rinaldo Cappellini, along with newcomer activist Thomas Maloney and stalwart labor priest Fr. John J. Curran. The battle this time included more than the UMWA and the large coal companies. A new type of lease-holding corporation also participated.

PART III

The Leasing System, Labor's Response, Anthracite's Decline

Miners Memorial, Plymouth, Pennsylvania

CHAPTER SIX

The Leasing System: Tenancy, Organized Crime, and Labor Resistance

[The Alleghany Corporation represented] the holding-company device pushed to its uttermost limits ... built of nothing but faith.

— Ron Chernow, *The House of Morgan*, 1990

[John L.] Lewis is not interested in the anthracite fields. ... We are going to organize the anthracite union and we are going to have a rump convention.

— Rinaldo Cappellini, 1933

Well, there were a lot of the Italian families, I mean, [who] were hooked up in this crime business. I'm trying to think. I don't want to make any comments on who it might have been. I wouldn't want to get myself in the ringer on that. But it was some of the Italian men who were high up in the mining circles [who secured coal leases].

— E. Stewart Milner, vice president, Pennsylvania Coal Company, 1994

Subcontracting and Leasing: Two Forms of Tenancy

The leasing system represented a logical extension of the subcontracting system. Both reorganized mineral rights' access from a large company to a smaller tenant through a legal agreement. Both allowed the lessor and the tenant to exert greater control over the workers and the workplace in order to achieve greater output and, ultimately, profits. Organized criminals secured many of the subcontracts and leases at the Erie Coal Companies. The United Mine Workers of America (UMWA) responded ineffectively to the grievances that resulted from both forms of tenancy. Finally, both schemes led to major labor rebellions.

Despite the similarities, there were also a number of differences. The subcontractors were unincorporated entrepreneurs who secured agreements as individuals, whereas the lessees were incorporated businesses that enjoyed limited liability and other corporate advantages. The subcontractors usually mined in one or two chambers and hired comparatively small work crews, while the leaseholders gained access to whole veins and eventually entire collieries where they hired from several dozen to a few thousand employees. The subcontracts usually ran for weeks or months, but leases remained in force for years and even decades. The subcontractors did not run the mine and were cognizant of the lessor's work crews and bosses. The lease holders, on the other hand, directed their own operations. Moreover, the presence of their bosses made it clear who had ultimate authority. Lastly, while the early lessees and the subcontractors sent their freshly mined coal to the lessor for processing and sales, by the late 1930s lessees began leasing, buying, or building their own breakers and developing their own distribution systems.

There were two types of leases. During the 1910s and 1920s, "obligated" leases required the holder to sell all raw coal to the lessor who processed and sold the final product. After 1935, the "free lease" became common whereby a lessee paid the lessor a flat annual amount for "rent" plus a royalty on each ton mined and was then free to sell the raw coal to any processor, or to process it himself if he owned a breaker.[1]

The Van Sweringens and The Pittston Company

In 1924, the Van Sweringen brothers, Otis P. and Mantis J., of Cleveland, gained control of the Erie Railroad and thereby the Pennsylvania Coal Company (PaCC) and the Hillside Coal & Iron Company (HC&I).[2] Within two years the brothers appointed another Clevelander, Michael Gallagher, to the board of directors. Gallagher joined the Van Sweringens in 1926 after having spent 18 years as president of the M.A. Hanna Company's bituminous operations.[3] The directors immediately dismissed board president Frederick D. Underwood, who had served since 1901, and elected Gallagher in his place. Eighteen months later, company president C.S. Goldsborough resigned in a letter that demurred: "From my understanding of your organization plans, as stated by you today, I feel that there is little justification for my retention in office and I hereby, with regret, tender my resignation. …"[4] The entire board then stepped aside and the Van Sweringens appointed a new group of directors and made Gallagher company president.[5]

Figure 1. Michael Gallagher co-hosted a banquet for President Warren G. Harding at the Willard Hotel, Washington, DC, March 3, 1921. President and Mrs. Harding are seated left; Mr. and Mrs. Gallagher are fifth and sixth persons left. (Courtesy of Chuck Yungkurth)

Figure 2. Pittston Coal Sign.
(Courtesy of William Ferri)

Why were Goldsborough and the board removed? A major reorganization was underway and Gallagher and more favorable board members were prepared to lead the way. The first step occurred on January 26, 1929, when the Van Sweringens created the Alleghany Corporation as a holding company for the Erie Railroad, the Nickel Plate Railroad, the Missouri Pacific Railroad, and other subsidiaries in their expanding business empire.[6] The second step came on January 11, 1930, when the brothers formed The Pittston Company (TPC) as a subsidiary of PaCC. The new corporation's main purpose was to "operate on a rental and royalty basis the anthracite properties in Lackawanna and Luzerne counties owned by the Pennsylvania Coal Company and those leased by the Pennsylvania Coal Company from the Hillside Coal & Iron Company."[7]

Accordingly, HC&I transferred all mines, facilities, and properties to PaCC, which in turn conveyed all of its holdings except one, the Ewen Colliery, to TPC. The collieries included HC&I's Forest City and Butler, and PaCC's No. 1, No. 5, No. 9, Erie, Ewen, Old

Figure 3. Night view, No. 9 Colliery, The Pittston Company, Hughestown, PA.
(Courtesy of Steven Lukasik)

Forge, and Underwood.[8] As part of the deal, TPC agreed to pay PaCC a rental fee of $90,000 per quarter plus a royalty of 12 cents per ton. The lease was set to run for 25 years, from January 1930 to December 1954. PaCC would retake possession of the properties in the event of an early termination.[9] TPC was incorporated in Delaware because of that state's favorable business laws, although virtually all operations were in the northern field around Pittston and Scranton. The transaction converted PaCC into a holding company within a holding company functioning to mine coal at the Ewen and collect royalties from TPC's leaseholders. Alleghany offered TPC's shares to Erie Railroad stockholders for $20 each.[10] Gallagher added the presidency of TPC to his titles.

Why did the Van Sweringens create another subsidiary rather than allow PaCC to mine coal and oversee subcontracts and leases as it had done for decades? Four explanations can be considered. First, the decision allowed the Erie Railroad to avoid antitrust action by separating railroad operations from the coal-producing subsidiaries.[11] Second, the bargain provided cash to help cover Alleghany's high debt, for the Van Sweringens had undertaken one of the most aggressive growth programs on Wall Street. Herbert Harwood Jr.'s book on the brothers' rise and fall, *Invisible Giants* (2003), referred to Alleghany as "the Crown Jewel of New York Finance."[12] Ron Chernow described it as a "super holding company atop [a] pyramid of debt."[13] Frederick Lewis Allen wrote that the firm was under great pressure to perform because of the massive liabilities that threatened to bring down the entire Van Sweringen empire, which encompassed 10 percent of the nation's rail traffic.[14]

Ohio farm boys who became the masters of the debt-financed leveraged buyout, the Van Sweringens used each additional purchase as collateral for more acquisitions in "an endless hall of mirrors."[15] They were supported by banks in Cleveland but especially by J.P. Morgan and his financial affiliates in New York City. *The New York Times* reported that Alleghany represented "the holding-company device pushed to its uttermost limits," a house of cards "built of nothing but faith."[16] When the stock market began to implode in October 1929, even the Morgan banks could not prevent a failure. The overly indebted enterprise lost 80 percent of its value, going from $56 to $10 a share in two months. It took another four years before the public knew the questionable circumstances under which Alleghany's stock had been offered.[17]

Figure 4. Van Sweringen brothers. (Thea Gallo Becker, Images of America: Cleveland, 1796–1929, *Chicago, IL: Arcadia, 2004)*

The third possibility was that the Van Sweringens concluded that leasing represented the profitable future. PaCC and HC&I owned vast tracts of coal land, some suitable for first minings but most for second and third minings where leaseholders could remove the remaining coal pillars. Prior to the reorganization, consultant Harry N. Taylor, president of the U.S. Distributing Corporation of Chicago, advised Gallagher that royalties from lessees' coal sales could increase by as much as 200 percent by 1931.[18] The consulting engineers at Coverdale & Colpitts also predicted solid returns.[19] Through leasing, therefore, TPC expected significant monetary rewards for minimal capital outlays.[20]

Figure 5. Van Sweringen Terminal Tower, Cleveland, Ohio. (National Register of Historical Places)

Fourth, when it came to labor, the restructuring largely freed PaCC from dealing with perhaps the most militant work force in hard coal. "The labor problem" became the responsibility of the lessees, as mining engineer and second-generation coal operator Alex Chamberlain put it: "It was a way of a company insulating itself from labor."[21] However, as argued later, leasing negatively affected labor-management relations and helped precipitate a labor war that was among the most discordant the region had ever seen.

Figure 6. Engine House for the Checker Slope, Butler Colliery, Pennsylvania Coal Company, Pittston Township, PA. (Courtesy of National Canal Museum Archives)

Leaseholders at the Pennsylvania Coal Company and The Pittston Company

As discussed in Chapter Two, PaCC had issued a small number of leases as early as 1909 and during the 1910s, primarily to miners with English, Welsh, and Irish surnames. Under Gallagher's direction, the system expanded impressively in the late 1920s and 1930s. British-named leaseholders remained most prominent, but Eastern Europeans and Italians began obtaining more of the agreements. By the late 1930s, the Italians were taking the largest number. Ethnic diversity was apparent in company names such as the Boyle & Flanagan Coal Company, Harry F. Ellis & Company, the Jablonski & Macarelli Coal Company, and the McLane & Damiani Coal Company. Other TPC leaseholders during the 1930s included The Standard Anthracite Mining Company, which secured a tract of land for a second mining in 1931; The Barnum Coal Company, which signed a lease in 1932 for second minings in portions of the Clark and Marcy veins in Duryea and Pittston, respectively, and, in 1934, took an agreement for a second mining in the Clark and Marcy veins in Duryea and Hughestown; The Rovo Coal Company, which garnered a lease in 1933 that permitted the removal of remaining coal in the Fernwood Slope, Jenkins Township; The Vespi Coal Company, which gained access to the coal in the pillars of the Clark vein in Olyphant in 1937; and the Joseph Volpe Coal Co., held by a relative of alleged crime boss Santo Volpe, which took six leases between 1933 and 1936 for coal in the Clark and Marcy veins at the Schooley Shaft in Exeter.[22]

Several leaseholders grew into relatively large operators. Pasquale Adonizio (aka Tony Rose) acquired 49 subcontracts and leases at various collieries between 1921 and 1933. The Berge-Rose Coal Company—formed on September 26, 1931, through a multi-ethnic partnership that included Pasquale Adonizio and his brother, Joseph Adonizio (Sicilians), William H. Berge (Welsh), and John Kehoe (Irish)—began with a lease from Erie's New York, Susquehanna & Western subsidiary. The lessor canceled the lease in April 1933 for "breach of covenants ... in the nonpayment of taxes and nonpayment of royalties," amounting to over $45,000.[23] The problem notwithstanding, the Adonizios secured a series of other leases, often through informal, family-based contacts.[24] William S. Jermyn and E.M. Green, owners of the Jermyn-Green Coal Company, entered into 34 lease agreements between 1937 and 1949. Three of the transactions involved full control of the No. 14 Colliery in 1937, the No. 6 Colliery in 1943, and the Butler Colliery in 1943. With an abundance of coal in pillars and un-mined seams, the enterprise began subcontracting and subleasing to even smaller independent operators. When Jermyn-Green went bankrupt in 1949, Samuel and Philip Gelso (see below) took over the No. 14 lease and processed the coal at the Underwood Colliery in Olyphant.[25]

Louis Pagnotti Sr. secured the largest number of leases—over 60—between 1937 and 1954. He began in the 1920s with numerous subcontracts. During the early 1930s, he formed a corporation to take advantage of the burgeoning opportunities in leasing.[26] Pagnotti's first significant TPC lease occurred on June 19, 1937, when he procured rights to the entire No. 9 Colliery through his Anthracite Coal Company of Pittston.[27] The lease was tenured for 17 years and seven months; however, the lessor terminated the agreement due to deficient royalty and tax payments.[28] In another lease Pagnotti gained control of the No. 1 Colliery in Dunmore. Starting in the 1930s, he joined his nephew and partner, James Tedesco, in leasing numerous properties from the Lehigh Valley Coal Company

Figure 7. Diamond Slope and Engine House, No. 14 Colliery, Jenkins Township, PA, leased to Jermyn-Green Coal Company in 1937. (Courtesy of National Canal Museum Archives)

Figure 8 (right). Louis Pagnotti Sr., subcontractor, leaseholder, and colliery owner. (U.S. Treasury Department)[29]

(LVCC), as that firm implemented a leasing system almost as extensive as that of TPC and PaCC.[30] They also leased the Clear Spring Colliery in West Pittston through their Sullivan Trail Coal Company. Pagnotti and Tedesco formed Pagnotti Enterprises in 1942, which grew into one of the region's largest coal producers by 1950.[31]

As with most tenants, when Pagnotti assumed control of an operation, the work process and pay system changed. For example, whereas PaCC and HC&I typically paid contract miners by the ton,

West Pittston Shaft Explosion. August 24, 1936.

Caption: "Mine blast victims are brought to the surface." Five miners were killed in West Pittston Monday when an explosion wrecked a shaft of the Sullivan Trail Coal Company mine, 120 feet underground. The bodies are being brought to the surface while other miners look on.

Figure 9. Disaster at the Clear Spring Colliery in 1938. (Philadelphia Ledger)

Figure 10. Harry E Colliery, Temple Coal Company, Swoyersville, PA, postcard 1905.

at Pagnotti's leased No. 9 Colliery the workers were paid a flat rate for a seven-hour shift with a quota of two rather large cars per man. In disregard of the UMWA agreement, company miners who were supposed to be paid $7.70 per day and company laborers whose rate was $6.60 per day were paid by the hour. When Pagnotti raised the quota to three cars per shift, the miners went on strike whereupon he closed the operation.[32] The company's robbing of pillars caused extensive subsidence in West Pittston, Pittston, and nearby towns.[33] Louis Pagnotti and partner James Tedesco were alleged by government authorities to have organized-crime dealings,[34] a proposition supported by oral history evidence.[35]

Organized Crime and the Leasing System

As they had done with subcontracting, alleged organized-crime figures gained a solid foothold in PaCC's and TPC's leasing business. They included Santo Volpe, the region's alleged top crime boss between 1908 and 1933, who began as a subcontractor at PaCC's No. 6 Colliery in 1913 (see Chapter Two) and grew to become one of the firm's premier leaseholders.[36] He acquired a total of 56 PaCC or TPC subcontracts and leases—second only to Pagnotti—between 1934 and 1948. He was serving as secretary of Sibley–Old Forge Coal Company on August 22, 1934, when he and company president Joseph Adonizio signed a lease with Erie's New York, Susquehanna, & Western subsidiary. The agreement gave Volpe the right to extract the remaining coal in certain veins in Old Forge Borough and Ransom Township, Lackawanna County. PaCC reassigned the lease solely to Volpe in June 1935.[37] Also in 1935, Volpe gained the lease for the Jermyn Colliery. A new Volpe Coal Company was incorporated on November 23, 1937, with principals Santo Volpe, Charles C. Buffalino, and Charles J. Buffalino.[38] The firm secured leases for the No. 6 Colliery and the Forest City Colliery in 1937, as well as the Butler Colliery in 1938.[39] When PaCC and TPC conducted an inventory of coal reserves held by 13 leaseholders in 1941, Volpe controlled 14,480,780 tons, more than any other tenant and second only to the 25,839,600 tons held by the parent company, PaCC, at the Ewen Colliery.[40]

The reputed strongman gained a reputation as a good manager who ran profitable, efficient, and disciplined enterprises.[41] Like many tenants, he emphasized output over safety, as

Figure 11. Sibley Breaker, Pennsylvania Coal Company, Old Forge, PA, built 1905. (Courtesy of Carl Orechovsky)

evidenced by fairly high accident rates, including an explosion at the Volpe Colliery (HC&I's former Butler Colliery) where ten men died on June 2, 1938.[42] Volpe served as one of three company representatives during each of the labor-management contract negotiations between 1939 and 1944.[43] In the 1930s, Volpe secured a gubernatorial appointment to the Pennsylvania Coal Commission, an agency designed to stabilize the market by allocating production quotas to each company. He enjoyed a friendship with UMWA president John L. Lewis, who backed him in securing a lease in the middle coal field.[44] In the 1950s Volpe became a business partner with former Governor John S. Fine and others in the Newport Excavating Company, a lease-holding tenant of the Glen Alden Coal Company. The company's shady dealings led to indictments against Fine and two other principals (Volpe had died before the indictments were issued) for personal and corporate income-tax evasion.[45]

Stefano LaTorre, Volpe's brother-in-law and the alleged co-founder of the area's crime family, was another subcontractor and leaseholder. He took 21 tenancy agreements between 1930 and 1936 for projects such as driving headings, mining pillars, and digging rock tunnels.[46] He went into partnership with Carlo Saporito and John Sciandra in the Saporito Coal Company, another PaCC lessee. In an apparent split within organized-crime ranks, he later claimed that he and his Saporito partners were "squeezed out" of the lease when it was given to John Sciandra and other principals in the Knox Coal Company in 1943.[47]

John Sciandra allegedly followed Volpe as regional mob boss in 1933.[48] Sciandra began with a rock-tunnel subcontract in 1937, and took another similar agreement in 1938.

Figure 12. Santo Volpe with UMWA officials in 1953.

Left to right: Thomas Kennedy, Vice President; Volpe; John L. Lewis, President; Martin Brennan, President of UMWA District 7; unknown.

(Courtesy of National Canal Museum Archives)

On March 21, 1939, Sciandra, LaTorre, and Saporito formed the Saporito Coal Company and obtained a lease for a section of the Schooley Colliery in Exeter. In 1943, Sciandra purportedly forced out Saporito and LaTorre and took over the mine after having established the Knox Coal Company with Louis Fabrizio. Between March 21, 1939, and January 27, 1943, Sciandra and Fabrizio obtained seven leases, including one for sections of the Pittston and Marcy veins at the Schooley Shaft. Sixteen additional leases and lease adjustments followed between 1943 and 1959 as Knox became one of PaCC's larger tenants.[49]

In 1954, the Knox company secured the lease for the River Slope mine near Pittston, site of the infamous Knox Mine disaster of January 22, 1959. The catastrophe occurred when the Knox Coal Company mined illegally under the Susquehanna River and caused

Figure 13. Knox Coal Company, River Slope mine, Port Griffith, PA.
(Courtesy of George Harvan)

Figure 14 (above left). Robert L. Dougherty at a Knox Mine Disaster trial. (Courtesy of Scranton Times)

Figure 15 (above right). Louis Fabrizio, owner, Knox Coal Company, testifying before the Joint Legislative Committee to Investigate the Knox Mine Disaster. (Courtesy of Scranton Times)

Figure 16 (right). August J. Lippi. (Courtesy of Special Collections, Indiana University of Pennsylvania)

a breach in the river bed that drowned 12 men, flooded numerous mines, and ended deep mining in most of the Pittston area. The continuous flow of water underground eventually led to the abandonment of all Wyoming Valley mines. In addition to Fabrizio and Sciandra, the other Knox principals were General Manager Robert Dougherty and, in an egregious violation of the Taft-Hartley labor law, District 1 President August J. Lippi. Fabrizio, Dougherty, and Lippi, along with two officers of L.U. 8005, the Knox local union, served time in federal prison as a result of the investigations into the Knox Coal Company.[50]

Louis Fabrizio obtained his first subcontract on June 22, 1933, with his partner, attorney Frank J. Flannery, to mine a tract in the Pittston Vein. Between 1933 and 1940, the pair secured three other agreements from PaCC and TPC.[51] Though not listed as a crime-syndicate member, Fabrizio had several anthracite dealings with supposed mobsters. As mentioned above, he took a lease in 1941, covering 212.38 acres in the Pittston Vein in Jenkins Township and Pittston Township, in partnership with John Sciandra in the Knox Coal Company. He served as president of the company for most of its history. He pleaded guilty and served a prison sentence for failing to pay tax on income from the Knox company.[52]

Robert Dougherty, who, in addition to being one of four stockholders in the Knox Coal Company and serving as president and general manager, secured his own leases from the Glen Alden Coal Company for his Avon Mining Company and Peeley Coal Company beginning in 1956. The leases included the Grand Tunnel Slope in the Avondale Colliery in Plymouth Township (Avon lease) and the No. 20 Tunnel at the Truesdale Colliery in Hanover Township (Peeley lease). He was acquitted of income-tax evasion in a case associated with Avon and Peeley; however, both companies were found guilty. Dougherty joined Fabrizio and Lippi as Knox Coal Company owners who were convicted and imprisoned for income-tax evasion.[53]

Charles Gelso and his sons Philip and Samuel secured several leases between the 1930s and 1950s. Charles signed a lease on March 12, 1934, for a first mining in the Red Ash Vein of the Schooley Shaft. Between December 1949 and September 1955, the Gelsos secured ten other leases. They established the No. 14 Coal Company and secured a lease for the No. 14 Colliery in Port Blanchard, which they ran between 1949 and 1961.[54] Samuel Gelso emphasized the importance of personal contacts in garnering the agreements: "If you knew the people who controlled the leases you had a better chance of getting in," he said. "If they knew you ran a good operation, one that could get the job done, get the coal out, then you had a better chance of getting the lease." Samuel and Philip Gelso were convicted of wage violations in the investigations that followed the Knox Mine disaster; they were given suspended sentences.[55]

Did TPC and PaCC officers and board members realize they were issuing hundreds of subcontracts and leases to persons with criminal associations? The question becomes especially relevant when considering that the board of directors voted to approve each lease at regularly scheduled meetings. Apparently, the tenants' criminal backgrounds remained an unspoken matter, according to PaCC's last surviving officer, Vice President E. Stewart Milner (ESM), who kept the lease records for many years. In an oral history interview he described the relationship:

Q: There was a reported influence of organized crime in the Knox Company and in the Pittston area. What could you tell us about that?

ESM: Well, there were a lot of the Italian families, I mean, [who] were hooked up in this crime business. I'm trying to think. I don't want to make any comments on who it might have been. I wouldn't want to get myself in the ringer on that. But it was some of the Italian men who were high up in the mining circles [who secured coal leases].

Q: Did the Pennsylvania Coal Company ever think about the possibility that some of these lessees were reputed mobsters?

ESM: I never heard any comment on it. No one really commented on it. I suppose it was. It was probably well-known throughout the region and everything. But you just kept your mouth shut or you're in big trouble.

Q: Was it mainly a business relationship?

ESM: Yea, that's right. That's exactly right. It was a business relationship, yea, yea.[56]

Figure 17. E. Stewart Milner, last vice president of the Pennsylvania Coal Company. (Northeastern Pennsylvania Oral & Life History Project Collection)

While federal and state investigators probed the corrupt inner workings of the industry before and after the Knox disaster, the District and International offices of the UMWA seemed to not recognize the problem. The news media had also been remiss in uncovering and reporting the corruption. Yet the mineworkers as well as legitimate operators had long understood the criminal element's influence. According to former Knox employee and co-author William A. Hastie:

> The [leasing] system was very fertile soil for corruption. The large companies were providing for the exploitation of natural resources otherwise unattainable by the smaller companies. And in truth many of the leases were granted to known criminals. The man that the Pennsylvania Crime Commission considers to have been the first *capo mafioso* in Pennsylvania held leases on two of the Pennsylvania Coal Company's largest collieries, the No. 6 and the Butler. He was Santo Volpe. In the Mafia he was known by the sinister title King of the Night.[57]

David Panzitta, who worked for his father's subcontracting and lease-holding firm, the Panzitta Coal Company, recognized the problem:

> There was really a lot of corruption involved. See, when the big companies ran the mines years ago, people worked there by the day. [Then] they start handing these [sub]contracts out, you know, [sub]contractors. Corruption [started] amongst your officials in the company. If you come up with the right price, you got a nice vein of coal. If you didn't, and they didn't want you, you got the scraps. ... When the corruption came in, they were removing the pillars, but leaving them around the shaft, you know, secure so your shaft wouldn't cave. When they started giving out [sub]contracts they were trying to make money and they did some things there, and little by little it just deteriorated. Then when the river came in [the Knox mine disaster] that was the final straw .... [58]

Figure 18. Panzitta Coal Company, leaseholder of the Schooley Shaft, Exeter, PA. (Courtesy of Steve Lukasik)

TPC and PaCC were not the only firms granting leases to reputed gangsters or men with criminal connections, although they took the lead.[59] The companies apparently suffered no negative legal or public-relations consequences despite decades of association with presumed criminals and, for that matter, with corrupt political leaders. Among the latter was John Kehoe, the political boss of Pittston and an important ally of Governor John S. Fine. Kehoe made a great deal of money bootlegging alcohol in the 1930s, much of which he invested in the coal business, including the lease-holding Berge-Rose Coal Company and the Kehoe-Berge Coal Company.[60] He wielded broad influence in local politics, media, and mining circles. He became the Luzerne County Tax Assessor; founder, publisher, and chief editorialist of Pittston's *Sunday Dispatch*; and a Pittston School Board member. One former PaCC superintendent referred to the relationship between Kehoe and the Erie companies: "Whoever was the [political] boss of Pittston had the inside of the door in Dunmore, which was the main office of the Pennsylvania Coal Company. In other words, you do the work, elect the people that you want, that you can trust, and we play ball."[61]

It is important to note that most Erie tenants were not criminally associated. Moreover, the vast majority of Italian mineworkers were honest, diligent, and reliable. Former PaCC mine superintendent Lewis Casterline, whose father and mother were born in northern Italy, emphasized that the Italians led the fight against the tenancy systems and the injustices associated with them. As discussed further in Chapter Eight, he emphasized that many fought organized gangs from Italy to New York to Pittston, and paid for it by enduring fear, violence, and even death.[62]

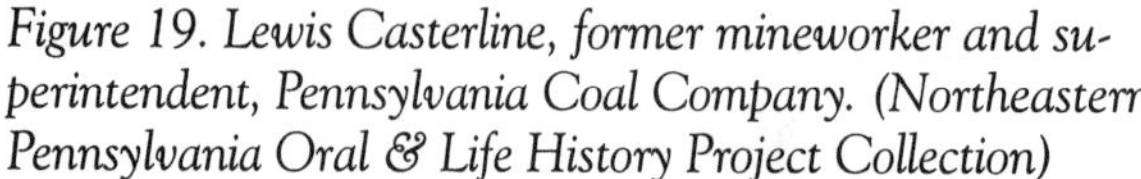

Figure 19. Lewis Casterline, former mineworker and superintendent, Pennsylvania Coal Company. (Northeastern Pennsylvania Oral & Life History Project Collection)

Figure 20 (below). William A. Colliery, Kehoe-Berge Coal Company, Old Forge, PA. (Saward's Journal *21 Jul 1934, 300; Courtesy of Jim Guthrie)*

Figure 21. Truesdale Colliery, DL&W/Glen Alden Coal Company, Hanover Township, Luzerne County, PA. (Courtesy of National Canal Museum Archives)

A New Union in the Northern Field: The United Anthracite Miners of Pennsylvania

In July 1929, in a remarkable turn of events, former District 1 President Rinaldo Cappellini reemerged as the leader of the very insurgents who helped drive him from office one year earlier.[63] With his energy and militancy revived, Cappellini ran for the District presidency against the incumbent, John Boylan. Cappellini and his vice-presidential running mate, Thomas Lavelle, built their candidacies around two issues: end subcontracting and establish an equalization program.[64] When the campaign failed and Boylan triumphed, Cappellini, Lavelle, and numerous other dissidents resigned from the UMWA and moved to establish a new labor organization called the Anthracite Mine Workers Union.[65] It represented the region's first dual or alternative labor union since the Industrial Workers of the World in the 1910s.[66] However, support for the organization flagged at least in part because the workers wanted to show unity for the approaching labor-management negotiations of 1930. The insurgents suspended their challenges and the District 1 labor wars subsided—at least for a time.

Figure 22. Rinaldo Cappellini, circa 1940. (Courtesy of Kim Cappellini)

The dialogues for the new contract commenced on July 1, 1930, in New York City. President Lewis headed a UMWA delegation that included vice president Philip Murray, secretary-treasurer Thomas Kennedy, as well as District presidents John J. Boylan, Michael Hart-

Figure 23. John Boylan, President, UMWA District 1. ("Debate Enlivens Miners' Session," Wilkes-Barre Record, *July 16, 1929)*

neady, and Martin F. Brennan of Districts 1, 7, and 9 respectively. Major W.W. Inglis of the Glen Alden Coal Company represented the operators along with A.J. Maloney of the Philadelphia and Reading Coal and Iron Company, Richard F. Grant of the Lehigh Valley Coal Corporation, E.H. Suender of the Madeira Hill Coal Company, J.B. Warriner of the Lehigh Coal and Navigation Coal Company, and Michael Gallagher of TPC.

The talks proved difficult. The union demanded equalization, better wages, adherence to established work rules, enforcement of seniority rights, the "check off," and an end to the special contract.[67] The operators opposed each demand and presented their own, including the elimination of wildcat strikes, weakened General Grievance Committees, and "an improved car of coal."[68]

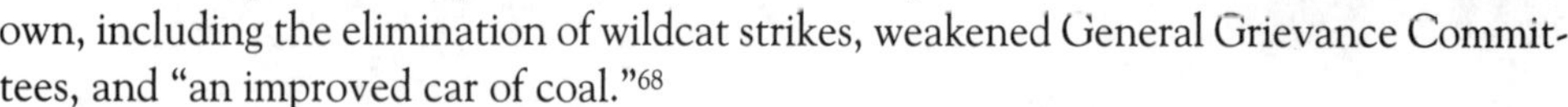

Despite tough negotiations, the parties came to terms on July 18, 1930. Set to run for an unusually long period of six years, the proposed contract prompted some observers to predict an era of industrial peace:

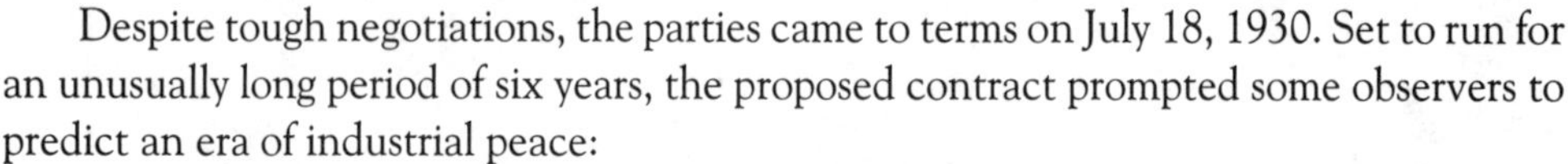

> Here is a real stabilization of the anthracite industry, for there has not been a [major] anthracite strike since 1926, and the new agreement will run to 1936, so we are assured of ten years of anthracite peace.[69]

The agreement froze wages and did not mention subcontracting or equalization. It allowed for a partial check-off, or payroll deduction for union dues, but brought no other advances for the workers. More important, it enhanced the workplace power of management. One article required the UMWA "to take active and affirmative steps to eliminate as far as possible strikes and shut-downs in violation of the agreement [i.e., wildcat strikes] and to eliminate group action designed to restrict output."[70] Another feature involved weakening General Grievance Committees, which had been important representative bodies since they were established in the 1912 agreement. Then there was the "improved car of coal," an undefined term that likely meant a higher topping and greater dockage for waste.

Conflict over the pact flared when delegates met at the UMWA Tri-District Convention in August 1930 to vote on approval. Thomas Maloney, president of L.U. 466 at Glen Alden's Stanton Colliery in Wilkes-Barre, led the critics. "I rise at this time in opposition to the entire agreement before this convention for ratification," he proclaimed, and added: "There isn't one demand, there isn't one resolution in the contract that was formulated in the [pre-negotiations] Tri-District Convention in Hazleton."[71] Lewis disagreed. He characterized the document as "a good agreement, constructive in its program, stabilizing in its scope, and intended and predicated upon the philosophy that we will live and let live."[72]

A storm of protests continued over the next four days. In his final address to the convention, Lewis made an impassioned appeal for a favorable vote. "My God, men," he began:

Figure 24. John L. Lewis, President, United Mine Workers of America, 1931. (Courtesy of United Mine Workers of America)

I don't mind fighting your adversaries and your opponents and your enemies, but I do shrink from being placed in a position where, after fighting them to exhaustion, I have to come back here and you force me to fight you ... I did sit here in this convention day after day and hear my honor impugned and myself and my associates humiliated because certain delegates felt secure in the body of the convention and they abused the privilege which had been given them. ... We have secured for you the best agreement possible under the circumstances confronting this industry and the country.[73]

The disapproval that came mainly from insurgent District 1 members failed to persuade the majority, and the agreement passed by more than a five-to-one-margin, 737 to 127.[74]

The overwhelmingly positive tally notwithstanding, the insurgents continued the protest under the leadership of Maloney. Immediately following the convention, they began mobilizing to challenge the Boylan forces in the District 1 election of May 1931. Maloney ran for president with Steve Hermann as the vice-presidential candidate. An end to subcontracting, a fair equalization policy, and an end to company violations of the 1930 wage agreement headlined the platform.[75] The candidates combed the northern field, attending dozens of rallies and speaking to thousands of mineworkers. An army of insurgents worked the grass roots on behalf of the ticket. To complicate matters for the UAMP, another rival union, the Communist-inspired National Miners Union, had been organizing in the northern field.[76]

Boylan warned the rebels about the consequences of fomenting a dual union.[77] In the face of a determined opposition, he ran a strong campaign that capitalized on the power of incumbency. Not only did he and his running mates, Michael J. Kosik for vice president, and Enoch Williams for secretary-treasurer, have the backing of John L. Lewis and the union hierarchy, but they had a clear advantage in personnel and monetary resources. Despite charges of voter fraud, a clear majority favored the incumbents, much to the satisfaction of Lewis. Not to be denied, the mutineers, with Maloney in the lead, contested the vote and staged a fight for control at the District convention in July.[78] When that strategy failed, the dissidents pulled out of the UMWA, formed a veritable dual-union organization, and called a series of selective strikes over the next several months. The UMWA joined with the operators to crush the movement.

With the civil war renewed, 44 of the District's 130 locals unions convened a rump convention in March 1932, despite warnings from the UMWA.[79] Maloney chaired the event, which was attended by workers' representatives from the field's major companies— Glen Alden, Hudson, Lehigh Valley, Susquehanna, PaCC and TPC—as well as from many independents. After a heated six-hour debate, the delegates voted for a general strike to begin on March 11, 1932.[80] Three familiar issues framed their demands: upholding established work rules and wage rates, ending the individual contract and the subcontracting system, and equalizing work.[81]

The shutdown's effectiveness varied from company to company and colliery to colliery. Early reports indicated that up to one-half of the northern field's pits were idled. TPC and PaCC employees were among the most committed to the strike, which they saw as an opportunity to confront Gallagher and his tenancy policies. They actually began the protest two weeks prior to the strike date when TPC closed ten collieries and kept only two working.[82] With Maloney present, some 4,000 of the firms' unemployed gathered at the Pittston Armory and endorsed the work suspension. Stanley Ocheski, president of L.U. 1703 at the No. 6 Colliery, presided over the assemblage, assisted by Feeney Buzzarelli, president of L.U. 1487 at the Ewen Colliery, and John Shipley and Lewis Casterline from L.U. 1581 at the No. 14 Colliery. As tensions rose again in Pittston, Mayor Ambrose Langan ordered police to patrol the collieries every morning to insure safe passage for men wanting to work. Police Chief J. Russell Taylor pledged to protect the safety of all workers. The strike was 90 percent effective in the Pittston area.[83]

Figure 25. The Von Storch Colliery, Penn Anthracite Coal Company, Scranton, PA. Breaker built in 1927, closed in 1947, and dismantled in 1948. (Courtesy of Charles Kumpas)

Five days into the strike, the insurgents reported that 32 collieries were shuttered while 20 were still working. At TPC, eight operations were closed and two worked; Glen Alden had ten facilities idle with eight working; Hudson saw three collieries working and six down; and Lehigh Valley had two mines idle with one operational. The independent Temple Coal Company had four inactive collieries while Penn Anthracite had two. Other independents experienced walkouts.[84] Violence erupted at several locations. Beatings occurred, and homes suffered damage from dynamite.[85] When Maloney learned that the operators were cooperating with the UMWA to break the strike, he sneered at the collusion: "They want war do they? Well, if they want it they can have it."[86] As he did on many other occasions, William W. Inglis, president of the Glen Alden Coal Company, sent private telegrams to Lewis expressing alarm about the strike:

INSURGENTS GAIN IN HARD COAL AREA

Strikers Prevent More Mines Operating in the District South of Wilkes-Barre.

PICKETS NOT MOLESTED

Troopers Are Less Harsh With Them After They Appeal to Pinchot and Local Officers.

Special to THE NEW YORK TIMES.

WILKES BARRE, Pa., March 24.—Insurgents made gains in the district below Wilkes-Barre today in their efforts to close every anthracite mine to bring about equalization of work and end wage cutting. In the upper end of Wyoming Valley, in Lackawanna and in the lower coal field the situation was unchanged, most of the miners remaining at their posts.

As the situation stands, the coal supply is not endangered, since the mines operating can take care of all orders.

Thomas Maloney, spokesman for the strikers, reported that all mines in the lower valley were idle. Other sources of information did not bear out this statement.

Auchincloss and Buttonwood collieries were idle for the first day because the workers, took strike votes. Pickets concentrated on the Truedale, largest of Glen Alden operations in Hanover township, and on the Maxwell in Ashley, but both operated, according to company, district and police officials. Four arrests were made at Ashley in clashes between troopers and pickets.

At Plymouth pickets made further inroads at the Gaylord mine of the Kingston Coal Company, but it was not crippled enough to halt operations. This is the only colliery operating in that field.

No effort was made to resume at the Stanton where Maloney was employed. A meeting of workers to take a vote broke up without action.

Pickets reported today that they were treated less harshly by troopers and deputy sheriffs after they had protested to Governor Pinchot, Captain William A. Clarke of Wyoming barracks and Sheriff Luther Kniffen.

It was charged that police cleared the streets of all persons, clubbing men and women who were not pickets and in instances merely churchgoers.

Tolay pickets in automobiles and trucks were not molested when they left Dupont for their posts at other collieries.

Taylor, in Lackawanna County, was the scene of an early morning disturbance, State police going to the rescue of local officers who were overpowered. Two arrests were made and there were two casualties.

As the miners usually do not work during the Easter holidays it is expected that quiet will prevail for the next few days. The companies will operate tomorrow if the miners will respond on Good Friday.

FINES SOVIET TRADE UNIT.

Istanbul Commerce Chamber Demands Bureau Become Member.

Wireless to THE NEW YORK TIMES.

ISTANBUL, March 24.—The Chamber of Commerce of this city has asked the Turkish Government to force the payment of 3,000 Turkish pounds [$1,620 at par] by the Soviet Trade Bureau established here in 1931. The chamber says the bureau should have complied with the law requiring all individuals and companies engaged in commerce in Istanbul to join the chamber and make an annual subscription, which in the case of the Soviet bureau was set at 1,000 pounds.

The bureau declines to join the chamber or pay a subscription on the ground that the bureau is a State institution. But because of its failure to join it is now called on to pay 2,000 more pounds as a fine. The Soviet Ambassador will intervene at Angora in behalf of the bureau.

Figure 26. Insurgent strike of March, 1932. (New York Times, *March 25, 1932*)

> In my mind this insurgent body that has always been a thorn in the flesh will not pay any attention to the officers in Number One District as long as they are in the state of mind daily exhibited. STOP. After you have thought this over may I ask you to call me at my residence tonight two seven four four nought. STOP.[87]

Despite the early support, large numbers feared the personal consequences and began complying with Boylan's command to resume work. The strike began to unravel.[88] The action officially ended on April 1, after three weeks, when the insurgents' governing council adopted a resolution ending the strike. The UMWA viewed the strike as nothing less than dual unionism.[89] The organization retaliated against Maloney and 34 others by banning them from the organization for 15 years. Because of his ardent dual unionism and long-standing militant disposition, John Torma of Nanticoke received a 99-year suspension.[90]

Maloney and his allies nevertheless maintained the movement. In May 1933, after a long hiatus, Rinaldo Cappellini reemerged yet again to help lead the insurgent movement. He joined with Maloney and others in canvassing the field to attend meetings and address rallies. The leaders bashed the UMWA for doing little or nothing to protect workers. During one event, Cappellini asked why the union did nothing about the wage cheating at the Ewen Colliery, where the workers were paid for five and seven days after having worked for twelve and fourteen. He censured other firms, such as Kehoe-Berge, for instigating similar schemes. Moreover, if workers complained about the cheating, they faced dismissal. James Musto of Pittston, a member of the Butler Colliery local union, also spoke at the rally.[91] Shortly before the District 1 convention of July 1933, Cappellini expounded on the alternatives to the *status quo*: "Lewis is not interested in the anthracite fields. ... We are going to organize the anthracite union and we are going to have a rump convention."[92]

Maloney worked as an outside laborer at Glen Alden's Stanton Colliery, where he served as L.U. 466 president and head of the grievance committee. In his speeches, he

Figure 27. Stanton Colliery, Glen Alden Coal Company, Wilkes-Barre, PA. (Courtesy of Underground Miners)

Figure 28. Thomas Maloney, elected UAMP president, August 7, 1933. (Courtesy of Times Leader*)*

criticized the layoffs, pay-rate violations, and intimidations, while advocating equalization. He also condemned the tenancy systems that had spread beyond TPC and PaCC to include his employer and other firms such as the Lehigh Valley Coal Company.[93]

The insurgents were, unsurprisingly, excluded from the District convention of 1933. The convention proved to be a relatively amiable affair with none of the rancor evident in 1932. Some delegates complained about unwarranted pay cuts and other unfair labor practices. Several others wanted to know when, and if, the expelled insurgents would be allowed to request reinstatement in the union. Overall, however, the rear-guard action among so many mineworkers received very little attention. Yet, President Boylan could not completely ignore the split within the organization: "In the last two years," he began, "as perhaps as never before in the history of our organization, the United Mine Workers of America has been on trial in the eyes of thousands of its members."[94]

The insurgents called another rump convention at Regal Hall in Scranton for August 7, 1933. Over 600 representatives from 66 local unions met to consider a course of action. Cappellini served as temporary chairman and greeted the delegates, after which Maloney took the gavel. In his first directive from the podium, Maloney asked Communist and other radicals to leave voluntarily lest they be physically removed.[95] In the day's most important decision, the delegates took the bold step of establishing a new union called the United Anthracite Miners of Pennsylvania (UAMP). Meant solely for hard-coal workers, the UAMP was intended to replace the UMWA as the bargaining agent of the northern field's 65,000 employees.[96] The insurgents hoped that the 86,000 workers in the southern and middle fields would eventually join. They elected Maloney as president, Willard Morgan as vice president, and Henry Schuster as secretary-treasurer. They also chose a full run of other officers, including six committeemen. Cappellini was elected state chairman, a symbolic office that nevertheless kept him involved as a principal in the movement.

In his featured address, Cappellini chastised the UMWA for neglecting the anthracite worker. "We actually had no contract [under the UMWA]," he offered. "If you had been treated right and if the contract had not been violated with reduction in rates there would not have been a rump convention. ..." He urged the conventioneers to champion the UAMP:

> You must remember we are no longer rumpers, [but] neither are we members of the United Mine Workers of America. We will have our own laws and we will establish the right [wage] rates. It is up to you to lend your support to your officers for as soon as the convention adjourns the coal companies and district officers will start to work and attempt to discourage the movement. It is up to you to show your colors and enlist the membership of every miner in the district. If you lose heart the old conditions will continue for many years to come"[97]

In Maloney's acceptance speech he underscored the legitimacy of the new association by pointing to the workers' right to organize under the recently passed National

Insurgents Elect Staff

Maloney Chosen President, Cappellini Chairman at Rump Convention

With the election of officers yesterday the rump convention of District 1, United Mine Workers at Scranton, laid the groundwork for the organization of a union to be known as the United Mine Workers of the Anthracite Region and took steps toward the extension of the separate union movement into Districts 7 and 9.

Rinaldo Cappellini, former district president, was drafted as the liason man to effect coordination between the three anthracite districts. Cappellini was elected temporary chairman of the tri-district union, the presidency to be permanent should he be elected if the separate anthracite union unaffiliated with the bituminous locals is organized.

Cappellini declined to consider any office during the session. When suggested as tri-district chairman, the office in effect a national presidency of the anthracite unions, Cappellini suggested that the proposal be reconsidered. Delegates stood to demand that Cappellini accept. The former district chief did so with the reservation that it was a temporary chairmanship.

Maloney District President

Thomas Maloney, temporary convention chairman, was unanimously elected district president after Stanley Rogers, Eynon; John Torma, Hanover; James Moleski, Duryea; James Mack, North Scranton and Cappellini declined to be plaqed in nomination.

Willard Morgan, Dickson City, won out in a three cornered contest with Moleski and Torma as his opponents for vice president.

Henry Schuster, South Wyoming Avenue, Scranton, was elected secretary-treasurer. Larry Enderline, Taylor, withdrew after his name was placed in nomination.

William Martin, Jermyn, was elected board member from the First Inspection district. He was opposed by Mack and Joseph Dougher, Archbald.

In the Second Inspection District, Bart Petrini, Old Forge, won out over Enderline and Michael Urbanowicz, South Scranton.

Torma Board Member

Torma was elected board member in the Fourth Inspection District after John Peters, Nanticoke; Stanley Gorkas Nanticoke, and Joseph Shovelin, Warrior Run, dropped out of the race.

Anthony Serafini, Exeter, was elected board member of the Third Inspection District, George Moleski, Pittston, and James Musto, also of Pittston, withdrew their candidacies.

The elections were held by roll call with delegates of the local unions voting for their respective choices. The newly elected officers, according to the program defined yesterday, are to function as district officers while locals will be requested to withhold payment of dues from the district offices in Miller building at Scranton.

Cappellini sounded the keynote of the convention in declaring that the purpose of the convention was to establish a separate anthracite union, retain counsel and frame a charter. By-laws and a constitution will be compiled and locals requested to divorce their affiliation from the United Mine Workers. The duly elected officers of District 1 will be requested to resign or be ousted. "This is a legal constitution under the laws of the United States of America. Before the convention adjourns we will elect organizers and field workers to whom you can take your grievances. Our organization will be based upon the principles of unity, harmony and brotherly love," Cappellini declared.

Cappellini Scores Lewis

Cappellini bitterly scored John L. Lewis, Philip J. Murray, Thomas L. Kennedy, John Boylan and member of the district executive board. He read an excerpt from an ar-

Figure 29. "Insurgents Elect Staff" excerpt. (Wilkes-Barre Record, *August 8, 1936*)

Industrial Relations Act (NIRA). The historic law—a cornerstone of the Roosevelt Administration's New Deal—contained the well-known Section 7(a), which guaranteed the workers' right to collective bargaining under any union of their choice, without fear or intimidation. In ridiculing the tactics of the UMWA, Maloney scoffed, "Boylan threatened to evict you if you came here. You have come and I want to say we are going to have a miners' union in this district without Boylan. ... We intend to give what Roosevelt gave to the people of this country, a square deal."[98] Before the convention ended, the delegates passed resolutions on the most serious issues:

1. Pay Rates: Enforcement of the negotiated pay rate sheets at all collieries.
2. Subcontracting: Abolishment of the subcontract mining system.[99]
3. Equalization: Opening all collieries except those permanently abandoned or otherwise inoperable so that each worker could have employment.[100]
4. Wages: Elimination of "starvation rates" at numerous mines.
5. Work Rules: Prohibition against the companies using salaried men for maintenance when mines are closed or on idle days.
6. Output: A limit of mineworkers loading two cars per man per shift.[101]

The mutineers built the UAMP upon a foundation of grass-roots, cross-ethnic, multi-company solidarity, coupled with a disdain for the company-friendly policies of Lewis and the UMWA.[102] The Communists and their National Miners Union supported the insurgents' strikes and they now backed the formation of the UAMP. However, when Maloney rejected a proposed alliance, the NMU condemned him and the new union.[103] The combination of UAMP and Communist agitation, unemployment at 30 percent, and widespread economic distress produced a chaotic situation in the northern field that the UMWA tried to blame on other groups.[104]

Although Cappellini played a pivotal role in establishing the union, his involvement waned and eventually ended following a series of personal and legal entanglements that landed him in jail.[105] The UAMP carried on with Maloney at the helm and with the backing of one-third to one-half of the northern field's workers. Along with a core of dedicated officers and committee members, the organization also enjoyed the support of Maloney's close advisor and pastor, 75-year-old Msgr. John J. Curran of St. Mary's Church, Wilkes-Barre. Curran lent the type of moral backing and tactical advice he had shown in numerous other labor-management confrontations since the strike of 1902. Lewis Casterline attended meetings with Maloney, Cappellini, Curran and others. He referred to Curran as "a very good friend of mine," and acknowledged the cleric's dedication:

> Every movement that was ever here, we'd go down to him. In fact, I went down there one time with poor Tom Maloney and crowd. There was [*sic*] 12 of us there from Lackawanna and Luzerne counties. Curran said, "Now look, if anybody's a Communist, that's your belief. But if there's anyone that is a Communist, I will not play. I will not go along with the idea." So we said, "Give us a chance." There was Tom Maloney and there was Jimmy Musto, [and] me. I don't remember all the names that there was.[106]

The UAMP precipitated an intense labor war that continued for two years.[107] The organization called a number of paralyzing strikes between 1933 and 1935. The often violent actions attracted the attention of Pennsylvania governors Gifford Pinchot and

Left to right: Figure 30. Msgr. John J. Curran, Pastor, St. Mary's Church, Wilkes-Barre, PA. (Courtesy of Agnes Kupstas)

Figure 31. Governor Gifford Pinchot.

Figure 32. Senator Robert F. Wagner of New York. (United States Senate Portrait)

George Earle, New York Senator Robert F. Wagner, and members of the Roosevelt Administration, among others. The selective strikes precipitated fights, beatings, and shootings that sent hundreds to the hospital and even caused a number of deaths. In the shutdown against Penn Anthracite's Scranton-area collieries in 1933, UAMP members were accused of dynamiting the tracks of the New York, Ontario & Western Railroad leading to the company's Capouse Colliery.[108] The Scranton vicinity experienced 24 dynamitings between August and mid-October, 1933. Both sides—UMWA and UAMP—were alleged to have participated in the destructive acts but the new union received most of the blame.

NIRA or not, Lewis and Boylan cooperated with coal executives to quash the dual threat. The UMWA used harassment and expulsions while the companies resorted to intimidation, lay-offs, and firings. The UAMP retaliated with more selective strikes against PaCC, TPC, the Pennsylvania Anthracite Coal Company, the Hudson Coal Company, and the Glen Alden Coal Company. The organization called a general strike in January 1934, one that Maloney termed "one of the most important events recorded in the history of our Wyoming and Lackawanna Valleys."[109]

Drawing on some of the same diplomatic skills he had honed in the 1902 strike, Curran began traveling to Washington D.C. during the fall of 1933 for discussions with Senator Wagner that continued into 1934.[110] The priest's efforts were complimented by a group of local business leaders who also traveled to the nation's capital urging the involvement of the federal government.[111] Although UAMP leaders had hope for a special investigative committee, the body was never appointed. Instead, Wagner and the National Labor Board (NLB) asked the Anthracite Board of Conciliation (ABC) to report on the situation. The ABC review proved most unsatisfactory to the UAMP and the strikes continued in late 1934 and well into 1935.[112] When Maloney, Vice President Schuster, and 27 other UAMP members defied a court order to end a strike in March 1935, Luzerne County Judge W.A. Valentine sent them to jail.[113]

Labor-Management Relations at PaCC and TPC

Soon after assuming the presidency of the Alleghany Corporation and TPC in 1929, President Gallagher set a negative tone for labor-management relations by ordering wage cuts and refusing to participate in the recently negotiated check-off plan with the UMWA.[114] John L. Lewis and John Boylan objected to the unilateral decision. The UMWA tried to negotiate with Gallagher and finally had to threaten a strike before the parties came to terms. The crux of the problem was that Gallagher wanted a separate, informal agreement for lower wages at some collieries because many workers had agreed to take a lower pay rate. In an effort to encourage "flexibility," Lewis permitted Boylan and Gallagher to pursue a quiet "local arrangement," which included substandard wages. However, many workers learned about the machinations against the negotiated rate sheet and refused to allow the reductions.[115]

As employment relentlessly fell in the early 1930s, equalization and tenancy-related grievances precipitated a strike at the Erie companies on June 20, 1930. One-half of the 14,000 workers employed at PaCC- and TPC-leased collieries stayed home for ten days. The same workers threatened an equalization strike in 1931.[116] Despite the company's attempt to partly equalize work time in 1928 (see Chapter Four), the goal remained elusive because both the managers and Lewis opposed the idea.[117]

Another murder occurred on January 4, 1931. PaCC employee Charles Calamara was hit with eight bullets as he walked up Railroad Street soon after returning from a year's trip to Montedoro, Sicily. He had accompanied a prominent subcontractor and they quarrelled over tenancy and other workplace matters. Calamara received threats while abroad and authorities speculated that the ongoing dispute led to the homicide.[118] A few months later, yet another person met his end. Three men emerged from a park car and shot Charles (Sam) Licata near his home in Pittston. Licata and Calamara were friends who had been unwavering promoters of the insurgent demonstrations at PaCC.[119]

The workers managed to avoid violence in March 1931 when 13 local unions at TPC-leased collieries protested the closing of the Butler and the No. 14.[120] They charged management with installing practices that had been used in the bituminous industry (where Gallagher had been a chief executive) such as reducing wages in violation of the negotiated agreement.[121] The mineworkers at PaCC and TPC were among the most active supporters of the aforementioned insurgent strike of 1932, and vowed to keep it going "by our lonesome if necessary."[122] PaCC, which operated only the Ewen Colliery, fired scores of dissidents and refused to rehire them when the strike ended. Many of the dismissed workers —including Ross Alba, Catal Giordano, George Kietyanna, Frank Gallo, John Asakavage, and hundreds of others—filed appeals with the Anthracite Board of Conciliation. They argued that the strike settlement required the companies to treat strikers fairly in rehiring decisions. Maloney defended his colleagues at one hearing where 26 complainants accused the company of discrimination. In another inquest, two workers who claimed discrimination heard the company's answer: "... both men took part in an illegal strike at the operation in March, 1932, that new men had been hired, and that the company had no right to discharge faithful employes [*sic*] to replace men who attempted an unsuccessful tie-up of the colliery."[123]

Representatives of some UMWA local unions at TPC and Ewen participated in the District 1 convention of July 1933. However, many locals did not send delegates because they were dominated by the insurgent faction.[124] The insurgent convention, in August

1933, generated broad enthusiasm among the dissidents. Shortly before the rump gathering, some 4,000 Erie and other workers cheered Cappellini at a rally in Albert West Park, Pittston, where he attacked the UMWA's District and International leaders and policies. He chided them for ignoring the plight of the mineworker and doing little to address the ever-present violations of the collectively bargained agreement. He cited instances where some men fortunate enough to have jobs were toiling 12 to 14 days but getting paid for only four or five.[125]

In preparation for the convention, the Ewen Colliery experienced infighting when the UMWA faction denied a group of insurgents admission to a regularly scheduled meeting. Over 1,000 protestors marched to an alternative hall and elected temporary officers including Sam Loquasto as chairman, and Vincent Scalzo as secretary. The members then selected seven delegates for the insurgent convention including Anthony Attardo, Sam Facciponte, Americo Pazello, Frank Kuma, Anthony Falzone, Clarence Nelson, and Edward Meroski.[126]

Other Erie locals enthusiastically supported the conclave. Although Michael Mundernar, president of L.U. 1719 at the No. 9 Colliery, submitted his resignation when some members questioned his leadership, the larger body convinced him to stay and preside over the election of 14 delegates to the rump assembly. The No. 6 Colliery quickly selected five delegates. The same non-confrontational atmosphere prevailed in the delegate selection at the Butler Colliery.[127]

When the UAMP formed, the Erie men were among its staunchest supporters, with the Italians among the most active.[128] In many ways, the UAMP represented a capstone to decades of upheaval against the companies whose policies originated the insurgent movement in the 1910s and produced many of its leaders.[129]

The Demise of the UAMP

Despite an initially broad base of support, the UAMP could not stand against the combined resources of the UMWA, the coal companies, the lessors, the ABC, various community leaders and institutions, organized crime, and even the Roosevelt Administration, which viewed the union as a detriment to New Deal recovery programs. By the fall of 1935, after four years of internecine warfare, the UAMP had been defeated.[130] Legitimate as well as crime-related lessees enjoyed greater freedom over pay rates, work rules, safety procedures, and other labor practices now that the company-friendly UMWA stood as anthracite's only viable labor organization. The defeat marked a triumph for criminally inclined leaseholders for, as indicated, operators such as Santo Volpe significantly expanded their enterprises after 1935.

The UAMP came to a fatal symbolic end on April 10, 1936. Thomas Maloney, age 44, and his four-year-old son were murdered by a cigar-box bomb evidently mailed by a disgruntled former UAMP member. On the same day, an arsonist firebombed St. Mary's Church rectory in Wilkes-Barre. The assaults caused Msgr. Curran to suffer a heart attack from which he never recovered.[131]

John Boylan resigned his position in July 1935, partly as a move by Lewis to placate the insurgents. The new District 1 president, Michael Kosik, worked with Lewis and the companies to prevent unauthorized strikes. The labor-management contract of May 7, 1936, included a "Violations of the Agreement" clause that placed the burden on UMWA officials "to use every power vested in them to require that grievances be taken up in the

TIMES LEADER

FINAL MARKET

LATEST NEWS CLOSING PRICES

FRIDAY EVENING, APRIL 10, 1936

1 KILLED, 4 HURT BY BOMBS HERE, 3 MORE DEATH PACKAGES SEIZED- SIXTH BOMB FOUND AT HAZLETON

MALONEY MAIMED BY 'EASTER PRESENT' RECEIVED IN MAIL

Ex-Mine Leader May Lose Arm From Bomb Sent In Cigar Box—Warned Daughter Against Opening Package—Kitchen Wrecked By Blast.

Thomas Maloney, president of the former United Anthracite Miners of Pennsylvania, and his two children were [illegible]ously injured today when a nitro-glycerine bomb—mailed [in] the guise of an Easter present—exploded in the kitchen of [his] home at Georgetown, Wilkes-Barre Township.

[Fu]ll force of the blast struck Maloney, 44, and his son, [illegible] 4, as they stood in front of a porcelain-topped [illegible] he had placed a cigar box delivered

MALONEY FAMILY VICTIMS OF MAIL PACKAGE

THOMAS MALONEY, SR. May lose arm — THOMAS MALONEY, JR. Serious — MARGARET MALONEY Serious — MRS. TOM MALONEY Sick

KITCHEN OF MALONEY HO[ME AF]TER BOMB EXPLODED

CEMETERY OFFICIAL KILLED BY BOMB; OTHERS ARE FOUND

Michael Gallagher, Hanover Township School Director, Meets Death—Bombs Addressed To Judge Jones, Sheriff Kniffen and Judge Fine's Relative Are Seized.

Nitro-glycerine bombs—disguised as Easter gifts—[killed] one and injured four others here today as terrorists [used] Wilkes-Barre post-office to deliver at least a half dozen [death-]dealing packages.

Up until 4 o'clock this afternoon, all [of the] bombs had been accounted for, but only [illegible] [ex]ploded

Figure 33. Package bomb kills Thomas Maloney and Thomas Maloney, Jr., Wilkes-Barre Township, PA. (Times Leader, *April 10, 1936)*

manner provided in the Agreement and to prevent illegal strikes in violation thereof."[132] The clause stipulated severe disciplinary actions against all wildcatters. A resolution adopted by the ABC on May 20, 1938, also prohibited unsanctioned walkouts.[133]

With the UAMP destroyed and the possibility of strikes greatly diminished, tenancy spread rapidly during the second half of the 1930s and into the 1940s and 1950s. Dozens of independent operators rushed to secure the hundreds of leases offered by the mineral-rights controlling companies. While Volpe represented the mob-connected lessee, the Moffat Coal Company of Taylor, Lackawanna County, stood as one of the largest legitimate firms. Robert Moffat began the enterprise in 1937 as a relatively small leasing operation but, by the mid-1950s, managed to purchase all of the Glen Alden Coal Company's collieries in the Lackawanna County area. Moffat mined and processed much of the coal but also issued several subcontracts and leases to smaller operators.[134]

Whether the lessees were crime related or legitimate, few supported the wage sheets, work rules, and other stipulations of the labor-management agreement. Many were not opposed to illegally robbing pillars and barriers and otherwise pillaging the coal. A deep and pervasive culture of corrupt mining and labor practices continued to engulf the northern field. Although the UMWA still represented the workers, many operations became union in name only.[135]

Figure 34. Taylor "concrete" breaker, Delaware, Lackawanna & Western Railroad, Taylor, PA, 1914. (Anneman Photo, Courtesy of National Canal Museum Archives)

Figure 35. Taylor "concrete" breaker, Moffat Coal Company, Taylor, PA, circa 1960. (Carl Corlesen, Buried Black Treasure, *Dansville, NY: F.A. Owen, 1954)*

PaCC Returns to the Leasing Business

TPC made money from leasing even as industrial relations suffered. In 1930 and 1931, the firm met all rent and royalty obligations to its parent, PaCC, and paid a stock dividend of $1.90 and 27 cents per share, respectively. However, earnings began to slip and, in 1932, showed a deficit of $1.43 per share. Over the next two years the shortfall hovered around one dollar per share. PaCC's board of directors began raising doubts about the subsidiary's viability. The board eased the financial burden in 1934 by eliminating TPC's quarterly rent payments of $90,000 and instead demanding a simple 32 cents per ton royalty.[136] The new terms did not stem the negative cash flow. TPC began borrowing up to $2 million annually to stay afloat. The 1935 deficit topped two dollars per share and stayed at one dollar over the next four years. TPC made royalty payments between 1935 and 1938.[137] By late spring 1938, the royalty arrears totalled nearly $3 million.

It remains difficult to assess what combination of poor management, market conditions, corruption, and labor problems led to TPC's collapse. Certainly the market conditions during the 1930s did not help. In 1935, the anthracite industry suffered a net income loss of more than $10 million.[138] The profit squeeze meant that producers worked fervently to cut costs and stabilize prices. At a Stockholders and Executive Committee meeting in Cleveland during the summer of 1938, PaCC's board of directors voted to dissolve the eight-year compact with TPC.[139] Beginning in January 1939, PaCC resumed control of all properties that TPC held and began issuing and managing its own subcontracts and leases.

As the Van Sweringens' domain continued to crumble following the stock-market crash, J.P. Morgan and Company and the Morgan Guaranty Trust foreclosed on PaCC's parent company, the Alleghany Corporation, in 1935. Indebtedness to Erie, which had provided loans with Morgan's backing, topped $5 million. Total debt stood at $9.5 million. Alleghany's stock fell to 37.5 cents per share following the foreclosure. The firm lost $2.25 million in 1937.[140] It was sold for only $3 million, creating a loss of $9 million each for both the J.P. Morgan and the Guaranty Trust banks.[141]

Through it all, Erie retained an unwavering commitment to tenancy. PaCC operated only two collieries by the late 1930s, the Ewen and the Underwood, as eight others had been either leased or closed.[142] In 1939, one year after TPC's bankruptcy, PaCC held agreements with 55 leaseholders who paid $1.3 million in rent and royalties. In 1941, the company's own production stood at 1.34 million tons, mainly at the Ewen, while leaseholders produced 4.1 million tons.[143] In the same year, rent and royalty income totaled $1.2 million, bolstered by the wartime demand for coal.[144] Between 1938 and 1959, PaCC issued leases to an average of 44 operators per year which, together, produced an annual average of over 1.5 million gross tons worth about $500,000 of income to PaCC. To remain competitive, the other major producers created and/or expanded their own tenancy systems.[145]

It was apparent that PaCC's policy toward workers in general and organized labor in particular had not changed substantially since the strike of 1902. With a corps of leaseholders, the company gained a powerful new weapon in the drive to discipline workers, control the workplace, and enhance profitability. Corrupt lessees could also prove useful by mining in off-limit, but highly productive, coal seams that PaCC itself did not want to exploit. Moreover, the leaseholders colluded with District 1 union officials, including August J. Lippi who served as president between 1951 and 1965, to insure "labor peace" through graft.[146]

The UMWA Reacts to the Leasing System: Too Little, Too Late

Although opposition to the individual contract had been a demand since the strike of 1902, the union made no progress in eliminating the structure. After doing little to address over three decades of tenancy-related grievances and disorder, Lewis finally spoke ambivalently about leasing in an address at the Tri-District Convention of March 1939 in New York City:

> Some of the great coal companies have followed the practices of leasing out their coal lands and their collieries to individuals and companies of doubtful responsibility. They have failed to include as a stipulation in those leases that those lessees should be responsible to the degree that they will pay the rates under the industry, and as a result the chaos has been intensified and large operating companies have found themselves in the unique and paradoxical position that they have been unable to compete in the open market, with the lessee mining coal upon their own property and paying them a royalty for the right to mine.[147]

Thomas Kennedy, UMWA vice president and former District 7 president, also lectured the gathering. A native of Larksville, near Wilkes-Barre, he spoke more critically about leasing and its consequences:

> We have a condition where the parent that created the child, which might be known as the lessee[,] is being destroyed by its own creator, due to the fact that there has been no understanding had with respect to these leases as they affect the payment of wages and the maintenance of working conditions or the stabilizing of the price of that commodity in the consuming market.

Kennedy described the leasing problem as endemic to the industry. "This is not only true of one company but it is true of most companies in the anthracite industry. ..." He condemned the structure as a "problem that has done more to bring about instability and chaos in the industry than anything else I know of."[148]

The conventioneers conducted a spirited discussion on tenancy. District 1 delegates led the complainants. Those from Districts 7 and 9 emphasized the pressure that the northern field's leasing practices were having in their areas. District 7 President Hugh V. Brown complained that operators in his section wanted tenancy's lower wages, higher output, and other concessions. "Give me the conditions that are in District 1," he cited them as insisting.[149] The upshot was that the delegates approved two relevant demands: the familiar one demanding the abolition of the individual contract, and a new one calling for tighter regulation on lessors and lessees to ensure compliance with the labor-management compact.[150]

*Figure 36 (left). John L. Lewis, 1938. (*Time*, February 7, 1938)*

Figure 37 (right). Thomas Kennedy, Vice President, UMWA, 1939.[151] *(Courtesy of UMWA)*

Figure 38. John L. Lewis, Thomas Kennedy, Msgr. John J. Curran. (Courtesy of Luzerne County Historical Society)

District 1 officials later passed a resolution requiring all lessors to notify the president of the appropriate UMWA district when a new lease was issued.[152] However, without any effective enforcement, the operators ignored the request. Later in 1939, pressure from some union members led to a resolution that was jointly accepted by the operators and the presidents of the three anthracite districts, calling for the complete elimination of subcontracting:

> RESOLVED, that special and individual contracts in the mining of coal where now in use shall be eliminated. Steps shall be taken by the contracting party to have them superseded by proper agreements negotiated under the terms of the general wage agreement, providing for the payment of contract rates thereunder. ... Such agreements superseding special and individual contracts shall be consummated as rapidly as practical before April 30, 1940, with negotiations to continue thereafter if questions remain open on that date.[153]

The resolution was mainly symbolic and the companies paid little attention to it, as indicated by Kennedy's criticism of the subcontract in testimony before the National War Labor Board hearings in 1943.[154] By this time, however, the issue was moot because leasing had superseded subcontracting as the main form of tenancy.

Despite the internecine conflicts that permeated the industry and its workers over the previous three decades, the UMWA still had no effective strategy to deal with tenancy and its consequences. Lewis's business unionism philosophy meant that the union concentrated on "stability" and modest pay and benefit increases for an ever-declining workforce. Many workers chastised the president for his apparent indifference to the anthracite worker's plight.[155] Others correctly viewed District 1 President Lippi as a corrupt appendage of the operators and organized crime. Militant employees turned to localized actions, believing they could have no real effect on the International or District unions' officers.

While tenancy's impact on labor-management relations in the northern field seemed clear, questions remained regarding its real effects on productivity and output. To what extent did it boost tonnage? Did it prolong the life of a severely declining industry? Or did it contribute to the decline? Did the true costs outweigh the economic benefits? Hard coal travelled a very difficult road during the first half of the twentieth century and Chapter Seven will argue that subcontracting and leasing figured prominently in the final decades.[156]

Miners Monument, Nanticoke, Pennsylvania

CHAPTER SEVEN

Tenancy Systems and the Decline of the Northern Anthracite Industry

Many of these mines, in the judgment of the Commission, do not even pretend to comply with any part of the mine laws.

— Investigation Commission, Volpe Colliery Mine Disaster, 1938

It is with a great deal of satisfaction that your officers inform you that they have succeeded in abolishing the system of special or individual contracts in District No. 1.

— District 1 Official Statement, 1952

But up until today, I don't like that sneaky [thing], the special contract. Well, who the hell approved that special contract? I didn't, only I heard about it. And I'm still talkin' about it today! ... That still bothers me 'til today.

— John Mikulski, motor runner and local union officer, Avondale Colliery, Plymouth Township, PA, 2001

Why Did the Anthracite Industry Decline?

In the historiography of anthracite's final decades a common argument has been that the industry expired for market-related reasons deriving from competition from other fuels. Bituminous, petroleum, and natural gas enjoyed cost, convenience, and supply advantages that drew customers away from anthracite, so the argument has gone.[1]

To be sure, fuel oil sales did surpass anthracite in the northeastern U.S.—hard coal's main market—in 1947, and natural gas took the lead after 1955 (Figure 1). The question remains—why? Was it simply a matter of competitiveness or were other factors involved? In fact, other fossil fuels did enjoy certain competitive advantages over anthracite, some of which the federal government artificially supported. For instance, in 1916, Congress passed a bill allowing special tax-amortization benefits for oil companies so they could expand production, a concession extended to coal producers ten years later in 1926. In the 1920s, federal lawmakers approved an income-tax deduction for petroleum companies known as the Oil Depletion Allowance that climbed to 27.5 percent by the time it was abolished in the late 1970s. Anthracite producers received a similar concession for only 10 percent.[2]

In spite of such advantages, and contrary to popular belief, fuel oil and natural gas did not have a cost lead over hard coal until the 1960s. For example, in a study based on home heating in New York City during the late 1950s, the U.S. Bureau of Labor found that the price of energy produced by a specific quantity of fuel oil amounted to $23.49, while the

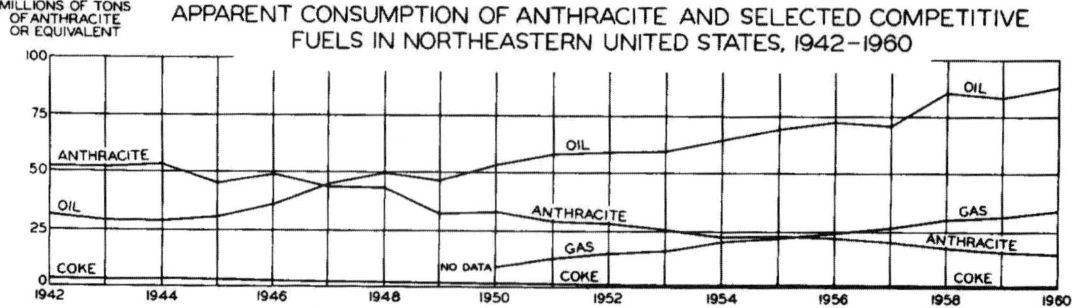

Figure 1. Consumption of anthracite, oil, natural gas, and coke, 1942–1960. (George F. Deasy and Phyllis R. Griess, Atlas of Pennsylvania Coal Mining, Part II—Anthracite, University Park, PA: Penn State College of Mineral Industries, 1963, 117)

cost for a comparable amount of anthracite was $19.90. For natural gas the figure was nearly three times as high, $58.74.[3] Therefore, while oil and gas were cleaner, more convenient, and more effectively marketed, they were not cheaper.

Bituminous, on the other hand, held both output and price advantages over anthracite. Part of soft coal's superiority resulted from greater productivity. Carter Goodrich reported that anthracite increased output from 2.17 net tons per employee per day in 1929, to 2.53 net tons in 1934 (a period that marked the beginning of the leasing system; see below). And yet, he added, "Despite the recent gains, the output per man is still hardly any greater than in 1899 (2.50 tons), whereas in bituminous mining it is now about 4.40 tons (seven-hour day), as against 3.05 in 1899 (eight and nine hour days)."[4] Lower productivity, in turn, related to anthracite's geological, technological, and other disadvantages.

Rather than deteriorating solely as a result of impersonal market forces, this chapter will argue that a more accurate analysis of hard coal's decades-long slide requires a study of four interrelated factors: (1) industrial relations involving managers, labor leaders, and workers; (2) investment decisions, especially regarding technology and innovation; (3) operational limitations related to geology, production, and overproduction; and (4) tenancy, including its associated corruptions and conflicts. Because industrial relations have been detailed in numerous other research volumes and in earlier chapters here, the following discussion will focus on the other three elements.

Figure 2. Mule and workers underground, circa 1905. (Courtesy of William Ferri)

Figure 3. Mineworkers propping the "roof" in a Scranton, PA, mine, circa 1905. (Courtesy of William Ferri)

Investment Decisions: Technology and Innovation

The anthracite corporations often lagged behind the competition in applying new technologies and innovations. Hard coal executives formed three organizations during the first third of the twentieth century to bolster competitiveness and enhance public relations: the Anthracite Bureau of Information in 1915, the Anthracite Coal Service in 1925, and the Anthracite Institute in 1929. However, the paybacks from the agencies were much less than anticipated. Technological innovation was the main purpose of the Manhattan-based Anthracite Institute and, toward this end, it created the Anthracite Laboratories to develop new coal-burning furnaces and approve existing ones. Between 1936 and 1953 the Laboratories' budget amounted to $3.4 million ($28.6 million in 2011 dollars, or an average of $1.58 million per year).[5] Yet, its most important invention, the stoker—a furnace with an automatic feeder—did little to boost demand. Knowledgeable observers judged the agency as "totally unsuccessful" in stemming the transition away from coal toward fuel oil and natural gas.[6] The Coal Service's inability to effectively market and innovate beyond home heating into electrical generation pointed to another failure.[7]

Figure 4. Link-belt automatic stoker advertisement, circa 1937.

When it came to mining technologies, anthracite trailed bituminous in most, though not all, areas. The soft coal industry saw greater improvements in cutting, drilling, and blasting devices. By the end of the nineteenth century, bituminous workers were using pneumatic drills to mine at the coal face, and by 1910 they were employing undercutting and short-wall mining machines. Within two decades, long-wall mining apparatuses provided even greater productivity.[8] Pneumatic drills and jackhammers were not used in anthracite until after 1900, and the industry rarely employed short- and

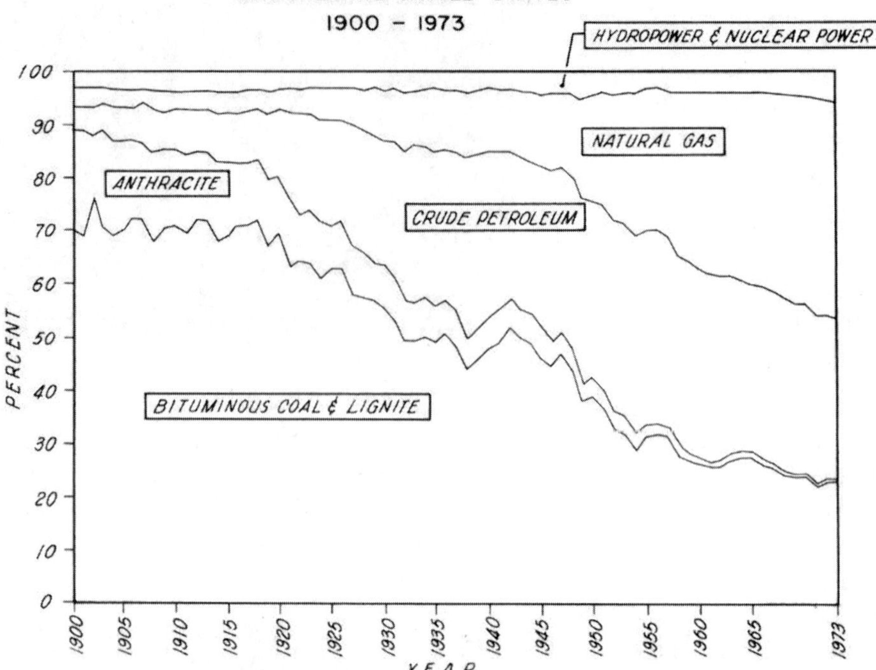

Figure 5. Distribution of energy production in the continental U.S., 1900–1973. (U.S. Bureau of Mines)

long-wall technologies because of geological faults, thin seams, steep pitches, and other physical irregularities.[9]

Both industries introduced undercutting machines early in the twentieth century, and more modern cutters and loaders in the 1920s. However, geology again hindered their full utilization in anthracite.[10] In 1929, for example, the Pennsylvania Coal Company (PaCC) and Hillside Coal & Iron Company (HC&I) used only 67 cutting machines in 12 collieries, most of them at the Ewen (13), Underwood (24), and Forest City (15) operations. By 1934, the number had fallen slightly to 64.[11]

Geology provided fewer constraints for anthracite producers when it came to newer and more portable loading machines. The two most popular were the shaker conveyor (or shaker chute) and the scraper loader. In 1929, PaCC and HC&I did not use any shaker chutes and had only four scraper loaders, but by 1934 the numbers had grown to 128 and 76 respectively.[12] The devices gained wide acceptance in the industry and represented one of the few instances where anthracite outpaced bituminous: by 1934 anthracite workers were loading 19.1 percent of the raw product by machine compared to 12.2 percent in bituminous.[13] The first electric locomotives (or "motors") were used to replace mules as the source of underground transportation in the late 1880s, but they were constrained in anthracite by the aforementioned geological problems. In the area of coal-processing technologies, the anthracite managers invested significant amounts into the late 1930s; however, hard coal still struggled to compete because it required much more elaborate and costly breakers than soft coal.[14]

Figure 6 (upper right). Hand-powered mechanical loading machine, circa 1905. (www.miningartifacts.org)

Figure 7 (lower right). Jeffrey Company sectional loader, circa 1920. (Ohio Memory Project, Ohio Historical Society)

Figure 8 (below). Menzies Cones advertisement, Huber breaker, Glen Alden Coal Company, Ashley, PA. (Coal Age, 44, April 1939)

The new equipment undoubtedly contributed to anthracite's modest productivity increases that began in the 1930s (see below).[15] However, management's strategy to lower the largest production cost, namely labor, through the tenancy systems and their often-corrupt practices also played a part. As discussed in Chapter Five, PaCC, HC&I, and the Hudson Coal Company preferred that subcontractors operate the new cutting machines in order to surmount the stint and gain maximum output. At the Penn Anthracite Collieries Company in Scranton during the early 1930s, subcontractor Joe Costa used a Jeffrey cutting machine and a mechanical loader, along with two laborers, to produce as many as 32 cars in a single shift. Needless to say, the regular contract miners did not welcome Costa's extraordinary yield because it violated the stint and meant less work for them.[16]

MENZIES CONE 1939

The Menzies Cone, developed in 1934 by W.C. Menzies, was a coal washing machine utilizing an upward current of water to affect a separation between coal and its impurities. The machine worked on the principle that coal has a lower specific gravity than its contaminants; such as slate, rock, bone, or clay. So, once the lighter coal was fed into the cone, the upward current of water carried it over the top, leaving the impurities behind.

Raw coal was fed into the cone with an agitator at the top. Rows of water nozzles were spaced in horizontal rings from top to bottom in the cone. The raw coal entered a well around the vertical shaft of the agitator in the cone's center. The pressure of the rising water, controlled by valves, and the movement of the agitator stratified the raw feed in levels depending on the various specific gravities. The lighter coal rose over the top and flowed out of the cone to dewatering and sizing screens and then on to the loading area. The impurities fell through the rush of water to the bottom of the cone into a refuse conveyor.

In the early 19th century, miners cleaned coal by hand. By the early 20th century mechanical cleaning equipment, including pickers, jigs, concentrating tables, and a wide range of cones, was introduced in the anthracite region. Experiments with machines using an upward current of water to separate coal from refuse had occurred since 1916. W.C. Menzies, a mechanical engineer from Scranton, Pa., installed an experimental cone unit at the Nottingham colliery, owned by Glen Alden Coal Company, in Plymouth, Pa. in 1932. In 1934, he formed the Menzies Separator Company and began selling the cones commercially. The fourteen units installed at the Huber breaker in 1939 treated each coal size separately, and were the latest improvement in coal washing equipment in the anthracite region. Twelve cones were 9'4" in diameter and two were 7 feet. Each had a capacity to treat one ton per hour per square foot of area at the top of the cone. They used 8,200 g.p.m. of circulating water.

The Menzies cones were the principal coal washing units at Huber until the late 1950s when new technology complemented their use. Ten original cones remain in the breaker.

This drawing is a reconstruction based on original drawings located in the Blue Coal Corporate Archives, historic photographs, and personal interviews with former Blue Coal employees. See the HAER historical report for an annotated list of original drawings used.

Spring Board
Fork
Dewatering Shaker
Agitator
Slush Valve Handle
Recovery Tank
Refuse Conveyor
Undercut Gate
Pump

Menzies Cone Flow Section

AGITATOR DRIVE
DEWATERING SHAKER
FEED
REFUSE
CLEAN COAL
REGULATING VALVES
RECOVERY TANK

5 10 15 20
Feet 3/16" = 1'-0"
1 2 3 4 5
Meters 1:64

4 Feet
30° 30°
Scale: 1/2" = 1'-0"
1 Meter

Figure 9. Menzies Cone detail, Huber Breaker, Glen Alden Coal Company, Ashley, PA. (Library of Congress, Historic American Engineering Record, Lazlo Gellar, delineator. Details described in Robert Janosov, "Glen Alden's Huber Breaker," Canal History and Technology Proceedings Vol XI, 1992, 124–126)

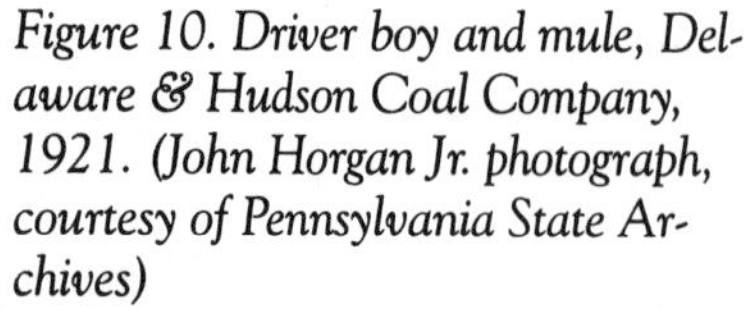

Figure 10. Driver boy and mule, Delaware & Hudson Coal Company, 1921. (John Horgan Jr. photograph, courtesy of Pennsylvania State Archives)

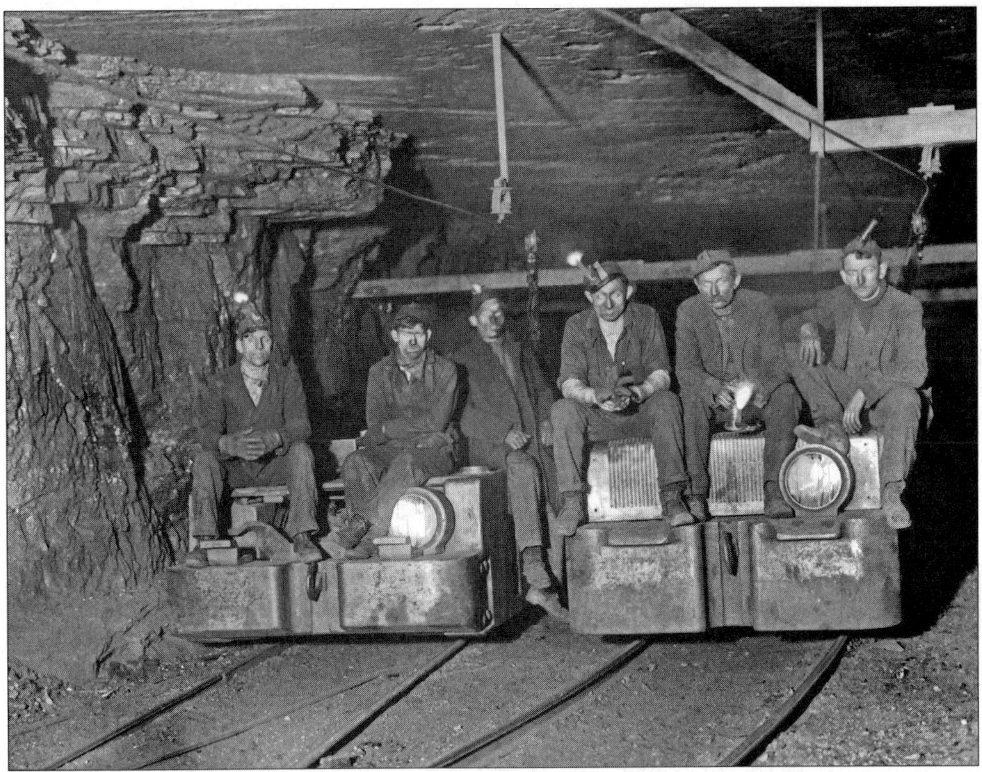

Figure 11. Electric mine locomotives, Delaware, Lackawanna, & Western Railroad, circa 1910. (Bunnell photograph, courtesy of National Canal Museum Archives)

Operational Limitations: Geology, Production, and Overproduction

Geology presented other critical limitations for anthracite. Unlike their bituminous counterparts, the anthracite companies could not find new deposits in other states or regions. Because substantial quantities of hard coal were found only in the ten-county area of northeastern Pennsylvania, each company had to dig deeper shafts to access new seams. However, as a shaft went further into the earth, it accessed thinner veins. The average thickness of the anthracite regions' mineable coal seams declined from 158 inches in 1872 to 80 inches in 1922, while their average depth increased from 235 feet to 415 feet.[17] Such constraints added significantly to production costs.[18] (Figure 12)

Year	Average Thickness (inches)	Average Depth (feet)
1872	158	235
1882	130	310
1892	108	360
1902	93	390
1912	84	400
1922	80	415

Figure 12. Thickness and depth of anthracite veins, 1872–1922. (Barger and Schurr, The Mining Industries, 1899–1939, A Study in Output, Employment and Productivity, *New York: National Bureau of Economic Research, 1944, 184)*

Figure 13. Anthracite coal veins separated by rock strata. The number and thickness of veins varied among regions and the various areas within. (Ohio State University eHistory)

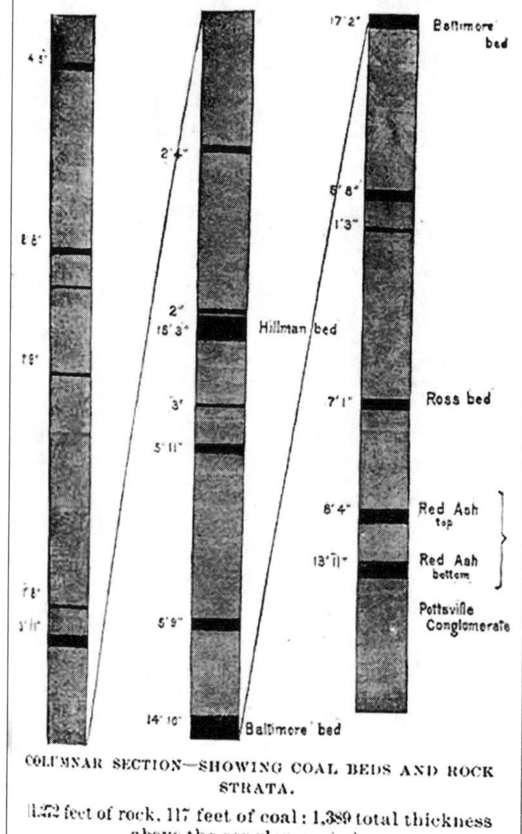

Geology related to another problem: overproduction. The escalating costs of mining deeper veins encouraged the companies to recover their investments through greater output, which in turn led to overproduction. The market characteristics of coal also contributed to regular supply gluts. As labor historian David Brody pointed out, in its heyday years coal was like few other commodities in that demand was "inelastic," meaning that price increases did not significantly affect demand, at least in the short run, because consumers needed the fuel and could not easily switch to other sources. Supply, on the other hand, was "elastic" when demand increased, meaning that production readily expanded due to excess capacity among the large companies as well as the ease of entry by small firms, but "inelastic" as demand fell because many operators, especially the independents, were willing to take low or even no profits rather than reduce output or close down.[19]

To deal with the overproduction dilemma, as early as the 1850s some companies established short-lived coal "pools" that set quotas on the tonnage going to market.[20] In 1873, a number of railroad-owned companies inaugurated a plan that collapsed when the participants consistently overshot their allocations. Other operator coalitions formulated production-limiting agreements five more times between 1873 and 1900, and they all had little success.[21] As Priscilla Long observed of the schemes during the 1880s and 1890s: "When prices started rising [the companies] scrambled to produce more than their agreed-upon allotment. The system was constantly breaking down."[22] In 1898, a highly consolidated industry instituted a quota plan through its so-called "community of interest," or monopolistic position; however, there were still numerous independents who often produced as much as possible and disrupted the system.[23]

The coal pools had an unintended consequence for the size of a colliery, as Harold W. Aurand pointed out. Because the pools' allocations were based in each firm's potential output capacity: "Each company strived for a larger share of the total allotment by increasing its [physical] plant."[24] Thus, overbuilding became the other side of the overproduction coin.[25]

In 1935, the former Line companies joined with independent operators (including many leaseholders) to fashion a new production-and-pricing strategy. The compact assigned 60 percent of the tonnage to the large companies and 40 percent to the independ-

Figure 14. Truesdale Colliery, Delaware, Lackawanna, & Western Railroad, Hanover Twp., Luzerne County, PA. (Courtesy of National Canal Museum Archives)

ents. Each firm received a percentage based upon its potential production volume. The covenant worked reasonably well, and when it expired after five years, most operators upheld the quotas on a voluntary basis. The Pennsylvania Coal Commission eventually regulated the plan.[26] However, many independent and virtually all bootleg producers paid no attention to the policy.

The leasing system exacerbated the overproduction problem along two dimensions. With so many entrepreneurs seeking to gain a stake, intense competition encouraged the lessees to out-produce and undercut each other. Moreover, lessor companies such as PaCC took full advantage of a loophole that allowed them to process their lessees' coal and sell it through an agent, but not have it count against their quotas. For example, in 1939 PaCC received a share of 2.83 percent of total production. However, when adding the 6.71 percent given to its leaseholders, output from PaCC breakers amounted to 9.54 percent of total production—more than three times the official assignment.[27] Therefore, the production quotas functioned to encourage leasing by firms that controlled the mineral rights.

Tenancy and Decline: Business Corruptions and Labor Conflicts

As important as these factors may have been in anthracite's decline, the present volume has focused on a largely unexamined contributor: tenancy systems and their accompanying corruptions and conflicts. The corruptions included illegal mining, substandard wages, inspector deceptions, bribery, kickbacks, and other improprieties.[28] With regard to the conflicts, as these pages have shown, the state of affairs in northern anthracite led to some of the most bitter worker-management and worker-worker confrontations in American labor and business history.

Following the defeat of the United Anthracite Miners of Pennsylvania (UAMP) in 1935, the labor wars against subcontracting and leasing faded and northern-field tenancy expanded precipitously. At least 66 lease-holding companies mined coal in 1939, with few of them honoring the wage schedule negotiated by the United Mine Workers of America

(UMWA).[29] Most also ignored or paid scant attention to safety and other aspects of state and federal mining law. In 1938, a special commission appointed by Governor George Earle to investigate a disaster that took ten lives at the Volpe Colliery in Pittston (formerly HC&I's Butler Colliery), looked into 138 other lessee firms and concluded that the great majority "... do not even pretend to comply with any part of the mine [safety] laws."[30] Leasing nevertheless grew not only because of the economic rewards to the lessors and lessees, but also because organized worker resistance had virtually disappeared.[31] In 1942, the northern field claimed 104 coal companies, with the overwhelming majority working under leases from the large mineral rights controlling firms. In 1952, the number stood at 76.[32] Some leaseholders eventually purchased whole collieries, including the mineral rights, and began issuing subcontracts to even smaller entrepreneurs. This transformation meant that one of the most consolidated industries in the U.S. in 1902 had, in less than 50 years, become a highly decentralized enterprise when it came to the mining and processing of coal.

Illegal mining became another prominent aspect of the new order of anthracite production under tenancy. For example, The Pittston Company (TPC) found extensive pillaging during a survey of all company mines in 1938. Surveyors discovered multiple perforations in the supposedly sacrosanct barrier pillars of solid coal that separated adjacent mines. Required by state law, the 100-foot-thick blocks of anthracite were intended to prevent a flood, fire, or explosion in one pit from spreading to its neighbor.[33] In 1914 the U.S. Supreme Court upheld the state's right to require the barrier pillars.[34] However, especially in the 1920s and 1930s, and continuing into the 1940s and 1950s, the bulwarks had

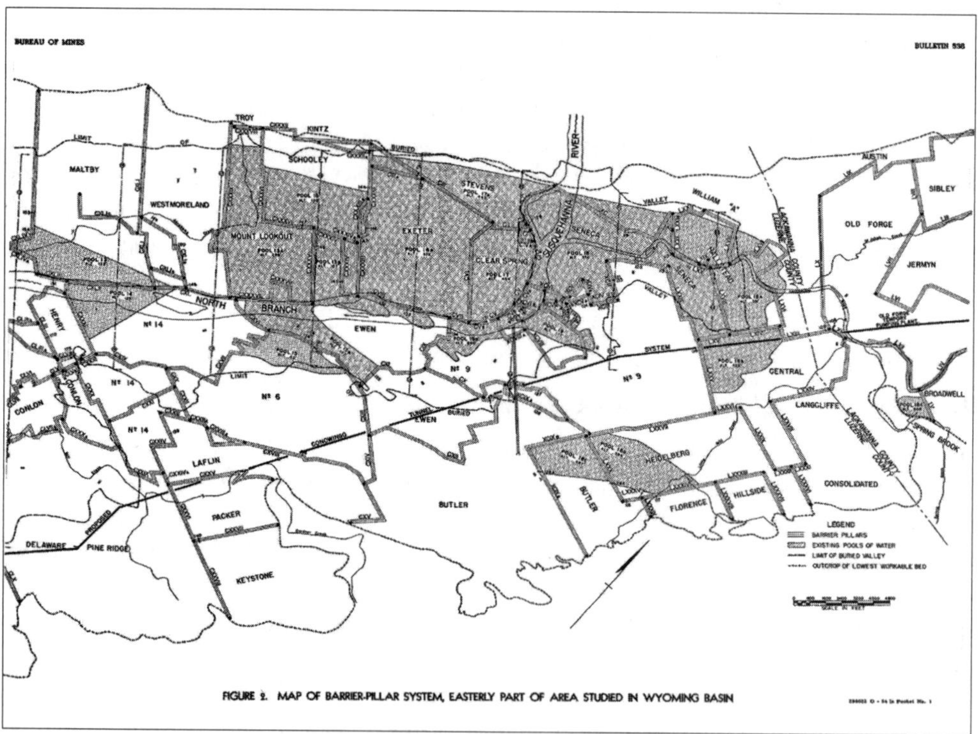

Figure 15. Barrier pillars in the eastern end of the Wyoming Basin around Pittston, PA. (Solomon H. Ash, "Barrier Pillars in Wyoming Basin, Northern Field," Bulletin 583, U.S. Bureau of Mines, Washington, DC: USGPO, 1954)

been plundered by several companies. TPC found that large quantities of coal had been removed from the barrier pillar between its No. 9 Colliery and the Heidelberg Colliery of the Lehigh Valley Coal Company (LVCC). The robbing meant that the Heidelberg's mines flooded whenever the No. 9's water level reached 535 feet from the bottom-most vein.[35] The Heidelberg was owned and operated by LVCC between 1881 and 1931, and then leased to the Heidelberg Coal Company in 1932, whose principal owner was James B. McDade of Scranton. It was not clear whether these firms or any of their subcontractors or leaseholders did the looting. Two other adjacent TPC-owned mines, the Central and Old Forge collieries, shared a similar problem.[36] On this occasion, it was obvious that TPC or PaCC mineworkers had looted the barrier at some point in time. As late as 1959, state mining officials discovered that LVCC's Henry Colliery, located next to PaCC's No. 14 Colliery, also had a partially mined-out barrier.[37]

Figure 16. Heidelberg Colliery, Heidelberg Coal Company, Avoca, PA. (Courtesy of Frederick W. Bartlett Estate, www.fbartcreations.com)

Figure 17. No. 9 Breaker, Pennsylvania Coal Company, Hughestown, PA. (Courtesy of Frederick W. Bartlett Estate, www.fbartcreations.com)

Figure 18. Old Forge Colliery, Pennsylvania Coal Company, Old Forge Twp., PA. (Courtesy of Frank Adams)

Government investigations and oral histories indicated that subcontractors and leaseholders commonly engaged in removing coal from the barrier pillars.[38] The pilfering not only broke the law but presented a potential safety hazard because the blocks of anthracite held back millions of gallons of so-called "make water" (or natural underground seepage) that had collected in abandoned mines. In 1935, the more than 100 barrier pillars in Lackawanna County—few of which were considered reliable—restrained a "staggering total" of ten billion gallons of water in 21 underground pools.[39] Geologist Solomon Ash found that Luzerne County's 224 barriers restricted an even larger volume in 1954.[40] If a thinned barrier gave way to the pressure of the underground reservoir behind it, a catastrophe of major proportion could strike the contiguous mine.[41]

Any discussion of corruption and illegal mining must include the Knox Coal Company. As mentioned earlier, Knox became a PaCC lessee in 1943 and secured rights to the River Slope lease in Port Griffith in 1954. The enterprise engaged in the illegal mining of support pillars as well as barrier pillars, but its most egregious violations occurred at the River Slope mine. Under

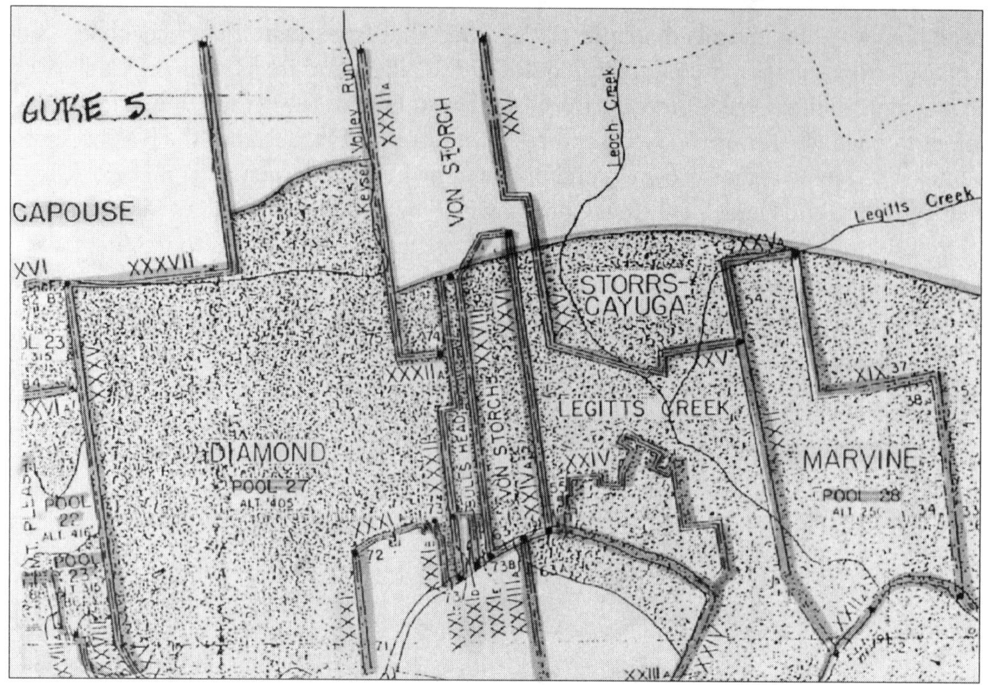

Figure 19. Barrier pillars under the City of Scranton, PA.
(U.S. Bureau of Mines)

Figure 20. Boiler house, Central Colliery, Pennsylvania Coal Company, Pittston Twp., PA.
(Courtesy of National Canal Museum Archives)

close managerial direction, workers removed coal in off-limit virgin seams under the Susquehanna River that came fatally close to the riverbed. The thin rock overburden gave way when the river level rose, causing a mine flood that took 12 lives, inundated numerous workings (some because the barrier had been robbed), caused widespread environmental damage, and hastened the industry's decline in Luzerne County.[42]

One of the most disturbing aspects of the incident involved the UMWA District 1 president, August J. Lippi, who was a secret and illegal partner in the company. The

Wilkes-Barre Record

12 Missing and 35 Rescued After River Breaks Into Mine Near Pittston

Cars Dropped In Pit to Try To Plug Break

Figure 21. The Knox mine disaster occurred on January 22, 1959. (Front page, Wilkes-Barre Record, *January 23, 1959)*

Figure 22. Mineworkers at the River Slope mine, Knox Coal Company, Port Griffith, PA, 1958. Left to right: Fred Costanzi, Steve Brek, John Angeli, unknown, unknown. (Courtesy of Carl Orechovsky)

disaster precipitated five federal and state grand jury investigations, which brought indictments against 22 individuals and four coal companies. The charges included manslaughter, conspiracy, personal income-tax evasion, corporate income-tax evasion, bank fraud, and labor- and safety-law violations.[43] Twelve persons and three companies were eventually convicted. Of the 22 indicted, nine were union officials and, of these, seven were found guilty. Six coal company owners were found guilty of income-tax or labor-law crimes.[44] In one of the darkest episodes in northern-field history, an astonishingly broad and entrenched culture of corruption had been exposed.[45]

Safety also suffered under tenancy. Like Knox, most subcontractors and leaseholders emphasized output over security. The drop in accident rates through the first half of the twentieth century related in large part to the decline of deep mining along with increased production from strip mines and reprocessed culm banks.[46] The official accident rate notwithstanding, oral-history evidence indicated that tenants regularly disregarded safety measures and put mineworkers at risk. Joe Stella worked as a surveyor for PaCC in the 1950s and described a general pattern:

Q: Are you saying that Knox was known as a pretty slipshod operation?

JS: It was a production outfit. In other words, that's all they were interest in was coal, coal, coal.

Q: Not safety?

JS: No.[47]

Mining engineer Alex Chamberlain witnessed the tenants' "rip and tear" mining practices firsthand when he worked for his father's Jermyn Green Coal Company in the 1940s.[48] Lax safety presented a constant source of anxiety for Charles Volpe (a distant relative of Santo) during the 1950s and, after two close calls, he decided to quit:

CV: I was hanging around the cigar store one day and a state mine inspector used to stop in there once in a while. He knew me well from the mines. He told me, "What's the matter Charlie, you're not working?" "No," I said, "Not working." He said, "What's the matter?" "Oh," I said, "They shut down and that's it." He said, "Now look, you're a foreman. They're crying for foremen down there at the Loomis [a Glen Alden colliery]." He says, "You want a job down there? I'll tell 'em. I'll send you right down then you can start down there tomorrow." That was Andy Wilson. I told him, "Andy, no more work for me." I said, "I've had it—fifteen years. That's it, I'm not going back no more."

Q: And you never regretted it?

CV: Nooooo! ... I had seniority if I wanted to work [for Pagnotti coal]. They came and visited me here at home. I could go back to work there. I told them, "I'm not going back no more." I said, "I got out, I'm lucky. I nearly got killed two different times." Two different times![49]

Another dimension of safety involved deception of federal, state, and company inspectors. Stella acknowledged that accurate mine examinations were rare because most inspectors announced their reviews in advance: "Soon as they [management] heard the state inspector was coming, they start fixing this up and that all over [the mine]."[50] As a surveyor, Stella's job was to assess whether a lessee stayed within the boundaries of a designated coal tract. He discovered systematic illegal mining among the lessees:

Figure 23. Strip mining in anthracite.

Q: Was it typical to find violations?

JS: Oh yes, very much. They'd always have some kind of excuse, you know. I had no power to do anything. My power was to go and check and report so then it'd be up to the higher ups [in PaCC management] to do something about it. They didn't even give me authority to stop [unsanctioned mining in] a place or anything like that.

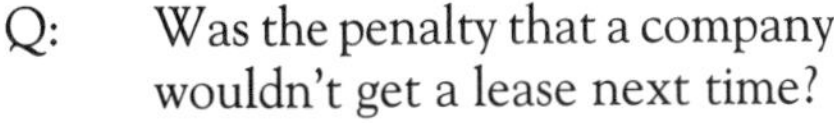

Q: Was the penalty that a company wouldn't get a lease next time?

JS: The [PaCC] boss, the engineer, would probably get mad at them [and say]: "No, nothing doing." Then, eventually they would ask for permission [to mine] in a different area. Sometimes they would get it sometimes they wouldn't.

Stella added that it was "normal procedure" for a tenant to steal, or "skip," eight to ten feet of coal from a support pillar even when it was not included in the subcontract or lease.[51]

After confirming that tenants often took "skips," subcontractor Joe Costa said that they often did not stop there:

> But what do you think they [eventually] did? They took it [support pillars] all! One time I drove a slope for Pagnotti up in the back here [Harry E Colliery in Swoyersville] and we robbed all the pillars and, when we heard an inspector was coming, the superintendent told me to blow it up. I said, "What about the shaker chutes, should we get them out?" And he said, "Forget those chutes and close it up." We closed it up [blew it up] so the inspector couldn't go down and see what we were doing.[52]
>
> When I was working for Pagnotti, we robbed lots of pillars. We even went up in the back and stole some beautiful coal that was owned by Haddock [Coal Company] from Luzerne. We'd take all the pillars, and we'd blow up the entrance, leaving the equipment inside to make it look good. The inspectors couldn't go in and check if we blew it up. I did this dozens of times.[53]

Costa also participated in distracting and essentially bribing the inspector:

> I remember Pagnotti one-time said to me, "Joe, here's a hundred dollars. Take the mine inspector out and show him a good time." So I took him fishing and we went out.[54]
>
> I used to get a hundred dollars or so and take the mine inspectors out for dinner and wine and dine them about every other month. They would only inspect around the outside of the mine if we kept them happy. It was a payoff for them.[55]

Wage violations presented another abuse. George DeGerolamo served as a local union officer at a TPC-leased mine during the 1930s (he did not indicate which one) and received a number of wage-related complaints from members. He took the grievances to Superintendent Donaldson and received a harsh greeting: "[He] jumped out of his swivel chair and poked his finger at me and told me that I was young at this game and that I had better leave

Figure 24. Harry E Colliery, Harry E Coal Company, Swoyersville, PA, circa 1950. (Courtesy of Frederick W. Bartlett Estate, www.fbartcreations.com)

his office and forget all about the grievances." DeGerolamo threatened to call a strike if the mistreatments were not rectified. In retaliation, Donaldson fired him. DeGerolamo requested the intercession of District 1 President John Boylan but received no support. The strike went forward; however, only the complainants appeared on the picket line. In further retaliation, the superintendent had DeGerolamo blacklisted from area mines. The anathematized union leader packed his car and drove his wife and baby to Detroit in search of employment in the auto industry.[56]

The tactics at Penn Anthracite Collieries Company, the former employer of John Boylan, exemplified the wage manipulations. Founded in 1931 upon the bankruptcy of the South Penn Collieries Company, Penn Anthracite leased and then bought mines from the Scranton Coal Company, the Hudson Coal Company, and others. The firm controlled several Scranton-area collieries, including the Capouse, Von Storch, Leggets Creek, and West Ridge.[57] The holdings became so extensive that Penn Anthracite began offering subcontracts and leases to smaller firms.

As was common among lessor-company officials, Penn Anthracite's general superintendent, John H. Harvey, encouraged his chief inspector, Oliver L. Davis, to push the tenants for higher yields and greater efficiency.[58] When it came to wages, the managers clearly ignored the UMWA's negotiated rate sheet. Penn Anthracite's due bills (or workers' pay statements) listed the regulation union wage; however, supervisors were instructed to pay less than scale so that the miners and laborers, as one official wrote: "... are still actually receiving some short rate for their work." Internal memoranda clearly indicated that short-changing employees stood as an intentional and widespread practice:

> If the full rate is actually paid to the men, experience indicates that there will be little or no profit to the T&E Company [Thomas & Edwards, a lessee] or to its subcontractor, Sokoloski & Blasi. Labor payrolls, compensation, insurance and other deductions made by the company, not including powder, timber and miscellaneous supplies, have generally exceeded the amounts of money earned under the [sub]contract.[59]

This anonymous memo, probably from the general manager to the president or from the chief mining engineer to the general manager, clearly showed that management had no qualms about the inferior wages, which they evidently believed were necessary to maintain a profitable business.

However, the managers did have legal concerns regarding the possibility of union officials discovering the conspiracy. Chief mining engineer T.S. Shoemaker wrote to Penn Anthracite President Charles Dorrance regarding the matter:

> At every colliery on our property there are signed rates which are not in use at the present time. By hard and tedious process the high or exorbitant rates have been gradually discontinued until there is little possibility of their revival except thru their discovery by such a [UMWA] committee making an examination of the col-

Figure 25. First Von Storch breaker, Delaware & Hudson Canal Company, Scranton, PA, 1865. Penn Anthracite Collieries Co. took over the new Von Storch breaker in 1931. (Johnson photograph, courtesy of National Canal Museum Archives)

> liery rate sheets as a guide to them in forming a rate sheet to be signed by some lessee on a part of the property. Such discovery of course would wipe out all the good effects [i.e., profits] which have been gained thru the years in which these rates have not been paid and might subject us to [officially filed] grievances for back-pay over the whole period.[60]

In 1936 the local union caught on and charged the aforementioned lessee, Thomas & Edwards, with paying lower wages than stipulated by the UMWA contract. The affected employees filed grievances with the Anthracite Board of Conciliation (ABC). They won the cases and received back pay. In response to the sanction, Penn Anthracite changed policies to require that all employees, including those in subcontracting crews, "get paid at the window on the regular colliery payroll and deductions made from the [sub]contractors for the amounts paid the men."[61] However, unlike the subcontractors, lessees such as Thomas & Edwards ran their own pay systems, which Penn Anthracite could not influence.

The outright stealing of coal presented another form of dishonesty. Joe Costa worked for a mob-affiliated subcontractor in the mid-1930s at TPC's No. 14 Colliery. He told how his job involved "stealing" unguarded cars of coal in a mine as they awaited transport to the breaker. His tactic was to change the written codes on the sides of the cars so as to attribute them to his boss's phantom employees:

> When I worked for this subcontractor, my job was to do the stealing. He gave me a little pot and I mixed the carbide and water together and he told me my job was to go around and change the markings on the loaded cars of coal to the markings of his "men." So we got paid for those cars because I changed the marking. I never got caught because nobody was ever watching.

> [Then on payday] I'd get a bunch of men off the street and pay them each five dollars. I'd give them a due bill [pay invoice] with some [phantom employee's] name on it and they'd go to the paymaster and pick up the envelope with the cash in it. I'd meet them outside and collect all the cash. That's when I'd give them their five dollars. The paymaster knew something was going on. Sure he did.[62]

This type of deceit prompted the 1939 Tri-District Convention to approve a resolution demanding the "maintenance of accurate records and payments for all cars loaded. ..." The resolution-makers offered as a rationale: "That cars are loaded and not paid for on account of inaccurate records."[63]

Costa believed that crooked practices such as stealing coal underground were fairly common. In the process, the subcontractors not only stole from each other, he said, but from the company's mining crews who also worked at No. 14: "So the company got stuck twice. First, they lost the coal that their own miners dug, then they had to pay the [sub]contractor a certain price for the same coal! The [sub]contractors were like pigs, stealing all the coal they could get."[64]

Coal pillaging and unsafe mining, along with wage, weight, and car cheating were hallmarks of the tenancy systems at PaCC, HC&I, D&H, LVCC, Knox, Penn Anthracite, and other companies, although the extent of the malpractice varied from operator to operator. The industry became engulfed in a culture of corruption where normal business dealings involved unethical and/or illegal actions. The dishonesty extended to the UMWA. DeGerolamo described the skullduggery he faced as a committeeman for the union local at his TPC-lessor employer: "I was a fairly good speaker," he wrote in his autobiography, "and very often would express my feelings against other UMWA officers who were leaning toward the coal company." It turned out that the officers of the local union did not appreciate his forthrightness:

> On a Sunday morning, the local president and two grievance committeemen called at my house and said that we had an important meeting with John Boylan. I got into the back seat of the car with the other two committeemen. After traveling about ten minutes in[to] a wooded area, I told them that this was not the way to the District office. The president stopped the car and one of the committeemen pulled out a .38 revolver and stuck it in my belly. He said that the next time I spoke of [criticized] the other officers I was going to get full of lead. I can't say I wasn't scared. I thought I was a "gonner." They left me off at my home. I didn't even say a word to my wife. I could never believe this would happen to me. ... At the next local union meeting, I asked to speak from the podium. The anger in me still possessed my better judgment. I exposed the gunman, who was at the meeting, and I said if anything happened to me, he was the man responsible for my life. Then I challenged him to fight American style, with fists. He was yellow and would not fight.[65]

Many local unions within District 1 participated in the connivance. The investigations into the Knox Coal Company provided one of the most striking examples: investigators found that L.U. 8005 officials Dominick Alaimo, Charles Piasecki, and Anthony Argo received thousands of dollars from management to maintain "labor peace" at Knox and essentially ignore the company's many illegal activities. The local union members at the No. 14 Coal Company, owned by the Gelso family, were similarly charged. Such chicanery beleaguered many mineworkers at Knox, No. 14, and other companies in District 1 even before August J. Lippi assumed the presidency in 1951. His reelections—including two after he had been indicted for labor law, income tax, and other violations—pointed to an *organized* mode of criminal behavior in the northern field's industrial relations.[66]

As this volume has shown, the workers vigorously resisted tenancy and its corruptions at PaCC and HC&I during the 1910s and 1920s. The rebellion expanded to employees at other firms in the late 1920s and early 1930s, as up to one-half of the northern field's workforce participated in the protest movements that culminated in the formation of the UAMP in 1933. When the region's most prominent dual union had been beaten and the coal companies expanded leasing, the UMWA became more concerned with "stability" and a hoped-for upsurge in demand for hard coal. However, the decline continued and tens of thousands of mineworkers left the area for work in Allentown, Bethlehem, Philadelphia, New Jersey, New York, and other locales, often commuting home on weekends.[67] Those who remained in mining continued to protest in small ways into the 1950s but, in reality, they had little choice but to accept the corruptions if they wanted to live locally and work in anthracite.[68] The ABC continued to adjudicate hundreds of wage and other grievances in the 1940s and 1950s, even though the filers risked blacklisting and various forms of retribution.[69]

The Economic Rationality of Tenancy

Subcontracting and leasing were originally intended to achieve two rational-economic objectives: reduce labor costs and increase productivity. The operators strategized that such changes would enhance competitiveness and thereby boost or at least stabilize profits. Greater control over the workers and the work process was integral to the strategy.[70] Whether purposeful or not, the anthracite operators followed the approach of Frederick W. Taylor who provided a general criticism of the American worker and workplace in his book, *The Principles of Scientific Management* (1911):

LIPPI IS NAMED TO WILLIAMS' PLACE IN MINERS' UNION

APR 21 1941

KOSIK ANNOUNCES SELECTION AND FORMAL APPOINTMENT WILL COME TOMORROW. CARDONI TO MOVE UP.

President M. J. Kosik, of the United Mine Workers of America in the Scranton district, today announced that August Lippi, Exeter, executive board member in the Pittston area, has been selected as the successor of the late Secretary-Treasurer Enoch Williams, Taylor. Explaining that there is urgent need for a secretary-treasurer being named with-

Figure 26. August J. Lippi, appointed secretary-treasurer of UMWA District 1 in 1941. (Courtesy of Scranton Times*)*

> . . . in most of the shops in this country, the shop was really run by the workmen and not by the bosses. The workmen together had carefully planned just how fast each job should be done, and they had set a pace for each machine throughout the shop, which was limited to about one-third of a good day's work.[71]

To enhance efficiency and decrease what he called "soldiering" by employees, Taylor argued that managers should be in charge: "It is only through *enforced* standardization of methods, *enforced* adoption of the best implements and working conditions, and *enforced* cooperation that this faster work can be assured." Who was responsible for the enforce-

Figure 27. Frederick W. Taylor, the founder of Scientific Management.

ment? "[T]he duty of enforcing the adoption of standards and enforcing this cooperation rests with the *management* alone," proclaimed Taylor.[72]

A generation earlier, in the 1880s, H. Martyn Chance studied anthracite and expressed views strikingly similar to Taylor's. In the *Second Geological Survey of Pennsylvania*, Chance wrote:

> Insubordination in the force employed underground should be treated with military severity. It can only be viewed in the light of a crime, and should be punished accordingly. ... Absolute and immediate obedience, unfailing adherence to all colliery rules, and faithful performance of his duties, are essential to the safety of those employed underground.

After noting that "our most prominent mine superintendents ... find it almost impossible to carry out, even in part, the policy they would endorse in dealing with underground employes [*sic*]," Chance stressed that: "The same difficulty has been experienced in other mining districts, and cannot be removed while artisans and laborers maintain their present attitude towards corporations and individuals supplying them with employment."[73]

Carter Goodrich applied Taylor's ideas to the American coal-mining industry. His assessment in *The Miner's Freedom* (1925) that "mining is an old industry with old traditions as well as in part a union industry with union traditions" led to the conclusion that any change in the organization of work and work culture would prove most difficult. He argued that the miner's independence, and the implied and real indiscipline central to the "culture and conditions" of the occupation, would hinder any Taylorist intentions.[74]

However, Goodrich envisioned technology as a powerful agent of change. In his survey of coal mining between 1890 and 1920, he concluded that modern managerial methods (meaning Scientific Management), coupled with new technologies such as high-speed cutters and loaders, would greatly reduce if not end the miner's independence (including the stint) and usher in a new era of factory-like efficiency.[75] Goodrich's predictions were effected more in bituminous than in anthracite. Neither the practices of modern management nor many of the latest mining technologies (such as short-wall and long-wall mining) were implemented to any significant degree in hard coal. Likewise for the analysis of Chris and Charles Tilly who concluded that between 1930 and 1950, U.S. coal production witnessed "high-priced labor and industrial consolidation [that] promoted a pronounced shift toward mechanization, time-discipline, and time-payment."[76] Again the conclusions applied more fully to bituminous than to anthracite.

In plotting a different course, the managers at PaCC, HC&I, and other anthracite companies used subcontracting and leasing as strategies to achieve the outcomes promoted by Scientific Management. That is, while the tenancy schemes brought significant changes to *who* mined anthracite and *how* and *where* it was mined, they more importantly altered the balance of power in the workplace. As such, they profoundly affected the work culture and the organization of work while also influencing output and productivity.

In support of management's policies, statistical evidence suggests that tenancy did, in fact, boost output. If 1913–1916 is taken as the period when subcontracting recommenced at PaCC, HC&I, and, to a lesser extent, at the Hudson Coal Company, and if 1933–1936 as the period the leasing system began to expand at PaCC, HC&I, and other major northern companies, it is clear that production surged after these dates. *Output per man hour* and *output per man day* increased most years, although there were some exceptions such as the strike years of 1922 and 1925–1926. The upturn in the categories reversed 15 years of declines following the 1902 strike. *Output per man* increased after 1902, remained relatively high until the mid-1920s (excepting the strike years of 1922 and 1925–26), fell between 1929 and 1932, and began rising to comparatively high levels after 1932. (Figure 28)

Northern-field production and employment between 1924 and 1948—prime years for subcontracting and leasing—showed that, after 1930, output remained fairly stable while employment plummeted more than 50 percent. (Figures 28 & 29) Both lessor and lessee companies introduced new technologies such as shaker chutes, jalopy loaders, and electric motors, while strip mining and reprocessing culm banks also supported output. However, at least some of the productivity—it is impossible to know exactly how much—can be associated with the efficiencies associated with tenancy.

For these reasons, it seems reasonable to conclude that northern-field tenants delivered two types of proficiencies. First, they put downward pressure on wages by often disregarding rate sheets, corrupting the local and district unions, and turning many certified

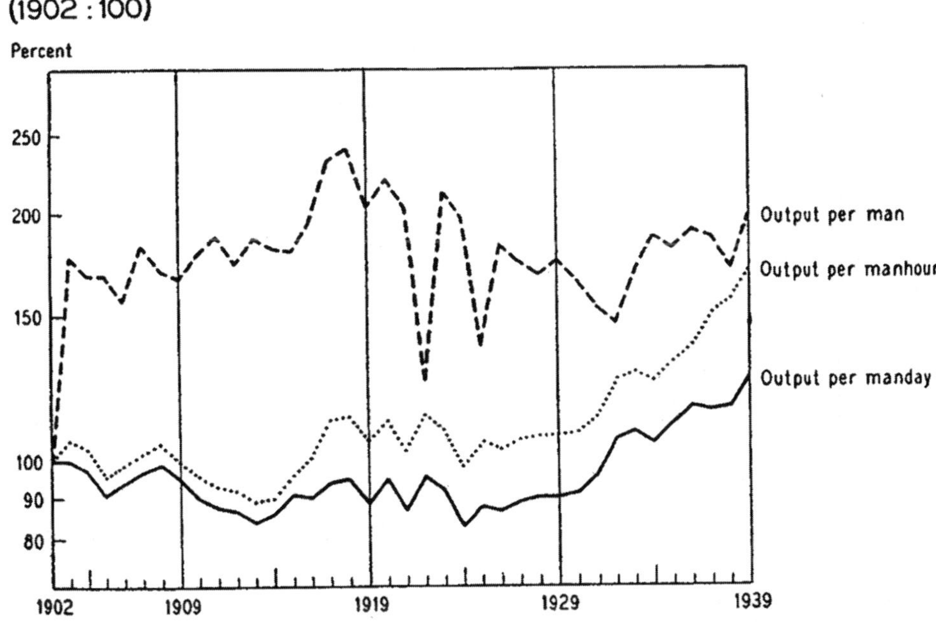

Figure 28. Anthracite Productivity, 1902–1939.
(Barger and Schurr, The Mining Industries, 1899–1939: A Study in Output, Employment and Productivity, *New York: National Bureau of Economic Research, 1944, 133)*

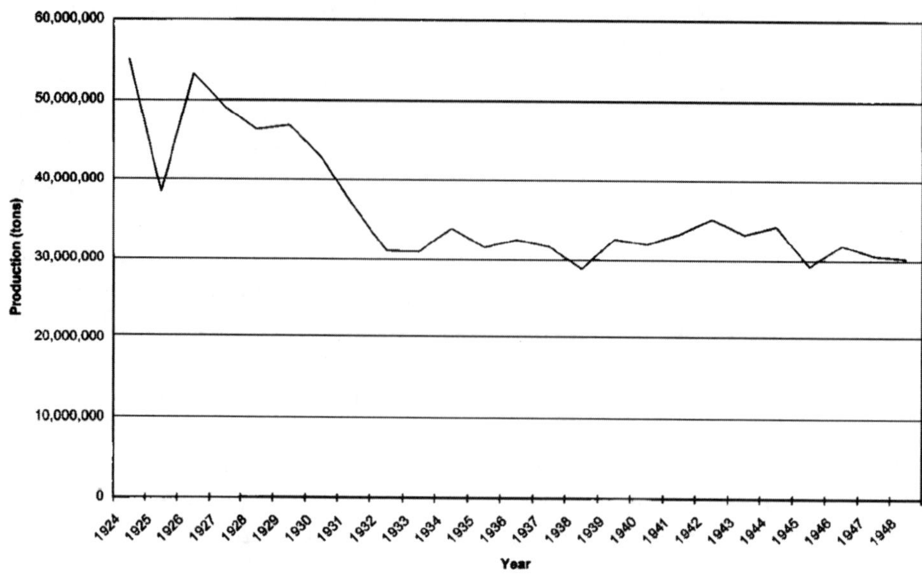

Figure 29. Northern Field Production, 1924–1948.
(Pennsylvania Department of Mines and Mineral Industries, Annual Report: Anthracite Division, 1963, *Harrisburg, PA: Commonwealth of Pennsylvania, 1963)*

miners into laborers. Indeed, it can be hypothesized that anthracite lagged in certain

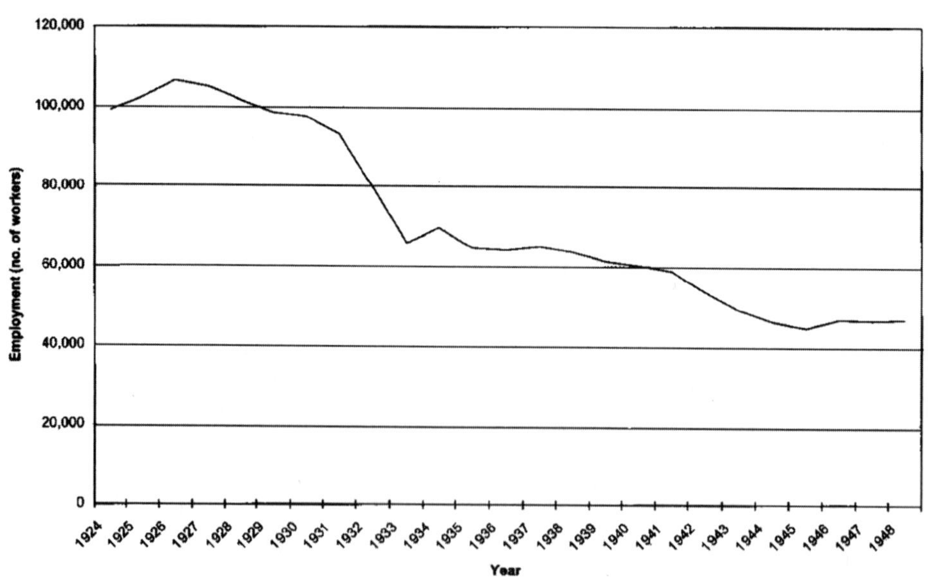

Figure 30. Northern Field Employment, 1924–1948.
(Pennsylvania Department of Mines and Mineral Industries, Annual Report: Anthracite Division, 1963, *Harrisburg, PA: Commonwealth of Pennsylvania, 1963)*

technological innovations in part because the tenants drove down the price of labor.[77] Second, productivity rose even as employment fell, partly because of technology but also because tenants pushed employees toward higher tonnage by eliminating or circumventing stints, instituting production bonus schemes, and taking illegal but "easy coal" in off-limit pillars and barriers.[78]

The "Irrational" Consequences of Tenancy

Despite the probable rational-economic paybacks, the two anthracite tenancy systems also yielded numerous "irrational" or detrimental outcomes. Their praetorian structure resulted in dishonesty, violence, conflict, and disaster. They threatened work, family, and community. In spite of the short-term economic remunerations, tenancy spawned a regime of subsidence and other environmental hazards, as well as injuries, bribes, kickbacks, bogus inspections, short-weightings, illegal mining, shorted wages, broken union agreements, and even murder. These were not merely irresponsible acts, or the results of irresponsible acts, by crooked individuals. They were predictable outcomes of *systematic* degradations initiated by the major anthracite corporations, their tenants (including alleged organized criminals), the mineworkers' union and its leaders, state regulatory agencies and inspectors, and, in certain cases, "coal hungry" mineworkers.

A main economic argument in support of the tenancy systems was that they kept investors investing, mineworkers working, consumers burning coal, and communities surviving.[79] The systems' cost-cutting and labor-disciplining efficacies—as evidenced by operators such as Santo Volpe, Louis Pagnotti, Robert Dougherty, John Kehoe, Morgan Bird,[80] and many others—permitted the industry to survive longer than the "laws" of economics would otherwise have allowed.[81] Although the argument has some basis in fact, it is also clear that the social, environmental, legal, and moral costs were exceedingly high. Few comparable instances of such pervasive and fatally consequential corruption and conflict can be found in the coal-mining industries of other Western nations.

Indeed, it can be argued that subcontracting and leasing themselves constituted a type of *organized* criminal activity. Legal (and ethical) principles were systematically violated by companies, tenants, union leaders, and, in some cases, workers. The lessors looked the other way as long as lessees delivered coal and/or paid royalties. While most lessees had no affiliation with organized crime, the tenancy *systems* nevertheless promoted various types of duplicitous behaviors during anthracite's final decades. The improprieties at the northern field's last major operator, the Blue Coal Corporation, which purchased most of the Glen Alden Coal Corporation's remaining holdings in 1966, were detailed by journalists Gil Gaul and Elliot Jaspin in a five-part Pulitzer Prize-winning series published in the *Pottsville Republican* between June 12 and June 16, 1978.[82] James Durkin, an alleged close associate of organized criminals, was one of the owners of Blue Coal.[83]

In his analysis of mine-working under Tayloristic management, David Brody concluded:

> In recent labor scholarship, Taylorism has been perceived primarily as a struggle for control of the workplace. By that definition, the coal miner remained singularly untouched. His control over the labor process went virtually unchallenged at the turn of the century and, indeed for at least another three decades.[84]

Figure 31. Blue Coal Company pockets for coal storage, circa 1950s. (Courtesy of Lackawanna Historical Society)

Figure 32. Burning culm bank, Lackawanna County, PA, circa 1950s. (Courtesy of Lackawanna Historical Society)

Figure 33. Fighting a culm bank fire, Marvine Colliery, Hudson Coal Company, Scranton, PA, 1968. (Courtesy of Lackawanna Historical Society)

Figure 34. Huber breaker in ruins, Ashley, PA, 2012 (the last standing breaker in the Northern Anthracite Field). (Robert P. Wolensky photograph)

Figure 35. John Mikulski, motor runner, Avon Coal Company, Plymouth, PA, holding a "sprag" used to stop coal cars underground. (NPOLHP Collection)

Although there was no evidence to suggest that most of anthracite's managers were formal advocates of Taylor, the present study has shown that the miner's freedom, and the work culture that supported it, were profoundly touched by tenancy during the first half of the century. PaCC, HC&I and other coal companies took precise aim at workers' control—including hours and rules, seniority rights, stints, and even collective bargaining. However, the primary vehicles of attack were not new management philosophies or innovative technologies, as Goodrich predicted and Brody implied. Among the prime vehicles of change were two tenancy systems that reorganized mineral rights' access. To confront what they saw as the systems' threat to their livelihood, work culture, community, and craft, the mineworkers fought a 35-year labor war.[85]

The End of the Individual Contract and the Subcontracting System?

Contrary to what some scholars have argued (see Chapter One), the individual or special contract and the subcontracting system did not disappear in the 1910s. The UMWA demanded an end to the policy in virtually every contract negotiation during the first half of the twentieth century. At long last, in the 1952 labor-management contract, the parties agreed to abolish the structure. An official District 1 pronouncement hailed the achievement:

> It is with a great deal of satisfaction that your officers inform you that they have succeeded in abolishing the system of special or individual contracts in District No. 1. The long sought after goal of the union was attained only after years of unrelenting efforts. As far back as 1920, an anthracite commission appointed by President Woodrow Wilson declined to eliminate the system. ... Hardly a convention or a joint conference passed, however, without the UMWA renewing its fight against the system which had such a baneful effect on our contractual structure ... .[86]

In reality, however, the arrangement remained in force at some collieries into the late 1950s. Motor runner and union local officer John Mikulski described his distress over the system's use at his workplace in 1958, the Avondale mine in Plymouth—operated by Robert L. Dougherty:

JM: I often said to my buddy, "How come they can have a special contract? We have a [union] contract of our own." He says, "Well, the men agreed on it."

The hell with the men. We're the representatives, we're the union guys, we're supposed to look into it. And here when I found out—all your big shots, your big officials [wanted special contracts], and I'm only the small fry. They got away with it for about a year or so and, finally, the mines all shut down on account of the Knox mine disaster. ... But they pulled one over onto us because unions is good if you go by union orders and all, like I just mentioned. When the big guys tell me what to do, well I gotta listen to them. But not all the time; sometimes you can spit back in their face. I found it out a little bit too late.

Q: Now, when you say "the big guys" wanted the special contract, do you mean the big guys at the company or the big guys at the union?

JM: Big guys in the union. Like there was Lippi. Dougherty [and Lippi] operated mines up in Pittston. They had their fingers in it. They were makin' the good buck. ... He [Dougherty] got Avondale and that's when they were going to have a special contract. They picked the bosses who knew who's the better miners, who can take coal and all. They got Billy Evans for one, and Big Mike, I don't know Big Mike's last name ... he was a laborer for Billy Evans. They're all dead and gone now. Jesus Christ, they [the men] worked their-selves to death. Like I mentioned before, two cars per man [was the standard]. Ya know, the mines wouldn't be a bad place to work if it was two cars for a man. ... But up until today, I don't like that sneaky [thing], the special contract. Well, who the hell approved that special contract? I didn't, only I heard about it. And I'm still talkin' about it today! And I'm damn sure Leonard [Statkiewicz] knowed about it because he was a higher [union] office holder than me. I'm only just the little guy. I'm only the guy with the bag holdin' the tools. That still bothers me 'til today. If you got a [labor-management] contract how the hell can you come in here and mine under this local, under this union, and forget about the two cars a man? They were loading three-four-five cars a man, which they were makin' a couple extra hundred dollars. That I know because I heard 'em sayin' it. They got a couple a hundred extra dollars. Well, Jesus Christ, why shouldn't you get a couple a hundred extra dollars? Look at the amount of cars you're loadin'... That's what we had unions for.[87]

The goal that had occupied three generations of anthracite colliers came as a hollow victory for another reason. Well before the 1950s, with PaCC leading the way, the large companies instituted a more complex putting-out plan, the leasing system, that broadened mineral rights' access to an even greater extent. Leasing became so widespread that even local governments issued leases.[88] Perhaps more surprising, the International office of the UMWA, with John L. Lewis's assured approval, provided the financial backing for the Coaldale Coal Company's lease of the

Figure 36. UMWA District 1 officials August J. Lippi and Leonard Statkiewicz going to trial for income-tax evasion, 1963. The defendants were found not guilty. (Courtesy of Scranton Times*)*

Figure 37. Coaldale breaker, Lehigh Coal & Navigation Co., Coaldale, PA. Leased to the Coaldale Coal Company in 1954. (Courtesy of George Harvan)

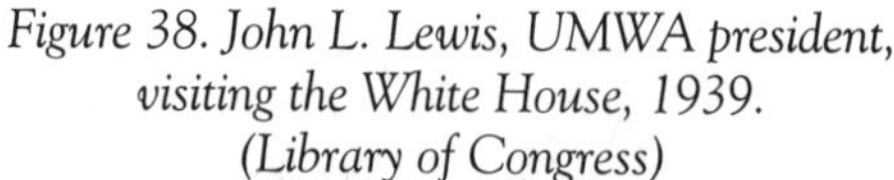

Figure 38. John L. Lewis, UMWA president, visiting the White House, 1939. (Library of Congress)

Coaldale breaker from the Lehigh Coal & Navigation Company in the middle coal field in 1954.[89] As leasing permeated northern-field anthracite and spread to the other fields, former subcontractors as well as investors, speculators, public officials, and persons with no mining background scrambled to garner the agreements. A prime example of the latter was evident in former Pennsylvania Governor John Fine's partnership with Santo Volpe and the Biscontini family in the Newport Excavating Company.[90] There was money to be made in leasing, and hundreds of stakeholders—big and small, famous and common, with or without mining experience—sought to gain a share.[91]

~ ~ ~

The fall of anthracite was clearly a complicated affair. A combination of factors related to investment decisions, technology, innovation, geology, production and overproduction, as well as tenancy and its corruptions, all combined with market conditions to bring down the hard-coal industry. For the workers, the powerful forces undergirding the decline were essentially beyond control. Theirs was, by necessity, a reactive position. They could flee, fight, or accept the policies and practices of a management system and corporate culture intent on attaining high profits in an industry beset with myriad problems.[92] Many chose to fight and their militancy was on full display in the northern field. Collective actions to include wildcat strikes, marches, rallies, alternative unions, and (on occasion) violent behaviors were their main responses to the hostile environment. The consternation among the colliers was clearly manifested in the field's many union movements. Between 1907 and 1939, the workers formed no fewer than five dual unions to challenge the UMWA, beginning with the Industrial Workers of the World and ending with the Progressive Miners Union.[93] (Figure 39)

- 1907–1916: Industrial Workers of the World (IWW)
- 1928–1931: National Miners Union (NMU)
- June–July 1928: Anthracite Mine Workers Union (AMWU)
- 1932–1935: United Anthracite Miners of Pennsylvania (UAMP)
- 1938–1939: Progressive Miners Union (PMU)

Figure 39. Five dual-union movements in the northern anthracite field, 1907–1939.

The northern field's Italian mineworkers were among the most vigorous participants in the dual-union movements. Yet, theirs is one of the untold stories of anthracite ethnic and labor history. Why has this been the case? The next chapter builds upon the analyses thus far presented about Italian mineworkers at the Erie Coal Companies and looks further into their participation in other anthracite conflicts, as well as their place in the larger hard-coal narrative.

PART IV

Postscript: Italian Mineworkers in Anthracite

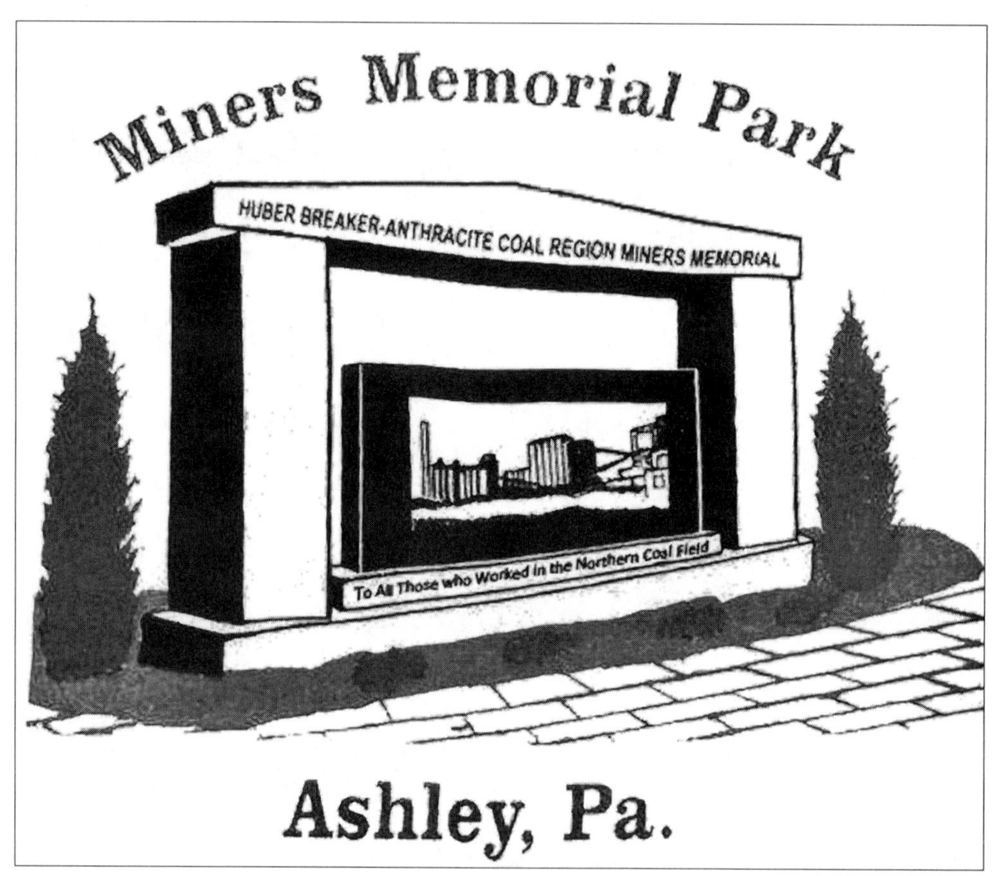

Miners Memorial at Ashley, Pennsylvania

CHAPTER EIGHT

The Italian Community on Strike: A Forgotten Chapter in Anthracite Labor and Ethnic History

Possibly the most dangerous element of the anthracite population dominated by this spirit [of unionism] are the lower classes of Sclavs [sic] and the Italians.

— Rev. Peter Roberts, *The Anthracite Coal Industry*, 1901

You are badly treated, my friend? I quite believe it; indeed, I can see it. Well, go to Argentina and sell potatoes, or to the mines of Pennsylvania. There you will grow rich, like the rest of your compatriots. Then return and send your sons to university; let them become advocati [legal advisors] and members of Parliament, who shall harass into their graves these wicked owners of the soil.

— Norman Douglas, *Old Calabria*, 1915

See now, the Italians, they knew this was all wrong, they knew it was going to ruin the coal company [PaCC]. ... The Italians tried to warn them [management and UMWA officials] but they wouldn't listen.

— Lewis Casterline, Mineworker and PaCC Superintendent, 1970

Introduction: The Slavic Community on Strike

Victor Greene, in *The Slavic Community on Strike: Immigrant Labor in Pennsylvania Anthracite* (1968), challenged many existing notions regarding the participation of eastern Europeans in the anthracite labor movements of the late nineteenth and early twentieth centuries.[1] He argued against the predominant view that Polish, Slovak, Ukrainian, Russian, and other Slavic immigrants (among whom he included Lithuanians and Hungarians, though they are not Slavs) hindered organizing campaigns because they did not support the unions or their strikes. After a review of numerous period documents he concluded: "... it is chimerical to ascribe the delay in unionization to the Slavs."[2]

While economist John R. Commons and many contemporary journalists maintained that Slavic immigrants were indifferent to, or attempted to undermine, the strikes of 1877–1878, Greene found no evidence to support the claim.[3] He pointed out that Knights of Labor President Terrence V. Powderly wanted to recruit the newcomers in the 1880s and was successful in organizing some Knights-sponsored assemblies such as the lodge in Shenandoah in the Lehigh Region, headed by Fr. John Wolansky, anthracite's first Ukrainian priest.[4] During the same period, the Miners and Laborers' Amalgamated Association showed a similar interest in the eastern Europeans. However, mainly because of feared economic competition, the Knights remained a mainly Irish association while British and German mineworkers dominated the Amalgamated.[5] Nonetheless, when the organizations called a strike late in the decade, the newcomers were as supportive as any other group. They showed even greater dedication to the United Mine Workers of America (UMWA), which arrived in the hard coal regions during the mid-1890s: Greene determined that the union succeeded because of "the predominant determination of the east European, rather than the Anglo-Saxon community."[6]

Harold W. Aurand reached the same judgment in his book, *From the Molly Maguires to the United Mineworkers* (1972). Like Greene, he contended that Slavic immigrants were not to blame for the defeat of the Knights and the Amalgamated unions and added that the Italians joined the eastern Europeans in holding the strikes of 1887–1888:

> Mine operators adopted two strikebreaking strategies—reopening and starvation. As labor leaders feared, the operators tried wooing the new immigrants back to work, but to the surprise of everyone, the Slavs and Italians supported the strike. Many immigrants even left the region rather than take part in strikebreaking.[7]

The operators proceeded to recruit workers from other lands who they hoped would be more compliant. The Pardee Coal Company in the eastern middle field brought in a group of Italians during the strike of 1887 but, as discussed later, the company's existing Italian labor force prevented the blacklegs from entering the mine.

The main argument regarding the Slavs' supposed indifference to unions related to what social scientists later called the *split labor market hypothesis*. The concept held that immigrants divided labor markets into an upper sector (established employees) and a lower sector (immigrants) because they were willing to accept lower wages and inferior working conditions.[8] Congregationalist minister and sociologist Peter Roberts articulated the idea in his book *Anthracite Coal Communities* (1904):

> These peoples furnished the cheap labor ... and supplied the operators with men willing to work under conditions which labor of a higher grade resented. The Sclavs [*sic*] represent possibly the lowest grade of European workmen that can be imported [and they have] displaced the Anglo-Saxon, the Celt and German.[9]

The idea implied that immigrants had no experience with collective action and were easily swayed by the operators who often paid their way to the coal fields and provided jobs and housing.[10]

Greene found no support for the hypothesis in anthracite. Although the Slavs encountered considerable prejudice and discrimination, they wanted the same wages and working conditions as fellow mineworkers from the established ethnic groups. They opposed the individual contract and supported collective bargaining because they recognized injustices when they saw them; they would not pas-

Figure 1. Imported Italian labor threatened established workers. (www.latinamericanstudies.org/immigration/italian-nativism)

Figure 2. John Mitchell, President, United Mine Workers of America. (John Mitchell, Organized Labor, *1903, frontispiece)*

sively accept the causes or consequences; they quickly learned from the British and German mineworkers who had long advocated labor's rights; and they viewed concerted action as a means to improve their situation. By 1900, President John Mitchell and his core of well-trained immigrant organizers had gained their trust and the UMWA became an integral part of the ethnic and work culture. Greene characterized their loyalty as based on "sociological motivation."[11] The acceptance of the union was seen in the location of Mitchell's picture next to images of Jesus and the Holy Family in the living rooms of Slavic mineworkers.[12]

Greene helped rewrite a research record that sociologist Michael A. Barendse described as having "an anti-Slavic bias among the investigators from 1890 to at least 1960."[13] Rev. Roberts was among the scholars who expressed such partialities. He cast aspersions on the new arrivals' moral and intellectual capabilities while advocating alcohol prohibition as a necessary way to improve their lot.[14] Although Roberts acknowledged the Slavs' role in building the UMWA, he drew different conclusions than Greene.[15] He affirmed that labor had a right to "combine" on behalf of their interests because the owners had done so; however, he was much less sanguine about the workers' tactics as well as the short- and long-term consequences of unionization. He decried the immigrants' violent propensities, especially when fueled by alcohol, and condemned all strikes as belligerent, selfish, and destructive. They threatened not only management's need to earn a profit, but the well-being of the whole community:

> The pecuniary loss from strikes is great, but that is limited and can be made good; the loss in moral sympathy and social harmony perpetuates itself and its miasmatic effect lingers on for a generation.[16]

In the final analysis, he viewed labor's crusades not as unified behavior in the pursuit of justice and fairness, but as anti-social agitation.[17]

Figure 3. Loading a coal delivery wagon, Lackawanna County, circa 1904. Wording on the side reads "The Hartt & Adair Co." and includes a Lackawanna Coal logo. (Robert J. Taylor photograph, courtesy of Wisconsin Historical Society)

Figure 4. Loading hard coal in Milwaukee, WI, circa 1915. (Robert J. Taylor photograph, courtesy of Wisconsin Historical Society)

Italians in Early Anthracite Labor History

What role did Italians play in the development of anthracite unions? Greene and Aurand devoted little attention to the subject. Yet between 1880 and 1920, Italy provided the largest number of immigrants to the United States, and second-largest numbers to the anthracite regions between 1890 and 1900 after the Poles.[18] According to the U.S. Census of 1900, the anthracite counties were home to 9,958 sons and daughters of Italy. In 1917, they comprised eight percent of the workforce, surpassed only by the Poles at 13 percent.[19] As discussed in previous chapters, the majority of Sicilian arrivals in Pittston and Dunmore had previously worked as miners and laborers in their homeland's sulfur industry.[20]

Because many of the first Italian immigrants entered the coal fields as strikebreakers, "Italian-born workers were the bane of union organizers everywhere in the United States, particularly in Pennsylvania."[21] John Siney, president of the Workingman's Benevolent Association, complained about them in the late 1860s, as Anthony F.C. Wallace noted: "Siney was very much concerned about the importation of 'Italian cutthroats' to break a miners' strike in western Pennsylvania, and his language was sprinkled with words like 'fight,' 'war,' 'outrage,' and 'death.'"[22] Of course, the Italians were not alone. Thousands of eastern Europeans began their mining careers as scabs. It was clear that the operators did not care which group extracted the coal:

> It made no difference to [the coal operators] *which* racial or national group worked in the mines. "We are determined to put an end to strikes," explained an operator. "[I]f the Italians fail we intend to import Swedes to do the work for us, and if they do not meet our expectations then we shall bring a lot of Negroes from the south."[23]

Historians Richard D. Grifo and Anthony F. Noto were not surprised by the result: "Is it any wonder that coal miners denounced Italian, as well as Polish and Hungarian workers, as degraded and servile lackeys who frustrated efforts to improve wages and working conditions?"[24]

However, the Italian and Slavic reputations as strikebreakers had been transformed by the 1890s. Certainly by the aforementioned strike of 1887, Italians had freely joined the picket lines with Slavs and other mineworkers.[25] When Italian employees at the

Figure 5. Coal delivery wagon crossing the Milwaukee River, circa 1904. (Robert J. Taylor photograph, courtesy of Wisconsin Historical Society)

Pardee Coal Company's Hollywood mine in the Lehigh Region learned that the bosses had contracted for 100 Italian strikebreakers, they so threatened the blacklegs when they arrived that only 14 showed up for work on the first day. Even this number angered the Italians. One news report quoted their response [in Italian-accented English]: "'We killa-them!' they cried, 'No blacka-leg us.'"[26] The next day, none of the recruits appeared at the mine. There were many other instances of Italian labor activism, including the potentially deadly Battle of Archbald in 1896, where a group of laborers called a strike over a subcontracting-related wage grievance.

Figure 6. Milwaukee Western Fuel Company delivery trucks, 1935. Wording on the door reads: "D L & W Scranton Blue Hard Coal." (Courtesy of Wisconsin Historical Society)

The Battle of Archbald

On December 5, 1896, 60 Italian laborers who were employed robbing pillars at a Forest Mining Company pit in Archbald, Lackawanna County, walked off the job. A *Wilkes-Barre Record* story headlined: "Italians Out on Strike; Exciting Scenes in Archbald on Saturday."[27] In an episode that could have led to the type of violence that took place 50 miles away in Lattimer the following year (discussed later), The Battle of Archbald grew out of the workers' refusal to accept an indirect pay cut.

The problem arose when the company reorganized production by replacing a single supervising subcontractor with three new ones. The original subcontractor had received his regular wage plus 6.5 cents per ton for the workers' output, but the company decided to give *each* of the new subcontractors (William Matthews, M. Baltus, and George Schemhell) 6.5 cents per ton, thus tripling the amount taken from the workers' pay. The result was a sizeable income loss for the laborers and they mobilized against the perceived injustice:

> The Italians stated ... that all they wanted was to have the middlemen removed from receiving that 6.5 cents, which those men were receiving for themselves. They claim that the [sub]contractors do nothing to entitle themselves to any fruits of their labor.[28]

The protestors rejected the company's assertion that all three subcontractors were necessary for supervision and safety. Indeed, they claimed that their new bosses often shirked responsibility for safety: "When there is any danger the Italians say these men run away and leave the Italians to load the coal."[29]

The mine foreman, Henry Chapman, and assistant foreman, Thomas J. Keilly, tried to settle the dispute but failed. A solution then rested with company superintendent E.S. Jones and head foreman E.A. Jones, who arranged a meeting with what the *Wilkes-Barre Record* termed "the mob" of strikers. It was a peaceable conference but the workers did not yield, prompting Superintendent Jones to conclude that "nothing but force would bring the Italians to their senses."[30]

Jones solicited the assistance of the Barrington & McSweeney detective agency and secured the appropriate warrants from Squire (Justice of the Peace) Gildea of Archbald. On Saturday, December 7, the superintendent joined Gildea, Constable McHale, and a large contingent of deputized agency men for the short train ride to the mine. Their purpose was to arrest the ringleaders and break the strike. Area newspapers backed the company's position. The *Wilkes-Barre Record* reported that the train stopped a short distance from the mine to unload a box of valuable cargo: "It was afterwards learned that the box contained a large number of repeating rifles, which constituted the perfect arsenal to be used on the strikers should they offer any resistance to the law's mandates."[31]

Upon arriving at the pit, "a swarm of about 200 men" confronted the posse. When Gildea tried to arrest one of the leaders, a scuffle ensued. The deputies reached for their revolvers, whereupon the boycotters scattered in all directions. A large number reassembled on a nearby hill and, within minutes, another confrontation occurred. With pistols drawn, Constable McHale led the Barrington & McSweeney men in a charge up the hill. The Italians held their ground and drove the attackers back with a barrage of sticks and stones. Out came the repeating rifles and the private army took aim at the dissidents. The sight of the weapons again caused the strikers to disperse, but not before three of the leaders—James

Most, Saboth Most, and Ralph Dominick—were caught and arrested. One of them possessed a bludgeon that he intended to use on the officials, according to news reports: "Constable McHale climbed the hill, drew his revolver, and confronted the leader. He knocked the bludgeon from his hand and, taking him by the collar, took him from his supporters."[32] Fortunately, bullets had not yet been fired by either side. The strikers regrouped at "the castle," a large company house where many resided with families.

Figure 7. "The Castle," a company-owned rooming house in an unknown anthracite town. (Peter Roberts, Anthracite Coal Communities, *1904, p. 128)*

Shortly thereafter, a group of irate Italian women paraded out of the dwelling, confronted the authorities, and berated them with threats, epithets, and brickbats. The clash did not lead to violence or injury and, after a short period, the women returned home. The troopers followed them and considered storming the house but cooler heads prevailed and they retreated. The officials and deputies planned to return in two days, on Monday, December 9, to secure the building and eject the residents. The community braced for another confrontation. One report fanned the flames of anxiety: "There is no telling what the outcome will be as they [Italians] are all armed, and judging from their conduct Saturday there certainly will be conflict."[33]

In the contest for public opinion, the company reemphasized the subcontractors' role in safety and highlighted the Italians' unreasonableness and intractability. The immigrants should have nothing to complain about, the managers said, because they had good jobs with good pay.[34] However, the Italians saw the issue not as a matter of thankful employment but of fair wages.

The company scheduled the mine to restart on December 9 at 6:00 a.m. Perhaps not by coincidence, it was at the same date and time that constabularies planned to siege the castle. When the officials and deputies arrived at the rooming house, they found no one present. The workers and their families had taken refuge in the woods. Still, the *Wilkes-Barre Record* feared "... that the quiet is a calm before an impending storm and the strikers may return for blood."[35]

The decisive battle never occurred. The powerless immigrants were "badly frightened, and, as they have lost their leaders, they were not likely to collect enough courage to face the men with the Winchester rifles, who are prepared to shoot them down."[36] Constable McHale later caught up with and arrested five other organizers and took them before Squire Gildea, who committed them to the county jail where they awaited trial and certain conviction.

Realizing their predicament, a group of strikers met with Superintendent Jones on December 9 and said they would resume mining if the charges against the leaders were dropped. Jones allowed 25 of the 60 men to work on the afternoon shift, which began at 3:00 p.m. However, he did not drop the charges and, when the situation calmed a few days later, most of the strikers were fired and thrown out of company housing.[37]

The Lattimer Massacre

The following year, on September 10, 1897, Italians actively participated in the strikes that led to one of the nation's most deadly labor massacres, at Lattimer in the eastern middle coal field near Hazleton, Luzerne County.[38] The first in a series of work-related conflicts preceding the tragedy began on August 12, 1897, when Gomer Jones, the recently appointed superintendent of the Lehigh & Wilkes-Barre Coal Company (L&WBCC)'s Honey Brook division, imposed new work rules for mule drivers at the Audenreid Colliery. The rule change meant that the drivers, most of whom were Italians and Slavs, had to spend two additional hours each day taking their animals to a centralized barn. Jones was convinced that mule-tending practices and numerous other inefficiencies plagued the workforce and he was determined to implement a more disciplined regimen. "When I give orders I expect them to be obeyed," he asserted.[39]

Two days later, the mule drivers at the company's Honey Brook Colliery went on strike. On the second day of the walkout, an irate Jones advanced upon the picketers with a crowbar and struck a young boy named John Bodan. The strikers leaped on Jones and beat him until a foreman pulled him away. Bodan filed charges and police arrested Jones for assault. He was set free on bail.

Figure 8. A mule barn at a colliery in Plymouth, PA. (Courtesy of Plymouth Historical Society)

News of the incident spread throughout the area. Two thousand employees at L&WBCC walked off the job in sympathy with the mule drivers and said they would maintain the strike until the company fired Jones. On August 16, nearly 350 protestors marched on six of the company's pits and shut them down, idling 3,000 men. The *Wilkes-Barre Times* reported that the Italians and the Hungarians (a label often used for Eastern Europeans) led the movement.[40] To plan for further demonstrations, the strikers elected a steering committee headed by Joseph Keshilla, a Hungarian, and Nille Duse, an Italian. They also sought assistance from the UMWA.

On August 19, the Hazleton *Daily Standard* indicated that, although the walkout began with Italians and Slavs, most other groups backed the action. The angry workers formed another committee with broader representation with Pasco Delecto representing the Italians, John Burska the Hungarians, Andrew Damian the Magyars, and Alex McMullen and William Hopkins the English speakers.[41] To protect their property, the company called in the state-sanctioned Coal & Iron Police whose members rode horses and carried pistols and Winchester rifles. In compliance with state law, the coal operators were required to arm, pay, and direct the police force.

The protestors resented not only Jones's policies and conduct, but also the general state of working conditions, wages, and living arrangements. Immigrants were especially dissatisfied with Pennsylvania's recently passed Campbell Act, which imposed a three-cents-per-day tax on non-citizen employees, a fee the operators passed on to the newcomers as payroll deductions. The men demanded not only an end to the tax but also higher wages, safer mines, and relief from patch-town burdens such as company stores and physicians. The boycott subsided when L&WBCC's head superintendent, Elmer Lawall, revoked Jones's decree and promised to investigate his conduct. He also raised workers' wages 10 cents a day.

On August 21, a new round of regional disorder commenced when 2,000 men at the Coleraine Colliery of the Van Winkle Coal Company struck over the Campbell Act. The strikers marched between the company's mines and shut them down. After learning that Mr. Van Winkle had offered a pay raise in exchange for a work resumption, 1,000 L&WBCC colliers demanded the same treatment. To reinforce their petition, they planned a demonstration for September 3.

On the appointed day, 5,000 workers staged one of the largest marches the area had seen. One of the foremost leaders was a "burly Italian" who addressed the gathering in a loud and enthusiastic voice. He threatened Superintendent Lawall and waved a large knife to show his determination. A *Wilkes-Barre Times* reporter who witnessed the proceedings wrote: "... the burst of Italian eloquence tended to invigorate the crowd."[42] The strikers prepared for further movement, whereupon, "The big Italian took matters in hand. [In Italian-accented English, he pronounced:] 'We getta do move on, and closa up the district.' The protesters responded eagerly and marched forward."[43]

When the column reached the No. 1 Colliery in Jeansville, the breaker whistle sounded a warning. At the sight of the strikers, the breakermen fled the premises. At that point, "the Italians hooked a plank to the whistle leaving it blowing to announce the victory to the surrounding countryside."[44] The phalanx soon dropped to 3,500 men and the leaders pondered whether they should continue to the next colliery in Hazleton about six miles away. Just then, a courier arrived and spoke to "the Italian leader with bared head":

> "Turn back! Turn back!" he shouted. "Why?" asked the leader. "You know not the danger awaiting you," exclaimed the breathless arrival. "Tell us!" shouted a thousand throats. "Tell us!" "Hark it then! Not over two miles from here on the outskirts of the city of Hazleton now stand the police force."... The leader thought. He slowly raised his head and with a voice as steady as the earth upon which he stood, exclaimed, "I gotta da right! I am-a American citizen. I have my papers. They cannot stoppa us. Forward!" He pulled his naturalization papers from his pocket and waved it aloft. The "army" revived. Enough! They were protected and with one throat the vast "army" yelled: "On to Hazleton."[45]

As the marchers approached the city, they were met by a police captain and three deputies who told them to disperse. In response to the captain, who maligned them as "agitators of the peace,"

> the leader of the strikers with a voice equally as resolute as the captain's exclaimed: "Getta outa de way. We noa stoppa." The undismayed captain replied: "Disperse or I'll run ye in!" "Ha ha," laughed the leader of the strikers as he pulled his papers from his pocket. "I am-a Americano citizen. I a defia you. We a goa through your-a city."[46]

The officer quickly assessed the situation and convinced the strikers to accept a police escort around the city so they would not get lost on the way to their destination, the Hazle Colliery. They agreed and proceeded to the outskirts under police direction.

Figure 9. Postcard of the Cranberry Colliery, Hazleton, PA, 1906.

The breaker whistle sounded as the throng approached the colliery. Mining operation immediately ceased, prompting cries of "Victory! The strikers again had won the day."[47] The campaigners' ranks continued to swell as they carried on to the Cranberry and Hazle Brook collieries where, again, they were effective in calling out the workers and halting production. In all, the picketers had closed four L&WBCC operations. By the end of the day, their ranks had surged to an estimated 10,000.

The growing labor rebellion alarmed not only L&WBCC managers, but also other middle-field producers who were mainly family-owned independents such as the Van Winkles, Pardees, Markles, Coxes, and Fells. They demanded that Luzerne County Sher-

Figure 10. Postcard of the Hazle Colliery, Hazleton, PA, 1915.

iff James Martin end the boycotts and establish order. Martin assembled a posse of 150 deputies with the great majority being of Anglo-Saxon heritage. Many had advanced education, including several engineers.[48] Martin and his advisors pondered a strategy.

On September 8, a "mob of young Italians and Huns" descended on the New Ebervale Coal Company's colliery near Hazleton. They assaulted Superintendent John Scott and ordered him to terminate operations. "I informed them this request would not be complied with," he later said, "and when they attempted to beat me, I drew my revolver." However, he immediately had three guns pointed at his chest. "About this time," he continued, "the real leader appeared and said they intended to win the strike if they had to apply the torch to every town in the region."[49] The same day, other militants forced closings at Van Winkle's Coleraine operation, Cox's Beaver Meadow mine, and other pits.

Figure 11. Hazleton-area strikers on the march. (Wilkes-Barre Times, *September 7, 1897)*

On September 9, a delegation of Lattimer Colliery men conferred with strikers at Pardee's Harwood Colliery, ten miles away, who had recently formed a UMWA local under the direction of organizer John Fahy. The Harwood village and colliery were populated largely by Slavs and Lithuanians, while Lattimer had mostly Italian residents and workers. The men said that if the Harwood men marched on their workplace they would walk out and thereby shut down the last of Calvin Pardee's working operations:

> After his father's death in 1892, Calvin Pardee, then 51 years old, took over the mines at Lattimer and Harwood, two distant colonies about 10 miles apart, and kept in stride with the other mine owners. He demanded more and more coal at less cost. He filled the [company] houses at Lattimer largely with Italian immigrants, and those in Harwood with Slovaks, Poles and Lithuanians. With a wholesale mixture of nationalities he felt that there would be less chance of a consolidation of the working men against his interests. The company stores thrived. Wages were steadily cut.[50]

Figure 12. Coal operators in the eastern middle field. Clockwise from upper left: George Markle, John Markle, Eckley B. Coxe, and Ario Pardee, coal operators in the eastern middle field. (Courtesy of Schuylkill County Historical Society)

The Harwood men agreed to the plan and the demonstration was set for the next day.

The morning of September 10 saw a corps of 400 Harwood strikers set out on the trek to Lattimer. When they reached Hazleton, Mayor Justus Altmiller refused to issue a parade permit so they had to circle around. Upon arriving in West Hazleton, they were met by Sheriff Martin and a group of 86 deputies "who saw the strikers as enemies of public order and the wellspring of lawlessness."[51] The lawmen harassed the picketers but could not halt the procession. Martin and his men then boarded a trolley for the short trip to Lattimer, where they formed a blockade across the road. As the demonstrators approached, the sheriff ordered them to stop. Although the front lines complied, those in the rear continued walking. In the midst of the pushing, shoving, and falling that resulted, the deputies opened fire. Many of their bullets entered the backs of the fleeing, unarmed marchers. The immediate death toll was 19, with 32 (some sources say 37) wounded. The deceased included six Poles, five Lithuanians, and eight Slovaks.[52] Six others later died in the hospital, bringing the total number of victims to 25. Lattimer was the bloodiest labor massacre of the nineteenth century.

Figure 13. Deputies firing on marchers at Lattimer. (Philadelphia Inquirer, *September 12, 1897)*

The slaughter at Lattimer was called "A Slavic Massacre" because virtually all of the victims were Slavs. More than a few anthracite mining tragedies have been associated with the ethnic groups that suffered the greatest losses. The Avondale disaster of 1869, in Plymouth Township, was termed "A Welsh Tragedy" because 69 of the 110 casualties were Welsh men and boys. The Twin Shaft mine collapse at Pittston in 1896, in which most of the 58 dead were of Irish ancestry, was deemed "An Irish Disaster."[53]

Figure 14 (left). The Lattimer breaker, circa 1892. (Carl Corlsen, Buried Black Treasure, *1954)*

Figure 15 (below). Identifying the bodies of Lattimer victims.

(Philadelphia Inquirer, *September 12, 1897)*

Figure 16 (bottom). Luzerne County Sheriff James Martin. (Philadelphia Inquirer, *September 12, 1897)*

IDENTIFYING BODIES IN THE STABLE OF UNDERTAKER BOYLE
Drawn on the spot by an Inquirer Staff Artist.

Sheriff Martin and 19 deputies were indicted for the murder of one victim, Mike Cheslak. The trial began on February 1, 1898, in the Luzerne County Court House in Wilkes-Barre. Nearly 200 witnesses testified over five and one half weeks. The jury rendered a "not guilty" verdict. The defendants were judged as not having malicious intentions in the conduct of their duty, and the "riotous condition" fostered by the strikers was viewed as a key mitigating circumstance. The decision triggered outcries from victims' families, fellow mineworkers, clergy, and ethnic newspapers. One protest came from *La Questione Sociale*, an Italian-American journal that editorialized its disbelief "… that the American middle class could reach the point to overlook and approve of infamy. It reached the point where the word justice is meaningless, brutality and cruelty are tolerable."[54]

Italians were not among the slain, although some may have been wounded.[55] The Italians joined other mourners in the many funeral ceremonies, including a procession for 12 victims in Hazleton:

> Through a fine, chilling drizzle and a heavy mist, a brass band playing the death march led the cortege. With muffled drums it preceded the three biers, the families of the deceased, and a thousand paraders in grocery and beer wagons past culm piles crowded with fascinated onlookers. Prominent in the snake-like procession wending its way along the road was a double line of men and boys of St. Joseph's Society wearing red, white, and blue sashes on their shoulders and crepe badges over their hearts. The brightly colored uniforms of the participating *Societa Italia-Americano* contrasted with the drab "In Memoriam" patches on their arms. Mine-workers and their families, dressed in their best, completed the train ride and, reaching the church, filled it to capacity. ... After the Requiem Mass the twelve were buried in St. Stanislaus' cemetery before a throng of six thousand.[56]

Although Lattimer was undoubtedly a Slavic tragedy, the historiography of the events preceding the catastrophe has devoted scant attention to the Italians' participation. The accounts provided by authors such as Greene, Aurand, Blatz, Miller and Sharpless, and Dublin and Licht illustrate the point.[57] In the few cases where they have been recognized, the Italians have been given a low profile and nowhere has their role been placed in a larger context of Italian and Italian-American labor activism during the 1890s (discussed later). Of course, it can be argued that a detailed analysis along these lines was beyond the scope of existing studies. Yet, the historical construction of Lattimer as a wholly eastern European tragedy, however understandable, illustrates a problem common to many conventional formulations: they can exclude important actors which, in turn, can obscure broader understandings.[58]

An example of the historiography's persistence was seen in the scholarly papers presented as part of the 100th anniversary commemoration of the massacre in 1997, as well as in the special issue of *Pennsylvania History* that published the papers in 2002. The two-day remembrance program in September, 1997—which included a re-enactment of the march, an historical-marker dedication, an eastern European dinner at which UMWA President Cecil Roberts spoke, and an academic conference—was sponsored by the Pennsylvania Historical and Museum Commission. None of the events or speakers used the word "Italian," and the same can be said of the research papers, although a few referred to "Southern Europeans."[59] Thus, the most recent commemorations and historical investigations have continued to exclude the ethnic group that helped instigate the period's protests, including the Lattimer march itself.[60]

Italian-American Narratives

The established canon raises two important questions: Why have Italians been neglected in telling the story of anthracite labor? Why has there been no scholarship equivalent to Greene's treatise on the Slavic workers?[61] One answer can be found in the concept of the *narrative*. A narrative is a story "constructed" by an author who uses interpretive strategies to convey messages that create particular meanings. Narratives are shaped by an author's personal values and ideologies as well as the broader historical circumstances and social conditions that form the basis of knowledge in any given period. The shifting ground between author, circumstance, and meaning explains why narratives can change over time. As sociologist Peter Berger observed: "The past is malleable and flexible,

Figure 17. Italian immigrants aboard ship. (www.latinamericanstudies.org/immigration/italian-nativism)

changing as our recollection reinterprets and re-explains what has happened."[62] Historian Raphael Samuels concurred: "History is a house of many mansions and its narratives change over time."[63]

A study of the narratives surrounding any group requires answers to four questions: What are the existing narratives? Who constructed them? Why have certain narratives been included? Why have others remained untold? The following section will argue that three narratives have dominated the contemporary understanding of the Italian-American experience in the U.S.: (1) *The Unfit Immigrant, Worker, and American*, (2) *Hard Work and Assimilation*, and (3) *The Organized Criminal*. All three narratives were constructed by numerous "authors" writing during the nineteenth and twentieth centuries, including some who wrote about anthracite's workers and communities.

A fourth narrative, however, has been largely excluded from the discussion: *Workers' Rights and Economic Justice*. Although some researchers have rediscovered the narrative in recent years, it has remained virtually forgotten within the Italian-American community, the larger society, and anthracite historiography. The following discourse has two main purposes. The first is to dissect the social, intellectual, and economic roots of the predominant narratives that have shaped (and often clouded) the ethnic and labor history of Italian-Americans and, by extension, Italian-American mineworkers. The second is to consider the fourth, obscure, narrative associated with workers' rights and justice. It is within the fourth narrative that a deeper understanding of the anthracite labor wars—along with the dogged involvement of Italian mineworkers—can be found.

Narrative I: The Unfit Immigrant, Worker, and American

A few years after the Lattimer Massacre, Rev. Peter Roberts reported that a single Italian immigrant facilitated the closure of a colliery in anticipation of the UMWA's first general strike in 1900:

> The workingmen in a shaft in Lackawanna County held a meeting to discuss the situation on September 8, ten days before the strike was ordered. The "foreigners" said "strike now"; the English-speaking element, which was in the minority, argued that there was no order issued, and tried to persuade them to keep at work until the order came. The question was decided by an Italian swinging a revolver around his head and shouting "Strike, Strike." The shaft was shut down a week before the general strike.[64]

On another occasion at an unnamed colliery, Roberts told how a group of Italians, purportedly members of a criminal gang, dominated the labor force:

> Twelve Italians, said to be members of the Mafia, held all the colliery in terror, and nothing could be done unless endorsed by them. Anglo-Saxons know how to slug a "scab," but the "foreigners" use the knife and revolver.[65]

Roberts further stated that when leaders called the strike, the workers—Italians and Slavs among them—rose at dawn and vowed to guard the mines and use force if anyone tried to enter.[66]

Rev. Roberts decided that the Italians and Slavs were "possibly the most dangerous element of the anthracite population" because they were "dominated by this spirit" of unionism. He told how the English-speaking workers had "respect for personal rights, even when unionism appears in its most rampant form, but some Slavs and Italians pass beyond all restraint."[67] As further evidence, he discussed one Luzerne County mine where the immigrants comprised the majority and, consequently, controlled the local union. Even though most of the English-speaking men endorsed the UMWA, they "held different views from those of the Sclavs [*sic*] and Italians. When the Anglo-Saxons expressed their opinions in a meeting of the union they were thrown out."[68] Roberts related similar stories of uncivil and mob-like behavior in testimony before the Anthracite Strike Commission of 1902–03.[69]

In his analysis of the early research on immigrant mineworkers Barendse concluded that Roberts was one of several scholars who relied more on preconceived notions than factual information:

> In other words, what the Americans "knew" about the immigrants was the result of expectations and assumptions internal to their own culture and having very little to do with the observable facts concerning the lives of these immigrants. ... [T]he English-speaking population seems, in this instance, to have assumed themselves once again to be culturally far superior to the immigrant workers.[70]

Barendse argued that the storylines composed by dominant group members such as Roberts, who was born in Wales, were based less on fact than on the prevailing anti-immigrant ideologies and prejudices.[71]

Because of persistent narrative biases, it is arguable that Italians have been among the most misrepresented of anthracite's immigrant groups. Some scholars have mis-

Figure 18. Italian immigrant family. (Lewis W. Hine photograph, Library of Congress)

takenly determined that they avoided mine work, with the implication that it was either too difficult or required skills that they did not possess or want to cultivate. In 1897, George O. Virtue wrote: "The Italians seem to have a dislike for underground work ..."[72] Roberts echoed the judgment a few years later: "The Italian is very rarely found working underground."[73] Ignoring the intervening decades of evidence to the contrary, historian Caroline Golab repeated the appraisal in 1977: "Italians consistently avoided underground work."[74] However, even

Figure 19. "Little Italy," West End Coal Company, Mocanaqua, PA. (Courtesy of Anthracite Heritage Museum)

Figure 20. "Worst class workers' housing," unknown anthracite town. (John Mitchell, Organized Labor, *1903, opposite p. 192)*

when they were recognized as mineworkers, the assessment was often negative. In Virtue's opinion: "They have all the vices of the Poles with but few of the virtues."[75]

The fact was that, by 1920, Italians constituted 20 percent of the American mining workforce.[76] As the present study has shown, Italians were employed at the Archbald mine and at numerous other workings in the Hazleton/Lattimer areas in the 1890s.[77] PaCC and its Erie affiliate, the Hillside Coal & Iron Company, together employed thousands of skilled and unskilled Italian workers during the first three decades of the twentieth century, and their sons and grandsons continued to mine coal until the northern field industry expired in the 1960s and 1970s. Thousands more toiled in different types of mines throughout the U.S. and in other countries during the nineteenth and twentieth centuries.[78]

The misrepresentations extended beyond mine working, however. The Italians' degenerated state as human beings became a central tenet of the narrative. The Pottsville *Daily Republican* wanted to purge the land of the debased sons and daughters of Italy as early as 1884: "It would be a good thing if there were some means of ridding America of this particular class of cattle."[79] Another affront came from the Pennsylvania Commissioner of Industrial Relations who, in 1884, disparaged Italian coal workers: "The illiteracy, turpitude, and degraded habits of this class of immigrants, innate and lasting as they are, stamp them as a most undesirable set."[80] An editorial in *The New York Times* around the same time expressed even stronger sentiments toward Sicilians:

> These sneaking and cowardly Sicilians who have transplanted to this country the lawless passions, the cutthroat practices and the oathbound societies of their native country, are to us a pest without mitigation. Our own rattlesnakes are as good citizens as they.[81]

Matthew Stanley Kemp, in his novel, *Boss Tom: The Annals of an Anthracite Mining Village* (1904), provided a derogatory literary contribution when he described the Italian quarter of a mining village as the most "dirty and squalid as also were the people and their habitations."[82] Such views harmonized with the pejoratives that had become part of everyday language and thought, as typified by a construction company manager who quipped following a fatal industrial accident in 1911: "There wasn't anyone killed except just wops."[83] Historian Robert F. Foerster sympathized with the victims of such insults:

> No one can follow the fortunes of the Italians abroad without being struck by a sort of contempt in which they are often held. "Dago," "gringo," "carcamano," "badola," "cincali," "macaroni"—how long the list of epithets might be![84]

When it came to becoming Americans, the landmark Dillingham Commission of 1911 emphasized the Italians' unsuitability for citizenship: "It is impossible to make any discrimination between races upon the score of their reputations as citizens. The fact is there are some respected citizens in all the races, except possibly the Italians."[85] The commission also highlighted the Sicilians' inveterate foreignness and their status as the most "unassimilable" of all groups. W. Jett Lauck, who conducted research for the commission on the anthracite work force, provided a similar assessment: "The Syrians and South Italians show the least tendency toward Americanization, with a slight advantage in favor of the former."[86]

The prejudices and stereotypes had ramifications for the Italians no less than for other European groups; however, the Italians experienced the most harmful and even

Figure 21 (left). A mob breaks into a New Orleans prison to lynch Italians, 1891. (E. Benjamin Andrews, History of the United States, *Vol. V, New York: Scribner's, 1912)*

Figure 22 (below). Italians being lynched in New Orleans, 1891. (www.latinamericanstudies.org/immigration/italians-lynched, 1891)

deadly consequences. Sicilian immigrants suffered 27 murders at the hands of Southern lynch mobs between 1886 and 1910, according to historian Clive Webb. Among them were 11 Italian prisoners (some allegedly Mafia members) in New Orleans who were lynched in 1891 by members of the White League, a Reconstruction-era terrorist group much like the Ku Klux Klan, after they had been acquitted of murdering the police chief.[87] The hangings brought a diplomatic protest from the Italian government, after which President Benjamin Harrison offered an apology in his state of the union address and offered $25,000 of U.S. State Department's funds to assist the victims' families.[88] In 1896, in Hahnville, Louisiana, three Italians were hanged by a mob after being jailed under suspicion of murder.[89] In 1899 in Tallulah, Louisiana, three shopkeepers and three bystanders, all Italian, were lynched by vigilantes because the shopkeepers were judged as having treated blacks with equality.[90]

The mistreatment was not confined to the south. Three Italians held under suspicion of murder were hanged by a mob of mineworkers in Walsenburg, Colorado, in March 1895.[91] Labor historian Herbert Gutman documented the killing of three Italian strikebreakers and the wounding of eight others in the bituminous fields of western Pennsylvania on November 29, 1897. The shootings occurred when the company brought in 200 scabs from the Italian Labor Company of New York. Four strikers were indicted for murder and inciting a riot. A jury acquitted them in a trial held in Greensburg, Pennsylvania, on May 10, 1875. Mine owner Charles H. Armstrong, along with the contract company superintendent, Frederick Guscetti, were indicted and convicted of a minor charge that carried a five dollar fine.[92]

Two other homicides occurred in the northern anthracite field during the Strike of 1902. Mineworkers Sistieno Castelli and his friend and brother-in-law, a Mr. Kiblotti,

Figure 23. Maltby Colliery, Lehigh Valley Coal Company, Swoyersville, PA, circa 1902. (Courtesy of Carl Orechovsky)

were slain on a foggy morning near the Maltby Colliery in Swoyersville, owned and operated by the Lehigh Valley Coal Company. The victims were mistaken for two of the strikebreakers the company had hired to keep the facility running. Castelli cried out when the assault began but to no effect as his attackers continued to beat him over the head with a club. Kiblotti tried to flee but fell to gunshots. Immediately following the killings, the assailants discovered that the men were card-carrying UMWA members employed at another colliery who were out on an early morning hunting trip.[93]

After reviewing the historical evidence, historians Stanley Feldstein and Lawrence Costello concluded: "The Italians were among the most maligned and ill-treated of the new immigrant groups."[94] Cristogianni Borsella argued that Italians have been the third-most-mistreated group in U.S. history, after Native Americans and African-Americans.[95] Fifteen different authors analyzed the problem in the edited volume by William J. Connell and Fred Gardaphe, *Anti-Italianism: Essays on Prejudice* (2010).[96]

The stereotypes and prejudices imposed physical as well as social, psychological, and economic injury.[97] Matthew Frye Jacobson documented several consequences of such enmity.[98] For example, as recently as 1992 the federal courts placed Italian-American faculty members on the list of "protected groups" at the City University of New York because of a long-standing pattern of bias in hiring and promotion. No other European group is on the list.[99] Stefano Luconi argued that persistent organized-crime stereotypes have hampered the political progress of Italian-Americans, particularly those running for elective office, such that they "are not yet completely integrated into the U.S. political system."[100] Other implications can be seen in the so-called "whiteness" literature, which has focused on the social and cultural construction of race and ethnicity. The title of the edited volume by Jennifer Guglielmo and Salvatore Salerno capsulized the question: *Are Italians White? How Race Is Made in America* (2003).[101] In another study on the topic, sociologist Eduardo

Bonilla-Silva found that "Italians seem to have a tenuous claim to whiteness, since many 'whites' mention them as examples of 'minority friends' in the interviews."[102]

Psychological repercussions were evident in autobiographical studies such as Maria Laurino's book *Were You Always Italian?* (2000). She wrote of an episode that occurred when former New York Governor Mario Cuomo pointedly asked her: " 'Were you always an Italian?' With childlike guilt, I shook my head no." She had been ashamed of her ethnic heritage and physical appearance even as an adult because of the ridicule she had suffered at the hands of friends and classmates in her youth. She remained repulsed by her own dark body hair and detested the "Guidos" or the Italian working-class boys depicted in films such as *Saturday Night Fever.* "I know all about ethnic self-hate," the Governor told her, an assessment with which she fully identified.[103] Like the women in Iranian author Azar Nafisi's best-selling book, *Reading Lolita in Tehran* (2003), Laurino had seen her personal identity and ethnic story damaged, not by a regime of fundamentalist clerics as in Iran, but by prejudicial society.[104]

Along these same lines, Salvatore J. LaGumina critically examined the social and psychological effects of films such as "The Godfather" and "The Untouchables," which promote the mobster stereotype. Sociologist Jerome Krase provided a similar critique of the animated film "Shark Tale," and the unsuccessful protest by Italian-American organizations to alter the production.[105] The popular MTV program "Jersey Shore" has been censured by Italian-American groups for promoting negative images to a younger generation.[106]

In 1914, well-known sociologist E.A. Ross of the University of Wisconsin described the low station of Italian immigrants and attributed it not to their poverty and social class but to inherent genetic and moral traits. Ross became one of many academicians urging

Figure 24. Immigrants arriving in New York circa 1900.
(John Mitchell, Organized Labor, *1903, opposite p. 96)*

the federal government to impose restrictions on the number of immigrants from southern and eastern Europe because they were diluting the nation's genetic stock and moral character.[107] University researchers, official investigations such as the Dillingham Commission, and articles published in the popular press all contributed to the passage of the Emergency Quota Act of 1921, the Immigration Act of 1924, and the National Origins Formula of 1929. The measures greatly reduced the entry of southern and eastern Europeans, as well as Asians, while favoring immigrants from northern Europe and Britain.[108]

The tradition of jaundiced research by social scientists can be seen in American sociologist Edward Banfield's study of a town he called Montegrano (Chiaromonte was its real name), located in the southern Italian province of Basilicata. In *The Moral Basis of a Backward Society* (1958), Banfield criticized the residents for numerous regressive traits including, most importantly, "amoral familism." To wit: "It is not too much to say that most people of Montegrano have no morality except, perhaps, that which requires service to the family." The pull of immediate kin so constrained the townsfolk, he contended, that it prevented them from "act[ing] concertedly or in the common good," and it became "a fundamental impediment to their economic and other progress."[109] Banfield, in effect, concluded that the villagers were hopelessly tribal and, consequently, their own worst enemies. Although he did not address organized crime, the leap from amoral and insular familists to organized crime "families" was quite easy.

Many other scholars, including Canadian political scientist Filippo Sabetti, have criticized Banfield's culturally biased perspective.[110] They have argued that he actually presented an ahistorical caricature of the town, the culture, and the people. Perhaps most significantly from an analytical perspective, he devoted insufficient attention to the fact that cultural patterns can result from political and economic conditions such as underdeveloped institutions, imbalanced power relations, unfair government policies, and systemic inequalities.[111] Moreover, research by historians such as Donna Gabaccia in *Militants and Migrants* (1988), Marcalla Bencivenni in *Italian Immigrant Radical Culture* (2011), and many others (discussed later), have documented a long history of peasants' and workers' organizations and militancy in southern Italy that required determined group action.

Banfield's conclusions have nevertheless become an established standard in social science. It informed Nathan Glazer and Daniel Patrick Moynihan who wrote *Beyond the Melting Pot* (1963), and Robert Putnam in *Making Democracy Work: Civic Traditions in Modern Italy* (1992). Both books contrasted the high civic values of northern Italians against the amoral individualism of the southerners.[112] The studies demonstrated that, despite numerous examples to the contrary, Banfield's research model had become one of the most influential statements on the culture and people of the Italian south. In the process, they have continued to buttress a narrative based on "unfitness."

A Digression: Ethnicity and the Ku Klux Klan in the Anthracite Region

Darker skins, foreign accents, and Catholic religion clearly marked the Italians as "strangers" to racist groups such as the Ku Klux Klan (KKK). Historian Philip Jenkins, in *Hoods and Shirts: The Extreme Right in Pennsylvania, 1925–1950* (1997), found that, although the "imperial heartland" of the KKK lay in the southwestern part of the state around Pittsburgh—with 60,000 members organized in 99 klaverns, or 23 percent of the state's total —hard coal country had the second-largest following in the 1920s:

Figure 25. Ku Klux Klan parade in Scranton, PA, circa 1930. (Courtesy of Lackawanna Historical Society)

> Fifty-seven of the state's 423 klaverns were located in the eight anthracite counties centered on Scranton, Wilkes-Barre, Pottsville, and Hazleton. At its height, this may have represented 30,000 to 40,000 members. Luzerne County was a stronghold, with no less than thirteen klaverns, while Schuylkill had eleven, so that these two counties had 10,000 members between them by 1924. Loucks describes Luzerne, Carbon, Lehigh, and Schuylkill Counties as a major regional center of Klan activity, an eastern parallel to the Pittsburgh region. The anthracite country became one of the eight provinces of the Invisible Empire.[113]

Wilkes-Barre, Kingston, and Carbondale had active klaverns that made regular public displays. Towns such as Scranton, Dallas, Nanticoke, Swoyersville, and Olyphant did not have klaverns but experienced various demonstrations. However, the KKK committed few, if any, overt acts of violence in the ten-county anthracite area. There were no lynchings, unlike the American South where about 85 percent of the 5,000 Klan-attributed murders in the 100 years following the Civil War occurred.[114] Rather, in northeastern Pennsylvania the organization relied on tactics such as cross burnings, nighttime rallies, and menacing parades directed mainly against immigrant Catholic groups.[115]

There was no evidence to suggest that Italians were singled out for special Klan persecution. They were principally viewed as persons from one of over two dozen undesirable nationalities.[116] Fear of and antagonism toward a generalized immigrant "other" motivated the KKK's mainly British-Protestant membership. The fear derived from the newcomers' for-

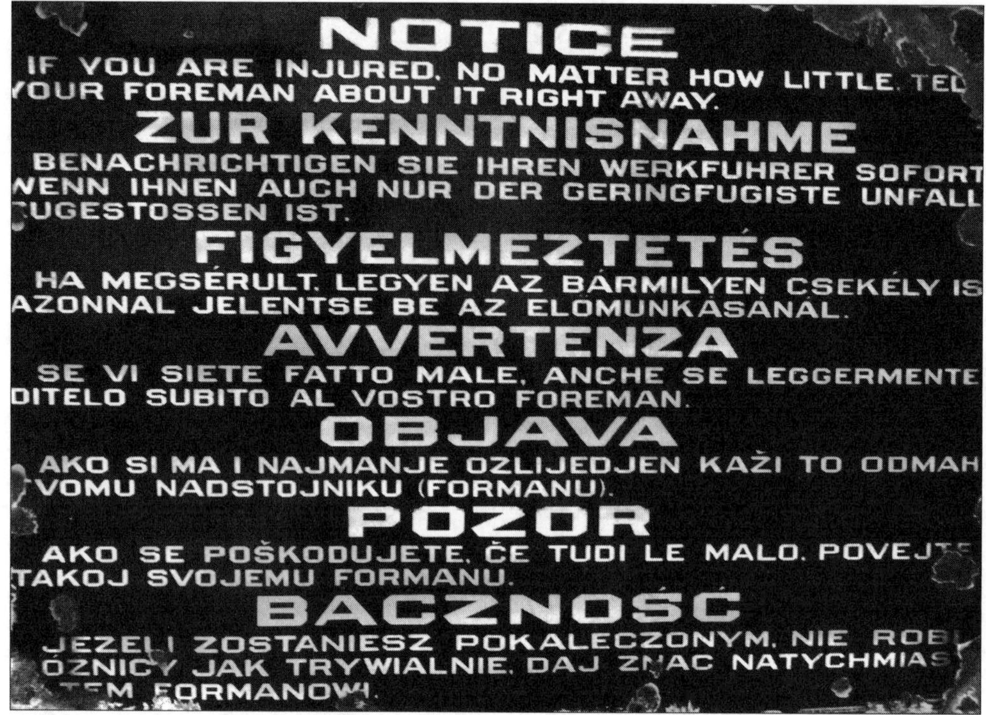

Figure 26. Mine safety sign in six languages, circa 1920.
(Courtesy of Luzerne County Historical Society)

eignness (including their religion), while the antagonism sprung from perceived labor market competition. As Jenkins wrote: "These conditions ... applied in the anthracite region in eastern and northeastern Pennsylvania, where there were long traditions of conflict between older Protestant groups such as the Welsh with Catholics like the Irish, Poles, Slovaks, and Italians."[117]

Klan recruitment relied heavily on the Freemasons and other fraternal orders.[118] Some solicitations were even made in Protestant churches.[119] Membership peaked between 1923 and 1925, fell during the latter part of the decade, and surged again such that "by 1930 Pennsylvania reported the largest Klan membership of any state."[120] No doubt the northern field's labor wars during the 1920s and early 1930s added fuel to the fires of KKK fear and mistrust. Numbers began to slide in the mid-30s but revived during 1938 and 1940, where the traditional animosities were expanded to include new obsessions with anti-unionism, anti-Communism, and anti-New Dealism.[121] By the 1940s, Paul M. Winter, the leader of Wilkes-Barre's Klavern No. 311 and a nationally prominent anti-Semite, established his base of operation in Shavertown, Luzerne County. He had spent much of his youth in Wilkes-Barre and had been active in one of the city's American Legion posts in the 1920s. He collaborated with American Nazi Robert E. Edmonson, publisher of the Nazi journal *American Vigilante Bulletin*, whose headquarters were moved from New York City to Stoddardsville, Luzerne County.[122]

Although the KKK and its members gained a justly deserved reputation for racism and bigotry, they managed to escape an organized-crime categorization despite their having exhibited behaviors that clearly qualified as such.[123] To be sure, however, while Italians were among the Klan's enemies, they were also among the perpetrators of bigotry particularly under the banner of Fascism between the 1920s and 1940s. Numerous sons of Italy, especially in larger cities such as Philadelphia, New York, and San Francisco, but also smaller cities such as Scranton and Pittston, joined the Blackshirts—who followed fascism, its policies, and leaders.[124] As the Blackshirts worked to buttress Italian nationalism among immigrants and support dictator Benito Mussolini, the group managed to avoid the level of public and governmental scrutiny that similar right-wing associations—such as the Nazis, KKK, Christian Front, and German-American Bund—experienced. In another facet of prejudice, northern Italians have long held discriminatory attitudes toward southern Italians. "Italy stops at Rome," has long been a Northern euphemism.[125] It is obvious that no single group has had a monopoly on intolerance.[126]

Narrative II: Hard-Work and Assimilation

The arrival of four million Italians in the U.S. between 1880 and 1920 (80 percent from southern Italy) led to numerous inquiries on the history, sociology, psychology, and economics of the immigrant experience. Employing a variety of theoretical and methodological approaches, scholars produced three common findings: the difficulties encountered (including prejudice and discrimination), the problems remaining (continued, though lessened, prejudice and discrimination), and the successes achieved (through hard work and assimilation). Narrative II emphasizes that latter aspect of the Italians' acculturation and integration into American society.[127]

Contrary to misrepresentations about evading hard work (including coalmining), Italians were actually employed on some of the country's most ambitious and arduous undertakings—certainly comparable to Irish and Chinese immigrants in erecting American's infrastructure, for example. By 1890, upwards of 90 percent of New York City's

Figure 27. Arano family picking berries, Philadelphia, PA, 1910. (Lewis Hine photo, Library of Congress)

Figure 28 (above). Italian-Americans working on a trolley road for the New Troy, Rensselaer & Pittsfield Electric Railroad, Lebanon Valley, NY, circa 1910. (http://www.everyculture.com/multi/Ha-La/Italian-Americans.html)

Figure 29 (right). Italian-American clam seller on Mulberry Bend, New York City, 1900. (Byron photograph, Library of Congress)

public works employees and 99 percent of Chicago's street workers were Italian. In the process of helping build (and defend) the nation in the twentieth century, the second narrative proclaims that, over the course of three generations, Italians have become an integral part of American society and have thereby shared in the American Dream.[128]

A renewed interest in the assimilationist narrative occurred with the "white ethnic" movement of the 1970s. Identity and pride, especially among citizens of southern and eastern European descent, held center stage. Founded partly as a backlash to the Civil Rights movement of the 1950s and 1960s, and the perceived social and economic gains by African-Americans and other minorities (including women), the movement fostered a greater appreciation for the first and second generations' efforts to become Americans.[129] Numerous accounts told the saga of the individuals who worked hard, "paid their dues," and climbed into middle- and upper-class life. The title of Linda Brandi Cateura's book, *Growing Up Italian: How Being Brought Up as an Italian-American Helped Shape the Characters, Lives, and Fortunes of Twenty-four Celebrated Americans*, articulates the premise.[130]

WORLD WAR II ENTHUSIASTS!

Join The Italian American Veterans Museum for

PRELUDE TO INVASION!

A WORLD WAR II RE-ENACTMENT

Saturday, October 13, 2012 ✪ 10 a.m. to 5 p.m. ✪ Casa Italia in Stone Park

Join the Allied Forces in England as they gear up for D-Day!
American, British and other Allied troops • Encampments • Weaponry • Military Vehicles
Drills & Maneuvers • Informational Displays • Tours of The Italian American Veterans Museum

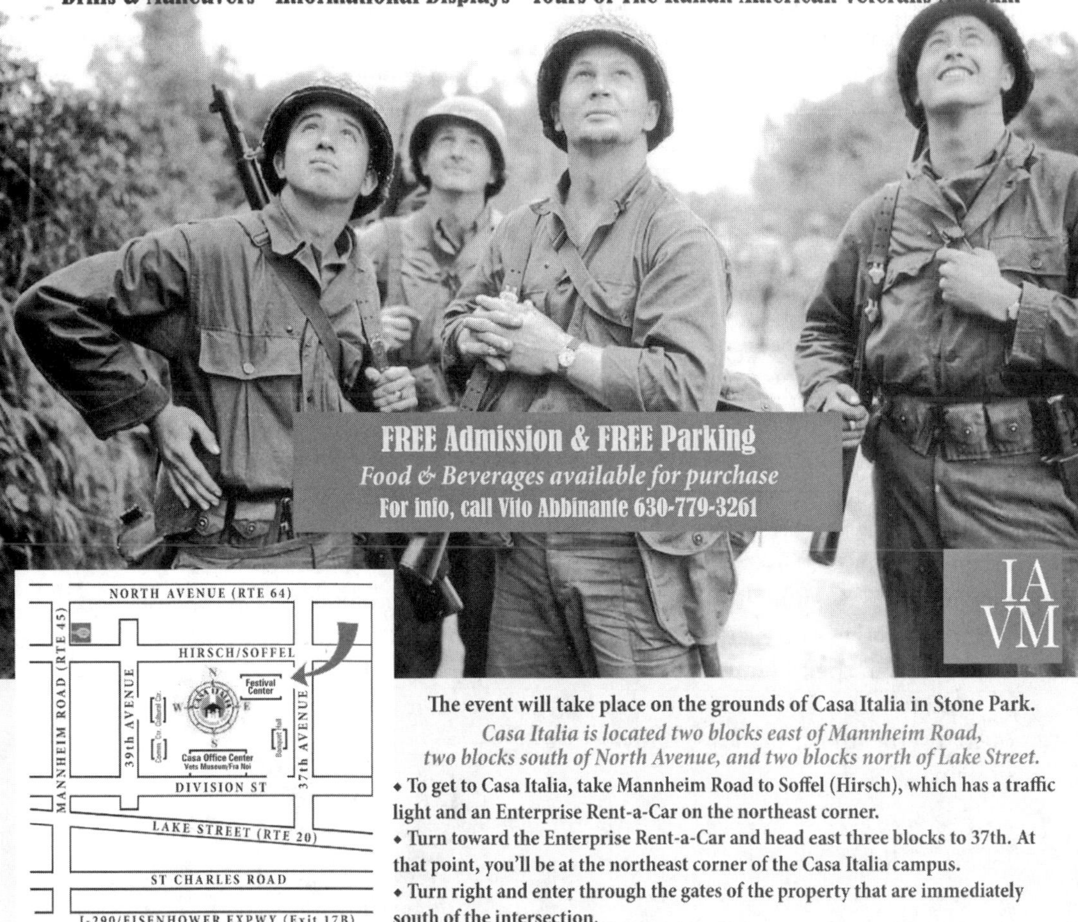

Figure 30. Poster from an Italian-American Veterans Museum Event, Chicago, 2012. (Italian-American Veterans Museum)

Other works have enumerated reasons why Italian-Americans should be proud of their heritage.[131] The pride extended to regional groups such as Sicilian-Americans who, many authors declared, should feel as gratified by the achievements of their fellows as any other ethnic body.[132]

The narrative hinged on the demonstrated accomplishments of individual immigrants and their offspring. The case of Antonio Martin Cogliandro, who left the Calabria region in the south of Italy for Scranton, was not uncommon. He came to America to find a better future and soon moved to San Francisco. He met and married Caterina Stephanie D'Acquisto, who was born in Palermo, Sicily. Mr. Cogliandro possessed strong entrepreneurial abilities and proceeded to establish the first restaurant at Fisherman's Wharf and, in succession, a spaghetti factory, an insurance firm, and a real-estate agency. His name and photograph appeared in *Attività Italiane in California*, by G.M. Tuoni and G. Brogelli (1929), a volume that highlighted dozens of prominent and successful Italians who had immigrated to the Bay Area around the turn of the century.[133]

Many similar stories catalogued the realizations of Italians in the anthracite regions. Two of the most successful non-mining businessmen were Andrew J. Sordoni, the twelfth child of Italian immigrants, who began with a team of borrowed horses in 1910 and built one of the largest construction firms in the East, with headquarters in Forty Fort, Luzerne County. He was elected to the Pennsylvania State Senate in 1926 and served as Secretary of Commerce under Governor John S. Fine in the 1950s. The other was Amadeo Obici, born in Oderzo, Veneto, Italy, who with fellow immigrant Mario Peruzzi established Planter's Nut and Chocolate Company in 1908. Now the Planter's Peanut Co., a subsidiary of Kraft Foods, Inc., the firm's headquarters were in Wilkes-Barre for most of its history.[134]

Several successful anthracite workers have been revealed in this volume. Sicilian sulfur miner Cologero Giarrtano, who emigrated from Serradifalco to Pittston in 1876, is cited in Chapter Three, note 53. He toiled in the coal industry (probably for PaCC) for many years, raised a family, and joined the middle class before succumbing to black lung disease in 1963.[135] George DeGerolamo, who was blackballed as a UMWA leader at a Pittston Company colliery in the 1930s but later returned to work in mining, wrote an important autobiography mentioned in Chapter Seven. Lewis Casterline, who has been quoted several times, represented another hard-work-and-assimilation story as he progressed from laborer, to miner, to foreman, to assistant superintendent, to superintendent at PaCC and other companies. The dozens of legitimate Italian-American coal operators who worked with leases from the large firms also achieved economic and social success in their lifetimes.[137]

As attractive as this narrative has been, however, it has been enormously overshadowed by the third, far less gratifying, storyline.

Figure 31. Rep. Nancy Pelosi of San Francisco, then Speaker of the U.S. House of Representatives, leading the Italian-American congressional delegation in a toast, 2011. (World News Network)[136]

Figure 32 (above left). "Underworld" organized crime images. Top, Al Capone (1931, U.S. Justice Department); bottom, Charles "Lucky" Luciano (1936, New York Police Department).

Figure 33 (above right). Fictional organized crime images. Top, from The Godfather; *bottom, from* The Sopranos.

Narrative III: Organized Crime

This narrative has dwelled on a gangster caricature that has accentuated the criminality of a few over the honest lives and labors of the great majority. Despite turn-of-the-nineteenth-century research documenting the low crime rate among Italian immigrants, as well as recent studies showing the same for contemporary Italian-Americans, the popular culture is rife with representations of the Italian criminal-mobster.[138] The phenomenon has been referred to as "The Godfather Industry," but it existed well before Francis Ford Coppola and Mario Puzo's extraordinarily popular film trilogy.[139] As one author opined about the most-watched American television series of the 2000s, the New Jersey-based gangster show, *The Sopranos*:

> That's right, I'm pinning the ethnic tail right on the depraved donkey. Defenders of the good name of Italian-Americans, the anti-defamation crowd ... will tell you that Mafia gangsters are just gangsters who happen to be Italian. Not for me. If they weren't Italian through and through, from the cut of their suit to the cut of their pasta, I wouldn't be interested. My guess is that without their Italian roots, they would have long ago been relegated to the dustbin of history along with the old Irish and Jewish "mobs."[140]

The writer composed the statement with tongue in cheek but it undoubtedly resonated with many, if not most, Soprano fans, as well as a movie-going public eager to patronize the omnipresent and always popular racketeer films and documentaries.[141] Although in recent years other ethnic gangs, including Russians, Mexicans, and Columbians, have entered the scene, the "stars" of the show have remained Italian. Well-publicized and often sensationalized government investigations such as the Kefauver hearings in 1950 and the McClelland hearings of 1957–1959, along with exposés such as the Valachi hearings (published as *The Valachi Papers*, 1969) and Vincent Teresa's book, *My Life in the Mafia* (1973), have helped reinforce an "Italians as criminals" typecast. The same can be said of popular depictions of the Apalachin crime meeting of 1957.[142]

The strength of the hoodlum stereotype did not merely overwhelm the other narratives, but also prevented an accurate understanding of organized crime itself. As Robert J. Kelly pointed out in *The Upperworld and the Underworld* (1999), the overemphasis on the Italians-as-gangsters scenario has encouraged far too much attention to an ethnic/ alien-culture interpretation of organized crime:

> The phrase *non-traditional organized crime* has become synonymous with *non-Italian crime*. Unfortunately, books and investigations examine non-Italian organized crime in the same one-dimensional fashion that Mafia-related crime has been approached. That is, the use of ethnicity as a descriptor of criminal activity is overemphasized and fails to shed light on the existence of the activity itself, and often comes perilously close to ethnic stereotyping.[143]

Kelly reasoned for another conceptualization that went beyond ethnic or racial attributions:

> The alternative is to describe organized crime not just in racial and ethnic identities but in terms of the criminal activity itself, and how and why various groups participate in certain activities. In this way, we can see that organized crime is very often the result of complex situational opportunities rather than just a problem or predilection of particular ethnic groups.[144]

Kelly concluded that organized criminals—from the first Wild West gangs who were overwhelmingly of British ancestry to later European, Jewish, Asian, and Hispanic offenders—have commonly worked hand in glove with established business and/or political leaders. Actors in the "upperworld" have often been as corrupt as their counterparts in the "underworld," and one could not have existed or advanced without the other. Sociologist Richard Alba contended that the key to understanding organized crime was not centralized control by ethnic bosses but "cultivated relationships involving reciprocal obligations" with numerous persons across a variety of businesses and government agencies.[145]

In a similar vein, historian Michael Woodiwiss proposed that organized crime should not be viewed as "bad people who corrupted government and business," because such a conspiratorial approach "[does] not take into account institutional corruption and unenforceable laws."[146] He urged a more sociological

Figure 34. Organized crime cartoon: Corruption in the "Upperworld".

Figure 35. "Upperworld" organized crime poster.
(Courtesy of Bob Mannseichner, adbusters)

orientation that avoided an ethnic-alien etiology and instead paid attention to values, laws, and institutions:

> The mistake that has always dogged U.S. organized crime control efforts was the misperception that organized crime was composed of conspiratorial entities that were alien and distinct from American life. A great deal of evidence suggests the opposite. High-level politicians and respectable members of business and professional communities have gained more from organized criminal activity than any other group and, for the most part, stayed out of prison.[147]

In *Gangster Capitalism* (2005), Woodiwiss dissected the extensive business and executive dishonesty that has plagued the U.S. from "the founders of America's industrial and commercial dynasties [who] showed little hesitation in bribing, stealing, cheating and using violence to further their businesses interests," to contemporary banking, investment, tobacco, medical, industrial, and other business corporations.[148] The overwhelming desire of law enforcement agencies, academic researchers, newspaper reporters, and the general public to focus on the ethnic or racial component has shifted attention away from organized crime's solid upperworld connections.

Indeed, a wide-ranging literature has shown that the upperworld is quite capable of criminal behaviors on its own. Criminologist Marshall B. Clinard, in *Corporate Crime* (1983) and *Corporate Corruption* (1990), documented the numerous illegal and unethical practices within some of America's largest corporations (auto, oil, pharmaceutical, and defense).[149] Sociologist Al Gedicks recognized the organized crime-like methods of mining and energy corporations as they have exploited natural resources in several parts of the world, particularly on lands owned by indigenous peoples.[150] Appropriately, in the 1990s several American tobacco corporations were found guilty of violating the federal

Racketeer Influenced and Corrupt Organizations Act (RICO), a law enacted in 1970 to facilitate Mafia convictions.[151]

The organized-crime characterizations recently attributed to Rupert Murdoch and his News International Corporation in the United Kingdom—where the corruption extended to the London Metropolitan Police, Scotland Yard, members of Parliament, and even prime ministers—and his denial of any knowledge regarding the law-breaking, prompted Member of Parliament Tom Watson to state that Murdoch was perhaps "the first Mafia boss in history who didn't know he was running a criminal enterprise."[152] In an article titled, "The Scam Wall Street Learned from the Mafia," which scrutinized the recent financial crimes committed by executives from GE Capital, Inc., journalist and political writer Matt Taibbi wrote: "[T]he crimes the defendants and their co-conspirators committed were virtually indistinguishable from the kind of thuggery practiced for decades by the Mafia."[153] Parallel arguments have been offered by Yves Smith, who wrote *ECONned: How Unenlightened Self-interest Undermined Democracy and Corrupted Capitalism* (2010), in which she proposed that contemporary business, especially financial corporations, have become "systematically predatory" on the consumer and the public and have caused a world-wide economic crisis.[154] The illegal behaviors perpetrated by the Massey Energy Company and its "notorious" chief executive officer, Don Blankenship, illustrate the problem in coal mining. The company was found guilty of systematically violating federal mining laws at the Upper Big Branch mine in West Virginia in April 2010, a felony that led to the deaths of 29 mineworkers. Massey agreed to pay a record $210 million fine to resolve the government's case against it.[155]

It is apparent that the Erie Railroad's (and, later, the Van Sweringen brothers') anthracite tenancy dealings exemplified the upperworld's cooperation with the underworld. For decades both PaCC and HC&I collaborated directly and indirectly with organized criminals in creating tenancy systems that undermined the workers' rights, defeated insurgent unions, corrupted the UMWA, circumvented safety laws, and degraded the environment. As discussed in Chapter Six, the corporation's board of directors approved each subcontract and lease at regular meetings, and PaCC's last vice president, E. Stewart Milner, acknowledged that the agreements were issued to alleged mobsters as part of simple "business relationships."[156] To reinforce the conclusions of authors such as Kelly and Woodiwiss, it is clear that Erie engaged members of the region's organized crime "family" (i.e., the Volpe/Sciandra/Bufalino organization) and their affiliates (e.g., Charles Consagra, Louis Fabrizio, Robert L. Dougherty, August J. Lippi) in the pursuit of rational economic goals. They did so without concern for the moral, social, safety, or environmental repercussions.[157] Because of its structure and operation, tenancy itself became a type of organized criminal activity.

Narrative IV: Workers' Rights and Economic Justice

Of course, organized crime has represented an undeniable aspect of Italian and Italian-American life. This remarkably long-lived narrative has endured for well over 100 years, even though it has applied to a relatively small number of people. More importantly, it has helped conceal another story: Italians as advocates for workers' rights and economic justice.

Lewis Casterline spoke directly to the fourth narrative in a commentary on the anthracite tenancy systems during the 1920s and 1930s. He extolled the Italian workers for their decades-long fight against the organized criminals who secured subcontracts and leases, the large companies that offered the agreements, and a union that was unable or unwilling to deal with the results:

> Now with this setup [tenancy], the way it came into Luzerne County, it came and the Italians fought that. They fought it and fought it. ... [Company officials would ask:] "What does the 'dagos' want around the mines? You have to stay there and talk to them until you got blue in the face." I'd say: now wait a minute. Now let's see what they want. They have established something that you never had [in the old country], self-respect. Now you got self-respect. So the Italians begin to say, "Here, we're giving our blood." In one month there was seven lives lost. Twenty-eight homes were dynamited ... Some of us fought this right to the end because we were caught right in the middle. ... See now, the Italians, they knew this was all wrong, they knew it was going to ruin the coal company [PaCC]. ... The Italians tried to warn them [management and UMWA officials] but they wouldn't listen. ... The [Wyoming] Valley lived fifty years behind the times in comparison with other sections of America. Not only did they [the company and organized crime] ruin the Valley but they ruined northeastern Pennsylvania.[158]

Edward Banfield and his followers notwithstanding, recent historical investigations have uncovered many similar examples of Italian activism. By using a comparative-internationalist perspective, historians have found that numerous Italian immigrants were dedicated to rights and justice causes well before their arrival in the U.S. Their commitments in this regard were based upon a series of politicizing and radicalizing experiences in Italy.[159] Bruno Cartosio, in "Sicilian Radicals in Two Worlds," reported that workers from various occupational groups developed a high degree of political astuteness and militancy even though most were from rural villages and could not read or write. Indeed, during the 1890s, Sicily had spawned a vigorous Socialist movement and maintained the largest percentage of unionized agricultural workers in Europe.[160] As part of the activism, a great number of sulfur workers mobilized against the *padrone*-like labor practices of the subcontractors who dominated production.

Jacquerie was the term used for the Sicilian peasants' and workers' revolts that occurred once or twice each generation between 1772 and 1950.[161] *Revenge* was the underlying motivation for the uprisings, but it was revenge within a broader context, according to historian Donna Gabaccia: "Scholars have seen *jacquerie* as the peasants' efforts to punish wealthy or powerful men who violated communal notions of fairness, decency, or traditions in some way."[162] The uprisings were, therefore, partly based on a moral foundation of *retributive justice* or "a form of collective violence designed to redress violations against a particular understanding of what was socially right and wrong."[163] However, the protests were not merely "acts of revenge" but also sought to redress rampant social inequalities and injustices. Hence, *distributed justice* was also a motivation. Moreover, the island's 14 sulfur mining towns experienced *jacquerie* more often than agricultural or wheat-producing settlements.[164] Approximately 86 percent of the sulfur-mining communities witnessed *jacquerie* or strikes after 1860, which was the highest count among the different types of municipalities.[165] An abortive miners' strike in 1900 encouraged large numbers to migrate to the U.S., some undoubtedly to the Pennsylvania hard coal fields.[166]

Sulfur miners were also among those who led and joined the *Fasci* or peasants' and workers' organizations formed mainly by Socialists in the early 1890s. The groups grew to more than 300,000 members from diverse occupational categories, among them sharecroppers, day laborers, industrial workers, and mineworkers. The *Fasci* led a wave of *jacquerie*-like strikes in dozens of towns during the period 1892–1894. The failure of the work suspensions again prompted many to emigrate.

It seems evident that the sulfur miners who arrived in northern anthracite towns such as Pittston, Dunmore, and Old Forge were well prepared to join fellow immigrants, as well as established American mineworkers, in the labor wars of the late nineteenth and early twentieth centuries. Gabaccia addressed some of the commonalities of workers in the two countries:

> In fact, the workplace experiences of wheat cultivators in western Sicily, cane harvesters in Louisiana, and railroad laborers in Chicago can be considered remarkably similar. All such jobs were seasonal, organized as gangs, controlled by middlemen, poorly paid, and abusively supervised. A visible enemy, usually a foreman, dominated each of these workplaces.[167]

Certainly, the Sicilian sulfur and Sicilian-American anthracite miners who toiled for their respective subcontractors and leaseholders can be added to the list. The tenant, as well as the foreman, served as "the enemy" in both cases. Consequently, while it was true, as Gabaccia wrote, that: "Peasants and wage earners had few options in both Sicily and the United States," it was also true that "they were never passive followers."[168]

Along another dimension of group mobilization, during the immigration period between 1880 and 1920, some Italian émigrés were the so-called *sovversivi* (subversives); they were a vanguard group that believed root-level—i.e., radical and systemic—changes were necessary to make progress on "the labor question" and in the larger society. They included Socialists, Communists, Syndicalists, Anarchists and, after World War I, Antifascists and Communist refugees.[169] Among them was Dr. Alberico Molinari, a physician and a Socialist who emigrated from the northern Italian city of Cremona to Scranton in 1903. According to historian Elizabetta Vezzosi, Molinari "won the warmth and esteem of the entire community." He was not only a very good doctor but represented "Socialism with a human face." A contemporary report characterized his organizing abilities: "In Wilkes-Barre, Old Forge, Plainsville, and in all the places Molinari goes, there are Socialists and Anarchists," where years before there had been few if any.[170] Molinari was one of many "radical ethnic brokers" who operated between the community and the society in an effort to encourage fellow immigrants in a leftist direction.[171] Their numbers included not only educated, higher ranking persons but also artisans, tenant farmers, farm workers, and sulfur miners. Rudolph Vecoli found that the foremost leaders among them were from southern Italy.[172]

What happened to the workers' rights and economic justice narrative? Why did it fade from the social memory of the ethnic group, the community, and the nation? Purges by the federal government had a discouraging influence within the Italian community, beginning with attacks on the Industrial Workers of the World (IWW) in the 1910s (see Chapter Three), continuing into the 1920s with the Red Scare—a decade that included the Sacco and Vanzetti executions in 1927—and carrying into the 1930s.[173] The immigration laws of the 1920s also had a negative impact, as did deportation and imprisonments in the 1910s and 1920s. A growing concern for assimilation and material gain further dampened activism. The promise of New Deal labor policies in the 1930s and 1940s turned many radicals into organizers for the Congress of Industrial Organizations and, eventually, Roosevelt Democrats.

Figure 36. Bartolomeo Vanzetti (left) and Nicola Sacco (right) arriving at Superior Court, Dedham, MA, for sentencing on April 19, 1927. They were executed for murder on August 23, 1927 (see note 173).

Within northern anthracite, the militancy among Italians survived the IWW's defeat in the late 1910s, and actually gained momentum through the 1920s and the several strikes at the Erie Coal Companies. The dissidence continued into the early 1930s with the United Anthracite Miners of Pennsylvania (UAMP). However, it began to fade in the mid-1930s following the demise of the UAMP and the triumph of the leasing system. No doubt military service in World War II, patriotism, and post-World War II anti-Communism in the 1940s and 1950s diminished any remaining radical elements. Still, a strong progressive and militant tradition endured in the area's International Ladies' Garment Workers' Union into the 1970s. The organization attained a membership of nearly 12,000 in the Wilkes-Barre area alone and they included thousands of Italian-American women and men.[174]

The decline in the insurgent movement and the triumph of tenancy brought about the virtual elimination of historical memory regarding the radical Italians and their multi-ethnic comrades. In discussing such lost or hidden memories, Luisa Del Giudice wrote: "... 'hidden' histories may result from omission or neglect ... or conscious avoidance on the part of those creating the official historical record ... but they may also result from a silence that is culturally imposed from within."[175] We do not believe that the story of "The Italian Community on Strike" in northern anthracite has been missing from the historical record because researchers have consciously avoided it, nor becasue the Italian community has instigated a self-imposed silence. Rather, our point is that the fourth narrative has remained "hidden" primarily because it has been overwhelmed by three other narratives—two of them quite negative toward Italians—within the ethnic and scholarly communities and, indeed, within the region, state, and nation.

Summary: The Italian Community on Strike

In broadening Victor Greene's findings about anthracite unionism, the evidence presented in this volume has shown that Italian immigrants were among the most dedicated to workers' rights and economic justice through collective action. Some served as leaders while others were among the participants in the strike of 1887–88, the Battle of Archbald in 1896, and the Lattimer-related protests of 1897. They joined the UMWA strikes of 1900 and 1902. They were integral to the wildcat shutdowns at the Erie collieries in 1905, 1907, 1908, and 1910, and spearheaded the IWW strike of 1916. In 1920, they were at the forefront in a work suspension that temporarily removed the subcontracting system from the Erie companies and, just as important, brought 12,000 workers into the UMWA. The Italians were enthusiastic contributors to the subcontracting strikes at PaCC and HC&I in 1924. Throughout 1928 they joined in the battles against management, organized crime, police, and even the UMWA in strikes that led to murder and chaos at the No. 6 Colliery. To be sure, the strikers often engaged in violent acts based on retributive justice, and persons of Italian extraction were among the perpetrators.[176]

Through it all, fairness and justice were the main goals for the Italians and their allies from over two dozen ethnic groups, including the English speakers who were the region's first mineworkers and union advocates. While wages were a concern, they were not paramount. Although violence occurred, it was not a universally avowed method. Relatively large numbers joined the IWW after 1907, and a small group joined the Communist, Socialist, and other radical groups in the 1920s and 1930s; however, the great majority eschewed radicalism. For better or worse, they chose not to follow the *sovversivi*, such as Dr. Alberico Molinari, or the Syndicalists and Socialists in the IWW, or other radicals in the UMWA, most of whom—contrary to the commonly accepted view—were not immigrants but established Anglo-Saxons.[177] By the 1920s, the Italians (and their militant colleagues) preferred the label *insurgents* (*Gli Insorti*) to define themselves and their protests. As such, they supported the UMWA and tried to push it in a progressive direction, but the organization's conservative leadership and policies turned them to broader protest movements and dual unionism. After the UAMP's defeat, they returned to John L. Lewis and his organization because there were no viable alternatives. Yet many remained critics, expressing their discontent by filing complaints with the Anthracite Board of Conciliation and by joining those who wrote hundreds of complaint letters to Lewis and other UMWA officials.[178]

It is important to recall that Italian mineworkers not only consistently mobilized on behalf of anthracite labor, but often did so in the face of threats and intimidation from management, the UMWA, and the small but powerful criminal element. The mourners in the funeral cortège of slain detective Sam Lucchino in July 1920 displayed admirable courage as they processed through the streets of Pittston to honor the victim and defy the perpetrators.[179] Of course, it must also be recognized that Italian émigrés were among the most prominent subcontractors and leaseholders at PaCC and HC&I. They included legitimate operators but also alleged gangsters such as Santo Volpe, Charles Consagra, Steven LaTorre, and John Sciandra. Other notorious Italians discussed in this volume include Antonio Puntario and Peter Erico, hired murderers of Sam Lucchino in 1920; and Ralph Melissari, Peter DeLucca, and Vincenzo Damiano, convicted assassins of Alex Campbell and Peter Reilly in 1928.

However, there were many more persons, both acknowledged and unsung, who participated directly or indirectly in the labor wars. Lucchino was one of the recognized heroes: a former Black Hander, he became a Pittston detective after working for the U.S. Secret Service and was gunned down while compiling evidence against subcontracting. Rinaldo Cappellini led the Erie insurgents in the early 1920s, was elected District 1 president in 1923, became a "Lewis man" between 1923 and 1928, and subsequently rejoined the insurgent movement to become a determined leader and advocate of the UAMP. James Musto was a participant in the anti-subcontracting protests of the 1920s and 1930s until he was blackballed by PaCC managers, whereupon he became a barber and a grocer and was later elected to Pennsylvania House of Representatives where, among other things, he served on the Joint Legislative Committee that investigated the Knox Mine Disaster and exposed a great deal of industry-wide corruption.[180]

Union activists and murder victims Charles Attardo (1914), Samuel Spachia (1925), Frank Bonita (1928), and Charles Licata (1931) should also be remembered among the unsung, along with the non-Italian victims of the No. 6 Colliery strikes in 1928: Alex Campbell, Peter Riley, Thomas Lillis, and Thomas Lewis.[181] They are among those listed in Appendix II.

Finally, it should be noted that the Italian contributions to anthracite history paralleled their engagement in various other American union movements. Labor historians have affirmed their work in the Industrial Workers of the World, the International Ladies' Garment Workers' Union, the Amalgamated Clothing Workers of America, and others.[182] Italians figured prominently in the industrial strikes in Lawrence, Massachusetts, in 1912 and 1919, the Paterson silk strike of 1913, the Mesabi Iron Range strikes of 1907 and 1916, the New York City Harbor strikes of 1907 and 1919, as well as numerous bituminous strikes. A good description of their role in various Pennsylvania labor movements can be found in historian Edwin Fenton's article, "Italians in the Labor Movement"

Figure 37. Mining in Scranton, Delaware & Hudson Coal Company, circa 1915. (John Horgan photograph, courtesy of Lackawanna Historical Society)

Figure 38. Mineworkers going down a shaft, Delaware & Hudson Coal Company, circa 1915. (John Horgan photograph, courtesy of Lackawanna Historical Society)

(1959).[183] "The Italian Community on Strike" in anthracite, as chronicled in the present volume, represents another chapter in a time-honored story.[184]

Revising Anthracite Labor and Ethnic History

Beginning in the 1960s, the "new social history" and the "new labor history" sought to overcome the biases inherent in historical studies of privileged groups—i.e., "history from the top." Accurate investigations, according to this perspective, require a more inclusive orientation. As a result, working people, minorities, ethnic groups, women, the poor, and other marginalized peoples have secured a place at the historical-narrative table.[185] There has also been an effort to include positive and constructive histories about often misrepresented and mistreated groups as an antidote to the many negative chronicles that abound.[186]

In addition to detailing the tenancy-related labor wars at PaCC and HC&I, another purpose of this study has been to recover, as historian Rudolph Vecoli put it, "ethnic histories of which we know little or nothing."[187] The story of the Italian anthracite mineworkers who, with colleagues from various other groups, resisted the corporate, gangster, and union forces aligned against them is one chapter of ethnic and labor history about which we have known very little.[188]

Mary Jo Bona found that storytelling among Italian-Americans has been of prime importance: "Through narrative, Italian American storytellers have constructed another space to revise hierarchical discourse, to give voice to those without power, to shape perceptions or invent alternative worlds." Moreover, she concluded that the stories were fundamentally *moral*—in the ethical (values and beliefs) and sociological (group solidarity) meanings of the term. They have served as guides to "a burdened people" in search of resistance, community, liberation, and healing. Themes central to peasant morality were apparent:

> [A]n insistence on equality and justice, despite arbitrary division of human beings; the desire to determine one's own fate, despite adversity; and, the belief that liberation can only be achieved, as [Italo] Calvino relates, "if we liberate other people, for this is the *sine qua non* of one's own liberation."[189]

The Italian anthracite mineworkers' role in the *Workers' Rights and Economic Justice* narrative can contribute to the storytelling and thereby help remedy a critical problem articulated by Vecoli: "Ethnicity is a form of memory and many Italian-Americans are suffering from amnesia."[190]

Like similar studies of amnesia and misrepresentation, perhaps this volume can help transform the often-cast villains of popular discourse into a position to which they have not been accustomed: that of heroes.[191] Without trying to romanticize the often violent and tempestuous situations they faced (and produced), perhaps an understanding of "The Italian Community on Strike" can encourage the descendants of Italian and other stereotyped groups—mineworkers and otherwise—to imagine "better stories and endings for [our]selves," our traditions, and our history.[192] For as sociologists Ronald J. Berger and Richard Quinney have asserted: "What is included or omitted from our stories makes plausible our anticipated futures."[193]

NOTES

Notes to the Preface

1. E.P. Thompson, *The Making of the British Working Class*, Harmondsworth, England: Penguin, 1980 [1963], 12.

2. Herbert G. Gutman, "Work, Culture, and Society in Industrializing America, 1815–1919," *American Historical Review*, 78 (Jun 1973), 531–558.

3. See, e.g., Alan Murray, *Holding the Line: A Narrative History of Australian Coal Miners and their Union in the 1980s*, Sydney, Australia: Coal Forestry Mining & Energy Union, 2009.

4. See, e.g., Donna R. Gabaccia, *Militants and Migrants: Rural Sicilians Become American Workers*, New Brunswick, NJ: Rutgers University Press, 1988; Bruno Cartosio, "Sicilian Radicals in Two Worlds," 117–128 in Marianne Debouzy (ed.), *In the Shadow of the Statue of Liberty: Immigrants, Workers, and Citizens in the American Republic, 1880–1920*, Urbana, IL: University of Illinois Press, 1992; Elizabetta Vezzosi, "Radical Ethnic Brokers: Immigrant Socialist Leaders in the United States between the Ethnic Community and the Larger Society," 121–138 in Donna R. Gabaccia and Fraser M. Ottanelli (eds.), *Italian Workers of the World: Labor Migration and the Formation of Multiethnic States*, Urbana, IL: University of Illinois Press, 2005; Marcella Bencivenni, *Italian Immigrant Radical Culture: The Idealism of the Sovversivi in the United States, 1890–1940*, New York: New York University Press, 2011.

5. For other reports on the region's organized crime gang see, e.g., Thomas Hunt and Michael A. Tona's investigative article, "Men of Montedoro," *Informer* (Apr 2011, informer-journal.blogspot.com), an account that our own parallel primary research largely corroborated; and Dave Janoski, "The Bufalino File: An Exclusive Look Inside the Massive FBI Paper Trail on Northeastern Pa.'s Most Notorious Mobster," *Citizen's Voice*, 17 Jul 2011.

6. Luisa Del Giudice, "Speaking Memory: Oral History, Oral Culture, and Italians in America," 3–18 in Luisa Del Giudice, *Oral History, Oral Culture, and Italians in Americans*, New York: Palgrave Macmillan, 2009. See also Paul R. Thompson, *The Voice of the Past: Oral History*, New York: Oxford University Press, 1988; and Allesandro Portelli, *They Say in Harlan County: An Oral History*, New York: Oxford University Press, 2010.

7. The co-authors have made numerous public presentations in eastern Pennsylvania based on our anthracite research: at the Anthracite Heritage Museum, the National Canal Museum Annual Symposium, the Huber Breaker Preservation Society, the West Pittston Historical Society, the Exeter Historical Society, the Plymouth Historical Society, the Greater Pittston Historical Society, the Lackawanna Historical Society, the Nanticoke Historical Society, the Shickshinny Historical Society, King's College, Wilkes University, and during several Mining History Week programs each January. We have also published articles in the *Citizens' Voice* (Wilkes-Barre), the *Scranton Times*, the *Pittston Dispatch*, and in the National Canal Museum Symposium *Proceedings* (see References).

8. Ronald J. Berger and Richard Quinney, *Storytelling Sociology: Narrative as Social Inquiry*, Boulder, CO: Lynne Rienner, 2005. From this volume see especially Robert P. Wolensky, "Working Class Heroes: Rinaldo Cappellini and the Anthracite Mineworkers," Chapter 18.

9. For a detailed analysis of anthracite culture see Harold Aurand, *Coalcracker Culture: Work and Values in Pennsylvania Anthracite, 1835–1935*, Selinsgrove, PA: Susquehanna University Press, 2003; however, this work devoted little attention to the culture of corruption and conflict discussed in the present volume.

10. In addition to labor movements in mining, garment manufacturing, shoemaking, and other industries (dating back to the 1840s in mining), there have been several notable examples of grassroots militancy in the northern anthracite field. The so-called "water gouge" protest of 1928–1936, called "probably the largest and longest [utility] rate case in the history of Pennsylvania," involved a successful coalition of citizens and local governments aligned against the ex-

cessive price increases of the newly formed water Scranton-Spring Brook Water Company, later called the Pennsylvania Gas & Water Company (Glenn Gooch, "Pennsylvania Gas and Water Company: A History," unpublished manuscript, PG&W, Wilkes-Barre, PA, 1976, 11) The unemployment councils of the early 1930s, organized mainly by Communists, held marches to pressure local and state political leaders to alleviate the deprivations brought by the Great Depression. With over 20,000 members in 1931, the councils were "authentic popular organizations that developed to meet real and urgent needs" (Donald L. Miller and Richard E. Sharpless, *The Kingdom of Coal: Work, Enterprise, and Ethnic Communities in the Mine Fields*, Philadelphia, PA: University of Pennsylvania Press, 1985, 319; see also Steve Nelson, James R. Barnett, and Bob Ruck, *Steve Nelson, American Radical*, Pittsburgh, PA: University of Pittsburgh Press, 1981). A third example can be seen in the citizens' movements following the Tropical Storm Agnes flood of June 22, 1972. The storm ravaged several Eastern states, but caused the most severe damage, over $1 billion, in the Wilkes-Barre/Wyoming Valley area. The decade-long recovery became a highly political and conflict-ridden affair that precipitated a social movement consisting of 48 newly founded citizens' groups, some of which acted singly while others formed coalitions. The groups wielded considerable influence over post-flood politics and decision-making (Robert P. Wolensky and Edward J. Miller, "The Everyday Versus the Disaster Role of Local Officials: Citizen and Official Definitions," *Urban Affairs Quarterly*, 16 [1981], 483–504; Robert P. Wolensky, Power Structure and Group Mobilization Following Disaster: A Case Study, *Social Science Quarterly*, 64 [Mar 1983], 96–110.) Another example involved the broad community mobilization in support of four labor unions at *The Times Leader Evening News* in Wilkes-Barre when Capital Cities, Inc., bought the company in 1978 and proceeded to break the unions. In the midst of an intense strike, the workers established and owned the *Citizens Voice*, which has remained the largest circulation daily newspaper in the area (Thomas Keil, *On Strike! Capital Cities and the Wilkes-Barre Newspaper Unions*, Tuscaloosa, AL: University of Alabama Press, 1988; in 2000, the workers sold the newspaper to Times-Shamrock, Inc., a private company based in Scranton, which publishes the *Scranton Times* and other newspapers). Yet another instance was the broad-based movement created by citizens groups that joined with local governments against the harmful economic development and environmental policies of the Pennsylvania Gas & Water Company in the 1980s (Robert P. Wolensky, "POWER: Collective Action and the Anthracite Region Water Crisis," 230–261 In S.R. Couch and S. Kroll-Smith (eds.), *Communities at Risk: Collective Responses to Technological Hazards*, New York: Peter Lang, 1991). A final case can be seen in the current collective action against the hydraulic fracturing or "fracking" used in extracting natural gas in the Marcellus Shale reserve throughout northeastern Pennsylvania (see, e.g., "Anti-drilling Activists to Hold Meeting this Morning," *Times Leader*, 4 Mar 2011; "Back Mountain Group Discusses Drilling," *Citizens' Voice*, 19 Mar 2010). This is to say that the workers' movements discussed in this volume drew upon, and contributed to, a long-standing tradition and "infrastructure" of citizen and community-based activism.

Notes to Chapter One

1. Geologists have classified coal into three categories: lignite, bituminous, and anthracite (with grades in between). As the highest grade, anthracite is almost pure carbon, burns most efficiently, and is nearly smokeless. For a review of Pennsylvania's coal industry, see E. Willard Miller (ed.), *A Geography of Pennsylvania*, University Park, PA: Penn State Press, 1995. See Appendix I, The Glossary, for the definitions of anthracite, bituminous, and other mining-related terms used in this volume.

2. Alfred D. Chandler, "Anthracite Coal and the Beginnings of the Industrial Revolution in the United States," *Business History Review*, 46 (1972), 141–181.

3. According to the U.S. Geological Survey, as late as 1913 anthracite employed slightly more workers within Pennsylvania than did bituminous (175,745 anthracite; 172,196 bituminous), even though soft coal had nearly twice the output (91,626,964 tons anthracite; 173,781, 271 tons bituminous) (U.S. Geological Survey, *Mineral Resources of the United States, Part II-Nonmetals*, Washington, DC: USGPO, 1913, *Coal Production in 1913*, 751; Frank Moore Colby et al., *The New International Year Book*, New York: Dodd & Mead, 1914, 532; "Coal and Coke News," *Coal Age*, 6 [19 Sept 1914], 482). The output differences reflected the more efficient production systems and technologies in bituminous. Also in 1913, Pennsylvania produced coal worth $388,220,933, with anthracite's share ($195,181,127) slightly higher than that of bituminous ($193,039,806). After reaching peak production of over 100 million tons in 1917, anthracite began a slide that continued through the next four decades. At the national level, soft coal claimed 84 percent of total U.S. production by 1900, a figure that climbed to 88 percent by 1920, and over 91 percent by 1925. The relative *value* of the two types of coal showed a similar

trend: they remained fairly even until the early 1880s with anthracite worth $42 million and bituminous $53 million. By 1890 the disparity had grown to $66 million for anthracite and $110 million for bituminous and, by 1925, to $328 million vs $1.06 billion respectively (U.S. Geological Survey, *Mineral Resources of the United States*, Part II-Nonmetals, Washington, DC: USGPO, 1910, 22; Samuel D. Warriner, "Reply of the Operators *[to the Miners' Demands]*," in *The Anthracite Strike of 1925–1926*, Philadelphia, PA: The Anthracite Bureau of Information, 1926, 6). Hard coal faced a similar threat from oil where its advantage had vanished by the mid-1910s. In 1920 anthracite's value stood at $434 million while petroleum's had grown to $1.36 billion; by 1925, following the recession of the early 1920s, the difference stood at $328 million vs $1.29 billion respectively (U.S. Bureau of the Census, *Historical Statistics of the United States: Colonial Times to 1970*, Washington, DC: USGPO, 1975, 590–593—coal production statistics, 582–583—coal value statistics, 582—petroleum value statistics). Although the value of natural gas increased throughout the first half of the twentieth century and certainly helped erode anthracite markets, the total value of the fuel would not exceed anthracite on a regular basis until the 1950s (*Ibid.*, 582). Frederick Saward (*The Coal Trade: The Year Book of the Coal and Coke Industry*, New York: published by the author, 1889, 90), showed anthracite's virtual confinement to the home-heating market by the late 1880s.

4. On the early anthracite industry see Clifton K. Yearley, Jr., *Enterprise and Anthracite: Economics and Democracy in Schuylkill County, 1820–1875*, Baltimore, MD: Johns Hopkins University Press, 1961; and H. Benjamin Powell, *Philadelphia's First Fuel Crisis: Jacob Cist and the Developing Market for Pennsylvania Anthracite*, University Park, PA: Penn State Press, 1978. On Wilkes-Barre's history see Henry C. Bradsby (ed.), *History of Luzerne County Pennsylvania*, Chicago, IL: Nelson, 1893; Oscar Jewell Harvey and Ernest Gray Smith, *A History of Wilkes-Barre*, Vol. 4, Wilkes-Barre, PA: Smith Bennett, 1929; Oscar J. Harvey and Harrison G. Smith, *A History of Wilkes-Barre*, Vol. 5, Wilkes-Barre, PA: Smith Bennett, 1930; and Edward J. Davies, *The Anthracite Aristocracy: Leadership and Social Change in the Hard Coal Regions of Northeastern Pennsylvania, 1800–1930*, DeKalb, IL: Northern Illinois University Press, 1985. On Scranton's history see Horace Hollister, *History of the Lackawanna Valley*, Philadelphia, PA: Lippincott, 1885; and Burton W. Folsom, Jr., *Urban Capitalists: Entrepreneurs and City Growth in Pennsylvania's Lackawanna and Lehigh Regions, 1800–1920*, 132, Baltimore, MD: Johns Hopkins University Press, 1981. For overviews of anthracite history see Donald L. Miller and Richard E. Sharpless, *The Kingdom of Coal*, Philadelphia, PA: University of Pennsylvania Press, 1985; and Thomas Dublin and Walter Licht, *The Face of Decline: The Pennsylvania Anthracite Region in the Twentieth Century*, Ithaca, NY: Cornell University Press, 2005.

5. On the development and longevity of the anthracite canals see Harvey and Smith, *History of Wilkes-Barre*, Vol. 4, 1886–1894; Ronald L. Filippelli, "The Schuylkill Navigation Company and its Role in the Development of the Anthracite Coal Trade And Schuylkill County, 1815–1845," unpublished M.A. thesis, Penn State University, 1966; Miller and Sharpless, *Kingdom of Coal*, Chapter 2; and Dublin and Licht, *The Face of Decline*, 13–18.

6. Davies, *Anthracite Aristocracy*; Folsom, *Urban Capitalists*. Wilkes-Barre and Scranton were part of Luzerne County until 1878, when, after a long dispute, the Scranton-Old Forge-Carbondale areas broke off to form Lackawanna County, which remains the youngest of Pennsylvania's 67 counties.

7. The Morgan-controlled anthracite railroads included the Erie; Reading; Lehigh Valley; Central Railroad of New Jersey; Delaware & Hudson; and Delaware, Lackawanna, & Western. To address the railroad's virtual monopoly on coal production and shipment, the Commodity Clause of the Hepburn Act of 1906 prohibited them from directly owning the coal they transported. The companies fought the law for years but eventually lost and were forced to spin off their mining subsidiaries (Eliot Jones, *The Anthracite Coal Combination in the United States*, Cambridge, MA: Harvard University Press, 1914, 186-212; Jules I. Bogen, *The Anthracite Railroads*, New York: Ronald, 1927; Scott Nearing, *Anthracite: An Instance of Natural Resource Monopoly*, Freeport, NY: Books for Libraries, 1915). Despite the law, the railroads' influence over their former mining operations continued in many instances.

8. Domination of the anthracite industry by New York, Philadelphia, and other urban capital interests resonates with the core-periphery concept in dependency theory, whereby "core" areas that control wealth and power invest in "peripheral" areas in order to develop and exploit natural resources and labor. On the concept of core-periphery see Appalachian Land Ownership Task Force, *Who Owns Appalachia: Landownership and Its Impact*, Lexington, KY: University Press of Kentucky, 1983; and Immanuel Wallerstein, *World-Systems Analysis: An Introduction*, Durham, NC: Duke University Press, 2004. On the core-periphery concept as applied to an-

thracite see Miller and Sharpless, *Kingdom of Coal*, 74–75; and Stephen R. Couch, "The Coal and Iron Police in Anthracite Country," 100–119 in David L. Salay (ed.), *Hard Coal, Hard Times: Ethnicity and Labor in the Anthracite Region*, Scranton, PA: Anthracite Museum Press, 1984.

9. The presidentially appointed U.S. Coal Commission of 1923 put the labor cost figure at 75 percent, as did Samuel D. Warriner, chairman of the Anthracite Operators' Association in the 1920s ("Reply of the Operators [to The Miners' Demands]," in *The Anthracite Strike of 1925–1926*, Philadelphia, PA: The Anthracite Bureau of Information, 1926). Miller and Sharpless (*Kingdom of Coal*, 291) used 70 percent as did the Hudson Coal Company (*The Story of Anthracite*, New York: Hudson Coal Company, 1932, 338). Theodore Bakerman (*Anthracite Coal: A Study in Advanced Industrial Decline*, New York: Arno, 1979 [1956], 4) put it at "more than two-thirds," while Irving Bernstein (*The Lean Years: The History of The American Worker, 1920–1933*, Boston, MA: Houghton Mifflin Company, 1960, 360) said it was "about two-thirds."

10. Carter Goodrich, *The Miner's Freedom: A Study of the Working Life in a Changing Industry*, New York: Arno, 1977 [1925].

11. The individual contract had been well established by the Long Strike of 1875, which signaled the end of the Workingmen's Benevolent Association, anthracite's first successful union. During that strike, northern field coal operator Charles Parrish "forced strikers from their company homes and refused to rehire them without individual contracts" (Victor R. Greene, *The Slavic Community on Strike: Immigrant Labor in Pennsylvania Anthracite*, South Bend, IN: University of Notre Dame Press, 1968, 68).

12. Anna Rochester (*Labor and Coal*, New York: International Publishers, 1931, 129) confirmed that "Contract miners of this petty boss type are employed only by certain operators in the northern part of the anthracite region." Testimony by company officials and mineworkers during the Anthracite Strike Commission hearings in 1902–03 also indicated that subcontracting existed almost exclusively in the northern field. However, historian Perry Blatz (*Democratic Miners: Work and Labor Relations in the Anthracite Coal Industry, 1875–1925*, Albany, NY: SUNY Press, 1994, 51) reported that one of the UMWA's earliest organizers, John Rinn, found it in the southern field in 1894. It is unclear why the subcontracting system advanced in the northern field, but the area's dominance by large, railroad-owned companies, who made major investments in large production facilities and therefore required substantial monetary returns, most likely contributed. Another factor may have been the corporate culture of the Erie Railroad and its two coal subsidiaries, the Pennsylvania Coal Company and the Hillside Coal & Iron Company, which were the leaders in developing the subcontracting system. Expectations for exceedingly high rates of return were part of the Erie corporate culture and PaCC and HC&I were historically among the most profitable of anthracite companies (see Chapter 2). A final reason, discussed in later chapters, could have been the presence within the companies of criminally associated miners from Sicily, who had worked as subcontractors in the island's sulfur mines and may have encouraged subcontracting at the Erie companies.

13. David Montgomery, *Workers' Control in America: Studies in the History of Work, Technology, and Labor Struggles*, New York: Cambridge University Press, 1979, 14–15. On the origins of the putting-out system in Britain see Eric Hobsbawm, *Industry and Empire*, New York: Pantheon Books, 1968, 15–16.

14. British engineers, managers, and miners built the Pennsylvania anthracite industry during the nineteenth century and brought in British mining practices. With regard to the individual contract, beginning in the eighteenth century, British companies required mineworkers to sign such agreements during the spring "binding time" committing them for one or two years of work (T.S. Ashton and Joseph Sykes, *The Coal Industry of the Eighteenth Century*, Manchester, England: Manchester University Press, 1964 [1929], Chapters 6 and 7). Anthracite likewise adopted the individual contract. McAlister Coleman presented three similarities between the British and American mining industries: (1) *The Truck System*: "whereby miners were forced to deal with company-owned stores and buy their own powder and tools as well as other necessities"; (2) *Piecework*: "the piecework system of payment by the ton mined by the skilled workers, which led in turn to the contract system in vogue in anthracite, whereby the older and more experienced miners hired and paid their own helpers—these were mainly taken over from English mine practice"; and (3) *Signing of Contracts*: "the tradition of signing contracts in April, a time disadvantageous to the miners facing the long summer slack season" (McAlister Coleman, *Men and Coal*, New York: Farrar & Rinehart, 1943, 35–36).

15. With some exceptions, especially in the Lehigh and Schuylkill regions, anthracite workers extracted coal using the room-and-pillar method in a first mining. Contrary to popular opinion, pillar removal (or robbing) was usually a legal activity in a second or a third mining especially after the courts exempted the companies from surface damage liability caused by subsidence or caving (see Chapter 2). As historian Robert Gordon ("Custom and Consequence: Early Nineteenth-Century Origins of the Environmental and Social Costs of Mining Anthracite," 240– 277 in Judith A. McGraw [ed.], *Early American Technology: Making and Doing Things From the Colonial Era to 1850*, Chapel Hill, NC: University of North Carolina Press, 1994) pointed out, nineteenth-century anthracite workers left behind 45 percent of the coal in pillars and barriers during a first mining. Sociologists Chris Tilly and Charles Tilly (*Work Under Capitalism*, Boulder, CO: Westview, 1998, 46) concluded that the room-and-pillar method involved a simplistic mining administration more conducive to subcontracting, while the more efficient long-wall mining—common in the bituminous industry—required a more elaborate division of labor, a complex payment system, and greater use of machines. On mining off-limit pillars see Chapter 7.

16. Harold W. Aurand ("Mine Safety and Social Control in the Anthracite Industry," *Pennsylvania History*, 52 [Oct 1985], 227–241) argued that the operators' growing concern for safety during the first two decades of the twentieth century was actually motivated by a desire for greater workplace control. On the politics of safety in anthracite mines during the nineteenth century see Anthony F.C. Wallace, *St. Clair: A Nineteenth-Century Coal Town's Experience with a Disaster-Prone Industry*, New York: Knopf, 1987, Chapter 5; and Mark Aldrich, "The Perils of Mining Anthracite: Regulation, Technology and Safety, 1870–1945," *Pennsylvania History*, 64 (1997), 361–383.

17. Tilly and Tilly (*Work Under Capitalism*, 34) identified three categories of mine-working men in the late 1800s: hewers (miners and laborers who dug the coal), haulers (who took the coal to the surface), and banksmen (who processed the coal). The hewers' main task was to cut the coal from the face, but they also participated in loading coal cars, timbering the "roof," and laying rails over which the run-of-mine (raw) product traveled to the shaft and then the surface. The term "hewer" derived from the early miners having to cut or chop the coal from the face with hand tools. The terms applied mainly to bituminous mining.

18. Pennsylvania Law 142, approved on May 9, 1889, stipulated that only certified miners could hold the title *miner* and fulfill the occupation's duties, which included, among other tasks, driving chambers, propping the roof, drilling holes for charges, firing the charges, and checking for safety (Pennsylvania Department of Mines, *Anthracite Mining Laws of Pennsylvania*, Harrisburg, PA: Pennsylvania Department of Mines, 1948). The coal operators opposed the law because it prevented the hiring of lower-skilled, cheaper workers, but also because they had difficulty hiring replacement miners during strikes ("Miners Prepare to Fight; Anthracite Operators Trying to Kill Certificate Law," *New York Times*, 12 Mar 1905).

19. Most northern-field mineworkers were paid by the piece, or the car of coal. The work culture encouraged a stint (or quota) of two or three cars per worker. Therefore, a work crew of one miner and two laborers would produce four to six cars per shift, with some variation from company to company and mine to mine depending on working conditions, local customs, and the size of the cars (*Proceedings of the Anthracite Mine Strike Commission, 1902–03*, Scranton, PA: *Scranton Tribune*, 1903, 84, 146). The contract-miner position actually resembled an amalgam of two forms of work organization: *inside contracting* and the *craft system*. The former predominated in nineteenth-century American manufacturing. Based on senior employees who secured company contracts to produce a quantity of goods within a specified time, inside contractors hired laborers and supervised the work process. On inside contracting and the craft system, see Dan Clawson, *Bureaucracy and the Labor Process: The Transformation of U.S. Industry, 1860–1920*, New York: Monthly Review, 1980, 71–125; and David M. Gordon, Richard Edwards, and Michael Reich, *Segmented Work, Divided Workers: The Historical Transformation of Labor in the United States*, New York: Cambridge University Press, 1982, 91–92.

20. Priscilla Long, *Where the Sun Never Shines: A History of America's Bloody Coal Industry*, New York: Paragon, 1989, 65. Chester Brozena from Plymouth, PA, who worked as a hard-coal miner between the late 1910s and the early 1940s, described the pride in his craft: "People underestimated the miner. The miner had to be an engineer on his own. He drove by lines [survey markers]. He didn't just go in there and dig coal. He had to drive by lines; he had to know where he was going, the width and all. Then he had to know his roof condition. He was his own engineer. He had to do this on his own. So he was sort of proud to be [a miner]" (Chester Brozena, audiotaped interview, 3 Dec 1988, NPOLHP). Historian Keith Dix's research on the bituminous miner-as-craftsman clearly applied to anthracite (Keith Dix, *What's a Coal Miner To Do?*

The Mechanization of Coal Mining, Pittsburgh, PA: University of Pittsburgh Press, 1988, 105). For a British perspective on the subject see Royden Harrison (ed.), *Independent Collier: The Coal Miner as Archetypal Proletarian Reconsidered*, New York: St. Martin's, 1978.

21. The Charter Master System and the Butty System were the main British subcontracting schemes. Both were built around a miner who secured a contract from a company, and hired laborers and sometimes other miners to assist in taking the coal. By the early 1800s, subcontracting had virtually disappeared from the northeastern English coal fields, although it persisted in Shropshire until the last quarter of the century, and in some places it continued well into the twentieth century. Workers protested the plans, and subcontracting facilitated the spread of labor unions during the early 1800s. On the Butty System see A.J. Taylor, "The Sub-Contract System in the British Coal Industry," 215–235 in L.S. Pressnell (ed.), *Studies in the Industrial Revolution*, London: University of London-Athlone, 1960; and Dave Douglass, "The Durham Pitman," 226–236 in Raphael Samuel (ed.), *Miners, Quarrymen and Saltworkers*, London: Routledge & Kegan Paul, 1977. On the Chartermaster System see Ashton and Sykes, *The Coal Industry of the Eighteenth Century*, 100; and Michael W. Flinn and David Stoker, *The History of the British Coal Industry*, Vol. 2, Oxford, England: Clarendon, 1984, 55–56, 377.

22. Rochester, *Labor and Coal*, 129.

23. Peter Roberts, *Anthracite Coal Communities: A Study of the Demography, the Social, Educational and Moral Life of the Anthracite Regions*, New York: Macmillan, 1904, 204.

24. Goodrich, *The Miner's Freedom*. Long (*Where the Sun Never* Shines, 138) discussed a mine inspector who described a similar production arrangement in a large bituminous mine in 1885. He reported that the pit functioned with 32 men who operated 32 undercutting machines, 32 boys who cleaned the cutters, and a third group of unspecified (but presumably large) size that blasted the coal and loaded it into cars. Economist David Gordon and his colleagues characterized such schemes as indicative of the "homogenization of the production process" initiated by companies in the late nineteenth century to control work and the worker (Gordon et al., *Segmented Work, Divided Workers*: 157–159).

25. Frank J. Warne, "The Effects of Unionism upon the Mine Workers," *Annals of the American Academy of Political and Social Science*, 21 (Jan 1903), 24.

26. Frank J. Warne, "The Anthracite Coal Strike," *Annals of the American Academy of Political and Social Science*, 17 (1901), 32; Anthracite Coal Strike Commission, *Report to the President on the Anthracite Coal Strike of May-October 1902*, 176.

27. "3,000 Striking Miners Cheer Wildly as Cappellini Closes Dramatic Plea for Support," *Scranton Times*, 7 Jan 1925.

28. Yearley, *Enterprise and Anthracite*, 170. In addition to the adult workers, all collieries employed unskilled boys above and below ground. Breaker boys were the most numerous above-ground workers, whose job it was to handpick waste (slate or rock) out of freshly mined coal. Within the mines, older boys drove mules that pulled coal cars while younger boys worked as "nippers" opening and closing the large doors that directed underground air sources. Roberts (*Anthracite Coal* Communities, 175) reported that 6,400 boys worked in and around the mines at the turn of the century. The National Civic Federation put the number at 12,000, while the Commonwealth of Pennsylvania stated that 8,100 boys under age 16 were employed in hard coal (Owen R. Lovejoy, "The Extent of Child Labor in the Anthracite Coal Industry," *Annals of the American Academy of Political and Social Science*, 29, [Jan 1907], 35–49). See also Lewis Hine, "The High Cost of Child Labor," *The Child Labor Bulletin*, 3 (1915), 63–67.

29. David Montgomery, *The Fall of the House of Labor: The Workplace, the State, and American Labor Activism, 1865–1925*, New York: Cambridge University Press, 1987, 334.

30. Victor R. Greene, "A Study in Slavs, Strikes, and Unions: The Anthracite Strike of 1897," *Pennsylvania History*, 31 (1964), 200.

31. The anthracite industry employed 143,824 men and boys on the eve of the strike of 1900.

32. Warne ("Anthracite Coal Strike," 16) put the capitalization figure at $161,784,473 in 1890. The 2011 figure is based on the Consumer Price Index value of $200 million in 1900, as compared to the same amount in 2011, computed at the website http://www.measuringworth.com/calculators/uscompare/index.php.

33. On the UMWA's founding see Selig Perlman, *Upheaval and Reorganization (Since 1876)*, Part 4, Vol. 2, in John R. Commons et al., *History of Labour in the United States*, New York: Macmillan,

1918, 487; and Maier Fox, *United We Stand: A History of the United Mine Workers of America, 1890–1990*, Washington, DC: United Mine Workers of America, 1990, Chapter 3. On the anthracite strikes between 1894 and 1897 see Greene, "Study in Slavs, Strikes, and Unions"; Perry K. Blatz, "Local Leadership and Local Militancy: The Nanticoke Strike of 1899 and the Roots of Unionization in the Northern Anthracite Fields," *Pennsylvania History*, 58 (1991), 278–297; and Blatz, *Democratic Miners*, Chapter 3. On the early anthracite unions see Harold W. Aurand, *From the Molly Maguires to the United Mineworkers: The Social Ecology of an Industrial Union, 1869–1897*, Philadelphia, PA: Temple University Press, 1971, Chapters 7 and 8; and Craig Phelan, *Grand Master Workman: Terrence Powderly and the Knights of Labor*, Westport, CT: Greenwood, 2000.

34. On Mitchell's life and legacy see John Mitchell, *Organized Labor*, Philadelphia, PA: American Book & Bible House, 1903; Elsie Gluck, *John Mitchell, Miner: Labor's Bargain with the Gilded Age*, New York: John Day, 1929; and Craig Phelan, *Divided Loyalties: The Public and Private Life of Labor Leader John Mitchell*, Albany, NY: SUNY Press, 1994.

35. *Proceedings of the Anthracite Mine Strike Commission*, 1902–03, 146. Tilly and Tilly examined the "intense conflict" over the empty coal cars (the free turn) because "the foreman's favorites got an extra car." The practice infuriated the workers: "In anthracite, where tonnage miners commonly hired laborers by the day and big subcontractors who did not themselves mine often stood between hewers and management, competition for cars became a bitter issue" (Tilly and Tilly, *Work Under Capitalism*, 49).

36. Andrew Roy found that England experienced one fatality for every 146,869 tons mined in 1873, while in all Pennsylvania mines the rate was one death for every 83,457 tons (Andrew Roy, *The Coal Mines: Containing a Description of the Various Systems of Working and Ventilating Mines*, Cleveland, OH: Robinson & Savage, 1876, 155, 159). David Rosner and Gerald Markowitz (*Dying for Work: Workers' Safety and Health in Twentieth Century America*, Bloomington, IN: Indiana University Press, 1989, xi) indicated that between 1900 and 1906 the U.S. mine-accident rate was four times that of France and two times that of Prussia. In a comparison of British and American mine-safety policy Aldrich concluded: "In Britain, where mining laws were a national matter, operator influence was diluted and regulations governed the main hazards of mining. American state law, by contrast, bore the strong imprint of operator influence" (Mark Aldrich, *Safety First: Technology, Labor, and Business in the Building of American Work Safety, 1870–1934* Baltimore, MD: Johns Hopkins University Press, 1997, 71). For a comparison of international mining death rates see also Aurand, "Mine Safety and Social Control," 228.

37. John Dorish, audiotaped interview, 31 Jul 1988, NPOLHP; William A. Hastie, audiotaped interview, 31 Jul 1989, NPOLHP; Alex Chamberlain, audiotaped interview, 11 Jul 1995, NPOLHP.

38. *Proceedings of the Anthracite Mine Strike Commission, 1902–03*, 17–18. On the Anthracite Strike Commission see Appendix I. Establishing culpability for accidents had always been a difficult subject in anthracite. When over 11,000 mineworkers died in work-related deaths between 1870 and 1900, another nearly 15,000 between 1901 and 1930, and a grand total of approximately 35,000 throughout hard coal's history, the companies and their officials, as well as the mine inspectors, remained virtually blameless. (On the fatality statistics see the Pennsylvania Department of Mines and Mineral Industries, *Annual Report: Anthracite Division*, 1961, 10). Instead, the workers were most often cited for having a careless disregard for personal safety. Peter Roberts (*The Anthracite Coal Industry: A Study of the Economic Conditions and Relations of the Co-operative Forces in the Development of the Anthracite Coal Industry of Pennsylvania*, New York: Macmillan, 1901, 154) suggested that up to 50 percent of all mining accidents were caused by employee carelessness. However, Frederick L. Hoffman ("Problems of Labor and Life in Anthracite Coal Mining," *Engineering and Mining Journal*, 74 [1902], 811) criticized Roberts and others for too easily blaming the workers: "Inspectors naturally are not likely to assume the responsibility for accidents which they can so readily shift upon the employees. ... The problem is not solved by shifting the responsibility upon men who must make their living by producing as much coal as possible in a given length of time, or whose very ignorance of the language precludes a thorough understanding of the rules and regulations intelligently framed for the health and safety of men employed in this occupation." Wallace (*St. Clair*, Chapter 5) argued that employers in the southern-field town of St. Clair similarly blamed workers first, but, in reality, the overriding problem was an inherently unsafe geology, business organization, and mining system.

39. Douglas Monroe, *A Decade of Turmoil: John L. Lewis and the Anthracite Miners, 1926–1936*, unpublished doctoral dissertation, Georgetown University, 1976, 81–82.

40. *Ibid.* On work culture and the working class see Herbert G. Gutman, "Work, Culture, and Society in Industrializing America, 1815–1919," *American Historical Review*, 78 (Jun 1973), 531–558.

41. Frederick W. Taylor, *The Principles of Scientific Management*, New York: Harper and Row, 1911, 48–49.

42. See Goodrich (*The Miner's Freedom*) on Taylorism and mechanization in coal mining. Taylor and Goodrich are again discussed in Chapter 7.

43. *Proceedings of the Anthracite Mine Strike Commission, 1902–03*, 17.

44. Montgomery, *Workers' Control in America*, 14–15.

45. William Graham, audiotaped interview, 11 Jul 1995, NPOLHP.

46. Warne, "The Anthracite Coal Strike," 32. On the practice of excessive shipping charges by the anthracite railroads see Pennsylvania Anthracite Coal Industry Commission, *Report of the Pennsylvania Anthracite Coal Industry Commission*, Harrisburg, PA: Commonwealth of Pennsylvania, 1938, 324.

47. "Report of the Anthracite Coal Strike Commission of 1902, Reprinted with Subsequent Agreements and the Majority and Minority Reports of the U.S. Anthracite Commission of 1920," Philadelphia, PA: Anthracite Bureau of Information, Oct 1920, 113.

48. Powers Hapgood and Mary Donovan, "Murdered Miners," *The Nation*, 14 Mar 1928, 293.

49. *Proceedings of the Anthracite Mine Strike Commission, 1902–03*, 14.

50. Elizabeth Gurley Flynn, *The Rebel Girl—An Autobiography, My First Life (1906–1926)*, New York: International Publishers, 1973 [1955], 121.

51. On the strikes of 1900 and 1902 see *Proceedings of the Anthracite Mine Strike Commission, 1902–03*; Warne, "The Anthracite Coal Strike;" Robert Cornell, *The Anthracite Coal Strike of 1902*, Washington, DC: Catholic University Press, 1957; Robert H. Wiebe, "The Anthracite Strike of 1902: A Record of Confusion," *Mississippi Valley Historical Review*, 48 (1961), 229–251; Greene, *The Slavic Community on Strike*, Chapter 8; Joe Gowaskie, "John Mitchell and the Anthracite Mine Workers: Leadership Conservatism and Rank-and-File Militancy," *Labor History*, 27 (1985), 54–83; Blatz, *Democratic Miners*, Chapter 4; and Dublin and Licht, *The Face of Decline*, Chapter 2.

52. In an attempt to eliminate the individual contract in 1900, the UMWA demanded "no more than one breast, gangway or working place for one miner at any time nor more than an equal share of cars and work for any miner" (*Proceedings of the Twelfth Annual Convention of the United Mine Workers of America*, Indianapolis, IN: Hollenback, 1901, 29) For a discussion of the 1900 and 1902 strikes in relation to the subcontracting system see Robert P. Wolensky, "The Contracting-Leasing System and Industrial Conflict in the Northern Anthracite Field," 67–94 in Lance Metz (ed.), *The Great Strike: Perspectives on the 1902 Anthracite Coal Strike*, Easton, PA: Canal History and Technology Press, 2002.

53. Selig Perlman (*A History of Trade Unionism in the United States*, New York: Macmillan, 1922, 177) described the early public support for the 1902 strike: "What distinguished the anthracite coal strike ... was that for the first time labor organizations tied up for months a strategic industry and caused wide suffering and discomfort to the public without being condemned as a revolutionary menace to the existing social order calling for suppression by the government; it was, on the contrary, adjudged a force within the preserves of orderly society and entitled to public sympathy." However, the public's sympathy waned as the strike carried into the cold months.

54. The company executives demanded that the commission's members include a military engineer, a mining engineer, a judge, an expert in the coal business, and an "eminent sociologist." They were willing to accept a union leader as the "eminent sociologist," and Roosevelt named E.E. Clark, president of the Order of Railway Conductors, to the position. Pressure by Catholics led to the appointment of Bishop John Lancaster Spalding of Peoria as the sixth member. The other members were: Brigadier General John M. Wilson; Thomas H. Watkins, mining engineer; E.W. Parker editor and mining expert; Judge George Gray, U.S. Circuit Court, chairman; and Col. Carroll D. Wright, the U.S. Commissioner of Labor, who was the last member appointed and served as the recorder.

55. Anthracite Coal Strike Commission. *Report to the President on the Anthracite Coal Strike of May-October 1902*, Washington, DC: USGPO, 1903, 270–296.

56. Cornell, *The Strike of 1902*, 155–157.

57. Pope Leo XIII, *Rerum Novarum*, "Rights and Duties of Capital and Labour: On the Conditions of Labour," New York: Paulist, 1939. Also available on-line at http://www.vatican.va/holy_father/leo_xiii/encyclicals/documents/hf_l-xiii_enc_15051891_rerum-novarum_en.html.

58. *Ibid.*, paragraph 60.

59. *Ibid.*, paragraph 45.

60. *Ibid.*, paragraph 45.

61. *Ibid.*, paragraph 49.

62 *Ibid.*, paragraph 49.

63. In trying to avert a strike, the UMWA originally proposed a five-member arbitration committee selected either by the National Civic Federation, or by a committee consisting of Archbishop John Ireland of St. Paul, Minnesota, Bishop Henry Potter of New York, and one additional member chosen by these two. The operators rejected both options.

64. Rev. John P. Gallagher, *A Century of History: The Diocese of Scranton, 1868–1968*, Scranton, PA: The Diocese of Scranton, 1968, 212.

65. On Curran's participation in the 1902 strike settlement see Sr. Mary Annunciata Merrick, *A Case in Practical Democracy: Settlement of the Anthracite Coal Strike of 1902*, unpublished doctoral dissertation, University of Notre Dame, 1942; and Robert Cornell, *The Anthracite Coal Strike of 1902*, Washington, DC: Catholic University Press, 1957, 114–115 and 158–159. On other aspects of Curran's life see the audiotaped interview with his long-time secretary, Agnes Kupstas, 4 Dec 1988, NPOLHP. PHMC erected a historical marker near Holy Saviour Church, Wilkes-Barre, commemorating Curran's exertions on behalf of the strike settlement. Post-strike correspondence between Curran and Roosevelt can be found in the John Joseph Curran Papers, University of Notre Dame Archives. Other Roman Catholic clergy active during the 1902 strike were Fr. Malek of Plymouth, Fr. Szmoski of Plymouth, Fr. Schlosser of Wilkes-Barre; and Fr. (later Bishop) Hodur, who became the head of the Polish National Church in Scranton (Edward F. Hanlon, *The Wyoming Valley: An American Portrait*, Woodland Hills, CA: Windsor, 1983, 104–106). Curran remained a mineworkers' advocate over the next three decades, and will reappear in later chapters.

66. According to Morris J. MacGregor, Mitchell "became a primary exponent of the Church's teachings," and his speeches often contained "the philosophy and at times the exact wording of the encyclical [*Rerum Novarum*]" (Morris J. MacGregor, *Steadfast in the Faith: The Life of Patrick Cardinal O'Boyle*, Washington, DC: Catholic University of America Press, 2006, 11). See also Agnes Joy, "John Mitchell and Religious Leaders of the Period 1900–1910, unpublished M.A. thesis, Catholic University of America, 1954. Mitchell's wife was a Catholic and, in October 1907, she asked Fr. Curran to pray that her husband might convert to her faith. In December 1907, during an illness when he thought he might die, Mitchell summoned a priest and became a Catholic (Philip S. Foner, *The History of Labor in the United States: The Policies and Practices of the American Federation of Labor, 1900–1909*, New York: International Publishers, 1981 [1964], 120). Mitchell often dressed in black like a priest and wore what looked somewhat like a Roman collar (see figure 11).

67. Critics have argued that Leo XIII's encyclical was motivated as much by the fear of socialism as by the desire to improve the lot of the working class. The encyclical represented one of the first documents in what has been called Catholic Social Teaching. Other papal encyclicals have reinforced the right of workers to form labor unions, as did the American Catholic Bishops in a "pastoral letter" issued in 1986: "Labor unions help workers resist exploitation. ... The Church fully supports the right of workers to form unions or other associations to secure their rights to fair wages and working conditions" (United States Conference of Catholic Bishops, *Economic Justice For All: Pastoral Letter on Catholic Social Teaching and the U.S. Economy*, Washington, D.C.: USCCB, 1986, para 104). However, church support for unions is rarely practiced.

68. Thomas Larkin ("History of Labor Relations in the Anthracite Industry," 205–218 in *Transactions of the Fifth Annual Anthracite Conference of Lehigh University*, May 8–9, Bethlehem, PA, 1947, 207) categorized the UMWA's anthracite history into three periods following the strike of 1902: (1) 1903–1912—adjudication of grievances between employer and employee under a weak labor organization; (2) 1912–1923—grievance committees assistance in handling disputes to eventual complete union recognition; and (3) 1923–1947—grievance adjustments under business unionism with a closed shop and the check-off.

69. Warne, "The Anthracite Coal Strike," 51.

70. For a discussion of strikes as a response to management's desire to control labor in a highly competitive environment see Jon Amsden and Stephen Brier, "Coal Miners on Strike: The Transformation of Strike Demands and the Formation of a National Union," *Journal of Interdisciplinary History*, 8 (Sep 1977), 583–616.

71. *Proceedings of the UMWA District No. 1 Convention*, 1901, 7–9.

72. *Proceedings of the Anthracite Mine Strike Commission, 1902–03*, 131–132. The *Proceedings* are not clear on which Jermyn Colliery experienced the strike. There were at least three different Jermyn Collieries in the northern field in 1901. The Jermyn No. 3 Colliery in Manville was the only one operated by the D&H.

73. *Proceedings of the Anthracite Mine Strike Commission, 1902–03*, 132.

74. *Proceedings of the UMWA Tri-District (Districts 1, 7, 9) Convention of 1902*, 20–21.

75. *Proceedings of the Anthracite Mine Strike Commission, 1902–03*, 14.

76. *Ibid*, 146.

77. Montgomery, *Workers' Control in America*, 13.

78. Long, *Where the Sun Never Shines*, 69.

79. *Proceedings of the Anthracite Mine Strike Commission, 1902–03*, 146.

80. Lewis Casterline, audiotaped interview, 16 Jan 1970, interviewed by Clement Valetta and Mary Barrett, King's College Oral History Collections.

81. As early as the 1850s, the companies sought to stabilize prices by creating coal pools to limit the tonnage going to market (Yearley, *Enterprise and Anthracite*, 158–164). In 1868, numerous firms again agreed to reduce output by supporting a work stoppage sponsored by the Workingman's Benevolent Association in protest over the implementation of Pennsylvania eight-hour work day law (Marvin W. Schlegel, "The Workingmen's Benevolent Association: First Union of Anthracite Miners," *Pennsylvania History*, 10 [1943], 244). In 1873, the coal-carrying railroads inaugurated the first successful combination to stabilize supply and prices, but it lasted only a few years because members overshot their quotas. The same parties formulated similar agreements five more times before 1900 (Jones, *The Anthracite Coal Combination in the United States*, Chapter 3, and Appendix, Table 6). In 1898, a highly consolidated industry established a fairly effective production plan through a so-called "community of interest" among the mine-owning railroads. The nineteenth-century consortiums failed, said Long (*Where the Sun Never Shines*, 123), because "When prices started rising [the companies] scrambled to produce more than their agreed-upon allotment. The system was constantly breaking down." Warne ("The Anthracite Coal Strike," 21) discussed the demise of the 1898 consortium: "Among the operators it is known as 'an understanding among gentlemen. . . .' [but] the statistics of production show that the collieries rarely, if ever, keep within the allotments." The cartel-like machinations by the coal companies to manipulate supply and price was not limited to the U.S. In Britain several operators formed The Grand Allies, which functioned between 1710 and 1726 "for the purpose of restricting output and raising the price. As such it has much in common with other associations of owners, though for a time it was more successful than most" (Flinn and Stoker, *The History of the British Coal Industry*, 40).

82. *Proceedings of the Anthracite Mine Strike Commission, 1902–03*, 177.

83. *Ibid.*, 166.

84. Quoted in Goodrich, *The Miner's Freedom*, 43.

85. Quoted in Perry K. Blatz, "Workplace Militancy and Unionization: The UMWA and the Anthracite Miners, 1890–1912," Chapter 3 in John H.M. Laslett (ed.), *The United Mine Workers of America: A Model of Industrial Solidarity?* University Park, PA: Penn State Press & Penn State University Libraries, 1996, 67.

86. Anthracite Coal Strike Commission, *Report to the President*, Appendix A, 112.

87. Warne, "The Effects of Unionism upon the Mine Workers," 24.

88. Goodrich, *The Miner's Freedom*, 130.

89. Montgomery, *The Fall of the House of Labor*, 334.

90. Fox, *United We Stand*, 177.

91. The Anthracite Coal Controversy, Demands of 1906, Scranton Public Library Donation Boxes, Anthracite Heritage Museum, uncatalogued information sheets.

92 *Proceedings of UMWA Tri-District (Districts 1, 7, 9) Convention, 1916*, 70.

93. On the Padrone System in the U.S. see Herman Stump, J.H. Senner, and Edward F. McSweeney, *Report of the Immigration Investigation Commission to the Honorable Secretary of the Treasury*, Washington, DC: USGPO, 1895, 36. See also Carroll D. Wright, "Padrones in Chicago," *Bulletin of the Department of Labor*, 691–727 in *Italians in Chicago: A Social and Economic Study*, Washington, DC: USGPO, Jan 1897; Carter Goodrich, "Contract Labor," 342–344 in Edwin R.A. Seligman (ed.), *Encyclopaedia of the Social Sciences*, Vol. 4, New York: Macmillan, 1931; Humbert S. Nelli, "The Italian Padrone System in the United States," *Labor History*, 5 (1964), 153–167; Caroline Golab, *Immigrant Destinations*, Philadelphia, PA: Temple University Press, 1977, 58–63; and Gunther Peck, "Reinventing Free Labor: Immigrant Padrones and Contract Labor in North American, 1885–1925," *Journal of American History*, 83 (1996), 848–871. On the Irish padrone-type system in Philadelphia see Dennis J. Clark, *The Irish in Philadelphia: Ten Generations of Urban Experience*, Philadelphia, PA: Temple University Press, 1973. The Padrone system shared some similarities with Scotland's *Collier-Serf* system, which provided the country's hereditary mining workforce from 1606 until 1799 (Ashton and Sykes, *The Coal Industry of the Eighteenth Century*, Chapter 5; Long, *Where the Sun Never Shines*, 14–15; and Flinn and Stoker, *The History of the British Coal Industry*, 358–361). Golab (*Immigrant Destinations*, 58) traced the Padrone system to ancient Mesopotamia and found that in more recent times "the Italians perfected the system." The Italian involvement during the nineteenth century derived from a combination of high national unemployment, a strong work ethic, and other countries' demand for cheap laborers as well as skilled workers such as masons and stone carvers. By the 1890s, Italian immigrants had replaced the Irish as America's main source of unskilled labor. The U.S. Congress legalized labor contracting in 1864 and allowed brokers to garner laborers' wages for one year in payment for a work contract. However, the law was repealed in 1868 because of the exploitation it engendered. Labor contracting nevertheless continued because of the high demand from businesses. The Italian government enacted a law in 1873 prohibiting the inveigling of children by Padrones. Numerous other immigrant groups were subjected to the labor-contracting system including Greeks, Mexicans, and various Slavs and Asians (Peck, "Reinventing Free Labor," 1996). For a personal account of work under a Padrone see Pascal D'Angelo, *Son of Italy*, New York: Arno, 1975 [1924]. On the role of Padrones in Italian emigrations to various parts of the world in the nineteenth century see Steven Colatrella, *Workers of the World: African and Asian Migrants in Italy in the 1990s*, Trenton, NJ: Africa World Press, 2001, 65–70. The New York Society for the Prevention of Cruelty to Children criticized the Padrone system in *The Italian Padrone Case: U.S. Against Antonio Giovanni Ancarola*, New York: Styles & Cash, 1880.

94. The two stages of the Padrone system were discussed by Luciano John Iorizzo, *Italian Immigration and the Impact of the Padrone System*, New York: Arno, 1980 [1966], Chapter 4. On the unfairness of an Italian Padrone system operating in Philadelphia in the 1920s see Richard A. Varbero, "Workers in City and County: The South Italian Experience in Philadelphia, 1900–1950," 16–32 in Richard N. Juliani and Philip V. Cannistrararo (eds.), *Italian Americans: The Search for a Usable Past, Proceedings of the 19th Annual Conference of the American Italian Historical Association*, Staten Island, NY: American Italian Historical Association, 1989, 18–19.

95. Powderly as quoted in Stump et al., *Report of the Immigration Investigation Commission to the Honorable Secretary of the Treasury*, 36. See also Wright (*Italians in Chicago*, 691–727) who reported that nearly 22 percent of the 1,860 Italians surveyed were employed by Padrones. The Padrones in Wright's study took a commission on the workers' salaries and forced them to purchase food stuffs and supplies at inflated prices.

96. Greene, *The Slavic Community on Strike*, 11.

97. Michael La Sorte, *La Merica: Images of Italian Greenhorn Experience*, Philadelphia, PA: Temple University Press, 1985, 70. Lewis Casterline (audiotaped interview, 16 Jan 1970) spoke of Mrs. Liboranti, owner of a drinking establishment in Jessup, near Scranton, who "was paid by the management of the coal companies and she'd go to Italy and recruit these young boys to come over here. They called her *regina*, which means queen. She'd go over there and get all these young fellows to come over. [She'd tell them]: 'Don't you worry, there's a bed for you and there's a job.' And the ticket was paid. Most of them were Gubinis from Gubio. They were all Peruginis [from the province of Perugia]." On the durability of the Padrone system in the U.S. see Nelli, "The Italian Padrone System in the United States.

Notes to Chapter Two

1. Quoted from the *Laws of Pennsylvania*, 1838, No. 73, in Eliot Jones, *The Anthracite Coal Combination in the United States*, Cambridge, MA: Harvard University Press, 1914, 24.

2. On PaCC's history see Wyoming Coal Association and Washington Coal Association, Resolution of the Managers of the Wyoming Coal Association for the transfer of real estate to the Pennsylvania Coal Company, 8 Nov 1848, 1989.343, Folder 1, NCMA; Washington Coal Company, Minutes of the meeting of the directors of the Washington Coal Co. at Carbondale, Pennsylvania, 25 Jan 1848, 1989.343, Folder 1, NCMA; and Wyoming Coal Association, Agreement between the Delaware & Hudson Canal Co. and the Wyoming Coal Association, 31 Aug 1847, 1989.343, Folder 1, NCMA. See also "The Pennsylvania Coal Company," *Scranton Directory*, 1867; J.A. Clark (ed.), *The Wyoming Valley, Upper Waters of the Susquehanna, and the Lackawanna Coal-Region*, Scranton, PA: J.A. Clark, 1875; Hudson Coal Company, *The Story of Anthracite*, New York: Hudson Coal Company, 1932, 101; *History of Luzerne, Lackawanna, and Wyoming Counties*, New York: Munsell, 1880, 81–82; George H. Minor, *The Erie System: The Organization and Corporate History*, Published by the Erie Railroad, 1938, 429–436; Spiro Patton, "Delaware & Hudson Company vs. Pennsylvania Coal Company During the 1850s," 7–24 in Lance E. Metz (ed.), *Canal History and Technology Proceedings*, Vol. XI, Easton, PA: Canal History and Technology Press, 1992.

3. Mary Theresa Connolly, *'The Gravity': History of the Pennsylvania Coal Company Railroad, 1850–1885*, Olyphant, PA: Barrett, 1972; Edward Steers, "The Pennsylvania Coal Company's Gravity Railroad," 155–228 in Lance E. Metz (ed.), *Canal History and Technology Proceedings* Vol. 1, Easton, PA: Canal History and Technology Press, 1982; Michael Knies, "The Pennsylvania Coal Company: New Insights from the James Archbald Papers, 1852–1853, 'We Are Now at War,' " 45–76 in Lance Metz (ed.), *Canal History and Technology Proceedings*, Vol. XIII, 2004.

4. Among the PaCC board of directors between 1851 and 1853 were William R. Griffith, Isaac L. Platt, William Harvemeyer, Mr. Stout, Daniel Parish, John Ewen, Irad Hawley, William H. Falls, Moses Taylor, Robert B. Minturn, and George A. Hoyt. PaCC's ownership of large land tracts allowed the company to sell surface rights for home and other building use. At least some (and possibly most or all) of the land deeds granted the surface rights to the owner but retained the mineral rights for the company. Moreover, the deeds held the company blameless for any damage caused by mining. The deed issued to William A. Hastie's great-grandfather in 1864, for a property in Pittston, contained the hold-harmless clause.

5. Burton W. Folsom Jr., *Urban Capitalists: Entrepreneurs and City Growth in Pennsylvania's Lackawanna and Lehigh Regions, 1800–1920*, Baltimore, MD: Johns Hopkins University Press, 1981, 92.

6. "The Pennsylvania Coal Company," *Scranton Directory*, 3–4.

7. *Ibid.*, 8.

8. Oscar Jewell Harvey and Ernest Gray and Smith, *A History of Wilkes-Barre*, Vol. 4, Wilkes-Barre, PA: Smith Bennett, 1929.

9. "The Pennsylvania Coal Company," *Scranton Directory*, 5–6.

10. Steers, "Gravity Railroad," 162.

11. T. Addison Busbey (ed.), *The Biographical Directory of Railway Officials of America*, Chicago, IL: Railway Age, 1901.

12. Folsom, *Urban Capitalists*, 86, 94.

13. Edward J. Davies, *The Anthracite Aristocracy: Leadership and Social Change in the Hard Coal Regions of Northeastern Pennsylvania, 1800–1930*, DeKalb, IL: Northern Illinois University Press, 1985, 70–71.

14. Immigration continued to the Pennsylvania coal fields from Scottish towns such as Leadhills, Wanlockhead, Carsphairn, and the ironstone and coal-mine towns of Ayrshire. Five men left Shotts in North Lanarkshire and eventually worked for PaCC as late as the 1940s: Robert Groves, William Sinclair, Jock Williams, Andrew Wilson, and James Jamieson. They all later joined Shotts native Charles Irvin as employees of the Knox Coal Company, which leased mineral rights from PaCC beginning in 1943. Groves was the Knox superintendent at the time of the infamous Knox Mine Disaster (Robert P. Wolensky et al., *The Knox Mine Disaster: The Final Years of the Northern Anthracite Industry and the Effort to Rebuild a Regional Economy*, Harrisburg,

PA: PHMC, 1999). Some Scottish bosses were known to have created tensions with the newer European immigrants. For example, Portuguese miner Joe Costa remembered with anger a Scottish superintendent at the Anthracite Collieries Company in Scranton who demanded a kickback for a subcontract in the 1930s (Joe Costa, unrecorded interview, 11 Oct 1993, NPOLHP).

15. Harold W. Aurand, *Anthracite Heritage Museum and Scranton Iron Furnaces, Pennsylvania Trail of History*, Mechanicsburg, PA: Stackpole, 2002, 26.

16. On the strike of 1869 see Harold W. Aurand, *From the Molly Maguires to the United Mine-workers: The Social Ecology of an Industrial Union, 1869–1897*, Philadelphia, PA: Temple University Press, 1971, Chapter 7; and Robert P. Wolensky and Joseph M. Keating, *Tragedy at Avondale: The Causes, Consequences, and Legacy of the Pennsylvania Anthracite Coal Industry's Most Deadly Mining Disaster, September 6, 1869*, Easton PA: Canal Museum and Technology Press, 2008, Chapter 3.

17. J.A. Dacus, *Annals of the Great Strikes in the United States,* Chicago, IL: L.T. Palmer, 1877, 151–53; Wolensky and Keating, *Tragedy at Avondale*, Chapter 3.

18. The new Erie had been the former New York & Erie Railway Company, first incorporated in 1833, and reorganized after bankruptcy in 1859.

19. Pennsylvania Anthracite Coal Industry Commission, *Report of the Pennsylvania Anthracite Coal Industry Commission*, Harrisburg, PA: Commonwealth of Pennsylvania, 1938, 398.

20. PaCC's collieries included the Dawson, Gypsy Grove, Law, Stark, and Nos. 1, 2, 4, 5, 6, 7, 8, 9, and 14 (Colliery Books, Pennsylvania Coal Company, 1983.177, Folders 1–12, Boxes 1 & 2, PaCC Papers, NCMA; Minor, *The Erie System*, 1938, 435; www.northernfield.info/).

21. The other five major northern anthracite companies were also controlled by railroads: Delaware, Lackawanna, & Western Railroad; Delaware & Hudson Coal Company; Lehigh Valley Coal Company; Lehigh & Wilkes-Barre Coal Company; and Susquehanna Collieries Company (Jones, *The Anthracite Coal Combination*, 82–84).

22. On Erie's history see Stuart Daggett, *Railroad Reorganization*, Washington, DC: Beard, 1999 [1908]; Edward H. Mott, *Between the Ocean and the Lakes: The Story of Erie*, New York: Ticher, 1908; Jones, *The Anthracite Coal Combination*, 122–134; and Matthew Josephson, *The Robber Barons*, New York: Harcourt, Brace & World, 1962 [1934], Chapter 6; Minor, *The Erie System.*

23. Daggett, *Railroad Reorganization*, 73.

24. Minor, *The Erie System*, 219–226. See the HC&I Papers, MG 282, PHMC, PSA. The authors are very grateful to Jim Guthrie and Frank Adams for consultation on Erie's many problems, incarnations, strategies, and purchases.

25. Minor, *The Erie System*, 225.

26. Pennsylvania Bureau of Mines, *Report of the Bureau of Mines*, Department of Internal Affairs of Pennsylvania, Harrisburg, PA, 1899. See also Frank J. Warne ("The Anthracite Coal Strike," *Annals of the American Academy of Political and Social Science*, 17, [1901], 18), who listed the percentage of output by the large Line producers in 1899: Philadelphia & Reading (20.3 percent; 9,684,000 tons); Lehigh Valley (15.9 percent; 7,589,000); Delaware, Lackawanna & Western (13 percent; 6,372,000); D&H (13.5 percent; 6,430,000); and Central of New Jersey (11.1 percent; 5,393,000); PaCC (4.9 percent or 6,851,000 tons). All other coal companies, including the HC&I, produced 20.7 percent, or 9,851,000 tons.

27. Because NYS&WCC produced no coal of its own but contracted for the output of several independents, this study focuses on PaCC and HC&I only, which together employed a workforce approaching 12,000 by 1910.

28. Minor, *The Erie System*, 429.

29. Jones, *The Anthracite Coal Combination*, 74–75, 88–90; Robert Cornell, *The Anthracite Coal Strike of 1902*, Washington, DC: Catholic University Press, 1957, 36.

30. The Temple Coal Company, which Morgan had purchased on 1899, took control of the Simpson & Watkins Company, but Morgan's Guaranty Trust actually held the stock (Jones, *The Anthracite Coal Combination*, 79; see also "Companies Merge," *Pittston Gazette*, 10 Mar 1899).

31. The 2011 figure is based on the Consumer Price Index value of $32 million in 1901, as compared to the same amount in 2011, computed at the website www.measuringworth.com/ppowerus/.

32. Josephson, *The Robber Barons*, 413; Jones, *The Anthracite Coal Combination*, 84, 124; Cornell, *The Anthracite Coal Strike of 1902*, 36; Pennsylvania Anthracite Coal Industry Commission, *Report*, 398.

33. Josephson, *The Robber Barons*, 412.

34. Jones, *The Anthracite Coal Combination*, 24. PaCC's pattern of high rates of return began early according to H.C. Bradsby (*History of Luzerne County*, Chicago, IL: Nelson, 1893, 287). He reported that the favorable transportation terms PaCC secured from the D&H canal and from the Erie Railway "has given the company important advantages over rival companies. Without the heavy cost of locomotive railroads, owned or leased, or large indebtedness to draw interest from its treasury, it has been able to make dividends which sent its stock up to two hundred and eight percent, while other stocks were below par in the markets." Warne ("The Anthracite Coal Strike," 20) cited another advantage: "This company, however, has the exceptional advantage that its coal lands were bought many years ago at a small cost and it is not obliged to pay heavy royalties or interest on extravagant first-cost of property, as many of its competitors do."

35. In addition to PaCC's 2.9 million tons of output in 1902, HC&I produced 1.74 million tons, while the NYS&WCC had contracts for 1.33 million tons. In contrast, the largest anthracite producer in 1902, the Philadelphia & Reading, harvested 9.33 million tons (Anthracite Coal Strike Commission, *Report to the President on the Anthracite Coal Strike of May–October 1902*, Washington, DC: USGPO, 1903, Appendix J, 254). Following Erie's purchase of PaCC, the coal contracted by NYS&WCC was turned over to PaCC and sold under its label (Jones, *The Anthracite Coal Combination*, 124).

36. In June, 1907, The U.S. Justice Department filed an anti-trust action to dissolve the "anthracite combination." The government charged that, since 1892, there had been a conspiracy to monopolize the industry. Central to the government's case was the Morgan interests' defeat of PaCC's competing rail line and Erie's purchase of PaCC. On December 8, 1910, the U.S. Circuit Court ruled against the conspiracy charge. The government appealed to the U.S. Supreme Court, which ruled on December 6, 1912, that the competing line's collapse followed by Erie's purchase of PaCC did not constitute illegal practices (Jones, *The Anthracite Coal Combination*, 215). See also George O. Virtue, "The Anthracite Combination," *Quarterly Journal of Economics*, 10 (Apr 1896), 296–323.

37. The dividend payments and stockholder returns were listed in the U.S. Coal Commission, *Report*, Part II–Anthracite, Washington, DC: USGPO, 1925, 930, 964. Among the major producers, the Lehigh & Wilkes-Barre Coal Company distributed annual gains of about 19 percent. For a lower calculation of HC&I's profitability see Jones, *The Anthracite Coal Combination*, 123.

38. *Proceedings of the Anthracite Mine Strike Commission*, 1902–03, Scranton, PA: Scranton Tribune, 1903, 285; Anthracite Coal Strike Commission, *Report to the President*, 183–184.

39. *Proceedings of the Anthracite Mine Strike Commission*, 136. While all large producers opposed the UMWA, some independents were more accommodating. The Kingston Coal Company was one firm that reached an early settlement with the UMWA, according to Alex Chamberlain, whose family held interests in the company (Alex Chamberlain, audiotaped interview, 11 Jul 1995, NPOLHP). The Kingston Coal Company's papers are held by the Luzerne County Historical Society.

40. *Proceedings of the Anthracite Mine Strike Commission*, 139.

41. Anthracite Coal Strike Commission, *Report to the President*, 112–113. The date in the statement is inaccurate. The UMWA first came to anthracite in 1894 and organized several locals in the Schuylkill Region.

42. *Proceedings of the Anthracite Mine Strike Commission*, 145.

43. *Ibid.*

44. *Ibid.*

45. *Proceedings of the UMWA Tri-District Convention of 1902*, 20–21; emphases added.

46. *Ibid.*

47. *Proceedings of the Anthracite Mine Strike Commission*, 146.

48. PaCC's subcontracts and leases are located at PHMC, Harrisburg, PA, MG 282.

49. Warne, "The Anthracite Coal Strike," 5.

50. In 1911, PaCC operated ten collieries; six were located in Luzerne County (No. 6, No. 9, No. 14, Ewen, Central, and Barnum) and four in Lackawanna County (No. 1, No. 5, Old Forge, and Gipsy Grove) (see Colliery Books, 1983.177, Folders 1–12, Boxes 1 & 2 PaCC Papers, NCMA; Minor, *The Erie System*, 1938, 435; and www.northernfield.info/).

51. Boland's contracts and leases were written between October 1909 and September 1930; the contracts included No. 408, recorded in Deed Book 246, 192, and No. 12056, Contract File Index, Pittston Co. & Canceled, PaCC Papers, PSA. The Executive Committee deliberated and approved a Boland lease at a meeting on October 23, 1918: "President May asked for authority to execute an agreement dated October 24th, 1918, with John J. Boland leasing to the latter pillars in the Third Dunmore Vein underneath the same surface where he is now mining in the Clark and First Dunmore Veins; the royalty to be 75 cents a ton for prepared sizes, 45 cents for pea, 30 cents for buckwheat, and 22½ cents for sizes smaller" (Stockholders and Executive Committee Minutes, Folder: 20 May 1915 to 26 Nov 1919, Minutes, 1838– 1971, No. 6, Doc No. 1026, 203, PaCC Papers, PSA).

52. Contract File Index, Agreement No. 301, PaCC Papers, PSA.

53. UMWA leader Henry A. Schuster put 1918 as the date when PaCC expanded the subcontracting system ("Peace Restored and Miners Will Start Back Today," *Wilkes-Barre Record*, 10 Feb 1934). District 1 President Michael Kosik offered 1916 as the date of re-initiation "under several companies," including PaCC and Hudson (*Proceedings of the UMWA Tri-District Convention, 1939*, 71). Luzerne County Historian Charles A. McCarthy confirmed that subcontracts were offered to McHale and Mitchell in 1913 at the No. 6 Colliery, they turned down the offer, and Santo Volpe and Charles Consagra, who are discussed again in Chapter 4, accepted the agreements (Charles A. McCarthy, "1920 Mine Strike Included Murder of Pittston Detective," *Wilkes-Barre Record*, 26 Oct 1965; see also Charles A. McCarthy, "Pittston Montedoro Society Golden Jubilee to Be Sunday," *ibid.*, 7 Feb 1966). However, as indicated, it is clear that PaCC began offering tenancy agreements as early as 1909, with the earliest subcontractors and leaseholders being English-speaking "Americans." There is little evidence to support Selig Perlman's claim in his investigations for the Commission on Industrial Relations that, prior to 1916, "The [sub]contractor is more often a Pole than he is an English-speaking person," notwithstanding the fact that there were Polish subcontractors (Perlman as quoted in David Montgomery, in *The Fall of the House of Labor: The Workplace, the State, and American Labor Activism, 1865–1925*, New York: Cambridge University Press, 1987, 334).

54. Subcontracting was also present in the Lehigh Region in 1913 at the Oak Hill Colliery. At the UMWA District 9 convention that year, delegate Dando from L.U. 1640 complained: "At our colliery, Oak Hill, where I am working in, about half the work done in that colliery is done under big [sub]contracts. Men have refused work there just because they can't get a breast heading. I got to work for somebody else if I want to be a Union man. If I want to be a Union man, I got to either work for them or do company work. Is it right that we should be put down like that?" (*Proceedings of the Fourteenth Annual Convention*, UMWA District 9, 1913, 186–187).

55. Stockholders and Executive Committee Minutes, Folder: 20 May 1915 to 26 Nov 1919, Minutes, 1838–1971, No. 6, Doc. No. 1026, PaCC Papers, PSA.

56. Contract File Index, Pittston Co. & Canceled, Agreement No. 172. Stockholders and Executive Committee Minutes, Folder: 20 May 1915 to 26 Nov 1919, No. 6, Doc. No. 1026, PaCC Papers, PSA. The Executive Committee debated and approved the 1919 lease at a meeting on January 22, 1919, and offered "some pillars in the Clark Vein at No. 1 Colliery in about 2.35 acres of ground amounting to 13.334 tons of coal, to Messrs. Carney and Brown ... at a price of 75 cents per ton for pea and above, and 30 cents per ton for all sizes smaller. The coal in question cannot be reached to advantage by this company" (pp 216–217).

57. The exact number of workers in the subcontractors' crews cannot be known because records were not kept. If the 30 subcontractors were employed and each hired 30–35 workers (miners and laborers), the total would be approximately 1,000 employed in subcontracting.

58. "Report Ready for Mine-Cave Committee," Editorial, *Scranton Times*, 2 Jun 1910; "Mine-Cave Protection Plan Will Be Unfolded Tonight," *ibid.*, 10 Jun 1910; "Commission Will Devise Remedies for Scranton Mine-Cave Problem," *ibid.*, 11 Jun 1910.

59. *The Colliery Engineer*, 25 (Aug 1914–Jul 1915), 391.

60. Stockholders and Executive Committee Minutes, Folder: 20 May 1915 to 26 Nov 1919, Minutes, 1838–1971, No. 6, Doc. No. 102, PaCC Papers, PSA.

61. Philip V. Mattes, "The Mine-Cave Struggle," in Thomas Murphy (ed.), *Jubilee History of Lackawanna County, Pennsylvania*, Topeka, KS: Historical Publishing, 1928, 371. See also Philip V. Mattes, "The Mine Cave Problem," and "Kohler and Fowler Acts," Pp. 15–28 in *Tales of Scranton*, Scranton, PA: Published by the author, no date; and Charles P. Kumpas, "The Peoples Coal Company and the Law," *The Miner's Lamp*, 16 (1999), 1, 4.

62. Ellis W. Roberts, *A History of Land Subsidence and its Consequences Caused by The Mining of Anthracite Coal in Luzerne County, Pennsylvania*, unpublished doctoral dissertation, New York University, 1948. The authors are indebted to the late Ellis W. Roberts for information and consultation on the history of the northern field's subsidence problems.

63. "Pittston Ready to Aid in Fighting Mine Caves," *Scranton Times*, 8 Feb 1921.

64. In the early years of mining, the companies sent the waste from processed coal back underground to support the interior structure of the mine. However, as output increased over time it became cheaper to pile the waste in culm banks on the colliery grounds. Moreover, the accumulation of waste underground hindered second and third minings, which became a main source of income in later years.

65. Kohler Act, 27 May 1921, P.L. 1198, Section 1.

66. Mahon v. Pa. Coal Co., 118 A. 491 (Pa. 1922), Pennsylvania Supreme Court.

67. Pennsylvania Coal Co. v. Mahon, et al, 260 U.S. 393, 67 L Ed 322, 43 S Ct 158, 28 ALR 1321. See also Lawrence M. Friedman, "A Search for Seizure: Pennsylvania Coal Co. v. Mahon in Context," *Law and History Review*, 4 (Spring 1986), 1–22; and Evan B. Brandes, "Legal Theory and Property Jurisprudence of Oliver Wendell Holmes, Jr. and Louis D. Brandeis: An Analysis of Pennsylvania Coal Company v. Mahon," *Creighton Law Review*, 38 (2005), 1179.

68. In an unrecorded interview on the Penn Coal Case (18 Jan 1995, NPOLHP), the late U.S. Court of Appeals Judge Max Rosenn of Kingston, PA, stated that the Mahons likely initiated the suit at the request of PaCC and possibly other coal companies. He surmised that the companies wanted a friendly plaintiff to challenge the law. Many years prior to his appointment to the federal bench by President Richard Nixon in 1970, Rosenn defended the Lehigh Valley Coal Company in a subsidence case. The authors are grateful to Judge Rosenn, and also to Attorney Hopkin T. Rolands, for consultation on the Penn Coal Case.

69. *Keystone Bituminous Coal Association, et al., Petitioners v Nicholas DeBenedictis (1987)* 480 US 470, 94 L Ed 2d 472, 107 S Ct 1232. See also Susan Manges McMichael, "Mahon Revisited: *Keystone Bituminous Coal Ass'n v. DeBenedictis*, 480 U.S. 470 (1987)," *Natural Resources Journal*, 29 (1989), 1067–77. Nicholas DeBenedictis, the chief defendant in the case, was Secretary of the Pennsylvania Department of Environmental Resources when the suit was filed.

70. Edward J. Davies, *The Anthracite Aristocracy*.

Notes to Chapter Three

1. During the late 19th and early 20th centuries, the American coal industry (anthracite and bituminous combined) experienced more strike action than any other sector of the economy. Between 1881 and 1905, while 74.4 workers per thousand in all industries went on strike, the number for mining was 196 per thousand. The tobacco industry had the next-highest strike activity with 100 per thousand (Chris Tilly and Charles Tilly, *Work Under Capitalism*, Boulder, CO: Westview, 1998, 50).

2. "Conciliators at Last Agree to Judge Gray's Conciliation," *Wilkes-Barre Times*, 20 Aug 1904; "Gray to Settle Old Dispute," *Harrisburg Patriot*, 20 Aug 1904. The authors found no evidence of Gray having made a final decision on the matter. The Anthracite Strike Commission, set up by President Theodore Roosevelt to settle the 1902 strike, established the ABC to arbitrate workplace grievances. Despite perennial workers' complaints against it, the ABC continued for over five decades and celebrated a fiftieth anniversary in 1953, at a banquet attended by 1500 persons. John L. Lewis addressed the assembly in a speech that emphasized mining productivity and efficiency ("Increased Production and Efficiency Urged in Address by Lewis," *The Record American*, 2 Oct 1953).

3. Joe Gowaskie, "John Mitchell and the Anthracite Mine Workers: Leadership Conservatism and Rank-and-File Militancy," *Labor History*, 27 (1985), 64, 73–82.

4. In one of the earliest studies on Italians in the labor movement in Pennsylvania, Edwin Fenton ("Italians in the Labor Movement," *Pennsylvania History*, 26 [Apr 1959], 143) concluded that many joined unions soon after arriving in America, but when the union did not deliver either in material benefits or workplace improvements, they tended to withdraw, "retreating to their mutual benefit society." The same pattern was apparently true of the Italians and other ethnic groups at the Erie coal companies, and at other coal companies following the strike of 1902.

5. Joe Gowaskie, "John Mitchell and the Anthracite Mine Workers," 70.

6. "Coal Men Break, Strike Ordered," *New York Times*, 30 Mar 1906; Perry K. Blatz, *Democratic Miners: Work and Labor Relations in the Anthracite Coal Industry, 1875–1925*, Albany, NY: SUNY Press, 1994, 194.

7. In 1910, PaCC operated ten collieries: the Barnum, Central, Ewen, Gypsy Grove, Old Forge, and Nos. 1, 5, 6, 9, 14. HC&I had four: the Butler, Consolidated, Forest City, and Erie. The facilities were located in the Pittston–Avoca areas of Luzerne County and the Scranton–Old Forge areas of Lackawanna County, except for the Forest City Colliery in Susquehanna County.

8. Proceedings of the Twelfth Annual Convention of District No. 1, 10.

9. "Men Refuse to Work Despite the Ruling," *The Evening News*, 25 May 1910.

10. "Women Lead a Fighting Mob; Workman and Trooper Hurt," *The Evening News*, 24 May 1910. On the tradition of women participating in strikes and otherwise supporting husbands, sons, fathers, brothers, and neighbors see Anne Hard, "Anthracite Country," *Nation*, 121 (25 Nov 1925), 2–3; Harold W. Aurand, *From the Molly Maguires to the United Mineworkers: The Social Ecology of an Industrial Union, 1869–1897*, Philadelphia, PA: Temple University Press, 1971, 92; Donald L. Miller and Richard E. Sharpless, *The Kingdom of Coal: Work, Enterprise, and Ethnic Communities in the Mine Fields*, Philadelphia, PA: University of Pennsylvania Press, 1985, 309; and Thomas Dublin and Walter Licht, *The Face of Decline: The Pennsylvania Anthracite Region in the Twentieth Century*, Ithaca, NY: Cornell University Press, 2005, 64, 78.

11. "Foreman is Accused of Shooting Striker," *The Evening News*, 25 May 1910.

12. *Ibid.* The Pennsylvania State Police were established in 1905, largely as a result of the 1902 anthracite strike. Beginning in 1865, the coal companies had state authority to hire private law enforcement armies called the Coal and Iron Police. Because of abuses and misuses by the companies, the legislature created the State Police (Stephen R. Couch, "The Coal and Iron Police in Anthracite Country," 100–125 in David L. Salay [ed.], *Hard Coal, Hard Times: Ethnicity and Labor in the Anthracite Region*, Scranton, PA: Anthracite Museum Press, 1984).

13. "Rodda Prevented Riot at Central Colliery," *The Evening News*, 26 May 1910. See also Patrick M. Lynch, *Pennsylvania Anthracite: A Forgotten IWW Venture, 1906–1916*, unpublished M.A. thesis, Bloomsburg State College, 1974, 33; and Blatz, *Democratic Miners*, 211–214. No doubt adding to the IWW's perceived anthracite threat were the union's activities in Philadelphia and in Western Pennsylvania steel towns. In 1908, the Wobblies led steel workers on strike in McKee's Rocks and New Castle. In 1913, the union spearheaded a strike of thousands of longshoremen in Philadelphia (Perry K. Blatz, "Titanic Struggles, 1873–1917," 83–148 in Howard Harris and Perry K. Blatz [eds.], *Keystone of Democracy: A History of Pennsylvania Workers*, Harrisburg, PA: PHMC, 1999, 138–140).

14. Editorial, *Wilkes-Barre Record*, 4 Jun 1910.

15. Locally based Italian immigrant gangs were given the Black Hand moniker because of their signature black handprint on threatening letters. Their stock-in-trade was extortion perpetrated on fellow immigrants. According to Thomas M. Pitkin and Francesco Cordasco (*The Black Hand: A Chapter in Ethnic Crime*, Totowa, NJ: Littlefield, Adams, 1977, 228–229), in 1884, U.S. Secret Service agent William A. Flynn thought that the counterfeiters he had been investigating in New York City were affiliated with the Sicilian Mafia. He at first believed the gangs were well organized under a single boss but, when it turned out that they were not, the term Mafia faded in favor of Black Hand, which was in use until the 1920s. With Prohibition and vast new markets in alcohol, the Black Handers began to fade and the Mafia gained ascendancy. Pitkin and Cordasco concluded that the Sicilians did not penetrate American organized crime, which had been dominated by Irish and Jewish gangs, until the late 1920s. Only with the Congressional investigations of the 1950s did the term Mafia come back into wide use, along with the term *La Cosa Nostra*. Most researchers now argue that organized crime, including the Mafia, is more decentralized than previously thought. For critiques of organized-crime research in the U.S. see Robert J. Kelly, *The Upperworld and the Underworld: Case Studies of Racketeering and*

Business Infiltrations in the United States, New York: Kluwer Academic/Plenum, 1999; Michael Woodiwiss, *Organized Crime and American Power*, Toronto: University of Toronto Press, 2001; Michael Woodiwiss, *Gangster Capitalism: The United States and the Global Rise of Organized Crime*, New York: Carroll & Graff, 2005; and David Critchley, *The Origin of Organized Crime in America: The New York City Mafia, 1891–1931*, New York: Routledge, 2009. See Chapter 7 for a critical review of organized-crime research.

16. Gowaskie, "John Mitchell and the Anthracite Mine Workers," 68.
17. "District President McEnaney Appealed to the National Executive and Special Organizers Have Been Sent to This Vicinity," *The Evening News*, 25 May 1910.
18. "Watchman Shot," *Wilkes-Barre Record*, 1 Jun 1910; "Miners' Union Orders Strikers Back to Work," *Scranton Times*, 2 Jun 1910.
19. "Many Mine Workers Are Resuming Work," *The Evening News*, 27 May 1910.
20. "Sheriff Connor Called to Quell Trouble in Dunmore," *Scranton Times*, 3 Jun 1910.
21. Philip Cannistraro and Gerald Meyer (eds.), *The Lost World of Italian-American Radicalism*, Westport, CT: Praeger, 2003. An example of Italian labor activism around this time can be seen in the "Uprising of the 20,000" in New York City, in November 1909, where Italian and Jewish women shirtwaist makers staged a massive but unsuccessful strike on behalf of the union shop (Philip S. Foner, *Women and the American Labor Movement*, New York: The Free Press 1982, Chapter 7). See also Sally M. Miller, *The Radical Immigrant*, New York: Twayne, 1974, 97–112.
22. "Watchman Shot," *Wilkes-Barre Record*, 1 Jun 1910; "Settle the Strike," *Scranton Times*, 1 Jun 1910. Besides Captain May, the other members of PaCC's board of directors in 1910 were J.G. McCullough, G.F. Baker, Charles Steele, N.B. Reath, G.A. Richardson, F.D. Underwood, G.F. Brownell, and J.C. Stuart ("Pennsylvania Coal Company Elected Set of Directors," *ibid.*, 4 Jun 1910). Erie and PaCC shared some of the same board members.
23. "Says Captain May Renewed Offer to Take up Grievances," *Scranton Times*, 1 Jun 1910; "Union Lessens Efficiency," *New York Times*, 15 Jan 1903.
24. "Shooting Down Near Pittston," *Scranton Times*, 1 Jun 1910.
25. *Ibid.*
26. *Ibid.*
27. "Strike at Avoca," *Scranton Times*, 2 Jun 1910.
28. "Avoca Miners Join Strikers' Ranks Yesterday," *Wilkes-Barre Record*, 2 Jun 1910.
29. "Strikers' Meeting to Talk Over Peace Proposition," *Scranton Times*, 7 Jun 1910.
30. "Striking Miners Insist on a Settlement Before Resuming Work," *Scranton Times*, 6 Jun 1910.
31. "Strikers Meet To-Day," *Wilkes-Barre Record*, 7 Jun 1910.
32. *Ibid.*
33. "Strike is Settled," *Wilkes-Barre Record*, 8 Jun 1910.
34. *Ibid.*
35. "The Strike Ends," editorial, *Scranton Times*, 8 Jun 1910.
36. "Union Lessens Efficiency," *New York Times*, 15 Jan 1903.
37. Anthracite Board of Conciliation, Grievance 1126, Vol. 11, 82, 1920, Pennsylvania Coal Co. vs. Employees, Contained in Digest of the Decisions of the Anthracite Board of Conciliation, Wage Agreements, Carl A. Peterson Papers, Paterno/GST/Z/04.4 Box 1, Folder 3, HCLA, PSU.
38. The ABC heard a related case and rendered a split decision on the payment collection issue. An umpire and a judge ruled that it was indeed the responsibility of the workers to collect the funds used to pay the checkers ("Dead Mine Chief Told Wife Who His Killers Might Be," *Scranton Times*, 29 Feb 1928).
39. "The Strike Ends," Editorial, *Scranton Times*, 8 Jun 1910.
40. *Ibid.*; "If Men Return to Work Company Will Take Up All Grievances in 48 Hours," *Scranton Times*, 3 Jun 1910. Lynch (*Pennsylvania Anthracite*, 22–23) reported that even though the UMWA claimed 10,000 new members following the strike, the figure seemed "absurdly high"

considering that District 1's membership ranged between 4,174 and 9,366. Blatz (*Democratic Miners*, 214) put the number at 5,000 new members, and added that many stayed for only a short time.

41. Peter Roberts, *The Anthracite Coal Industry: A Study of the Economic Conditions and Relations of the Co-operative Forces in the Development of the Anthracite Coal Industry of Pennsylvania*, New York: Macmillan, 1901, 190. Roberts resided in Mahanoy City in the Schuylkill anthracite region. On mineworkers' solidarity, Tilly and Tilly (*Work Under Capitalism*, 50) wrote: "The solidarity of mining communities showed up in the deep and sometimes violent confrontations with mine managers, private police in the mines' employ, and sometimes state or federal military forces."

42. "Men Formulated Grievances," *Wilkes-Barre Record*, 11 Jun 1910; "Miners' Grievances," *ibid.*, 22 Jun 1910.

43. "Want Uniformity: Check Weighmen Discussed Important Mining Question," *ibid.*, 21 Jun 1910.

44. "Miners' Grievances," *ibid.*, 22 Jun 1910. The "card of recommendation" involved a practice whereby a worker dismissed at one PaCC or HC&I colliery had to remain unemployed while his former bosses sent his employment card to the main office where his dismissal was noted and the card returned to the former workplace. The process could take weeks and the workers saw the policy as both a punishment and an economic hardship.

45. PaCC and HC&I were not the only firms experiencing labor-related strife during the spring of 1910. For example, a strike closed the Lackawanna Colliery of the Temple Coal & Iron Company in Blakely when some 1,000 Polish and Italian miners protested the unjust firing of 34 miners and laborers. As with numerous other wildcat actions, the protestors spurned the UMWA when the organization tried to intercede. In another instance, the largely immigrant workforces at the Sunnyside Colliery of the Humbert Coal Company in Jessup and at the Richmond No. 3 Colliery of the Scranton Coal Company walked out over pay and work-rule objections ("Miners' Union Orders Strikers Back to Work," *Scranton Times*, 2 Jun 1910; "Avoca Miners Join Strikers' Ranks Yesterday," *Wilkes-Barre Record*, 2 Jun 1910; "Sheriff Connor Called to Quell Trouble in Dunmore," *Scranton Times*, 3 Jun 1910).

46. Melvyn Dubofsky, *We Shall Be All: A History of the Industrial Workers of the World*, Chicago, IL: Quadrangle/New York Times, 1969; Fred Thompson, *The I.W.W.: Its First Seventy Years (1905–1975)*, Chicago, IL: Industrial Workers of the World, 1976 [1955]; Salvatore Salerno, *Red November, Black November: Culture and Community in the Industrial Workers of the World*, Albany, NY: SUNY Press, 1989; Lynch, *Pennsylvania Anthracite*.

47. Lynch, *Pennsylvania Anthracite*, 11–13, 30–32. Ettor quit the IWW later in 1916 in a conflict over a strike on the Mesabi Range in northern Minnesota. That strike commenced on June 2, 1916, when an Italian immigrant miner, Joe Gruni, protested substandard wages caused, in part, by a local version of the subcontracting system. According to Philip S. Foner (*History of the Labor Movement in the United States*, Vol. 4: *The Industrial Workers of the World 1905–1917*, International Publishers, 1972, 494), Gruni reacted strongly: " 'To hell with such wages,' he cried; he threw down his pick, and decided to quit. To his surprise, the entire shift in the underground mine went along with him. Gruni and his coworkers went from stope to stope in Aurora, crying: 'We've been robbed long enough. It's time to strike.' " The strike failed although the companies later raised wages. See also Dubofsky, *We Shall Be All*, 322–330.

48. Walter T. Howard (ed.), *Anthracite Reds: A Documentary History of Communists in Northeastern Pennsylvania during the 1920s*, Vol. I, New York: iUniverse, 2004; 10; see also Lynch, *Pennsylvania Anthracite*, 90–91.

49. Salvatore Salerno, *Red November, Black November*; Salvatore Salerno, "*I Telitti Della Razza Bianca* (Crimes of the White Race): Italian Anarchists' Racial Discourse as Crime," 11–23 in Jennifer Guglielmo and Salvatore Salerno (eds.), *Are Italians White? How Race Is Made in America*, New York: Routledge, 2003; Joseph R. Conlin, *A Study of the Industrial Workers of the World before World War I*, unpublished doctoral dissertation, University of Wisconsin, 1966.

50. In 1916, PaCC's collieries included the Barnum, Central, Ewen, Old Forge, Underwood, and Nos. 1, 5, 6, 9, and 14. HC&I operated the Butler, Forest City, Consolidated, and Erie.

51. The Pennsylvania Department of Mines, *Annual Report, 1916 (Anthracite Division)*, Harrisburg, PA: Commonwealth of Pennsylvania, stated that Italian-born mineworkers constituted 25.5

percent of the Pittston area's mining workforce, for which PaCC and HC&I were the main, but not the only, employers.

52. Lynch, *Pennsylvania Anthracite*, 65, 151.

53. John W Briggs, *An Italian Passage: Immigrants to Three American* Cities, New Haven, CT: Yale University Press, 1978, 8–11. See also Clement Valetta, " 'To Battle for Our Ideas': Community Ethic and Anthracite Labor, 1920–1940," *Pennsylvania History*, 58 (Oct 1991), 311–329. The case of Cologero Giarrtano was typical. Born in Serradifalco in 1876, he worked in the sulfur mines beginning at age eight. He emigrated to the U.S. and eventually ended up in Pittston. He began working in the coal mines there because, he said, that was the work he knew. He was happy to find a large contingent of fellow immigrants from the same town and from other Sicilian towns living in Pittston. He died of black lung disease in 1963 (Joan Saverino, " 'Domani Ci Zappa': Italian Immigration and Ethnicity in Pennsylvania," *Pennsylvania Folklife*, 44 [Autumn 1995], 2–22).

54. Authorities charged Ettor, Giovannitti, and Caruso as accomplices in the death of Anna LoPizzo, a participant in the Lawrence textile strike of 1912. They were detained for the duration of the strike. They were acquitted on November 12, 1912.

55. Salvatore Lupo (*History of the Mafia*, New York: Columbia University Press, 2009, x) described the subcontracting system in the Sicilian sulfur industry: "The crews were small, independently operating entities, and they stood to profit by their own production, as long as a sizeable proportion was rendered to the owner of the mine." He also tied the structure to organized crime (p x). John Dickie (*Cosa Nostra: The History of the Sicilian Mafia*, New York: Palgrave McMillan, 2004, 72–82) described the Mafia's involvement in the sulfur industry from at least the mid-19th century. Until the early 20th century, Sicily had a virtual monopoly on international sulfur sales. The valuable commodity was used in the vulcanization of rubber and had other industrial uses. The Sicilian industry collapsed in the early 20th century with the development of mines in Louisiana and Texas. British citizens invested in the sulfur industry during their country's occupation of Sicily between 1806 and 1815 (Arnold George D. Maran, *Mafia—Inside the Dark Heart: The Rise and Fall of the Sicilian Mafia*, New York: Thomas Dunne, 2010). Anthracite mineworker Charles Volpe's father was among the Sicilian sulfur miners who worked for PaCC (Charles Volpe, audiotaped interview, 22 Jan 1998, NPOLHP). On Sicily's sulfur mines see also Eric J. Hobsbawm, *Primitive Rebels: Studies in Archaic Forms of Social Movement in the 19th and 20th Centuries*, New York: Norton, 1959, 5, 36, 41, 45.

56. Wilkes-Barre and Scranton had a combined total of 1,500 Italian-born residents in 1900, while the entire anthracite region had about 10,000. In Pittston (population 12,556) about one-third of the residents were Italian, with the largest group from Sicily, mainly Montedoro and San Cataldo. Dunmore also had a large Italian population, with one-fourth from San Cataldo (U.S. Bureau of the Census, Thirteenth Census of the U.S., 1910; and Richard D. Grifo and Anthony F. Noto, *Italian Presence in Pennsylvania*, University Park, PA: Pennsylvania Historical Associations, 1990, iii, 7–8). The population growth of southern and eastern Europeans in eight of the ten anthracite counties was accompanied by a decreasing proportion of English, Welsh, Scots, German, and Irish during the last two decades of the 19th century. In 1880, the "old" groups constituted 94 percent of the population; in 1890, 68 percent; and in 1900, 49 percent. In Luzerne County in 1900, the old groups numbered about 100,000, while the newer immigrants had about 89,000 residents.

57. "Blame Contract System for Pittston's Wave of Crime," *Scranton Times*, 17 Aug 1920. Subcontractor and alleged mobster Charles Consagra was charged and later freed in the murder of Charles Attardo ("Consargro [*sic*] A Prisoner in Murder of Lucchino," *Scranton Times*, 14 Aug 1920). It is not known whether the culprit was ever brought to justice.

58. Blatz, *Democratic Miners*, 66.

59. "Coal and Coke News: Harrisburg, Pennsylvania," *Coal Age*, 10 (28 Oct 1916), 733.

60. Lynch, *Pennsylvania Anthracite*, 75.

61. Luigi Galleani edited this principal Italian anarchist newspaper between 1903 and 1918, whereupon it was shut down by the U.S. Government under the Sedition Act of 1918.

62. Lynch, *Pennsylvania Anthracite*, 98–99. The IWW claimed 97 members at the Old Forge Colliery in 1916 (*ibid.*, 35).

63. Economist and social critic Scott Nearing referred to "The Anti-Social Nature of the Anthracite Wage" because the companies did not pay a living income. He reported that food prices be-

tween 1902 and 1914 rose by nearly 39 percent, and clothing prices by one-third, yet the mineworkers' pay lagged well behind inflation (Scott Nearing, *Anthracite: An Instance of Natural Resource Monopoly*, Freeport, NY: Books for Libraries Press, 1915, 137, 123, 143). Frank J. Warne found that, in 1900, a typical coal miner earned $30 a month, and a laborer $20. After costs for rent, clothing, household goods, physicians and medicines, and other essentials, a miner with four dependents had $14 remaining for food, or a little over three cents per person for three meals a day; the laborer's family had two cents. The figures for 1915 would probably not differ greatly given the loss in the buying power due to scant wage increases compounded by inflation (Frank J. Warne, "The Anthracite Coal Strike," *Annals of the American Academy of Political and Social Science*, 17 [1901], 40). In the economically robust year of 1917, economists W. Jett Lauck and Edgar Sydenstricker wrote that the occupational category of coal miner was among the lowest paid in the nation, with a wage that constituted a smaller portion of family income than any other working-class household head. The authors also found that coal-mining families had among the lowest total incomes and the largest number of children earning income (W.J. Lauck and Edgar Sydenstricker, *Conditions of Labor in American Industries*, New York: Arno, 1917, 252).

64. Selig Perlman, *Upheaval and Reorganization (Since 1876)*, Part 4, Volume 2, in John R. Commons et al., *History of Labour in the United States*, New York: Macmillan, 1918, 523–524. For an assessment of labor relations following the 1902 strike, particularly as related to grievance procedures, see Edgar Sydenstricker, "The Settlement of Disputes Under Agreements in the Anthracite Industry," *The Journal of Political Economy*, 24 (Mar 1916), 254–283.

65. "The Old Forge Salient," Editorial, *Scranton Times*, 9 Sept 1916.

66. "Ettor Forms I.W.W. Branch Here with 500 Men Enrolled," *Pittston Gazette*, 22 Mar 1916.

67. Luigi Galleani was on the FBI's investigative list for subversive activities (Mark Naison and Maurice Isserman [eds.], *Research Collections in American Radicalism*, Department of Justice, Investigative Files, Part II, The Communist Party, University Publications of America, Section 5, Previously Restricted Materials. 1921–1930, 90 pages, found at http://library.lexis nexis.com/ksc_assets/catalog/10836.pdf]). On the IWW see also Howard Zinn, *A People's History of the United States*, New York: Harper Perennial, 1980, 372.

68. "IWW Demands," *Scranton Republican*, 15 Sept 1916.

69. "Union Men at Greenwood Will Return to Work Monday Next; I.W.W. Members Will Stay Out," *Scranton Times*, 26 Feb 1916; "Deadlock in I.W.W. Strike at Greenwood," *ibid.*, 13 Mar 1916; "Industrial Workers of the World Join Union," *Coal Age*, 9 (4 Mar 1916), 419.

70. "Pennsylvania Anthracite: Pittston," *Coal Age*, 9 (29 Apr 1916), 775.

71. "I.W.W. Swarms Old Forge; Heavily Armed and With Women Cheering, They Take Possession," *Philadelphia Inquirer*, 24 May 1916.

72. "Twelve I.W.W. Men Prefer Prison to Payment of Fines," *Wilkes-Barre Record*, 31 Aug 1916; "IWW Placards in Old Forge," *Scranton Times*, 31 Aug 1916.

73. *Ibid.*

74. Marybeth Van Winkle, "The Other Side of the Story," *The Searcher*, 13 (2010), 6; "The Old Forge Salient," editorial, *Scranton Times*, 9 Sept 1916. The missive could have been a "plant" to discredit the IWW in what had become a vicious labor war.

75. *Ibid.*

76. "317 I.W.W. Caught in Secret Session By State Troopers," *Philadelphia Inquirer*, 15 Sept 1916. The Minnesota reference was to the IWW-led strike on the Mesabi Range, which resulted in several arrests. See also "I.W.W. Meeting Raided," *Gazette Telegraph* (Denver), 15 Sept 1916. The number of persons first reported as arrested varied between 250 and 400; however, news reports eventually put the number at 267. The authors are indebted to Patrick M. Lynch for consultation and criticism on this and other aspects of the IWW and the strike of 1916.

77. *Ibid.*; "I.W.W. Meeting Raided," *Gazette Telegraph* (Denver), 15 Sept 1916; "Miners Repudiate I.W.W. Organization," *News Tribune* (Duluth), 19 Sept 1916; "May Deport Men Arrested in Raid," *Harrisburg Patriot*, 19 Sept 1916.

78. "Leader of I.W.W. Put under Arrest as Spy of Kaiser; Organizer Joseph Graber Held at Scranton," *Idaho Statesman*, 5 July 1917.

79. "Sheriff Will Prevent All Meetings of I.W.W.," in James Bussacco, *Pittston's Coal Mining Era*, Pittston, PA: published by the author, 1995, 196; "Police Gather in Saloon Assembly," *ibid.*, 194; "500 Laborers Are Driven From Region by the I.W.W. Strike," *ibid.*, 191; Charles A. McCarthy, "Industrial Workers Caused Havoc in County," *Sunday Independent*, 29 Aug 1976.

80. Graber was arrested in the government's move against the IWW. On his arrest, conviction, and eventual release see Mark Naison and Maurice Isserman (eds.), *Research Collections in American Radicalism*, Department of Justice Investigative Files, Industrial Workers of the World, No. 753.PT1, Reel 11, Andruss Library, Bloomsburg University; "Leader of I.W.W. Put Under Arrest as Spy of Kaiser; and "Organizer Joseph Graber Held at Scranton," *Idaho Statesman*, 5 Jul 1917. Graber was convicted of espionage and sentenced on August 30, 1918, to serve five years at the federal penitentiary in Leavenworth, Kansas. His sentence was commuted on January 23, 1923. Three other IWW members from Scranton were among the 95 Wobblies detained in 1917: Albert B. Prashner, Salvatore Zumpano, and Stanley Dembicki (Charles F. Clyne, U.S. Attorney, to George B. Thatcher, U.S. Attorney General, 15 Oct 1917, in Mark Naison and Maurice Isserman [eds.], Department of Justice Investigative Files, Industrial Workers of the World, No. 753.PT1, Reel 8, Andruss Library, Bloomsburg University).

81. Melvyn Dubofsky, *We Shall Be All*, 326; "I.W.W. Would Drag in Italian Government," *Duluth News*, 10 Sept 1916.

82. "Eleven I.W.W. Men are Sent to Jail," *Scranton Republican*, 2 Apr 1919.

83. The IWW suffered various defeats during the fall of 1916 besides the anthracite strike and the aforementioned strike on the Mesabi Range in Minnesota. A major failure came in the State of Washington on October 30, 1916, when 40 IWW members from Seattle arrived by boat in Everett to support a five-month-long strike by shingle workers. They were greeted at the docks by local deputies who proceeded to assault and arbitrarily arrest them. On November 5, upwards of 250 IWW supporters from Seattle decided to sail to the town and fight for the strikers, for free speech, and for the right to organize. The sheriff and a large number of deputized citizens greeted the Wobblies when they landed, and forbade them to enter the city. Words were exchanged and shots rang out mainly, but not only, from the police. The boat nearly capsized as the Wobblies rushed to one side, and some may have fallen overboard and drowned. After ten minutes of gunfire, the toll was at least two deputies dead with up to 20 injured, while the Wobblies lost up to 12 dead with about 30 injured. IWW leader Bill Haywood and American Federation of Labor leader Samuel Gompers asked the federal government to protect the rights of workers in Everett. See Walker Smith, *The Everett Massacre*, Chicago, IL: IWW Publishing Bureau, 1918; Norman H. Clark, *Mill Town: A Social History of Everett, Washington*, Seattle, WA: University of Washington Press, 1970.

84. The northern field was not the only anthracite area that experienced wildcat strikes in the latter part of the 1910s. In September 1918, for example, some 25,000 workers in District 9 walked out in a wage dispute. The protestors then rejected the incumbent district president and replaced him with a fellow dissident (Dublin and Licht, *The Face of Decline*, 50).

Notes to Chapter Four

1. Enoch Williams to John L. Lewis, 4 Jan 1919; Enoch Williams to John L. Lewis, 13 Jan 1919; Enoch Williams to John L. Lewis, 18 Jan 1919; John L. Lewis to Enoch Williams, 7 Jan 1919; John L. Lewis to Enoch Williams, 22 Jan 1919, International Executive Board Correspondence, Box 1, Folder 1, AX/B40/HCLA/01083, UMWA Papers, HCLA, PSU. As often occurred, Lewis received private communications from a person knowledgeable about various affairs within District 1; with the grass roots problems in the late 1910s, he received letters from national UMWA organizer James Jones, who was based in Kingston, near Wilkes-Barre. See, for example, James Jones to John L. Lewis, 18 Dec 1918, *ibid.*

2. In a speech delivered to striking PaCC miners in 1925, District 1 President Rinaldo Cappellini stated that PaCC had no more than 100 UMWA members in the early 1920s ("3,000 Miners Cheer Wildly as Cappellini Closes Dramatic Plea for Support," *Scranton Times*, 7 Jan 1925).

3. *Proceedings of the UMWA District No. 1 Convention, 1921*, 156–157.

4. "Dead Mine Chief Told Wife Who His Killers Might Be," *Scranton Times*, 29 Feb 1928. See also Chapter 5.

5. "6,000 Miners Out Today," *New York Times*, 16 Jul 1920; "Eight Thousand Men Out in Pittston Mine District," *Scranton Times*, 16 Jul 1920; "8,500 Quit Scranton Mines," *ibid.*, 17 Jul 1920.

6. "Thousands of Miners Still on 'Vacation,' " *ibid.*, 7 Sept 1920.

7. The Pennsylvania Department of Mines *Annual Report*, 1916, indicated that Italian-born mineworkers constituted about 25.5 percent of the Pittston area's mining workforce. PaCC and HC&I were the largest employers in the area. Ben Selekman ("Miners and Murder: What Lies Back of the Labor Feud in Anthracite," *Survey Graphic*, 60 [1 May 1928], 155), reported that 36 percent of PaCC's workforce in 1920 were Italian-born, the majority from Sicily. It is not entirely clear how and why so many Sicilians were employed by the Erie companies. Erie might have sent recruiters to Sicily and/or perhaps Santo Volpe, Stefano LaTorre, and their colleagues played some role in recruitment.

8. "Mine Strike Still Unsettled," *Wilkes-Barre Record*, 27 Jul 1920. George E. Stevenson (*Reflections of an Anthracite Engineer*, published by the author, 1931, Chapter 8) maintained that the miners usually received payment for the fines at those companies that paid by the ton. Therefore, PaCC and HCI may have been among the exceptions.

9. "Young Man Dominant Figure in Troubled Anthracite Field, But Revolt Threatens Power," *Evening Herald* (Shenandoah, PA), 20 Mar 1928.

10. Sydney G. Tarrow (*Power in Movement: Social Movements and Contentious Politics*, New York: Cambridge University Press, 2011, 9) defined social movements as "collective challenges, based on common purposes and social solidarities, in sustained interaction with elites, opponents, and authorities." See also Charles Tilly, *From Mobilization to Revolution*, Reading, MA: Addison-Wesley, 1978.

11. "Cappelini [*sic*] Jailed in Pittston," *The Evening News*, 13 Jan 1923. Individual colliery grievance committees, along with company-wide General Grievance Committees, were allowed under the labor-management agreement of 1912, although because of antipathy toward the UMWA, PaCC colliery workers did not establish the committees until 1920. On the General Grievance Committees see Anthracite Board of Conciliation, *Award of the Anthracite Coal Strike Commission, Subsequent Agreements, and Resolutions of the Board of Conciliation*, Hazleton, PA: Anthracite Board of Conciliation, 1916, 19.

12. According to the U.S. Census of 1910, the Cappellini family consisted of the mother, three brothers, and a sister. The brothers worked in the mines, the two older ones as car runners and Rinaldo as a driver boy. The father's name was not listed. On Rinaldo Cappellini's biography see Perry K. Blatz, *Democratic Miners: Work and Labor Relations in the Anthracite Coal Industry, 1875–1925*, Albany, NY: SUNY Press, 1994, 250–260; Richard Stanislaus, "The Role of the Italian Immigrant in the Anthracite Coal Industry of Northeastern Pennsylvania During the 1920s," unpublished manuscript, 1997; Perry K. Blatz and Robert P. Wolensky, "Rinaldo Cappellini, the Knox Mine Disaster, and the Decline of the Anthracite Coal Industry," 154–157 in Howard Harris and Perry K. Blatz (eds.), *Keystone of Democracy: A History of Pennsylvania Workers*, Harrisburg, PA: PHMC, 1999; and Robert P. Wolensky, "Working Class Heroes: Rinaldo Cappellini and the Anthracite Mineworkers," Chapter 18 in Ronald Berger and Richard Quinney (eds.), *Storytelling Sociology: Narrative as Social Inquiry*, Boulder, CO: Lynne Rienner Publishers, 2004.

13. Ann H. Roller, "Wilkes-Barre, An Anthracite Town," *The Survey*, 55 (Feb 1926), 537.

14. Marie Cappellini, audiotaped interviews, 8 and 29 Dec 1988, NPOLHP.

15. There is no evidence that Campbell or Cappellini were involved in the IWW-led strike of 1916 (see Chapter 3), or that they supported the Wobblies, although both were clearly PaCC employees at the time. Therefore, they would have known about the IWW, the strike, and its purposes, and they could have participated.

16. David Montgomery, *The Fall of the House of Labor: The Workplace, the State, and American Labor Activism, 1865–1925*, New York: Cambridge University Press, 1987, 122.

17. Selekman, "Miners and Murder," 152.

18. Lewis Casterline, audiotaped interview, 12 Dec 1988, NPOLHP.

19. "Detective Lucchino Assassinated at the Door of Residence," *Pittston Gazette*, 22 Jul 1920; "Assassins' Bullets End Sam Lucchino's Career," *Wilkes-Barre Record*, 22 Jul 1920; "Lucchino Murder Puzzling Police," *ibid.*, 24 Jul 1920; "Contract System Cause of Loss to Company, Enoch Williams Says," *Scranton Times*, 24 Jul 1920; Charles A. McCarthy, "1920 Mine Strike Included Murder of Pittston Detective," *Wilkes-Barre Record*, 26 Oct 1965.

20. "Murder a Month with One to Spare in 1920, Record in County," *ibid.*, 1 Dec 1920.

21. "Claim Miner Hired Gunmen," *Wilkes-Barre Times*, 14 Aug 1920; "Make Fifth Arrest in Luchino [*sic*] Murder," *Wilkes-Barre Record*, 14 Aug 1920; "Blame Contract System for Pittston's Wave of Crime," *Scranton Times*, 17 Aug 1920; "Mine Contractor is Released from Jail," *Wilkes-Barre Record*, 20 Aug 1920; "Dying Detective is Quoted as Blaming Wealthy Contractor," *ibid.*, 11 Nov 1920.

22. "Black Hand Suspects Get One Year in Jail," *Wilkes-Barre Times*, 11 May 1907; "Light Sentence for Black Hand Men," *New York Tribune*, 12 May 1907. Pietro Lucchino, Sam's brother, was among those convicted. Santo Volpe was one of two persons acquitted by the jury.

23. The Lucchino family had long-standing ties to organized crime. They were related to the LaTorre family, Stefano LaTorre having married Rose Lucchino. Samuel's brother, Pietro, was among those convicted for conspiracy in 1908. His mother, Pasqualina, was charged in 1911 with larceny and receiving stolen goods ("Woman Lacked Legal Counsel," *Wilkes-Barre Times*, 29 Sept 1911). His father, Tranquilo, likely committed suicide on August 10, 1911, allegedly in despair over his son's work for the Secret Service and Pittston police; however, some suspected that the elder Lucchino was murdered because of his son's activities ("Father of Blackhand Victim Died of His Own Hand," *ibid.*, 11 Aug 1911). Known Black Handers facilitated Sam Lucchino's U.S. citizenship applications, and he facilitated others.

24. William J. Flynn, *The Barrel Mystery*, New York: James A. McCann, 1919, Chapter 4. On the conviction of the last Lupo-Morello gang member see "Federal Jury Convicts Boscarino," *New York Times*, 11 Dec 1910.

25. Thomas M. Pitkin and Francesco Cordasco, *The Black Hand: A Chapter in Ethnic Crime*, Totowa, NJ: Littlefield, Adams, 1977, 135.

26. Flynn, *The Barrel Mystery*, Chapter 4. Flynn wrote that Lucchino "never explained to me further just what his grievance against the 'Black Hand' was."

27. "Father of Blackhand Victim Died of His Own Hand," *Wilkes-Barre Times*, 11 Aug 1911.

28. Charles Consagra was charged with and acquitted of another murder, that of Joseph Casteleno of Pittston in 1905 ("Two Men Charge Each Other with Pittston Murder," *ibid.*, 23 Mar 1911; "Alleged Blackhander Acquitted of Murder," *Philadelphia Inquirer*, 22 Apr 1911). Samuel Lucchino served as one of the chief witnesses against Consagra, for which he received threatening letters ("Luchino [*sic*] Receives Another Threat from Black Hand," *Wilkes-Barre Times*, 11 May 1911).

29. It may not have been a coincidence that, after spending nearly ten years in prison, "Lupo the Wolf" Saietta was granted a conditional parole by President Warren G. Harding in June 1920, one month prior to Lucchino's assassination. See "Consagro [*sic*] A Prisoner in Murder of Lucchino," *Scranton Times*, 14 Aug 1920; Charles A. McCarthy, "1920 Mine Strike Included Murder of Pittston Detective," *Wilkes-Barre Record*, 26 Oct 1965; David Critchley, *The Origin of Organized Crime in America: The New York City Mafia, 1891–1931*, New York: Routledge, 2009, 50.

30. "Pittston at Bier of Murdered Detective," *Scranton Times*, 24 Jul 1920; "6,000 Miners in Attendance on Lucchino Burial," *Pittston Gazette*, 24 Jul 1920.

31. "Lucchino Was Actually Secret Service Agent," *Sunday Independent*, 1 Aug 1982.

32. "Consargro [*sic*] A Prisoner in Murder of Lucchino," *Scranton Times*, 14 Aug 1920.

33. G.L. Hostetter, G.L. and T.Q. Beesley ("20th Century Crime," 49–58 in Gus Tyler [ed.], *Organized Crime in America*, Ann Arbor, MI: University of Michigan Press, 1962 [1933], 51–52) described racketeering as the "exploitation of commerce and the public through circumscribing the right to work and do business." They also saw it as "the product of an age of ruthless business competition and lax or corrupt law enforcement … [which is] in turn the outgrowth of individual and collective indifference and of absorption in purely material affairs." The subcontracting system and organized crime's involvement focused on competition and prices, specifically controlling labor and corrupting mining and pay practices in order to keep costs down and profits up. Criminally controlled subcontracting clearly qualified as a form of labor and work racketeering.

34. William A. Hastie, audiotaped interview, 31 Jul 1989, NPOLHP.

35. *Ibid.*

36. During the trial, Mrs. Lucchino, along with mineworker Orra R. White, served as the key witnesses. White provided an eyewitness account and identified Puntario and Erico as the perpe-

trators. White had been a member of UMWA L.U. 1417 ("Eyewitness to 1920 Pittston Murder Dies," *Wilkes-Barre Record*, 18 Oct 1968). Mrs. Lucchino received at least four Black Hand death threats after her testimony ("Pittston Woman is Threatened by Black Hand," *Wilkes-Barre Times*, 15 Jul 1922; "Mrs. Sam Lucchino Receives Another Threatening Note," *ibid.*, 22 Oct 1922). Appeals by Puntario and Erico ended in the Pennsylvania Supreme Court, which heard the cases on January 3, 1922. Under threats of violence, a relative of Mrs. Lucchino was forced to go to police and discredit her testimony but, in the final analysis, he decided to tell the truth and support her account ("Witness Says He Was Threatened," *Wilkes-Barre Record*, 29 Jan 1921). The guilty verdicts were upheld (Court of Pennsylvania, May and October Terms, 1921, January Term, 1922, Commonwealth v. Puntario, Appellant, 1922, 501–08; Commonwealth v. Erico, Appellant, 1922, 508–511).

37. On the trials see "Ponterari [*sic*] Found Guilty of Lucchino Killing," *Scranton Times*, 5 Oct 1920; "Erico Fails to Testify," *Wilkes-Barre Record*, 15 Nov 1920; and "Erico Found Guilty of First Degree Murder," *ibid.*, 15 Nov 1920.

38. "Charley Consagra Reported Missing by Authorities," *Wilkes-Barre Times*, 2 Nov 1922; "Man Sought in Grecio Shooting Surrenders," *ibid.*, 21 Feb 1928.

39. "Contract Miner Offers to Quit the Business," *Scranton Times*, 3 Aug 1920.

40. "Strikers Vote to Remain Idle," *Wilkes-Barre Record*, 6 Aug 1920.

41. Stephen Fox, *Blood and Power: Organized Crime in Twentieth Century America*, New York: Penguin, 1989, 64. See also Thomas Hunt and Michael A. Tona, "Men of Montedoro," *Informer* (Apr 2011), 7, informer-journal.blogspot.com.

42. Fox's UMWA reference is to the mob-related corruption surrounding District 1 President August J. Lippi who assumed the high office in June 1951. Lippi was allegedly organized crime's man within the district possibly since his first election to the District Executive Board in 1927. As discussed in Chapters 6 and 7, Lippi was a secret and illegal partner in the Knox Coal Company. On labor racketeering in District 1 see also Robert P. Wolensky, Kenneth C. Wolensky, and Nicole H. Wolensky, *The Knox Mine Disaster: The Final Years of the Northern Anthracite Industry and the Effort to Rebuild a Regional Economy*, Harrisburg, PA: PHMC, 1999; and Robert P. Wolensky, Nicole H. Wolensky, and Kenneth C. Wolensky, *Voices of the Knox Mine Disaster: Stories, Remembrances, and Reflections on the Anthracite Coal Industry's Last Major Catastrophe, January 22, 1959*, Harrisburg, PA: PHMC, 2005.

43. Pennsylvania Crime Commission, *A Decade of Organized Crime*, Harrisburg, PA: Commonwealth of Pennsylvania, 1980, 50. Oral history evidence suggests that Cataldo Bufalino might have been the region's first crime boss (William Hastie, audiotaped interview, 31 Jul 1989, NPOLHP).

44. Frederick Sondern Jr. (*Brotherhood of Evil: The Mafia*, New York: Farrar, Straus & Cudahy, 1959, 22) referred to Volpe's "King of the Night" moniker and placed him at the head of a triumvirate that controlled the criminal rackets in northeastern Pennsylvania into the 1930s. The Pennsylvania Crime Commission's *Report* in 1984 called Stefano LaTorre "the oldest member of the Bufalino crime family," and said he was one of the three founding members of the region's organized-crime family, the other two being Santo Volpe and John Sciandra. All were born in Montedoro, Sicily. Volpe emerged as top boss and was allegedly followed by Sciandra in 1933, who in turn was followed by Montedorans Joseph Barbara in 1950 and Russell Bufalino in 1959 (Pennsylvania Crime Commission (Pennsylvania Crime Commission, *A Decade of Organized Crime*, 50, 59; Pennsylvania Crime Commission, *Report on Organized Crime, 1970*, Harrisburg: Office of the Attorney General, Commonwealth of Pennsylvania, 17–18). Bufalino began as laborer at PaCC's No. 6 Colliery in the 1920s. In the late 1950s, the McClellan Committee of the U.S. Senate identified him as the head of the Pittston crime gang and characterized him as "one of the most ruthless and powerful leaders of the Mafia in the United States" (*Hearings Before the U.S. Senate Select Committee on Improper Activities in the Labor or Management Field*, [McClellan Hearings], Eighty-Sixth Congress, 30 Jun, 1–3 Jul 1958, Washington, DC: USGPO, 1958, Part 32, 1225–26). Because of his connections with the Teamsters Union, Bufalino became a prime suspect in the 1976 disappearance of ex-Teamsters president James R. Hoffa. On Bufalino ties to Hoffa's disappearance see Charles Brandt (*I Heard You Paint Houses: Frank "The Irishman" Sheeran and the Inside Story of the Mafia, the Teamsters, and the Last Ride of Jimmy Hoffa*, Hanover, NH: Steerforth Press, 2005), a book that is expected to be made into a movie, *The Irishman*, directed by Martin Scorsese and starring Robert DeNiro, Al Pacino, and Joe Pesci ("Robert DeNiro Says Martin Scorcese's 'The Irishman' Still Happening," www.slashfilm.com/robert-de-niro-martin-scorsese-irishman-happening/).

45. McClelland Hearings, Appendix, "Partial Mafia (Syndicate) Relationship Study."

46. Humbert S. Nelli, *The Business of Crime*, New York: Oxford University Press, 1976, 136.

47. "Committee of Strikers See Coal Co. Directors," *Scranton Times*, 4 Aug 1920. On the strike of 1920 see also "Urges Miners to Call Off Strike," *Wilkes-Barre Record*, 21 Jul 1920; "Miners Are Driven from Hall by Strikers," *Scranton Times*, 12 Aug 1920; "Pennsylvania Strike is Settled," *ibid.*, 10 Oct 1920; and Blatz, *Democratic Miners*, 242–243.

48. The U.S. Department of Justice investigated the shutdown as a conspiracy by the company and/or the unionists to create a coal shortage, but found no evidence to support the claim ("Government Agents Investigating Strike," *Wilkes-Barre Record*, 22 Jul 1920).

49. "Miners Urged to Return to Work," *ibid.*, 12 Aug 1920; "Strikers Should Heed Wise Counsel," editorial, *ibid.*, 13 Aug 1920. On the Vacation Strike see Blatz, *Democratic Miners* 239–249, and Thomas Dublin and Walter Licht, *The Face of Decline: The Pennsylvania Anthracite Region in the Twentieth Century*, Ithaca, NY: Cornell University Press, 2005, 53–54.

50. *Report, Findings, and Award of the Anthracite Coal Commission of 1920*, Washington, DC: USGPO, 1920, 19–20.

51. *Ibid.*, 20. Emphasis added.

52. Neal J. Ferry, "Minority Report" in *Report, Findings, and Award of the Anthracite Coal Commission of 1920*, Washington, DC: USGPO, 1920, 312–323.

53. "Mineworkers Quit, Defying Leaders," *New York Times*, 2 Sept 1920; "Coal Peace Treaty Faces Opposition by Radical Miners," *Philadelphia Inquirer*, 5 Sept 1922; "Nearly All Local Miners Working Except for Pennsylvania Coal Co. Operations," *Scranton Times*, 20 Sept 1920.

54. "5,000 Striking Mine Workers to Remain Idle," *ibid.*, 9 Sept 1920.

55. "Striking Miners Threaten Violence," *New York Times*, 13 Sept 1920. See also "Another House is Dynamited," *Pittston Gazette*, 8 Sept 1916.

56. "Pittston Strike Over," editorial, *Scranton Times*, 24 Sept 1920.

57. "Pennsylvania Strike Is Settled; Long Tie-up Is Ended by Agreement of Company to Eliminate Contract System," *ibid.*, 7 Oct 1920.

58. J. Louis Engdahl, "The Yellow Streak in Coal," *The Liberator*, 9 (Sept 1923), 23–25.

59. "Joyce Now Labor Leader Stirs Pittston Strikers," *Scranton Times*, 6 Oct 1920.

60. "Contractors Giving Up to Restore Mine Peace," *ibid.*, 5 Oct 1920.

61. *The New York Times* put the number of PaCC subcontractors at 30 at all operations ("Mine Union Leader Guilty of Killing," *New York Times*, 15 Apr 1928). Lewis Casterline, a PaCC foreman and superintendent, recalled a higher number for the early 1920s: "We had 34 [sub]contractors under the Pennsylvania Coal Company" (Lewis Casterline, audiotaped interview, 12 Dec 1988, NPOLHP).

62. Charles Tilly, "War Making and State Making as Organized Crime," 169–186 In Peter Evans, Dietrich Rueschemeyer, and Theda Skocpol (eds.), *Bringing the State Back In*, Cambridge, England: Cambridge University Press, 1985, 183.

63. "Butler Colliery Strike," *Wilkes-Barre Record*, 22 Jan 1921; "Threaten Strike in Pittston Area," *ibid.*, 26 Jan 1921.

64. Dynamite struck James Joyce's home and business establishment on January 16, 1921. Apparently, a striker or group of strikers who did not welcome his attempts to compromise with management set the charge. In Joyce's view: "The dynamiting is a deliberate attempt at murder, and if the radical miners think they are going to scare me, they are mistaken. I have obtained for these men what officials of the United Mine Workers tried in vain to secure. Is this gratitude for what I am doing? I'm trying to help the poor coal miner keep bread on his table for him and his family" ("Blast Destroys Joyce Building," *Pittston Gazette*, 31 Jan 1921; see also "Radicals Dynamite Union Chief's Home," *New York Times*, 17 Jan 1921).

65. Another, more legendary, strike occurred in 1921, in Logan County, West Virginia. In a battle over unionization during August and September, over 10,000 armed colliers fought county, state, and private police, even the U.S. military, in the infamous Battle of Blair Mountain (David A. Corbin, *Life, Work, and Rebellion in the Coal Fields: The Southern West Virginia Miners, 1880–1922*, Champaign, IL: University of Illinois Press, 1981; Robert Shogan, *The Battle of Blair*

Mountain: The Story of America's Largest Labor Uprising, Boulder, CO: Westview, 2004; and Lon Savage, *Thunder In the Mountains: The West Virginia Mine War, 1920–21*, Pittsburgh, PA: University of Pittsburgh Press, 1990). Events surrounding the battle have been fictionalized by Denise Giardina in *Storming Heaven* (New York: Ballantine, 1987) and by Jonathan Lynn in *Blair Mountain*, Boulder, CO: Flying Machine Press, 2005. For a video documentary see *Even the Heavens Weep: The West Virginia Mine Wars*, Lou Buttino (writer), Institute, West Virginia: The Humanities Council of West Virginia, WPBY, 1985. Also relevant is the John Sayles film *Matewan* (1987).

66. "Another Outbreak between Miners and Contractors," *Wilkes-Barre Times*, 26 May 1921; "Miners of No. 4 Strike Again Over Rock Contractor," *Pittston Gazette*, 1 Jul 1921.

67. "Miners Vote to Remain at Home," *Wilkes-Barre Record*, 10 Jan 1921; "Threaten Strike in Pittston Area," *ibid.*, 26 Jan 1921; "1500 Men Quit Labor at the Butler Mine," *The Evening News*, 9 Jan 1923; "Butler Strike Not Settled," *ibid.*, 10 Jan 1923;"1,600 Pittston Coal Miners Strike," *New York Times*, 31 Oct 1923.

68. *Proceedings of the Nineteenth Successive and Fourth Biennial Convention of District No. 1, 1921*, 156–157.

69. William J. Brennan to Rinaldo Cappellini, 19 Jun 1922, President/District Correspondence, Paterno/GST/206.02/, Box 1, Folder 3; William J. Brennan to Officers and Members of Local Unions in District 1, 30 Jun 1922, *ibid.*

70. Critic J. Louis Engdahl followed the publicity and drama surrounding the 1921 District 1 race between Brennan and Cappellini and observed: "A study of the columns of *The New York Times* during this period would create the impression that there was a campaign on for governor of Pennsylvania. Brennan stood for reaction in power; Cappellini was the champion of a rising militant spirit. Cappellini was elected overwhelmingly, a high water mark in militancy in the coal miners' union" (Engdahl, "The Yellow Streak in Coal," 1).

71. Lewis Casterline, audiotaped interview, 12 Dec 1988, NPOLHP.

72. "At Wilkes-Barre," *Time*, (30 Jul 1923). These comments by Cappellini are not contained in the official convention Proceedings (*Proceedings of the UMWA Tri-District Convention* (Districts 1, 7, & 9), 1923). On the convention's first day, Brennan's supporters submitted a resolution declaring Cappellini ineligible for office because of his radical predilections. The resolution was withdrawn before a vote was taken. On the second day, Cappellini's men put forth a resolution to reinstate their leader as a district organizer despite his election victory. The resolution, if successful, would have further injured Brennan's status because he had fired Cappellini as an organizer. This resolution was also withdrawn. In his speech on the first day of the convention Cappellini addressed the term "radical," which had been applied to him: "You have heard many a time the word 'Radical.' There is no word more abused and there is no man more crucified than the man who goes out and tries to do his bit in trying to better conditions, and if you don't believe me then try it yourself and you will soon find out that you are a radical" (*Proceedings of the UMWA Tri-District Convention* (Districts 1, 7, & 9), 1923, 50). Some delegates criticized the *Scranton Times* for using the term to describe Cappellini and his supporters (*ibid.*, 87–94).

73. Cecil Carnes, *John L. Lewis: Leader of Labor*, New York: Robert Speller, 1936, 104.

74. On the strikes of 1922 and 1925–26 see Harold Kanarek, "The Pennsylvania Anthracite Strike of 1922," *Pennsylvania Magazine of History and Biography*, 99 (1975), 207–225; "Anthracite Miners to Get Reply Today," *New York Times*, 2 Jun 1922; Robert H. Zieger,"Pinchot and Coolidge: The Politics of the 1923 Anthracite Crisis," *Mississippi Valley Historical Review*, 52 (1967), 566–581; "Will Not Arbitrate," *Coal Age* 28 (Oct 1925), 451–452; Samuel D. Warriner, "*Reply of the Operators' [to the Miners' Demands]*," in *The Anthracite Strike of 1925–1926*, Philadelphia. PA: The Anthracite Bureau of Information, 1926, 6; Robert H. Zieger, "Pennsylvania Coal and Politics: The Anthracite Strike of 1925–26," *Pennsylvania Magazine of History and Biography*, 93 (1969), 244–262; Harold Kanarek, "Disaster for Hard Coal: The Anthracite Strike of 1925–1926," *Labor History*, 15 (1974), 44–62. See also Gifford Pinchot, "Wages, Margins and Anthracite Prices," *Annals of the American Academy of Political and Social Science*, 111 (1924), 61–81.

75. Samuel P. Hays, *The Response to Industrialism, 1885–1924*, Chicago, IL: University of Chicago Press, 1959, 159; Paul Kens, *Locher v. New York*, Lawrence, KS: University Press of Kansas, 1990. On the commission's final award see Hunt et al, *What the Coal Commission Found: An Authoritative Summary by the Staff*, Baltimore, MD: Williams & Wilkins, 1925.

76. Anthracite Board of Conciliation. *Award of the Anthracite Coal Strike Commission, Subsequent Agreements, and Resolutions of the Board of Conciliation*, Hazleton, PA: Anthracite Board of Conciliation, 1923, 47.

77. "Hard-Coal Operators Prepared for Finish Fight in War against Future Strikes," *Coal Age*, 28 (Nov 1925), 737.

78. Zieger, "Pennsylvania Coal and Politics," 254–256; Kanarek, "Disaster for Hard Coal," 50–51; "Five-Year Agreement Ends Longest Strike in History of Anthracite Industry," *Coal Age*, 29 (Feb 1926), 268–270.

79. "Cappelini [*sic*] Jailed in Pittston," *The Evening News*, 13 Jan 1923. For an oral history on Cappellini's union leadership in the 1920s see Lewis Casterline, audiotaped interview, 12 Dec 1988, NPOLHP.

80. Marie Cappellini, audiotaped interview, 8 Dec 1988, NPOLHP.

81. Raphael Musto, audiotaped interview, 30 Jul 2001, NPOLHP. James Musto was elected to the Pennsylvania General Assembly in 1948 and served until his death in 1971. He served on the Joint Legislative Committee that investigated the Knox Mine Disaster of January 1959.

82. Walter T. Howard (ed.), *Anthracite Reds: A Documentary history of Communists in Northeastern Pennsylvania during the 1920s*, New York: iUniverse, 2004, Chapter 3.

83. "Local Communist Activities," editorial, *Wilkes-Barre Record*, 30 Jan 1924; "Ex-Service led by Healy and Deputized by Mayor Disperse 'Red' Flag Gathering," *ibid.*, 28 Jan 1924;"Mayor Bans Communists," *ibid.*, 29 Jan 1924; "Mayor Stands Pat No Red or Communist Meeting in Wilkes-Barre, Hart Reiterates," *ibid.*, 31 Jan 1924. It is not clear whether Ku Klux Klan leader Paul M. Winter, the head of Wilkes-Barre's Klavern No. 311 in the 1940s (and probably earlier), was among the instigators, or even present, at these events. However, he resided in Wilkes-Barre and was a member of the American Legion who held rabid anti-Communist attitudes. See Chapter 8 for more on the KKK and Winter's activities in the northern anthracite region.

84. *Albany Evening Journal*, 16 Feb 1924.

85. "Lay Strike to Communists," *Wilkes-Barre Record*, 13 Mar 1924.

86. "Boring from Within," *ibid.*, 11 Mar 1924.

87. "Strike Ranks Solid," *Times Leader*, 4 Dec 1924; "Striking Mine Workers Ask for Special Convention," *Scranton Times*, 4 Dec 1924; "Crisis is Reached in Union Affairs," *Times Leader*, 4 Dec 1924; "Miners' President Faces Crisis at Pittston Meeting," *ibid.*, 9 Dec 1924.

88. "Cappellini Arranges Miners' Mass Meeting to End Pittston Strike," *Scranton Times*, 6 Jan 1925.

89. "Main Powder House Wrecked at Mine in Strike Zone," *Times Leader*, 26 Dec 1924; "Powder House Blows Up; Blame Strikers," *The Evening News*, 26 Dec 1924; "Pittston Miners Told to End Strike," *Times Leader*, 29 Dec 1924.

90. "Father Curran's Effort to Bring Strike to End Fails," *ibid.*, 15 Jan 1925. While the PaCC and HC&I car rates were not mentioned, comparable rates were paid by the Hudson Coal Company under the UMWA Wage Agreement of September 19, 1923: "a per car rate of $2.3072 in the Diamond Vein, Clark Vein, Dunmore No. 3 Vein, Dunmore No. 4 Vein, New County Vein, Rock Vein, and 14 Ft. Top and Bottom Split Vein." These were all veins at the Marvine Colliery in Scranton (Carl Axel Peterson Papers, Paterno/GST/Z/04.4,Box 1, Folder 4, HCLA, PSU).

91. "Father Curran's Effort to Bring Strike to End Fails," *Times Leader*, 15 Jan 1925.

92. "3,000 Striking Miners Cheer Wildly as Cappellini Closes Dramatic Plea for Support," *Scranton Times*, 7 Jan 1925.

93. "Cappellini Has No Fear of Impeachment," *ibid.*, 12 Jan 1925.

94. "Bullet-Punctured Body of Mine-Union Official Found on Street Near Pittston Home," *ibid.*, 6 Jan 1925.

95. "Mine Union Officer Slain by Gunmen," *New York Times*, 7 Jan 1925.

96. "Mine Workers to Follow Body of Pittston Murder Victim to Grave," *Scranton Times*, 7 Jan 1925.

97. "General Strike Danger Passes as Locals Vote against Big Walkout," *ibid.*, 22 Jan 1925. For a comparison of anthracite wages with those in other industries around this time see Horace H. Dury, "Wages in the Coal Industry as Compared with Wages in Other Industries (With Special Reference to the Anthracite Situation)," *Annals of the American Academy of Political and Social Science*, 111 (1924), 314–343.

98. "Coal Strike Over," editorial, *Scranton Times*, 24 Jan 1925.

99. "Commission to Meet Cappellini," *ibid.*, 26 Jan 1925.

100. The growth and power of the UMWA in anthracite did not support the position of labor articulated by economist John R. Commons: "It does not seem likely, when a corporation has reached the position of a trust [monopoly], that unionism will get a footing, no matter how class-conscious the workmen have become" (quoted in Frank Tracy Carlton, *The Industrial Situation*, New York: Fleming H. Revell, 1914, 37).

Notes to Chapter Five

1. Irving Bernstein, *The Lean Years: A History of American Workers, 1920–1933*, Baltimore, MD: Penguin, 1966.

2. Pennsylvania Department of Mines and Mineral Industries, *Annual Report: Anthracite Division*, 1962, 10–11.

3. "A Decade of Anthracite Peace," *Literary Digest*, 120 (2 Aug 1930), 40.

4. Walter H. Voskuil, *Minerals in Modern Industry*, New York: Wiley, 1930, 28 table. On the decline of the industry from the early 1920s to the early 1930s see Richard Mead, *An Analysis of the Decline of the Anthracite Industry Since 1921*, unpublished doctoral dissertation, University of Pennsylvania, 1933.

5. Lewis as quoted from a speech to the Anthracite Cooperative Association, published in the *Wilkes-Barre Record*, 11 Oct 1927. See also Address of John L. Lewis to the Anthracite Cooperative Congress, Mt. Carmel, PA, 9 Nov 1927, International Executive Board Correspondence, AX/B40/HCLA/01146, Box 6, Folder 12, UMWA Papers, PSU; and Address of Thomas Kennedy to the Anthracite Cooperative Congress, Mt. Carmel, PA, 9 Nov 1927, *ibid.*

6. Anna Rochester, *Labor and Coal*, New York: International Publishers, 1931, 241.

7. "23 Anthracite Companies Join in Ad. Campaign," *Times Leader*, 2 Apr 1928.

8. "Anthracite Ass'n Seeks Membership," *Pittston Gazette*, 3 Jan 1928; "General Support Being Given Hard Coal Association," *ibid.*, 23 Jan 1928.

9. A newspaper report described the purpose of the Payne Coal Company's scatter discs: "When orange disc anthracite is delivered into a householder's cellar, a black pamphlet explains that the orange disc coal is of exceptional quality, is guaranteed, and can easily be duplicated on re-ordering from the same dealer. All dealers selling this coal are franchised, just the same as the dealer in Ford automobiles" ("Coal Company is Marking its Coal," *ibid.*, 13 Mar 1928). On the Menzies coal-cleaning technology see U.S. Bureau of Mines, *A Dictionary of Mining, Mineral, and Related Terms*, Washington, DC: USGPO, 1968. PaCC's Old Forge Breaker installed the first working model of a Menzies Cone Hydro Separator in 1915, to clean buckwheat coal in a demonstration by its inventor, William Menzies (www.oldforgecoalmine. com).

10. "Glen Alden Coal Is Made Distinctive by Being Colored," *Pittston Dispatch*, 7 Feb 1928. On July 31, 1930, CBS radio broadcast "The Shadow" as part of *The Detective Story Hour*. In September 1931, *The Blue Coal Radio Revue* continued the adventures of *The Shadow* under the sponsorship of the Glen Alden Coal Company, which remained the sponsor until September 1949. The program continued broadcasting with another sponsor until it went off the air in December 1954 ("The Shadow: A Short Radio History," http://www.old-time.com/sights/ shadow.html).

11. Co-author William A. Hastie did his part to promote hard coal during World War II: "As I fought with the U.S. Army in Northern Africa, Sicily, and the Italian Peninsula, whenever I had a chance I would use chalk and write, "Buy Burn Boost Pennsylvania Anthracite," on the sides of jeeps, tanks, buildings, walls—wherever I could. I was a supporter!" (William A. Hastie, unrecorded interview, 12 Jan 2012, NPOLHP).

12. Bernard Heller, "The Anthracite Industry," *The Nation*, 128 (27 Feb 1929), 56. Eleven years later, the Pennsylvania Anthracite Coal Industry Commission issued a similarly gloomy assessment: "In the aggregate [anthracite] is a very sick industry" (Pennsylvania Anthracite Coal In-

dustry Commission, *Report of the Pennsylvania Anthracite Coal Industry Commission*, Harrisburg, PA: Commonwealth of Pennsylvania, 1938, 20).

13. Carter Goodrich et al., *Migration and Economic Opportunity*, Philadelphia, PA: University of Pennsylvania Press, 1936, 429; Theodore Bakerman, *Anthracite Coal: A Study in Advanced Industrial Decline*, New York: Arno, 1979 [1956], 8–9.

14. "Great Industry, Anthracite Mining Highly Important in Pennsylvania," *United Mine Workers Journal*, 39 (1 Oct 1928), 17; George A. Schnell, "Anthracite's Impact on Northeastern Pennsylvania's Population and Environment, 1880–1995," *Journal of the Pennsylvania Academy of Science*, 72 (1998), 22–28. The Communist Party of the United States of America (CPUSA) assigned Steve Nelson to organize the anthracite unemployed in the 1930s. Nelson described the region's bleak circumstances: "Behind the very real beauty of the region lay a bitter poverty. The depression had started here in the twenties, and by the time it reached other areas, the eastern Pennsylvania mining towns were already devastated. ... The political repression was as bad as anywhere in Illinois, and the economic situation much worse. At the time my new assignment was considered to be one of the roughest outside the South" (Steve Nelson, James R. Barrett, and Rob Ruck, *Steve Nelson: American Radical*, Pittsburgh, PA: University of Pittsburgh Press, 1981, 94).

15. Carter Goodrich et al., *Migration and Economic Opportunity*, 430. See also "Four Lehigh Valley Collieries Here Idle Indefinitely," *Pittston Gazette*, 8 Feb 1928.

16. William M. Boal, "New Estimates of Paid-up Membership in the United Mineworkers, 1920–1929, by State and Province," *Labor History*, 47 (Nov 2006), Table 2. International UMWA membership plunged from 442,072 in 1921 to 156,700 in 1929.

17. On the American de-industrialization trend that grew from the 1980s see Barry Bluestone and Bennett Harrison, *The De-Industrialization of America: Plant Closings, Community Abandonment, and the Dismantling of Basic Industry*, New York: Basic, 1982; Paul D. Staudohar and Holly E. Brown (eds.), *Deindustrialization and Plant Closure*, Lexington, MA: D.C. Heath, 1987; and Steven Greenhouse, *The Big Squeeze: Tough Times for American Workers*, New York: Knopf, 2008. On contemporary downsizing see Don Lee Bohl, *Responsible Reductions in Force: An American Management Association Research Report on Downsizing and Outplacement*, New York: American Management Association, 1987, and W.J. Baumol, A.S. Blinder, and E.N. Wolff, *Downsizing in America: Reality, Causes and Consequences*, New York: Russell Sage Foundation, 2003.

18. Journalist William Serrin ("Historians See Lessons for Present in Fate of Dead Anthracite Coal Towns," *New York Times*, 29 Dec 1985) wrote about anthracite's decline during the first half of the twentieth century and compared it to the deindustrialization that hit the nation during the 1980s. Serrin's conclusion that "... the anthracite fields remain devastated — economically, ecologically, psychologically," did not apply as directly to the northern field as to the southern and middle fields which have remained more economically distressed. See also the detailed study on the subject by Thomas Dublin and Walter Licht, *The Face of Decline: The Pennsylvania Anthracite Region in the Twentieth Century*, Ithaca, NY: Cornell University Press, 2005.

19. John Bodnar, *Anthracite People: Families, Unions and Work, 1900–1940*, Harrisburg, PA: PHMC, 1983.

20. "Mine Workers to Air Grievance; Employees' Strike at Butler Colliery Because of Continued Use of Loader," *Wilkes-Barre Record*, 6 Jan 1928; "Miners Want Work Equalized," *ibid.*, 31 Jan 1928; "Equal Division of Work Sought for All Miners," *Pittston Gazette*, 15 Feb 1928.

21. Through most of his tenure as UMWA president, John L. Lewis did not support equalization because it "threatened the rationalized *modus vivendi* established between the coal operators and the union for stabilization of the workplace," according to Ronald M. Benson ("Commentary: The Family Economy and Labor Protest in Industrial American and the Coal and Iron Police in Anthracite Country," 120–25 in David L. Salay [ed.], *Hard Coal, Hard Times: Ethnicity and Labor in the Anthracite Region*, Scranton, PA: Anthracite Museum Press, 1984, 121). In addition to Cappellini, the other district presidents who supported equalization were Andrew Mattey of District 7 and Chris Golden of District 9 ("Equal Division of Work Sought for All Miners," *Pittston Gazette*, 15 Feb 1928).

22. Thomas Dublin, "The Equalization of Work: An Alternative Vision of Industrial Capitalism in the Anthracite Region of Pennsylvania in the 1930s," 81–98 in Lance E. Metz (ed.), *Canal History and Technology Proceedings*, Vol. 13, Easton, PA: Canal History and Technology Press, 1994; Dublin and Licht, *The Face of Decline*, 64–76.

23. On PaCC's equalization plan see "Nanticoke Men Back," *The Evening News*, 16 Jul 1928; "Butler Miners Return," *ibid.*, 19 Jul 1928.

24. District 1 President Rinaldo Cappellini referred to the No. 6 Colliery as the premiere mine in the Pittston area ("Will Reopen Mines to Avert Bloodshed," *New York Times*, 4 Mar 1928). The Pennsylvania Coal Company erected a wooden breaker at the No. 6 colliery in the early 1850s and built a new breaker in 1898. An immense structure for its time, the new breaker required about 1.5 million feet of timber to construct. According to the Pennsylvania Bureau of Mines *Report* in 1898: "Mechanically it is far superior to any breaker now in possession of the company in this district. The equipment is the most modern known to the anthracite coal business for the preparation of coal. After the coal leaves the car at the head of the breaker it is handled entirely by machinery until deposited in the pockets at the lower end of the breaker. An endless chain system is used for conveying the cars into the two patent Farrell dumps at the head, and as soon as the cars are emptied they pass over the tips, run up a short incline and switch themselves back to another set of chain conveyors. The cars are handled entirely by chain and gravity. In moving through the breaker the coal first passes over a set of bars through which the culm and fine coal find an opening and pass to the extreme bottom of breaker. The coal is then conveyed to the top of the breaker again and passes over a separate pair of screens, where it is crushed into sizes, re-cleaned by the patent Thomas slate pickers, and conveyed by the belt conveyors into the chutes. The larger coal passes over the grate bars and lands on a movable platform where it is cleaned as it passes over. The greater quantity of the coal after going through the rolls is elevated by three sets of elevators to the six main screens where it is separated into sizes. The culm is conveyed on belt conveyors to a pocket 100 feet from the breaker. The capacity of this breaker is estimated at 2,000 tons per day. The coal to be handled by it is mined from Shafts Nos. 5, 6 and 11" (Pennsylvania Bureau of Mines, *Report of the Bureau of Mines*, Harrisburg, PA: Department of Internal Affairs of Pennsylvania, 1898, 74).

25. The No. 6 Shaft within the No. 6 Colliery was completed to the Pittston Vein in 1853, and extended to the Red Ash Vein in 1898–99. The shaft also took coal from the Clark Vein and Babylon Vein (No. 6 Colliery Book, 1983.177, Folder 3, Box 2, 3–14 & 3–44, PaCC Papers, NCMA). On Michael Gallagher's tenure as president and board chairman see Chapter 6.

26. "Disagreement in No. 6 Local Over New Wage Scale," *Pittston Gazette*, 12 Jan 1928; "Warring Factions of Union Promise to Come Together," *ibid.*, 20 Jan 1928.

27. "Transcript of Proceedings, Board of Conciliation Meeting," 3 Mar 1928, International Executive Board Correspondence, AX/B40/HCLA/01146, Box 6, Folder 30, UMWA Papers, PSU; "Disagreement in No. 6 Local over New Wage Scale," *Pittston Gazette*, 12 Jan 1928.

28. "Transcript of Proceedings, Board of Conciliation Meeting," 3 Mar 1928, International Executive Board Correspondence, AX/B40/HCLA/01146, Box 6, Folder 30, UMWA Papers, PSU.

29. One of the subcontractors who gained access to a coal vein reduced the existing work crew from 40 to 20 men ("Warring Factions of Union Promise to Come Together," *ibid.*, 20 Jan 1928).

30. "Mr. Cappellini," *Time*, 9 Jul 1923.

31. The other deposed officers were Joseph Shimokonis and Charles Alba, vice presidents; Joseph Moleski and Philip Juliani, financial secretaries; Sam Seit, recording secretary; Andrew Spernatz, treasurer; Sylvester Morris and Matthew Hanahoe, trustees; and E. Aquilina, Sam Seit, Sam Latore, and James Antonello, grievance committee members ("Union Officials are Overthrown," *Wilkes-Barre Record*, 12 Jan 1929; "Disagreement in No. 6 Local Over New Wage Scale," *Pittston Gazette*, 12 Jan 1928). Ben Selekman ("Miners and Murder: What Lies Back of the Labor Feud in Anthracite," *Survey Graphic*, 60 [1 May 1928], 151, 154ff) confirmed that all of the officers were subcontractors, and Powers Hapgood and Mary Donovan ("Murdered Miners," *The Nation*, 14 Mar 1928, 293–94) also reported that Agati and the recently ousted officers were subcontractors. See also "Union Head Silent on Marianelli's Expected Coming," *Times Leader*, 13 Jan 1928.

32. On the 1928 No. 6 strike see Douglas K. Monroe, *A Decade of Turmoil: John L. Lewis and the Anthracite Miners, 1926–1936*, unpublished doctoral dissertation, Georgetown University, 1976, 81–92. In 1925, District 1 President Rinaldo Cappellini criticized the condition of the local unions at PaCC and HC&I collieries: "It is true they had a local of the organization at each of the collieries but with few exceptions the colliery superintendents directed the work of the local. They directed the officers and even went so far as to name the check-docking boss and check-weighman."

33. "Dead Mine Union Chief Told Wife Who His Killers Might Be," *Scranton Times*, 29 Feb 1928; "Pittston in Terror from Mine Killings," *New York Times*, 1 Mar 1928.

34. The *Philadelphia Inquirer* reported that 27 murders occurred in Pittston between 1914 and 1920 ("27th Murder Stirs Pittston Residents," *Philadelphia Inquirer*, 26 Jul 1920). On July 20, 1920, the *Wilkes-Barre Record* published a story referring to the "Epidemic of Murders" plaguing Luzerne County ("Murder Plains Main in Brutal Manner") and blamed "the foreign element" for the problem. Another article reported 13 homicides in the county in 1920 ("Murder a Month with One to Spare Is 1920 Record in County," *ibid.*, 1 Dec 1920). Selekman ("Miners and Murder," 151) reported that Pittston witnessed 30 murders between 1919 and 1928, many associated with the subcontracting system. The labor-related homicides continued into the late 1920s and early 1930s. Among the other victims were union activist Frank Bonita of Pittston, who was killed one week before the Save-the-Union convention in Pittsburgh in September 1928; Charles (Sam) Licata, a 38-year-old miner and union activist from Pittston, who was shot dead near his home on March 2, 1931; and Giuseppe Sperrazza, 53, a mining subcontractor from Pittston Township, found lifeless in his car with nine shots to the body on May 26, 1934 (Rochester, *Labor and Coal*, 223; *Wilkes-Barre Record Almanac*, 1931 & 1934; "Riddled by Bullets While Near His Home," *Trenton Evening Times*, 2 Mar 1931).

35. "Pittston Union Leader Found Slain," *The Evening News*, 19 Jan 1928; "Thos. Lillis, Well Known Mine Leader Slain from Ambush," *Pittston Gazette*, 19 Jan 1928. Two suspicious men had been visiting Pittston immediately prior to Lillis's demise, both wanted for murder in their respective cities of Buffalo and Philadelphia. According to Hapgood and Donovan: "The morning following the [Lillis] murder one of these men sent several hundred dollars home to his family. Both disappeared from town the day after the murder and have not been found. Their landlord admitted to police that [Frank] Agati [a former subcontractor and president of L.U. 1703] had been their visitor during their stay in Pittston" (Hapgood and Donovan, "Murdered Miners," 294). The *United Mine Workers Journal* stated that no person was convicted for Lillis's murder ("Frank Agati Slain by Gunmen in Offices of Union at WB," *United Mine Workers Journal*, 39 [1 Mar 1928], 17).

36. "Labor Leader Murder Victim," *Wilkes-Barre Record*, 20 Jan 1928; "See Causes of Murder in Pittston Mine Dispute," *ibid.*, 20 Jan 1928.

37. The other officers were Joseph Savage and Joe Costa, financial secretaries; Alfred Moleski, treasurer; Joseph Mullarkey, Thomas Griffin, and Frank Lenardo, trustees; Adam Moleski, Charles Verdine, and James Armat, committeemen ("Warring Factions of Union Promise to Come Together," *ibid.*, 20 Jan 1928). See also Selekman, "Miners and Murder," 155. Given Joe Costa's involvement with organized crime elements at this time, he may have served as a mob informant, although he never mentioned so during numerous interviews (see his interview listings in the References). See Chapter 6 for more on Mr. Costa, a Portuguese immigrant mineworker who was the first author's maternal uncle through marriage.

38. "Butler Men and Coal Officials Agree on Rate," *Pittston Gazette*, 3 Feb 1928.

39. For a moving account of child labor in Sicily's sulfur mines see Booker T. Washington, "Child Labor and the Sulfur Mines," Chapter 11 in his book, *The Man Farthest Down: A Record of Observations and Study in Europe*, Garden City, NY: Doubleday & Page, 1913. Washington described (p 214) the appalling working conditions in the sulfur mine for children and other workers: "I am not prepared just now to say to what extent I believe in a physical hell in the next world, but a sulphur mine in Sicily is about the nearest thing to hell that I expect to see in this life."

40. "Butler Men and Coal Officials Agree on Rate," *Pittston Gazette*, 3 Feb 1928.

41. "4 Collieries of Pa. System Work Monday," *ibid.*, 7 Jan 1928.

42. PaCC's 10 collieries included the Central, Erie, Ewen, Old Forge, Nos. 1, 5, 6, 9, 14, and Underwood collieries. HC&I had four collieries: Butler and three operations in Forest City (Pennsylvania Coal Company Colliery Books, 1983.177, Folders 1–12, Boxes 1 & 2, PaCC Papers, NCMA; and www.northernfield.info/).

43. "Conditions at No. 6 Discussed at Joint Meeting," *Pittston Gazette*, 14 Feb 1928; "Employes [*sic*] of No. 6 to Take Part in Peace Meeting," *ibid.*, 1 Mar 1928; "Pennsylvania Coal Workers Will Make Another Effort to Adjust Their Differences," *Scranton Times*, 5 Mar 1928; "Contract Mine System That Brought Strife to Pittston Region Before Conciliators," *ibid.*, 15 Mar 1928; "Idle Miners Want to Discuss Trouble with President of Pennsylvania Coal Company,"*ibid.*, 23 Apr 1928.

44. "Union Head Shot Down in Local Bank Building," *Times Leader*, 16 Feb 1928; "District Organizer Slain in Cappellini's Office," *Wilkes-Barre Record*, 17 Feb 1928; "Bonita, Accused Killer, is Caught at Pittston," *Times Leader*, 17 Feb 1928; "Pittston in Terror from Mine Killings," *New York Times*, 1 Mar 1928.

45. A story in the *United Mine Workers Journal* stated that District officials Anthony Figlock, August Lippi, and Frank Shiffko, as well as auditor Michael Gallagher, were also in the room when Agati was shot and they "just missed being struck by the shower of lead" ("Frank Agati Slain by Gunmen in Offices of Union at WB," *United Mine Workers Journal*, 39 [1 Mar 1928], 17.) President Rinaldo Cappellini was out of the office on other business.

46. Skeptics such as Rochester (*Labor and Coal*, 215) concluded that both Agati and Bonita drew guns and fired. Various reports indicated that five or six bullets were lodged in the office walls and one other in a window. One bullet hole was found in the wall near where Bonita stood. Bonita had obtained a permit to carry a gun following Thomas Lillis's murder. Bonita fled the scene with Mendola and Moleski but, one day later, surrendered to police in the company of Campbell and Reilly. Moleski went directly to the district attorney's office to explain the shooting, while Mendola was arrested some hours later. Hapgood and Donovan ("Murdered Miners," 293) wrote of Agati: "Eight miners to whom I [*sic*] talked the week-end after the shooting said Agati was always armed and often carried two guns conspicuously in his hip-pockets for the purpose of intimidation." See also "Testimony Tends to Prove Slayer Was Attacked," *Pittston Gazette*, 10 Apr 1928; "Miners' Leader Admits Wilkes-Barre Killing," *New York Times*, 12 Apr 1928; "Self-Defense, Expert Shows," *Daily Worker*, 12 Apr 1928; and "Dando Declares Death of Frank Agati in Wilkes-Barre Office 'Justifiable Killing,'" *Scranton Times*, 12 Apr 1928.

47. "Three are Held for Slaying," *Pittston Gazette*, 25 Feb 1928.

48. "Mine Union Leader Guilty of Killing," *New York Times*, 15 Apr 1928. Historian Walter Howard ("Anti-Labor and Anti-Radical Violence in Northeastern Pennsylvania During the Great Depression," 74–100 in Robert Mittrick [ed.], *Proceedings of the Eighth Annual Conference on the History of Northeastern Pennsylvania*, Luzerne County Community College, 6 Oct 1995, 76–78) pointed out that the CPUSA and its International Labor Defense (ILD) affiliate worked behind the scenes to raise money and organize a defense team for Bonita, Moleski and Mendola. The three accused union men faced high legal costs, as indicated by a partial payment to lawyers of $3,700 in May, 1928 ("Final Plans for Rump Convention Made at Meeting," *Scranton Times*, 5 May 1928). Expenditures for Bonita's defense were put at over $700, while "bail premiums and interest thereon" amounted to $1,500. Relief for the families took another undisclosed sum (Martin Abern, International Labor Defense Activities, 1 Jan–1 Jul 1928, www.marxists.org/history/usa/parties/cpusa/1928/ildactabern.htm). The ILD also provided financial assistance to the families while the men were in prison.

49. The jury in Bonita's trial deliberated for 43 hours before rendering a verdict of involuntary manslaughter, which recognized that Agati likely fired shots. However, Judge William S. McLean sent the jury back because there was no such count in the indictment. The jury's second verdict found Bonita guilty of voluntary manslaughter with the caveat that the court show mercy given the circumstances. McLean ignored the mercy suggestion and sentenced Bonita to the maximum sentence. Mendola and Moleski were convicted as accomplices in separate trials before Judge McLean who again handed out maximum jail terms. See "Mine Leader Guilty of Killing," *New York Times*, 15 Apr 1928; "Mendola Sentenced to Penitentiary," *The Evening News*, 17 Jul 1928; "M'Lean Sentences Bonita; Plea for Mercy is Ignored," *Pittston Gazette*, 18 Apr 1928; "Capitalist Justice," *Daily Worker*, 16 Apr 1928; Hapgood and Donovan, "Murdered Miners."

50. The letter was contained in "Union's Appeal is Firm," *Sunday Independent* (Wilkes-Barre), 1 Apr 1928.

51. "F. Agati Resigns as Organizer of the Miners' Union," *Pittston Gazette*, 16 Jan 1928; "Union Mine Head Brutally Slain in Wilkes-Barre," *ibid.*, 16 Feb 1928.

52. The 2011 figure is based on the Consumer Price Index value of $3,000 in 1928, as compared to the same amount in 2011, computed at the website www.measuringworth.com/calculators/uscompare/index.php.

53. "Band, 121 Automobiles in Agaty [*sic*] Funeral Cortege," *Pittston Gazette*, 19 Feb 1928.

54. Researchers Thomas Hunt and Michael A. Tona ("Men of Montedoro," *Informer* [Apr 2011], 5, informer-journal.blogspot.com), cited an FBI report dated 30 Oct 1967, linking Agati to the mob: "Frank Agati ...was a silent partner with Santo Volpe, Sr. and Steve LaTorre in one of their coal [sub]contracting businesses ... Frank Agati was believed to be a member of the Ma-

fia." Agati's son Guy later married Santo Volpe's daughter Euphamia ("Fanny"), further suggesting an organized-crime relationship. Agati was also linked to the murder of L.U. 1703 union leader Thomas Lillis (see note 35 above). Agati's affiliation with Cappellini indicated that the District 1 president had some ties to—or at least had to deal with—the organized criminal element. Cappellini 's name never appeared in official inquests or documents as a member or affiliate of organized crime.

55. "Pittston Mine Shooting Puts Area in Dread," *New York Times*, 20 Feb 1920; "Man Sought in Grecio Shooting Surrenders," *Wilkes-Barre Record*, 21 Feb 1928.

56. "Arrests Follow Finding of Guns at No. 6 Meeting," *Pittston Gazette*, 2 Feb 1928.

57. "Men Searched for Deadly Weapons, Three Arrested," *ibid.*, 1 Mar 1928; "Men Who Carried Pistols Indicted by the Grand Jury," *ibid.*, 10 Mar 1928. Pistols and other weapons were commonly carried in the mines around this time, said Lewis Casterline: "When I became superintendent, I had about 30 percent of the men [who] carried revolvers in the mines. Yeah, ... 500 men I had under me. That's Traders Coal Company, Ridgewood, here in Keystone section of [Plains Township]. Nineteen twenty-six. I was the superintendent there .... The Company was being 'taken' [cheated] by the workers. Not all of them, but a certain percentage. So I go up there and I met the grievance committee. I took a six-months leave of absence from the Pennsylvania Coal Company [to become superintendent at Traders]. And I says [to Trader's owners]: 'all right, I'll be up and see what I can do for you. And after I get through I'll give you my report.' [The Traders Coal Company was a PaCC lessee.] So as we were going along, I run into an obstacle. My foreman said to me: 'How the hell can you boss men when they got revolvers in the mine?' And [they] used those revolvers .... So I went up to the local [union] meeting, and after you suspend your meeting, now I'm not talking to the union, I'm talking to the membership ... I said: 'Now look boys, you can't have bosses in the mines if some of you guys carry revolvers. Now let's be fair about it. When you see a gun in front of you or a guy flashes it to you what are you going to do? You're going to scare that man and that man is going to scare somebody else, because he's going to tell the story to somebody.' 'Now,' I said, 'tell you what we do. You either take those revolvers and keep 'em home, or I'm going to buy a belt with a revolver on each side. And when I come around I'm going to have two pistols on each side.' We all started to laugh. So one of the guys got up and says [in broken English]: 'Well, we a taka the goddamn reboto [revolver] out.' And they took 'em out" (Lewis Casterline, audiotaped interview, 12 Dec 1988, NPOLHP).

58. Selekman, "Miners and Murder," 155. See also "Alex Campbell was Prominent in Mine Union," *Pittston Gazette*, 29 Feb 1928; and "Two Mine Leaders Cruelly Slain by Gunmen Last Night," *ibid.*, 29 Feb 1928.

59. "Strenuous Hunt for Assassins is Without Results," *ibid.*, 1 Mar 1928.

60. "Anthracite," *Time* (12 Mar 1928). The assassination weapons were not pump guns but automatic shot guns. They were found, along with a .38-caliber revolver and six empty cartridges, in the killers' blue-green abandoned car in the nearby borough of Moosic. Witnesses saw three men, presumed to be the murderers, jump on a moving freight train headed for Scranton ("Two Mine Leaders Cruelly Slain by Gunmen Last Night," *Pittston Gazette*, 29 Feb 1928). Joseph Piccillo and James Costanza of Pittston were young boys when the Campbell and Reilly shootings occurred across the street from their homes. Mr. Costanza was on the front porch and did not see the cars but heard the shots, whereupon an older sibling quickly ushered him inside. Mr. Piccillo saw the cars drive up the street and caught sight of the shotguns protruding from the windows. His cousin pushed him to the floor when the bullets started flying and they crawled inside. Mr. Piccillo remembered seeing the blood-splattered street and the large crowd gathered around the Campbell home. Piccillo and Costanza recalled the episode in unrecorded interviews on 6 Jul 1999, and 22 Jun 1999, respectively, NPOLHP.

61. "Campbell Chosen Check-Weighman at No. 6 Colliery," *Pittston Gazette*, 16 Jan 1928; "Alex Campbell and Peter Reilly, Miners' Leaders, Are Slain," *Wilkes-Barre Record*, 29 Feb 1928. The "Black Hand letter" refers to the type of threats written and delivered by local organized criminals (see Chapter 3).

62. "Dead Mine Union Chief Told Wife Who His Killers Might Be," *Scranton Times*, 29 Feb 1928.

63. Campbell's oldest daughter, Margaret, age 20, was about to receive a nursing diploma at the Pittston Hospital when news of her father's death arrived. She told one reporter: "I wish I could avenge my father's death. No one who knew my father, even though they disagreed with him, would have shot him down like that." She added that the family knew her father's life was in constant danger. According to the news story: "She steeled herself for the sad news that came to

her when she responded to an emergency call at the hospital last evening." The other Campbell children were Daniel, 18; Kathryn, 15; Alexander Jr., 12; Helen, 8; Anna, 5; and Marian, eight months. Mrs. Campbell was the former Anna Jenkins. One of Campbell's close friends stated: "There's little we can say. Alex, as everybody knows, was a strong union man. He served as an International organizer for a time in the other fields." The friend noted that Campbell was elected check-weighman at No. 6 on a number of occasions and often had difficulty gaining access to the colliery ("Dead Mine Chief Told Wife who His Killers Might Be," *ibid.*, 29 Feb 1928). Campbell spoke of the company's refusal to seat him as check-weighman following his 1928 election to the position in "Voice Protests at Mass Meeting," *Wilkes-Barre Record*, 3 Jan 1928. Charles McCarthy cited 1913 as the year when PaCC first disallowed Campbell to assume the check-weighman's duties at No. 6 (Charles A. McCarthy, "1920 Mine Strike Included Murder of Pittston Detective," *Wilkes-Barre Record*, 26 Oct 1965; see also Charles A. McCarthy, "Pittston Montedoro Society Golden Jubilee to be Sunday," *ibid.*, 7 Feb 1966).

64. "Alex Campbell and Peter Reilly, Miners' Leaders, Are Slain," *ibid.*, 29 Feb 1928; "2 Union Leaders Slain by Gunmen in Pittston 'War'," *The Scranton Republican*, 29 Feb 1928; "Pittston in Terror from Mine Killings," *New York Times*, 1 Mar 1928; "Two More Slain; Local Union Officers Meet Death in Pennsylvania," *United Mine Workers Journal*, 39 (15 Mar 1928), 16; "Death of Campbell and Reilly Laid to Contract Operators," *The Coal Digger*, 14 Apr 1928.

65. "Wilkes-Barre Local Denounces Assassins," *Pittston Gazette*, 1 Mar 1928. Campbell's wake and services took place at the family home under the direction of Rev. Richard A. Rinker, pastor of the First Presbyterian Church. Burial followed at the Pittston Cemetery. Reilly's was waked at the family home, followed by a Requiem Mass at St. Casimir's Lithuanian Church. He was interred at the church cemetery in Pittston ("Arrangements for the Funerals of the Tragedy Victims," *ibid.*, 1 Mar 1928).

66. "Cappellini Says He Will Not Quit as Miners' Head," *ibid.*, 1 Mar 1928.

67. "End This Reign of Terror," Editorial, *Times Leader*, 29 Feb 1928.

68. "Union Head Shot Down in Local Bank Building," *ibid.*, 16 Feb 1928. See Chapter 3 on the IWW.

69. Selekman, "Miners and Murder," 153.

70. "Alex Campbell and Peter Reilly, Miners' Leaders, are Slain," *Wilkes-Barre Record*, 29 Feb 1928. See also "Lucchino's Death Started Bloodshed," *Times Leader*, 20 Feb 1928. On Lucchino's murder see Chapter 4.

71. Walter T. Howard, *Forgotten Radicals: Communists in the Pennsylvania Anthracite, 1919–1950*, Landham, MD: University Press of America, 2005, 42.

72. "Police See Latest Pittston Shooting as Retaliation for Slaying of Frank Agaty [*sic*]," *Wilkes-Barre Record*, 29 Feb 1920; "Death of Campbell and Reilly Laid to Contractor Operators," *The Coal Digger*, 14 Apr 1928. See also "Union Head Shot Down in Local Bank Building," *Times Leader*, 16 Feb 1928; and "Alex Campbell and Peter Reilly, Miners' Leaders, Are Slain," *ibid.*, 28 Feb 1928.

73. Four other authors have examined the violence at No. 6: Selekman ("Miners and Murder"); Monroe (*A Decade of Turmoil*, 82ff); Perry K. Blatz (*Democratic Miners: Work and Labor Relations in the Anthracite Coal Industry, 1875–1925*, Albany, NY: SUNY Press, 1994, 258); and Howard, *Forgotten Radicals*, 39–45. None of the authors discussed organized crime's involvement in either the murders or the subcontracting system, an omission at least partly due to the difficulty in obtaining information. When Selekman researched the striking No. 6 workers in 1928, the structure and operation of organized crime was obscure; moreover, out of fear, the workers may not have wanted to discuss the criminal element with an outsider. Monroe, Blatz, and Howard, on the other hand, wrote after the structure and operation of organized crime had been studied and was well known; however, they relied almost exclusively on period documents that, unsurprisingly, contained scant information on the subject. The methodological problems associated with documentary research reinforces the value of oral history which, despite its own methodological problems, is often the only, or one of the only, sources of information available. Oral histories can also serve as a supplement to, or check on, written sources. On oral history see Paul R. Thompson, *The Voice of the Past: Oral History*, New York: Oxford University Press, 1988; Michael Frisch, *A Shared Authority: Essays of the Craft and Meaning of Oral History*, Albany, NY: SUNY Press, 1990; Alessandro Portelli, *The Death of Luigi Trastulli and Other Stories: Form and Meaning in Oral History*, Albany, NY: SUNY Press, 1991; Alessandro Portelli *The Battle of Valle Giulia: Oral History and the Art of Dialogue*, Madison, WI: University of Wisconsin

Press, 1997; Allesandro Portelli, *They Say in Harlan County: An Oral History*, New York: Oxford University Press, 2010; Luisa Del Giudice (ed.), *Oral History, Oral Culture, and Italian Americans*, New York: Palgrave Macmillan, 2009. The most recent major work on anthracite history by Dublin and Licht (*The Face of Decline*) completely ignored organized crime's involvement within the industry.

74. Transcript of *Proceedings*, Board of Conciliation Meeting, 3 Mar 1928, International Executive Board Correspondence, AX/B40/HCLA/01146, Box 6, Folder 30, UMWA Papers, PSU, quote is from p 19; Lewis Casterline, audiotaped interview, 16 Jan 1970, King's College Oral History Collections.

75. William Hastie, audiotaped interview, 31 Jul 1989, NPOLHP. Police charged Ralph Melissari of New Jersey, and Peter DeLucca and Vincenzo "Little Jimmy" Damiano of New York City with the homicides. A fourth person listed as "John Doe" was also indicted but the charges were evidently dropped; he was described as "an Italian, about 35 years old ("Four are Indicted for the Campbell–Reilly Murders," *Pittston Gazette*, 28 Nov 1928). Acquitted of Campbell's murder but convicted of Reilly's, Melissari was sentenced to life in prison. DeLucca was also convicted in Reilly's homicide and given a life sentence. Gunmen murdered Damiano before his trial began. On Melissari's attempt at a retrial see Commonwealth v. Ralph Melissari, Court of Quarter Sessions of Luzerne County, April Session, 1928. See also "Arrest Made in Pittston Double Murder by Trooper," *The Evening News*, 31 Mar 1928; "Police Say Publicity Has Delayed Arrest of Two Murder Suspects," *ibid.*, 2 Apr 1928; "Net is Closing in on Men suspected of Dual Slayings," *Pittston Gazette*, 2 Apr 1928; and *Wilkes-Barre Record Almanac*, 21 Nov 1928; 8 Feb 1932; 16 May 1932; 4 Jun 1932; 9 Jun, 1932; 13 Dec 1932.

76. "Employees of No. 6 to Take Part in Peace Meeting," *Pittston Gazette*, 1 Mar 1928. The scheduled speakers were Save-the-Union leaders Stanley Dziengelswski and George Papcun; labor activist Powers Hapgood; L.U. 1703 vice president Joseph Victor; L.U. 1703 financial secretary Joseph Savage; and the acting recording secretary of L.U. 1703, Charles Licata, who had assumed Peter Reilly's duties.

77. *Ibid.* Charles Siracuse, a miner and president of L.U. 452 at the Harry E Colliery in Swoyersville from 1927 to 1933, was the first author's maternal uncle. The first author's grandfathers, Oney Wolensky and Cologero "Carl" Siracuse, were also miners at the Harry E; his father, Nicholas Wolensky, and uncle, John Siracuse, worked in the breaker; uncles Frank Siracuse and Angelo Siracuse worked underground in the mines; and cousin Paul Wolensky was a mine laborer who perished in an explosion in 1941 (on the explosion see the interview with Paul Wolensky's co-worker, Peter Lazar, who survived; audiotaped interview, 25 Jun 1990, NPOLHP).

78. "Employees of No. 6 to Take Part in Peace Meeting," *Pittston Gazette*, 1 Mar 1928.

79. Mayor Asks for Peace," *Times Leader*, 23 Feb 1928. Mayor Gillespie's Herrin reference was to the 1922 massacre of 19 company men in Herrin, Illinois, at the hands of striking UMWA members (McAlister Coleman, *Men and Coal*, New York: Farrar & Rinehart, 1943, 114–25).

80. Telegram, William H. Gillespie to John L. Lewis, 28 Feb 1928, President/District Correspondence, 1920–1933, GST/2/06.02, Box 6, Folder 30, UMWA Papers, PSU. See also John L. Lewis to Hon. William H. Gillespie, 1 Mar 1928, *ibid.*; John L. Lewis to Rinaldo Cappellini, 1 Mar 1928, *ibid.*; and "Pittston in Terror from Mine Killings," *New York Times*, 1 Mar 1928.

81. "Employes [*sic*] of No. 6 to Take Part in Peace Meeting," *Pittston Gazette*, 1 Mar 1928; "President Lewis Has Declined to Come to Pittston," *ibid.*, 2 Mar 1929.

82. "Two More Slain: Local Union Officers Meet Death in Pennsylvania," *United Mine Workers Journal*, 39 (15 Mar 1928), 16.

83. "Official Letter to Local Unions in District No. 1," *ibid.*, 39 (15 Jun 1928), 7.

84. Robert H. Zieger, *John L. Lewis: Labor Leader*, Boston, MA: Twayne, 1988, 100. Lewis's red-baiting tactics were on display at the 1923 Tri-District Convention in Scranton, as discussed by Melvyn Dubofsky and Warren Van Tine (*John L. Lewis: A Biography*, New York: Quadrangle/New York Times, 1977, 99–100), and also by "Coal Miners Demand a 20 Per Cent. Raise," *New York Times*, 30 Jun 1923. See also Howard, *Forgotten Radicals*, 38–39.

85. "Hapgood and Wife Sent to Jail for Inciting Riot," *Pittston Gazette*, 5 Mar 1928; "Radicals Jailed for Stirring Up Feud Among Coal Miners; Hapgood and Wife Among Them," *Trenton Evening News*, 5 Mar 1928. Hapgood and his wife, Mary Donovan, were jailed and held without bail until April 12 when they were released by Judge Benjamin R. Jones for lack of evidence. Several attempts to post bail were denied. Local activist Charles Licata of Pittston was arrested at the

same time but freed on bail the next day. See also "Hapgoods Acquitted of Inciting Coal Riot," *New York Times*, 13 Apr 1928.

86. The bituminous men had been on strike for nine months during 1927 and 1928 when the Jacksonville wage scale agreement, which guaranteed workers a daily minimum wage of $7.50, expired and negotiations failed. Over 85,000 mineworkers in western Pennsylvania, Ohio, and West Virginia joined the strike, and many persons—including a subcommittee of the U.S. Senate that convened to investigate the matter—feared that the industrial action would spread to 24 other soft-coal states. Violence occurred at several pits. At a mine near Steubenville, Ohio, 300 union and non-union workers clashed at a Y&O Coal Company mine where 51 strikebreakers had been hired. In Walsenburg, Colorado, state police shot and killed a striking miner named Klemento Chavez, 42, who, police claimed, had fired at them from the meeting hall used by the IWW. Near Pittsburgh, the mining villages of Bruceton and Broughton experienced violence attributed to the hiring of a "negro strikebreaker" by the Pittsburgh Terminal Coal Company." On these incidents and other aspects of the strike see "Labor Factions Engage in Riots in Mining Town," *Pittston Gazette*, 13 Jan 1928; "State Police are Keeping the Peace in Colorado Town," *ibid.*, 13 Jan 1928; "Mine Villages Fear 'Night of Terror'," *ibid.*, 2 Feb 1928; "Coal Committee Fears Spread of Industrial War," *ibid.*, 28 Feb 1928; Hon. John J. Casey, *Miners' Strike in the Bituminous Coal Fields*, Washington, DC: USGPO, 8 Feb 1928; "The 1927–1928 Colorado Coal Strike," *Pacific Historical Review*, 32 (Aug 1963), 235–50; John L. Lewis, Philip Murray, and Thomas Kennedy to the Officers and Members of Local Unions, UMWA, 24 Apr 1928 and 31 Oct 1928, International Executive Board Correspondence, AX/B40/HCLA/ 01146, Box 6, Folder 16, UMWA Papers, PSU.

87. "Official Letter to Local Unions in District No. 1," *United Mine Workers Journal*, 39 (15 Jun 1928), 7. See also "Strike Closes Mine," *Times Leader*, 9 Mar 1928.

88. Howard (*Forgotten Radicals*, 33–34) listed the number of communist members in District 1 from various ethnic groups, the data having been gleaned from Communist Party records in Moscow following the end of the Cold War. In 1926, the CPUSA had 49 Lithuanians, 42 Italians, 10 Russians, 19 Ukrainians, 10 Poles, and 4 Croatian members. The breakdown of Italian membership by town was Luzerne (which could have included Swoyersville and the surrounding areas), 16; Old Forge (which could have included Pittston), 14; Jessup, 16; Plains (which could have included Pittston), 14; West Wyoming, 8; and Dunmore, 4. Interestingly, Pittston was not listed among the towns having members, meaning that the data were incomplete, or the city's large Italian immigrant population had little interest, or members did not want their membership known so they traveled to other towns for meetings.

89. "Communists in the Coal Region," *Wilkes-Barre Record*, 5 Oct 1928. On communist activism in Chicago during this period see Randi Storch, *Red Chicago: American Communism at Its Grassroots, 1928–35*, Urbana, IL: University of Illinois Press, 2009.

90. "Pittston Coal Miners Begin Anthracite Strike," *Daily Worker*, 21 Mar 1928. See also "Pittston Miners Support 'Save-the-Union' Group," *ibid.*, 22 Mar 1928.

91. On the Communist Party's unsuccessful attempt to organize northern-field workers see Howrd, "Anti-Labor and Anti-Radical Violence." in Northeastern Pennsylvania During the Great Depression." Communist influence reappeared in July and August 1928 with the Save-the-Union faction within the UMWA ("Save-the-Union Leader Hits at W.H. Gillespie," *The Evening News*, 4 Aug 1928). On the communist movement among mineworkers see also John Brophy (John O.P. Hall, ed.), *A Miner's Life*, Madison, WI: University of Wisconsin Press, 1964, 214–18; Dubofsky and Van Tine, *John L. Lewis*, 127–28; and Alan Singer, "Communists and Coal Miners: Rank-and-File Organizations in the United Mine Workers of America During the 1920s," *Science and Society*, 55 (1991), 143.

92. "Strike Closes Mine," *Times Leader*, 9 Mar 1928; "Miners' Interest Center in the No. 6 Controversy," *Pittston Gazette*, 9 Mar 1928. The statement was submitted to the press by L.U. 1703 officers Frank McGarry, Joseph Victor, James Lamarca, Joseph Savage, Nick Benefati, and James Kearney.

93. "Will Reopen Mine to Avert Bloodshed," *New York Times*, 4 Mar 1928.

94. For example, two blasts hit the Pittston area in 1927, one on October 19 at St. John the Evangelist Church, the Irish-Catholic house of worship, and the other on November 9 at the recently built West Pittston High School. Most previous explosives charges smashed homes and businesses, but these were the first to strike institutional buildings. On April 1, 1928, another charge destroyed the office of *La Voce Italiana*, an Italian-language newspaper. On April 2 alleged gang-

sters dymanited the Raymond Court House. On these bombings see "Pittston Church Dynamiting Stirs Authorities," *Times Leader* 19 Oct 1927; "Office of '*La Voce Italiana*' Wrecked by Bombers Late Last Night," *Scranton Times*, 1 Apr 1928; "Underworld Gangsters Use Dynamite to Wreck Raymond Court House," *ibid.*, 2 Apr 1928; "The Terror of the '20s," *Times Leader*, 27 Aug 2006.

95. "Brothers Pleaded Guilty in Scranton in Dynamiting Case," *Pittston Gazette*, 27 Mar 1928.

96. William A. Hastie, unrecorded interview, 18 Jan 2011. Miner Joe Costa confirmed the criminal element's use of "eyes" and "spies" around the mines: "They just knew how to hire men and use them to get the coal to make money. They had their watchdogs, too. You'd see their men watching all over, watching everything around the mines" (Joe Costa, untaped interview, 22 Jul 1995).

97. "Glen Alden Miners Urge Cappellini to Resign Office," *Pittston Gazette*, 5 Mar 1928.

98. "Will Reopen Mine to Avert Bloodshed," *New York Times*, 4 Mar 1928; "Lewis Sees Hope of Agreement," *ibid.*, 4 Mar 1928; "Reject Reopening of Pittston Mine," *ibid.*, 6 Mar 1928.

99. "Grievance Board Gives Support to Oust Contractors," *Pittston Gazette*, 10 Mar 1920.

100. "Guzior Elected to Head Local of Butler Colliery," *ibid.*, 25 Feb 1928. The other elected officers were John Paluski, vice president; Michael Tello, secretary; and Albert Strucke, treasurer. The deposed president was John Galick.

101. "Butler Mines Strike; Oppose Hiring New Men," *ibid.*, 17 Mar 1928; "Butler Miners Will Stand Firmly on Their Demands," *ibid.*, 21 Mar 1928.

102. "Butler Men Are Ready to Return to Work at Once," *ibid.*, 23 Mar 1928.

103. The other elected insurgent officers at the No. 9 Colliery were Frank Chiavinko, vice president; Charles Koval, financial secretary; Adam Dulkas, treasurer; Patrick Flynn, recording secretary; John Markowski and Joseph Dessoye, assistant secretaries; John Rudis, sentinel; Peter Flynn, Arthur Renfer and Frank Tross, auditors; and Edward Richards, Stanley Moskaitis, Geralds Hobbs, and Frank Weitz, trustees ("No. 9 Colliery Employes [*sic*] Expel local Officers," *ibid.*, 26 Apr 1928). On the conflict at No. 9 see also "No. 9 Men Honor New Officers at Regular Session," *ibid.*, 9 May 1928; "No. 9 Local Meeting in Charge of Old Officers," *ibid.*, 10 May 1928; and "Factional Fight in Miners' Union Gets into Courts," *ibid.*, 15 Jun 1928.

104. Transcript of *Proceedings*, Board of Conciliation Meeting, 3 Mar 1928, International Executive Board Correspondence, AX/B40/HCLA/01146, Box 6, Folder 30, UMWA Papers, PSU; "No. 6 colliery to be Reopened to Test Rates," *Pittston Gazette*, 3 Mar 1928.

105. "Plans to Resume Operation at No. 6 Mine Rejected by Men," *ibid.*, 5 Mar 1928.

106. Transcript of *Proceedings*, Meeting of Thomas Kennedy, Anthracite District Presidents, and Representatives of L.U. 1703, 15 Mar 1928, City Hall, Hazleton, PA, International Executive Board Correspondence, AX/B40/HCLA/01146, Box 6, Folder 30, UMWA Papers, PSU. See also "District Board Declares against Mine Contractor," *Pittston Gazette*, 6 Mar 1928; and "Miners Seeking Umpire's Ruling on Contractor," *ibid.*, 6 Mar 1928. Hudson Coal Company workers had filed a grievance with the ABC a month earlier because the company wanted to mine certain low veins using subcontractors ("Anthracite Board to Discuss Hudson Case," *ibid.*, 6 Feb 1928).

107. Transcript of *Proceedings*, Meeting of Thomas Kennedy, Anthracite District Presidents, and Representatives of L.U. 1703, 15 Mar 1928, City Hall, Hazleton, PA, International Executive Board Correspondence, AX/B40/HCLA/01146, Box 6, Folder 30, UMWA Papers, PSU, 17–18. Matty and District 9 President Chris Golden stated that coal companies in both of their districts tried to implement the subcontracting system but the men refused to work for the subcontractors, and the District organization and the local unions would simply not allow it (*ibid.*, 8–9). See Chapter 6, note 99, on this topic.

108. *Ibid.*, 19.

109. "Pittston Flares Up in a New Shooting," *New York Times*, 19 Mar 1928; "Boy Uses Shotgun to Kill Intruder; Father is Wounded by Two," *Pittston Gazette*, 19 Mar 1928. A summary of the incident in the *Wilkes-Barre Record Almanac* for 1928 indicated that the intruders were trying to extort money from the senior Alfili.

110. "Conciliators to Hear Grievance of No. 6 Miners," *ibid.*, 14 Mar 1928; "Contract Mine System That Brought Strife to Pittston Region Before Conciliators," *Scranton Times*, 15 Mar 1928;

"Conciliators Offer Slim Encouragement to No. 6 Sub-Committee," *Times Leader*, 16 Mar 1928.

111. "No. 6 Workers Reject '30-Day' Peace Plan," *Scranton Times*, 19 Mar 1928.

112. "Upholds Contract Mining," *New York Times*, 18 Mar 1928; "Contract Mining System Sustained by Umpire," *Scranton Times*, 17 Mar 1928. Two years earlier, in 1926, the ABC upheld the Hudson Coal Company's right to issue individual and special contracts for machine mining (Anthracite Board of Conciliation, *Report* [Grievances and Actions], Grievance No. 2035, Hazleton, PA: Anthracite Board of Conciliation, 1926, 16, 30–48). Crucial to the board's decision was the fact that two worker demands during the labor-management negotiations of 1916 (elimination of individual and special contracts, and a prohibition against contract miners having more than one working place) were not part of the final agreement. Moreover, the ABC concurred with Hudson's management that the company had conformed to another provision in the 1916 agreement requiring that no special contracts could be issued to individual employees at less than the prescribed wage rates. A District 1 letter to the membership in March 1928 stated that the Hudson Coal Company had been using subcontractors "for 15 or more years" (Rinaldo Cappelini and Enoch Williams To the Membership of the UMWA of District 1," International Executive Board Correspondence, AX/B40/HCLA/ 01146, Box 6, Folder 18, UMWA Papers, PSU).

113. On forced overtime without pay see "Neill's Decision on Overtime Pay Favors Miners," *Pittston Gazette*, 23 Apr 1928.

114. "3,000 Nanticoke Mine Workers Call Strike," *Times Leader*, 23 Feb 1928.

115. "Strike Closes Mine; 900 Mocanaqua Miners Walkout in 'Topping' Controversy," *ibid.*, 9 Mar 1928; "West End Coal Dispute Ends at Mocanaqua," *The Evening News*, 31 Jul 1928.

116. "700 Valley Miners Strike at Exeter," *Times Leader*, 4 Apr 1928; "600 Employes [*sic*] of Exeter Colliery on Strike Today," *Pittston Gazette*, 5 Apr 1929.

117. "3,000 Miners Strike," *Times Leader*, 9 May 1928.

118. The No. 20 Colliery strike is discussed in "Boylan Makes Known Plans to Head District," *The Evening News*, 23 Jul 1928.

119. "Brennan Heads Movement for Mine Meeting," *Pittston Gazette*, 19 Mar 1928.

120. "Cappellini Will Ignore Request for Convention," *ibid.*, 12 Mar 1928; "Conciliators Offer Slim Encouragement to No. 6 Sub-Committee," *Times Leader*, 16 Mar 1928.

121. All of the District's General Grievance Committees—one for each coal company—met weekly in preparation for the special convention ("Grievance Board Expels Papcun and Other Agitators," *Pittston Gazette*, 14 May 1928).

122. "Brennan Heads Movement for Mine Meeting," *ibid.*, 19 Mar 1928.

123. "Miners Demand Convention," *New York Times*, 19 Mar 1928; "Brennan Heads Movement for Mine Meeting," *Pittston Gazette*, 19 Mar 1928; "Miners Appeal to Lewis for a Convention Call," *ibid.*, 26 Mar 1928. A story in the *Pittston Gazette* put the number of attending locals at 60.

124. "Call for a Special Convention," International Executive Board Correspondence, AX/B40/HCLA/01146, Box 6, Folder 30, UMWA Papers, PSU. See also "Miners Appeal to Lewis for a Convention Call," *Pittston Gazette*, 26 Mar 1928.

125. *Ibid.* See also "Union's Appeal is Firm," *Sunday Independent*, 1 Apr 1928.

126. "Miner's Demand Denied," *Times Leader*, 22 Mar 1928; "Expulsion Threat by Cappellini is Met with Silence," *ibid.*, 23 Mar 1928; "Union Officials Say Trouble Due to Petty Politics," *Pittston Gazette*, 23 Mar 1928. The officers of the committee were William Davis (Glen Alden), chairman; David Jones (Glen Alden), vice chairman; Edward McCrone (workplace unknown), secretary. The executive committee members were Frank Sobers, John Cavanaugh, Charles Kaschinsky, John Orr, Frank McGarry, James Gallagher, Elliott Miller, Michael Kuratz, Adam Minch, Edward Hogan, and William J. Brennan.

127. In mid-March 1928, the Lehigh Valley Coal Company restarted three collieries that had been idle for several weeks: the Seneca in Pittston, William A. in Duryea, and Broadwell in Moosic. Another of the firm's collieries, the Heidelberg in Avoca, was expected to resume work the following week ("Three Collieries of Lehigh Valley Will Resume Work," *Pittston Gazette*, 15 Mar 1928; "30 Anthracite Collieries are to Resume Work," *ibid.*, 2 Apr 1928; "Penna. Mines to Work on Friday," *ibid.*, 5 Apr 1928).

128. "50 Mine Locals Pass Resolution for Convention," *ibid.*, 23 Apr 1928.

129. "Anti-Cappellini Group Will Gather Tonight," *Scranton Times*, 21 Apr 1928; "Miners Plan Session to Oust Cappellini," *New York Times*, 23 Apr 1928; "Final Plans for Rump Convention Made at Meeting," *Scranton Times*, 14 May 1928.

130. "Meet Call is Branded Dual Move," *ibid.*, 23 Apr 1928.

131. "Miner's District Executive Board Called to Capitol," *Times Leader*, 23 Mar 1928. The executive board consisted of John Kmetz, Nanticoke; August Lippi, Exeter; John Boylan, Scranton; and George Dorsey, Scranton. Dorsey was appointed following the death of James Gleason, and Lippi took Dorsey's seat with the biennial election in 1927.

132. "'Return to Work' Lewis' Message to No. 6 Miners," *Pittston Gazette*, 28 Mar 1928. See also "Lewis Understood to Have Adopted Hands-Off Policy," *Times Leader*, 27 Mar 1928.

133. "Lewis Sends Sealed Note to Local 1703," *ibid.*, 29 Mar 1928; "Official Text of Lewis Letter Received Here," *Pittston Gazette*, 29 Mar 1928. Along with McGarry, the other officers of L.U. 1703 were Joseph Victor, vice president, James Kearney, secretary, and Frank Licata, treasurer.

134. "Union's Appeal is Firm," *Sunday Independent*, 1 Apr 1928.

135. "No. 6 Miners Reject Proposal of Leaders for Resumption of Work," *Scranton Times*, 5 Apr 1928.

136. "Accept Relief Plan for Men of No. 6 Colliery," *Pittston Gazette*, 27 Mar 1928.

137. "Lewis' Proposal to No. 6 Employes [*sic*] Rejected Today," *ibid.*, 5 Apr 1928.

138. "Mediator Davis May Intervene in No. 6 Controversy," *ibid.*, 6 Apr 1928; "Mediator Davis to Address Men of No. 6 Colliery," *ibid.*, 16 Apr 1928.

139. "No. 6 Committee Invited to Confer with Union Leaders," *ibid.*, 2 Apr 1928.

140. Telegram, Rinaldo Cappellini to Michael Gallagher, 5 Apr 1928, and telegram, Michael Gallagher to Rinaldo Cappellini, 6 Apr 1928, President/District Correspondence, 1920– 1933, GST/Z06.02, Box 1, Folder 11, UMWA Papers, PSU. See also "Cappellini Wants Company Head to Meet Idle Miners," *Scranton Times*, 5 Apr 1928; and "Gallagher Refuses to Meet No. 6 Men Unless They Return," *Times Leader*, 11 Apr 1928.

141. "Pittston Men Hold Stormy Mine Meeting," *The Evening News*, 2 Aug 1928. See also "McGarry Takes Campbell's Job at No. 6 Mine," *Times Leader*, 7 Mar 1928.

142. Selekman, "Miners and Murder," 193–94.

143. "Brennan Outlines Underlying Trouble in Pittston Mine War," *Times Leader*, 28 Apr 1928.

144. The insurgents barred deposed L.U. 1703 president Patsy Paglioca, who had also been a subcontractor, from participating in the emergency meeting; however, he was allowed to sit in the rear of the room with other non-member observers. On the return to work see "Pittston Miners to Resume Work," *New York Times*, 20 Apr 1928; "No. 6 Colliery Did Not Reopen this Morning," *Pittston Gazette*, 26 Apr 1928; "No. 6 Colliery Started Up Today Without a Hitch," *ibid.*, 30 Apr 1928; and "Pittston Miners in No. 6 Colliery are Back at Work," *Scranton Times*, 30 Apr 1928.

145. "Save-the-Union Committee Protests Police Action in Preventing Meetings," *ibid.*, 1 May 1928.

146. John Brophy to the membership of the UMWA, "John Brophy Appeals to International Executive Board for Honesty and a Square Deal," International Executive Board Correspondence, 28 May 1927, Box 6, Folder 12, AX/B40/HCLA/01146, UMWA Papers, PSU. See also Brophy, *A Miner's Life*, Chapter 15.

147. Minutes, National Conference, Save-the-Union Committee, April 1, 2, 3, 1928, Labor Lyceum, Pittsburgh, International Executive Board Correspondence, Box 6, Folder 18, AX/B40/HCLA/01146, UMWA Papers, PSU. Howard listed a "small cadre" of communist workers in anthracite who participated in the Save-the-Union movement: Charles (Sam) Licata, Joseph Dougher, George Papcun, Stanley Dziengielewski, and Ed Falkowski. Other anthracite Communists included Joseph Dougher, Frank Vrataric, and Alex Zarek. The District 1 members attending the Save-the-Union convention in Pittsburgh included George Papcun, Stanley Dziengielewski, Mike Zaldokis, and Anthony Ricci, among others. They were also supporters of the National Miners Union, which achieved no real success in anthracite (Howard, *Forgotten Radicals*, 35; Walter T. Howard, "The National Miners Union: Communists and Miners in the

Pennsylvania Anthracite, 1928–1931," *Pennsylvania Magazine of History and Biography*, 125 [Jan/Apr], 104–05). On the bituminous strike see "Save-the Unionists Claim 10,000 Answered Strike Call," *Pittston Gazette*, 17 Apr 1928.

148. Steve Nelson, James R. Barrett, and Rob Ruck, *Steve Nelson: American Radical*, 174. Nelson reported that the Communists had over 200 members in the four anthracite fields by the mid-1930s.

149. "Miners Determined to Oust Lewis Machine," *The Daily Worker*, 2 Jun 1928; "Mine Group is Opposed to New District Head, National Miners' Leaders Issue Statement, Attacking John Boylan," *The Evening News*, 15 Jul 1928; "Fist Fight Between 2 Officers Ends Local Meeting at Pittston; Police Stop Melee that Results at Ewen Union Meeting," *ibid.*, 3 Aug 1928; and "Save-The-Union Leader Hits at W.H. Gillespie," *ibid.*, 4 Aug 1928.

150. "Campbell's Widow Urges Solidarity," *Times Leader*, 28 Mar 1928. Mrs. Campbell also wrote to John L. Lewis and asked for an investigation of the Cappellini administration. Lewis expressed sympathies for the family's loss but declared the matter closed (Mrs. Alex Campbell to John L. Lewis, 9 Jun 1928; John L. Lewis to Mrs. Alex Campbell, 19 Jun 1928, President/District Correspondence, 1920–1938, GST/Z06.02, Box 1, Folder 12, UMWA Papers, PSU).

151. "Leaders Warn Miners to Keep Away from Grievance Group," *Scranton Times*, 25 Apr 1928.

152. Michael R. Bussel, *From Harvard to the Ranks of Labor: Powers Hapgood and the American Working Class*, University Park, PA: Penn State Press, 1999. Another case where a UMWA district official suffered consequences for illegal strike activity involved Alex Howat, District 14 president, who was sent to jail by authorities for calling a wildcat strike in Kansas, which had a no-strike law (James P. Cannon, "The Story of Alex Howat," *The Liberator*, 4 [Apr 1921], 25–6).

153. "Ask Lewis to Preside at Rump Convention," *Pittston Gazette*, 5 May 1928; "Mine Bosses Charged with Resorting to Intimidation to Keep Men from Convention," *Scranton Times*, 21 May 1928.

154. "Green and Davis are Invited to Miners Session," *Pittston Gazette*, 19 May 1928.

155. "Plans for Special Miners' Convention Being Carried Out," *ibid.*, 7 May 1928.

156. "No 6 Colliery Started Up Today Without a Hitch," *ibid.*, 30 Apr 1928.

157. "Alex Campbell is Eulogized at Mine Convention," *ibid.*, 22 May 1928.

158. "'Save-the-Union' Man Ejected from Convention," *Scranton Times*, 22 May 1928.

169. "Mine Convention Needs 7 Locals for a Majority," *Pittston Gazette*, 25 May 1928; "Charge District Officials Failed to Help No. 6 Men," *ibid.*, 26 May 1928; "Mine Convention Making Plans to Assume Control," *ibid.*, 29 May 1928.

160. "Fist Fight Occurs As Ewen Employees Meet in Local Hall," *ibid.*, 2 May 1928.

161. "Three New Locals Send Delegates to Convention," *ibid.*, 23 May 1928; "Mine Convention Lacks 12 Locals for a Majority," *ibid.*, 24 May 1928; "Mine Convention Needs 7 Locals," *ibid.*, 25 May 1928.

162. "Victory Parade as Convention Gets a Majority," *ibid.*, 26 May 1928.

163. The other District 1 officers ousted were Michael Kosik, vice president; Enoch Williams, secretary-treasurer; John Boylan, George Dorsey, August Lippi, John Kmetz, board members; and Dennis Brislin, International Board member.

164. "District Offices of Miners' Union Declared Vacant," *ibid.*, 30 May 1928.

165. "Insurgent Miners Elect Officers," *New York Times*, 2 Jun 1928. The other officers were Joseph Dougher, inspection board member, first inspection district; John Whitey, inspection board member, second inspection district; Edward Hogan, inspection board member, third inspection district; and Frank Sobers, inspection board member, fourth inspection district. John Bellfield joined the board at a later date ("Miners Choose New Officials in District 1," *Daily Worker*, 4 Jun 1928).

166. "McGarry Elected," *New York Times*, 1 Jun 1928.

167. "Frank M'Garry is Chosen President of Miners' Union," *Pittston Gazette*, 1 Jun 1928. See also "Recognition Denied Insurgent Miners," *New York Times*, 8 Jun 1928.

168. Frank McGarry et al., to John L. Lewis, 2 Jun 1928, President/District Correspondence, 1920–1933, GST/Z06.02, Box 1, Folder 12, UMWA Papers, PSU.

169. John L. Lewis to Walter Harris, 8 Jun 1928, President/District Correspondence, 1920–1938, GST/Z06.02, Box 1, Folder 12, UMWA Papers, PSU. The three-page letter was also sent to the District 1 membership. See also John L. Lewis to the Officers and Members of District 1, UMWA, 8 Jun 1928, *ibid.*

170. "Miners Convention Making Plans to Assume Control," *Pittston Gazette*, 29 May 1928. See also "Says Meeting Is Illegal," *Scranton Times*, 21 May 1928.

171. Rinaldo Cappellini and Enoch Williams to Officers and Members of UMWA District 1, 26 Jun 1928, President/District Correspondence, 1920–1933, GST/Z06.02, Box 1, Folder 12, UMWA Papers, PSU.

172. "Cappellini Out," *Scranton Times*, 20 Jul 1928; "Cappelini [*sic*] to Debate Insurgent Leaders at Sugar Notch, Is Plan," *The Evening News*, 13 Jul 1928. On the Save-the-Union Convention see the "National Conference, Save-the-Union Committee, Pittsburgh, Pa., April 1, 2, 3, 1928, Labor Lyceum, Miller St., Pittsburgh, PA," International Executive Board Correspondence, Box 6, Folder 28, AX/B40/HCLA/01146, UMWA Papers, PSU; and "Miners Plan to Win Strike and Save Union, *The Coal Digger*, 14 Apr 1928.

173. "Expel Leaders of Mine Convention, Cappellini Plan," *Pittston Gazette*, 6 Jun 1928.

174. "New Rates Plan at No. 6 Colliery Is to be Submitted," *ibid.*, 14 Jun 1928; "No Agreement on Mining Rates at No. 6 Colliery, *ibid.*, 21 Jun 1928.

175. "Pittston Miners to Resume Work," *New York Times*, 20 Apr 1928.

176. "Pittston Miners Reject Lewis Peace Proposal," *Times Leader*, 5 May 1928.

177. "Pittston Local Ousts Officers," *Wilkes-Barre Record*, 6 Jun 1928; "Grievance Body of Miners Meet," *ibid.*, 9 Jun 1928; "Cardoni, Head of Grievance Board, Will Be Ousted," *Pittston Gazette*, 14 Jun 1928; "Insurgents Take Over Committee," *Wilkes-Barre Record*, 16 Jun 1928; "M'Garry's Friends Elected to Office in Ewen Colliery," *Pittston Gazette*, 20 Jun 1928; "Cappellini Men of Ewen Local Name Officers," *ibid.*, 21 Jun 1928; "Miners Continue Factional Fight," *Wilkes-Barre Record*, 22 Jun 1928; "3,000 Attend the Mass Meeting of Miners at Armory," *Pittston Gazette*, 25 Jun 1928.

178. "Police Break up Miners' Meeting," *ibid.*, 28 Jun 1928.

179. "No. 6 Men Accept Old Proposal," *Scranton Republican*, 17 Jul 1928; "Pittston Mine Unions Would Stop Meetings," *The Evening News*, 17 Jul 1928; "Butler Miners Return," *ibid.*, 19 Jul 1928.

180. The 1916 labor-management accord stated: "The right of operators to use such machines shall be unquestioned and the method employed shall be at the option of the Operator" (*Award of the Anthracite Coal Strike Commission, Subsequent Agreements, and Resolutions of the Board of Conciliation*, Hazleton, PA: Anthracite Board of Conciliation, May 1916, 25). McGarry spoke for the men when he said: "I am not in favor of the mechanical loader." He reluctantly admitted that the labor-management agreement allowed for the devices, "but it does not say you have to work [with them]" ("No. 6 Men Accept Old Proposal; Return This Week," *Scranton Republican*, 17 Jul 1928; see also "Lewis Sees Hope of Agreement," *New York Times*, 4 Mar 1928).

181. "No. 6 Men Accept Old Proposal; Return This Week," *Scranton Republican*, 17 Jul 1928.

182. "No. 6 Miners Return Under Police Guard," *The Evening News*, 30 Jul 1928.

183. "Pittston Men Hold Stormy Mine Meeting," *ibid.*, 2 Aug 1928.

184. The No. 6 Colliery grievance committee consisted of James Kearney, Nick Belfonte, Adam Moleski and Frank Licata.

185. "Molesky [*sic*] Retained Head of Ewen Local," *Pittston Gazette*, 21 Jan 1928; "Moleski Ousted as Chairman of Ewen Mine Local," *ibid.*, 6 Jun 1928. The other new officers were Daniel Cavanaugh, vice president, and Samuel Latona, financial secretary. All other officers were retained including Thomas Cometa, recording secretary; Michael Lelack, treasurer; John Hidock, secretary-treasurer of the sinking fund; and Charles Greco, sentinel. The new officers appointed a new grievance committee.

186. "Mine Power Station is Dynamited," *The Evening News*, 13 Jul 1928.

187. "Union Factions in Test of Strength at Pittston Meet," *Scranton Times*, 16 Jun 1928; "Cappellini Men Victorious in No. 14 Mine Local," *Pittston Gazette*, 18 Jun 1928; "More Officers Are Elected by Pittston Local," *The Evening News*, 12 Jul 1928; "Frank Cardoni Field Worker in District 1," *ibid.*, 18 Jul 1928; "Factional Strife at No. 14 Closes Colliery," *ibid.*, 23 Jul 1928.

188. "Trouble Spreading in Pittston District as More Miners Strike," *ibid.*, 26 Jul 1928; "Work Resumed At No. 9 Mines After Strike," *ibid.*, 13 Aug 1928.

189. "Police Break up Meeting of Ewen Miners," *ibid.*, 8 Aug 1928.

190. "Cappellini Signs Expulsion Dates," *ibid.*, 27 Jun 1928.

191. "Frank Cardoni Field Worker in District 1," *ibid.*, 18 Jul 1928.

192. "3000 Nanticoke Miners Idle by Strike Action," *ibid.*, 11 Jul 1928; "Frank Cardoni Field Worker in District 1," *ibid.*, 18 Jul 1928.

193. "Florence Mine Troubles Lead to Strike Vote," *ibid.*, 10 Jul 1928; "Insurgent Officers Are Credited with Victory at Florence," *ibid.*, 13 Jul 1928; "Dupont Miners Quit Posts in Rate Dispute," *ibid.*, 8 Aug 1928.

194. Members of L.U. 1635 filed charges against Cappellini for receiving pay for work he did for the American Federation of Labor while president of District 1. However, the District Executive Committee ruled that the position was not salaried so there was no violation (Enoch Williams to Thomas Kennedy, 7 Jul 1928, President/District Correspondence, 1920–1933, GST/Z06.02, Box 1, Folder 12, UMWA Papers, PSU). In another case, members of a local union in Plymouth filed seven charges against Cappellini including "inefficient, misappropriation of District funds" and "violation of District and International constitutions." Again investigators found no violations ("No Action Is Taken on Cappelini's [*sic*] Expulsion When Charges Are Made," *The Evening News*, 6 Jul 1928).

195. "False Alarm Says District Head of Story," *Scranton Republican*, 19 Jul 1928. See also "Cappelini [*sic*] Will Resign, Says Rumor; Denial is Not Taken Seriously," *The Evening News*, 19 Jul 1928.

196. Both the *Wilkes-Barre Record* and *Scranton Times*, on 20 Jul & 21 Jul 1928, respectively, discussed Cappellini's loss of control over District 1. Monroe (*A Decade of Turmoil*, 85–86, 91) viewed Cappellini as a scapegoat for the District's problems and argued that the resignation was at least partly grounded in Lewis's never having fully accepted the dynamic and outspoken leader. Monroe accurately concluded that Cappellini found himself caught between the *status quo* intentions of Lewis and the reformist desires of many rank-and-file workers. Cappellini later explained his resignation to a group of insurgents who were planning a rump convention for August 1933: "Why did I resign? When the people lose confidence in a man he should be big enough to resign his job. When my friends in Pittston, Carbondale, Dunmore, Jessup, Exeter and all over the District, with the money that was paid to spread propaganda against me, refused to look me in the face, I resigned" ("Rump Meeting Chiefs Named," *Wilkes-Barre Record*, 22 Jul 1933). See also "Cappellini Out," *Scranton Times*, 20 Jul 1928.

197. "Resigns as President of Miners in District No. 1," *ibid.*, 20 Jul 1928.

198. According to union by-laws, the presidency should have fallen to Vice President Michael J. Kosik. However, Lewis orchestrated a questionable maneuver whereby Kosik resigned his post at the executive board meeting on July 20, whereupon the board appointed Boylan as vice president and Kosik as a board member. Cappellini then tendered his resignation and Boylan assumed the presidency. Kosik was then reappointed as vice president ("John Boylan, Scranton, is Elected President of Miners in District; Resignation of Cappelini [*sic*] From Office Accepted," *The Evening News*, 20 Jul 1928; "Boylan is President of Miners in District No. 1, Cappellini's Resignation is Accepted and Board Member is Chosen as His Successor," *Scranton Times*, 20 Jul 1928; "Boylan to be Elected President," *Scranton Republican*, 29 Jul 1928; "Elected President; John J. Boylan Becomes Head of District No. 1," *United Mine Workers Journal*, 39 [1 Aug 1928], 7).

199. Lewis also received regular private insider messages from W.D. Davis, who was an auditor for the UMWA. Dozens of his handwritten letters to Lewis can be found in the President/District Correspondence, 1920–1938, GST/Z06.02, UMWA Papers, PSU.

200. "Stand by Cappellini Removal," *Scranton Times*, 17 Jun 1928.

201. "Election of Boylan Fails to Halt Anti Squabble in District," *Scranton Republican*, 21 Jul 1928. See also "Anti Officers Declare Fight Will Continue," *ibid.*, 21 Jul 1928.

202. The foreman transfers included Richard Pollard from the Fernwood Slope of the Butler Colliery to the No. 7 Shaft of the Ewen Colliery; Alfred King from the No. 7 Shaft of the Ewen Colliery to the Thomas Shaft of the Butler Colliery; Charles Birbeck from the Thomas Shaft to the Fernwood Slope of the Butler Colliery; Richard Williams from the No. 10 Shaft of the No. 9 Colliery to the Sax Drift of the No. 14 Colliery; and Norman Smiles from the Sax Drift of the No. 14 Colliery to the No. 10 Shaft of the No. 9 Colliery ("Official Changes at the Collieries Took Place Today," *Pittston Gazette*, 16 Apr 1928; "Mediator Begins Survey into Coal Dispute at No. 6," *Times Leader*, 9 May 1928).

203. "John Brydon is Successor to Jennings," *The Evening News*, 14 Jul 1928. Erie management combined with local business and community leaders to fete Jennings at a testimonial banquet held at the Hotel Casey in Scranton ("Men of Penn'a Co. Paid Tribute to Joseph Jennings," *Pittston Gazette*, 19 Jun 1928).

204. "Reply Made to Suggestion for Separate Union," *Scranton Republican*, 17 Jul 1928.

205. "McGarry Warns Union Battle Will Continue; Recognition Is Only Thing That Will Satisfy Rump Officials," *ibid.*, 20 Jul 1928. See also "Election of Boylan Fails to Halt Anti Squabble in District," *Scranton Republican*, 21 Jul 1928.

206. "Insurgents Deny They Seek Compromise with District Union Heads," *The Evening News*, 31 Jul 1928.

207. "Lewis Revokes Charter of Local Union at Old Forge When 400 Mine Workers Accept Wage Reduction," *ibid.*, 1 Aug 1928. See also "M'Garry Denies He Asks Walkout for Personal Reasons in District 1," *ibid.*, 4 Aug 1928.

208. "Committee of 81 Meeting Called," *Wilkes-Barre Record*, 31 July 1928; "Insurgents Plan Meeting at Scranton," *The Evening* News, 31 July 1928.

209. "Favoritism Charged By District Insurgents; General Strike Threat," *ibid.*, 28 Jul 1928.

210. "Strike Action Off as Result of New Plans," *The Evening News*, 7 Aug 1928; "M'Garry Forces Will Not Pay Tax to Old Officers," *Pittston Gazette*, 18 Jun 1928. On August 16, 1928, the International UMWA office listed 33 local unions in District 1 as "not in good standing" because their dues were in arrears (International Executive Board Correspondence, AX/B40/ HCLA/ 01146, Box 6, Folders 32 & 33, UMWA Papers, PSU).

211. "Penna. Coal Co. Will Not Meet Rump Leaders," *The Evening News*, 11 August 1928.

212. Walter T. Howard, "Anti-Labor and Anti-Radical Violence in Northeastern Pennsylvania during the Great Depression," 81. The authorities also ransacked Minerich's boarding-house room and confiscated CPUSA, Save-the-Union, and NMU documents. Minerich was held several days until the American Civil Liberties Union and the International Labor Defense secured his release (Walter T. Howard, "The International Labor Defense and Antiradical Repression in The Pennsylvania Anthracite, 1928–1929," *Carver*, 15 [Spr 1998], 41–6).

213. "Testimony is Taken in Mine Union Dispute," *The Evening News,* 6 Aug 1928; "Penna. Coal Co. Will Not Meet Rump Leaders," *ibid.*, 11 Aug 1928; "Work Resumed at No. 9 Mines After Strike," *ibid.*, 13 Aug 1928.

214. "Fist Fight Between 2 Officers Ends Local Meeting at Pittston," *The Evening News*, 3 Aug 1928; "Police Break Up Meting of Ewen Miners," *ibid.*, 8 Aug 1928; "Pittston Men Hold Stormy Session," *ibid.*, 2 Aug 1928.

215. "Mine Engine House and Dwelling Bombed," *Trenton Sunday Times Advertiser*, 18 Nov 1928.

216. "Settle the Mine Trouble," *Wilkes-Barre Record*, 16 Nov 1928.

217. *Ibid.*

218. John Boylan and Enoch Williams to the Officers and Members of Local Unions in UMWA District No. 1, 18 Sept 1928, President/District Correspondence, 1920–1933, GST/Z06.02, Box 1, Folder 11, UMWA Papers, PSU.

219. "Mine Board to Fight System of Contracts," *The Evening News*, 28 Jul 1928. See also "Union Head Hits Back at Critics," *ibid.*, 8 Aug 1928; and Monroe, *A Decade of Turmoil*, 102.

220. The complaint read: "The '[sub]contract system,' whereby an individual [sub]contractor takes over a section of a mine, or several working places, and employs a number of men to do the work, is a system, to say the least, that is a menace to the peace and stability of the anthracite region. It breeds discontent, favoritism, graft, chaos and in some instances has resulted in physical

violence and murder. Its continuation in any section of the anthracite region is a challenge to the intelligence and constructive ability of the mining industry. Its baleful influence has been such as to bring about its downfall and elimination in practically every section of the anthracite region, with the exception of certain parts of District No. 1. In the sections now affected by this practice it has given every justification for its total elimination" ("President Boylan Files Complaint against Contract Mining with Board," *United Mine Workers Journal*, 39 [15 Sep 1928], 10; see also "No Split in Insurgent Ranks," *The Evening News*, 14 Oct 1928).

221. "Peace Parley Held by Miners," *Wilkes-Barre Record*, 16 Nov 1928.

222. John Boylan and Enoch Williams to the Officers and Members of Local Unions in UMWA District No. 1, 19 Sept 1928, Containing a "Resolution Expelling Insurgent Officers," President/District Correspondence, 1920–1933, GST/Z06.02, Box 1, Folder 12, UMWA Papers, PSU. See also Membership Restoration Appeals—1930, *ibid*., Box 1, Folder 20, UMWA Papers, PSU; W.D. Davis to John L. Lewis, 21 Jun 1929, *ibid*., UMWA Papers, Box 6, Folder 31, UMWA Papers, PSU; "Appeal of Certain Members of Local 898, former members of Local No. 6, Anthracite Miners of Pennsylvania, for restoration of all rights and privileges," Roll Call and Resume of Board Meetings, 1930–1931—International Executive Board Correspondence, AX/B40/HCLA/01146, Box 8, Folder 3, UMWA Papers, PSU; "District No. 1 Executive Board Expels Group Identified in Forming Dual Organization," *United Mine Workers Journal*, 39 (1 Oct 1928), 5; and "No Split in Insurgent Ranks Now," *The Evening News*, 14 Oct 1928.

223. Others expelled were John Hermansen, John Bellfield, Edward Hogan, Vincent Trettis, William Murnin, and Ray Delaney. Another group consisting of L.U. 699 members from Edwardsville were expelled: Michael Goodlavage, Mike Wallo, Andrew Campbell, Ed Gavonis, Zigmund Great, Walter Great, Ed Great, Joseph Bellas, John Barral, Stanley Scovitch, George Uncavage, Andrew Petrowski (Michael Goodlavage to John L. Lewis, 10 Jan 1929, International Executive Board Correspondence, AX/B40/HCLA/01146, Box 6, Folder 31, UMWA Papers, PSU; John Kmetz to John L. Lewis, 23 Jan 1929, *ibid*.). See also "New 'Baloney' Organization in Anthracite Region Will Seek Assistance from Anti-Union Employers," *United Mine Workers Journal*, 39 (15 Nov 1928), 7; and "Roundly Denounced by Murray and Kennedy at John Mitchell Memorial Meeting," *ibid*., 39 (15 Nov 1928), 10.

224. "Contract Mining System Abolished at Underwood Colliery in District No. 1," *ibid*., 39 (15 Dec 1928), 5.

225. Lewis Casterline, audiotaped interview, 16 Jan 1970, King's College Oral History Collections. Casterline further stated that organized criminal elements worked diligently to deliver votes for Fine and the Republican machine. Other interviews that linked Fine to organized crime included Lewis Casterline, audiotaped interview, 12 Dec 1988, NPOLHP; and William A. Hastie, audiotaped interview, 31 Jul 1989, NPOLHP. On the spoils system and other forms of corruption during Fine's rule in Luzerne County see former Democratic State Representative Fred Shupnik, audiotaped interview, 11 Jan 1995, NPOLHP. On Fine's political history see Robert Mittrick, "John S. Fine: The Rise and Fall of a Political Boss," unpublished manuscript, Luzerne County Community College, Nanticoke, PA; and www.nanticokecity.com/famous.htm#fine. For an electoral instance where the Italian community reacted against false pre-election allegations of mob affiliation directed at an Italian-American political candidate see Stefano Luconi, "'Mobsters' at the Polls: The Mafia Stereotype of the Media and Italian-American Voters in Philadelphia in the Early 1950s," 38–50 in Mary Jo Bona and Anthony J. Tamburri (eds.), *Through the Looking Glass: Italian & Italian/American Images in the Media*, Selected Essays from the 27th Annual Conference of the American Italian Historical Association, 10–12 Nov 1994, Staten Island, NY: American Italian Historical Association, 1996.

226. Angelo Siracuse, audiotaped interview, 19 Jul 1989, NPOLHP. On Lorenzo Ferraro as the alleged organized crime leader in Swoyersville see also Joe Costa, unrecorded interviews, 1 Aug 1992 and 24 Mar 1995; and Sam Mullay, audiotaped interview, 22 Jan 2005, NPOLHP.

227. Regarding the coal companies' control of local politics before the 1920s, State Senator Martin L. Murray (Democrat, Luzerne County) said: "If you weren't a Republican you couldn't get a job in the mines, in the utilities, in the banks. The special interests controlled everything. The [John S.] Fine organization even gave property assessment breaks to registered Republicans, and the coal companies didn't pay taxes. What we had was a fear built up among the people" (Senator Martin L. Murray as quoted in Paul Beers, *Pennsylvania Politics, Today and Yesterday*, University Park, PA: Penn State Press, 1980, 179. See also Edward F. Cooke and Edward G. Janosik, *Pennsylvania Politics*, New York: Holt, Rinehart & Winston, 1965, 14–15).

228. *Wilkes-Barre Record Almanac*, 7 Nov 1928.

229. Edward F. Hanlon, *The Wyoming Valley: An American Portrait*, Woodland Hills, CA: Windsor, 1983, 145, 150.

230. Scranton has been more reliably Democratic than Wilkes-Barre in twentieth-century municipal elections (Robert P. Wolensky, Thomas J. Baldino, and John V. Hepp IV, "Remaking Municipal Government? Charter Reform in Wilkes-Barre, Pennsylvania, 1968–2001," *Pennsylvania History*, 73 [Autumn 2006], 446–79). Political corruption has recently wracked Democratic and Republican elected officials in both Luzerne and Lackawanna counties. Extensive collusion between government and private-sector individuals has led to numerous indictments and convictions. Archived material on what has been described as a widespread "culture of corruption" can be found at the websites of the *Citizens' Voice*, *Times Leader*, and *Scranton Times*. See also Robert P. Wolensky and Kenneth C. Wolensky, "Corruption and Pennsylvania: A Perfect Combination?" *Harrisburg Patriot*, 14 Feb 2010.

Notes to Chapter Six

1. Secundo Sebastianelli of Jessup, near Scranton, described his father's preference for the free lease at his lease-holding Sebastianelli Coal Company: "A free lease is where a company [the lessee] can sell the coal wherever he wants to. If he wants to sell it to Joe Blow's breaker, or Tom's breaker, or Andy's breaker, either way. My father always liked free leases [because] he could sell the coal wherever he wanted and people weren't cheating him. Cheating him by weight. When the coal went up to the breaker, if you didn't have a free lease, he [the lessor] will give you whatever he wanted to pay for the tonnage" (Secundo Sebastianelli, audiotaped interview, 29 Jan 1999, NPOLHP).

2. As they were assembling their railroad empire, the Van Sweringens purchased the Erie, Nickel Plate, and Pere Marquette railroads in 1924 (Herbert Harwood Jr., *Invisible Giants: The Empires of Cleveland's Van Sweringen Brothers*, Bloomington, IN: Indiana University Press, 2003). See also "Three Erie Stocks Attain New Highs," *New York Times*, 15 Nov 1923.

3. "Gallagher Quits to Go with Vans," *Plain Dealer* (Cleveland), 21 Jul 1926.

4. Board of Directors Meeting, 28 Jun 1929, Minutes, 1847–1930, Vol. 8, 13 Jan 1926–14 Feb 1930, Box 1, PaCC Papers, PSA. The quotation is from p. 49. See also Stockholders and Executive Committee Minutes, Box 1, Minutes, 28 Mar 1930–29 Dec 1939, Doc. No. 102, Board of Directors Meeting, New York City, 28 Mar 1930, 1–4, PaCC Papers, PSA.

5. The board members' resignation letters are in Board of Directors Minutes, 1838–1971, Box 8, Folder: 13 Jan 1926–14 Feb 1930, PaCC Papers, PSA. The new board consisted of Michael Gallagher, chairman, plus M.P. Callaway, C.R. Nash, D.W. Cooke, H.T. Peters, D.E. Pomeroy, H.A. Knapp, and A.W. Putnam (Minutes, Board of Directors Meeting, PaCC, 9 Jan 1930, Board of Directors Minutes, 1838–1971, Box 8, Folder: 13 Jan 1926–14 Feb 1930, PaCC Papers, PSA). The executive officers of PaCC were Michael Gallagher, president; J.C. Brydon, vice president; C.H. Fredericks, secretary & treasurer; and C.J. Westlake, auditor. The members of the Executive Committee were Gallagher (ex-officio), M.P. Callaway, H.T. Peters, A.W. Putnam, and G.T. Slade (Minutes, Board of Directors Meeting, 28, Jun 1929, Board of Directors Minutes, 1838–1971, Box 8, Folder: 13 Jan 1926–14 Feb 1930, PaCC Papers, PSA).

6. Because the Interstate Commerce Commission denied the Van Sweringens' attempt to merge their railroads (Erie; Nickel Plate; Pere Marquette; Chesapeake & Ohio; New York, Chicago & St. Louis; Chicago & St. Louis; Toledo, St. Louis & Western; Lake Erie & Western; and Buffalo, Rochester & Pittsburgh, among others) into one entity, the brothers formed the Alleghany Corporation as a holding company for the railroads' stock. They issued $35 million in 5 percent convertible bonds, par 100, which sold quickly through J.P. Morgan & Co., National City Co., First National Bank, and Guaranty Co. at a 10 point premium. The brothers also hoped to sell 250,000 shares of preferred stock, as well as 3.5 million shares of common stock. As a holding company, Alleghany was allowed to purchase additional railroad stock but it had "no power to operate railroad properties or to engage in banking" ("Allegheny [*sic*] Corporation," *Time*, 1 [1 Feb 1929]). For a comprehensive history of the Alleghany Corporation, see Harwood, *Invisible Giants*, Chapter 14.

7. *Report of the Pennsylvania Anthracite Coal Industry Commission*, Harrisburg, PA: Commonwealth of Pennsylvania, 1938, 399. See also *Moody's Transportation Manual, 1960*: 774–75. As part of its incorporation, TPC also acquired control over other Van Sweringan companies including the United States Distributing Corporation, a holding company that owned United States Trucking Corporation; Pattison & Bownes, Inc., a wholesale coal distributor; Independent Warehouses, Inc.; and a bituminous mining company in the state of Wyoming.

8. The collieries were listed in the original lease between PaCC, the Erie Railroad, and TPC, 1 Jan 1930, 3–7 (Board of Directors Minutes, 1838–1971, Box 8, Folder: 13 Jan 1926–14 Feb 1930, PaCC Papers, PSA). The transfer also included a small number of bituminous mines in western Pennsylvania.

9. "Plan of Reorganization of the Mining and Distribution of Anthracite Coal From the Properties of Pennsylvania Coal Company and Hillside Coal & Iron Company," Minutes, 1838–1971, Box 8, Folder: 13 Jan 1926–14 Feb 1930, PaCC Papers, PSA; Minutes, Board of Directors, 28 Jun 1929, Minutes, 1838–1971, Box 8, Folder: 13 Jan 1926–14 Feb 1930, PaCC Papers, PSA; Minutes, Special Meeting of the Board of Directors, 9 Dec 1930, Minutes, 1838–1971, Box 8, Folder: 13 Jan 1926–14 Feb 1930, PaCC Papers, PSA; Minutes Special Meeting of the Board of Directors, 31 Jan 1930, Minutes, 1838–1971, Box 8, Folder: 13 Jan 1926–14 Feb 1930, PaCC Papers, PSA.

10. "The Pittston Story," *Mechanization*, 21 (Oct 1957), 60–94.

11. Adele Hast argued that Alleghany established The Pittston Company because "Competition in the hard-coal industry had intensified in the late 1920s [and] anti-trust laws prevented Erie Railroad from entering new markets" (Adele Hast [ed.], "The Pittston Company," in *International Directory of Company Histories*, Vol. 4, Chicago, IL: St. James, 1991, 180–182).

12. Harwood, *Invisible Giants*, 164.

13. Ron Chernow, *The House of Morgan: An American Banking Dynasty and the Rise of Modern Finance*, New York: Simon and Schuster-Touchstone, 1990, 309.

14. Frederick Lewis Allen, *The Lords of Creation*, New York: Quadrangle, 1966 [1935], 293–303.

15. Ron Chernow, *The House of Morgan*, 307.

16. *Ibid.*, 310.

17. *Ibid.*

18. Harry N. Taylor to Michael Gallagher, 8 Jan 1930, Board of Directors Minutes, 1838–1971, Box 8, Folder: 13 Jan 1926–14 Feb 1930, PaCC Papers, PSA.

19. Coverdale & Colpits to Michael Gallagher, 27 Dec 1929, Board of Directors Minutes, 1838–1971, Box 8, Folder: 13 Jan 1926–14 Feb 1930, PaCC Papers, PSA.

20. *Report of Pennsylvania Anthracite Coal Industry Commission*, 308–310; and "Empire for Sale," *Time* 6, (23 Sept 1935).

21. Alex Chamberlain, audiotaped interview, 11 Jul 1995, NPOLHP.

22. The leases were listed in the PaCC Royalty Records, 1883–1968, PaCC Papers, PSA; and in PaCC Contract File Index, Pittston Co. & Canceled, PaCC Papers, PSA. The Standard Anthracite Mining Company lease was No. 69; the Vespi Coal Company lease was No. 231; the Barnum Coal Company lease was No. 3007; the Rovo Coal Company lease was No. 132; the Joseph Volpe Coal Co. lease was No. 64.

23. Adonizio's original lease was No. C-165, Contract File Index, Pittston Co. & Canceled File, PaCC Papers, PSA. On the cancellation of the 1933 lease see "Notice of Forfeiture of Coal Lease, to Berge-Rose Coal Company," Contracts, Secretary's and Treasurer's Files, 1867–1969, No. C-165, Box 5, Folder 13, PaCC Papers, PSA.

24. Charles Adonizio Jr. (audiotaped interview, 14 Mar 1996, NPOLHP) recalled the tenancy days of his father and uncles in an interview that revealed the importance of informal relationships in securing the agreements.

25. Contract File Index, Pittston Co. & Canceled File, Contracts 1867–1969, Box 6, PaCC Papers, PSA. The Jermyn-Green lease for the No. 14 Colliery was No. 220. On Jermyn-Green's bankruptcy and Gelso's securing the lease see Stockholders and Executive Committee Minutes, 1838–1971, 24 Oct 1945–28 Jan 1953, Doc. No. 1021, PaCC Papers, PSA.

26. Pagnotti's parents emigrated from northern Italy to Old Forge where he was born Luigi Pagnotto in 1894 (see Appendix II). On Pagnotti's rags-to-riches story see Liberato Trotta, *Biography of Louis Pagnotti*, unpublished manuscript, Pagnotti Enterprises Archives, Wilkes-Barre, PA, 1981; "Louis Pagnotti Dies in Phila. Hospital," *Standard-Speaker* (Hazleton, PA), 11 Apr 1966; Chick Feldman, "Lou Pagnotti's Initials Spelled 'Love People'," *Scranton Tribune*, 13 Apr 1966; and "L. Pagnotti, Mine Owner, Builder Dies," *Anthracite Tri-District News*, 29 Apr 1966, 1.

27. Contract File Index, Contracts 1867–1969, No. 226, Box 6, PaCC Papers, PSA.

28. Pagnotti failed to make property tax payments and ran royalty deficits for 15 years, amounting to $55,089 in 1939, $81,033 in 1950, and $191,043 in 1954. After TPC folded in 1938 and PaCC resumed control of the leases, Pagnotti protested to the board of directors that the No. 9 Colliery proved insufficiently remunerative. Perhaps for reasons related to the size of his operations, the board members remained patient as one of its largest tenants made irregular royalty payments and built a substantial debt. They lowered Pagnotti's royalty rates on more than one occasion but finally terminated the agreement in the mid-1950s. The board discussed Pagnotti's deficit at meetings on 25 Sept 1944; 27 Sept 1944; 28 Jun 1950; 25 Jul 1951; and 22 Dec 1954 (Stockholders and Executive Committee Minutes, 1838–1971, 28 Mar 1930–29 Dec 1939; Doc. No. 1021, PaCC Papers, PSA). According to PaCC surveyor, Earl Stanton, the employees believed that PaCC president R.A. Lambert resigned or was forced out in part because he allowed Pagnotti to carry the deficit for such a long period (Earl Stanton, audiotaped interview, 15 Mar 1999, NPOLHP).

29. The photo of Louis Pagnotti, Sr. is from U.S. Treasury Department, *Mafia: The Government's Secret File on Organized Crime*, Foreword by Sam Giancana, New York: Collins, 2007, 714.

30. James Tedesco, audiotaped interview, 8 Oct 1996, NPOLHP. The Lehigh Valley Coal Company (LVCC) engaged in leasing almost as extensively as PaCC in the 1930s and into the 1940s, as apparent from a review of the company's 18 volumes of Lease Books. LVCC issued approximately 50 leases annually between 1932 and 1952 (Lehigh Valley Coal Company Papers, Lease Books, Pagnotti Enterprises Archives, Wilkes-Barre, PA).

31. Pagnotti and Tedesco purchased three LVCC collieries between the mid-1930s and mid-1940s: the Harry E in Swoyersville; Franklin in Wilkes-Barre, and Clear Spring (renamed the Sullivan Trail Colliery) in West Pittston. They purchased all of LVCC's remaining holdings in the mid-1950s. In 1958, the pair also secured a lease from the Glen Alden Coal Company for the Loomis Colliery in Nanticoke (John L. Dorris to Thomas Kennedy, 3 Apr 1959, President/District Correspondence, D.1 1954–1960, 26/5, Folder: Jan–Jun 1960, UMWA Papers, HCLA, PSU). In 1956, Pagnotti became a major shareholder and president of the Old Forge Discount and Deposit Bank. Tedesco served as the bank's chairman, president, and chief executive beginning in 1966 until his death in 2008 at age 98.

32. George DeGerolamo, *Autobiography*, 1980, published by the author, no date, 34, 40–41. William A. Hastie closely followed Pagnotti's coal operations in West Pittston and elsewhere: "In 1932, Pagnotti acquired the Clear Spring Colliery in West Pittston, which had been long-idled following an underground flood thought to have been caused by the nearby Susquehanna River invading the pit. However, Pagnotti and his crews determined that the real source of the water was that, as the workings expanded, the pumps could not keep up with additional 'make water' or the natural flowage underground through cracks and crevices in the overburden. Pagnotti rectified the problem by installing larger and more powerful pumps. Under his management, the Clear Spring soon became known as a 'slaughter house' when fatal accidents mounted. The worst of these occurred in August 1936, when five young mineworkers were killed in a methane blast. The colliery was also known for paying wages well below union rates." (William A. Hastie, unrecorded interview, 24 Apr 2012).

33. William A. Hastie has lived in West Pittston for over eight decades and has personally experienced the cave-ins caused by Pagnotti's mining:

"Wherever Pagnotti mined, he left a wide swath of destruction above ground through subsidence. The excessive pillar robbing caused serious cave-ins, which resulted in damage to houses, schools, churches, factories, and businesses. Brick and masonry buildings were the most susceptible as they lacked the elasticity of wooden frame buildings. In 1940, mining at the Clear Spring Colliery caused widespread subsidence in West Pittston. Nearly all of the colliery's mines underlay a land enclosed by a bend in the Susquehanna River that contained many of the mansions and fashionable homes of wealthy residents, as well as two schools and the sandstone Presbyterian Church. None of these buildings was spared severe damage. Worse for the future, the subsidence caused the entire area to fall lower on the surface in relation to the Susquehanna, thus worsening the effects of future flooding by that waterway. For example, flood waters that might have previously filled the cellars of homes now flowed through the living rooms of the same homes. Even the recent flood of September 9–10, 2011, caused by Tropical Storm Lee, which brought the most severe flooding in West Pittston's history, was much worse than it had to be because Pagnotti's caving lowered the street levels decades earlier.

"Louis Pagnotti was not the only mining subcontractor or lessee to destroy surface property—indeed, every one of them was guilty. However, Pagnotti was the most prolific in this regard. The vastness of the workings of PaCC's No. 9 Colliery, of which he was the lessor by the

1930s, ensured that. Those mines extended from Hughestown Borough on the northeast fringe of Pittston to Welsh Hill and Oregon Heights in the southern and southwestern extremities of the city. No section of Pittston escaped residential damage. Entire neighborhoods were left uninhabitable. Pittston High School was rendered idle for years in the 1940s while awaiting funds for repairs. My own wife couldn't go to the high school because of it; she had to attend school in another district. The badly tilted Welsh Congregational Church was fated to the wrecking ball as a result of the cave-ins—a tragedy and a grim partner of the devastation. In perhaps the most infamous case, in the summer of 1943, two and one-half-year-old Jule Ann Fulmer was toddling a step or two behind her aunt as they made their way down the long steep hill of Mill Street in north central Pittston. The aunt turned around to check on her young ward only to find that the child had disappeared from sight. The sidewalk had opened up and Jule Ann had been cast into a void of an old mine. It would require the removal of 40 truckloads of overburden to retrieve the little body. Some residents of Mill Street are still haunted to this day by memories of that day, and are still plagued by regular mine caves" (William A. Hastie, unrecorded interview, 24 Apr 2012). On the epidemic of cave-ins in West Pittston, Pittston, and elsewhere in the Wyoming Valley see Ellis W. Roberts, *A History of Land Subsidence and its Consequences Caused by the Mining of Anthracite Coal in Luzerne County, Pennsylvania,* unpublished doctoral dissertation, New York University, 1948.

34. Although not listed by state or federal investigators as a member of the region's organized crime family, the FBI concluded that Pagnotti had business dealings with alleged mobsters (U.S. Treasury Department, *Mafia: The Government's Secret File on Organized Crime*, 714). The Pennsylvania Crime Commission's 1980 report concluded that Pagnotti "has been identified as lending money along with Santo Volpe and Louis Consagra, deceased [Russell] Bufalino crime family members, and Angelo Polizzi, deceased close associate of Volpe, to a company owned by [regional mob boss Joseph] Barbara" (Pennsylvania Crime Commission, *A Decade of Organized Crime*, Harrisburg, PA: Commonwealth of Pennsylvania, 1980, 249). William D'Elia, the alleged current top boss since 1989 has had dealings with Pagnotti Enterprises; he been serving a prison sentence on federal charges of laundering $600,000 in illegal drugs money. Pagnotti's partner and nephew, James Tedesco, was also "linked to organized crime figures" (*Ibid.*, 249; see also 246–247, 249–250). The Crime Commission associated both Pagnotti and Tedesco with Joseph Barbara Sr., the head of the region's crime family prior to Bufalino, who convened the infamous gang meeting in 1957 on his estate in Apalachin, New York. Tedesco served as a member of Anthracite's Committee of Twelve in the early 1960s, an appointed body consisting of labor and management representatives. He was convicted, with Santo Volpe and others, of price fixing in 1944, and again with other persons in 1977 (*Ibid.*, 249).

35. Oral history evidence suggests that, while Pagnotti worked closely with alleged organized criminals, he was probably not a "made" member because of his Neapolitan birth. Mineworker Joe Costa had bootleg alcohol and other dealings with mobsters in the 1920s, and knew Pagnotti from the bootlegging business. According to Costa: "He wasn't in the mob at first but he had to work with them later as he got bigger" (Joe Costa, unrecorded interview, 24 Mar 1995, NPOLHP). Costa said that Pagnotti invested his bootleg alcohol profits in mining and other legitimate businesses. According to William A. Hastie: "Whether Louis Pagnotti was a member of organized crime or went it alone as a solitary racketeer is a matter of question. During prohibition he ran a sophisticated and efficient bootleg alcohol operation with few if any partners. He dealt solely in expensive whiskies, which eliminated the need for easily detected mass transportation, as did beer. He bought an eight-passenger "souped-up" limousine that he altered to pack with numerous cases of whiskey. I spoke to three of his drivers separately about this limo and the deliveries. Also, there were many reports that Pagnotti operated a string of brothels during prohibition. While these reports were all word of mouth, they were too persistent and widespread to be merely hearsay. The brothels would have been in competition with Scranton's infamous Raymond Court and Oxford Court, which were known as prostitution centers throughout the East Coast and beyond. After the Apalachin conclave in New York, the Pennsylvania Crime Commission reported that Pagnotti and Volpe financed Barbara's businesses in New York State. I should add that Pagnotti had many legitimate enterprises. As early as 1927, he had the Old Forge recreation building which was an imposing building with 10 bowling alleys, numerous pool tables, and other facilities. It was a three story building. He also promoted championship boxing matches. For example, in 1926, he promoted the fight where Mickey Walker lost his welterweight crown to Pete Latzo from Old Forge. Pagnotti promoted many fights in the Watres Armory in Scranton. He was a very popular and well-known coal operator. Louie was a T-totaler and an earlier riser (William A. Hastie, unrecorded interview, 24 Apr 2012).

36. In August 1932, Volpe was one of 14 persons arraigned in New York City for the murder of John Bazzano, the alleged crime boss of Pittsburgh. Bazzano was supposedly killed on order from a mob "board" because he had orchestrated the unauthorized murder of the three Volpe brothers (John, Arthur and James; no relation to Santo), also of Pittsburgh, in 1932. Angelo Colizza of Dunmore was also among those arraigned. All 14 pleaded not guilty. The charges were eventually dropped for lack of evidence ("Didn't Kill, Say 14, One Trentonian," *Trenton Evening Times*, 17 Aug 1932).

37. Volpe's leases were Nos. 42, 151 156, 188, 213, 218, 219, 230, 247, 248, 325, 2002, various boxes, Contract File Index, TPC & Canceled, PaCC Papers, PHMC, PSA.

38. *Scranton Times*, 15 Dec 1937.

39. An intriguing internal memo from E.D. Adair, TPC Legal Counsel, to Harry J. Connolly, TPC's vice-president and general manager, demonstrated the intense competition over the No. 6 Colliery lease in 1937:

    "Mr. Connolly:

    "Santo Volpe [came] in today [August 12, 1937] and executed the lease which has been drafted in as final form as I am able to get it .... I would not have bothered you with even this memo, if it were not for the story I got from Mr. Volpe today. On Tuesday evening, August 10th [1937], a man from Wilkes-Barre by the name of Gilligan, who represented himself to be a politician and a man who with others wanted to lease No. 6 Colliery, came to Mr. Volpe and inquired as to whether or not he was looking for this colliery. Mr. Volpe did not commit himself but Gilligan then told him that he and one Kirkendall of Wilkes-Barre (? former Revenue Collector) had been in to see Mr. Kirby on at least three occasions at which times he was referred to you, at which time you told Gilligan or his representative that No. 6 was not to be leased but the coal was to be delivered to The Pittston Company.

    "Apparently this was not a satisfactory arrangement to Gilligan nor his backers, one of whom was Major Lee White. Con McCole [future Wilkes-Barre mayor] was also represented to be one of the parties interested and Judge Fine's [future governor] name was mentioned and Mayor Langan [of Pittston] and [prominent local physician] Doctor Mundy [Chairman of the Luzerne County Democratic Party]. Gilligan apparently had been advised by one Mitchell—I believe Kirby's brother-in-law.

    "Finally, before the conversation was ended with Mr. Volpe, this man Gilligan said to him that he was authorized by Mitchell to say on behalf of Mr. Kirby that if Mr. Volpe would take them in with him—Gilligan and Kirkendall—the lease for No. 6 would be delivered to him within twenty-four hours. Otherwise he would never get it. Mr. Volpe then told him that the contract was in his safe and had already been delivered and he was not going to do anything, that his relations had been entirely with you and that he did not desire to talk with anyone else.

    "I wanted to get this information to you at once. I may be somewhat inaccurate in exactly what Mr. Volpe stated to me but I have endeavored to briefly say what he said to me as accurately as I can remember it. You will no doubt get Mr. Volpe's story directly from him when you return."

    The quotation is from the Santo Volpe File, PaCC Papers, memo dated 12 Aug 1937, E. Stewart Milner Collection, Anthracite Heritage Museum, Scranton, PA, uncatalogued. The names in the memorandum represented an impressive list of influential local people, all vying for a stake in the potentially lucrative lease. Their multi-ethnic surnames and occupational positions illustrated the diversity of persons interested in tenancy rights. The Mr. Kirby mentioned in the memo was probably Alan P. Kirby (son of F.M. Kirby who built the "dime store" fortune with F.W. Woolworth), who owned a large share of the Alleghany Corporation, which had become the owner of The Pittston Company. See note 141 below.

40. PaCC's *Annual Report* in 1941 ranked 13 leaseholders according to the net commercial tons each held as reserves. The top five were: (1) Volpe Coal Company: 114,480,780 tons; (2) Jermyn-Green Coal Company: 8,872,735 tons; (3) No. 9 Coal Company (Pagnotti): 5,242,930 tons; (4) Heidelberg Coal Company: 1,935,237 tons; and (5) Russell Coal Mining Company: 1,923,537 tons. The Ewen Colliery, operated by PaCC, had reserves of 25,839,600 tons while PaCC's Underwood Colliery held 5,242,930 tons. Another of the Volpe operations, the Daley & Volpe Coal Company in Lackawanna County, leased the Sibley Slope where, in 1939, the operation produced 41,272 tons (Pennsylvania Department of Mines, *Inspector's Report Anthracite Division*, 1939, 234). Volpe also had an interest in the Gateway Coal Company.

41. On more than one occasion in the early 1940s, Volpe expressed dissatisfaction with his PaCC leases. According to PaCC Stockholders and Executive Committee Minutes of December 23,

1942, PaCC President Harry J. Connolly told the board about a conversation he had with Volpe: "He was not too happy with the situation at the Volpe Coal Company operations at Butler and No. 6 Collieries; that he felt as compared with other lessees operating PaCC properties, that he was now being, and had been, discriminated against." In another conversation about a month later, Volpe again told Connolly "that his operations were unprofitable." Volpe requested a reduction in the royalties paid to PaCC, and the board consented to an independent audit to consider the matter (Minutes from 1838–1971, Stockholders and Executive Committee Minutes from 19 Jan 1940–16 Oct 1945, Doc. No. 102, Board minutes from 23 Dec 1942, 30 Jan 1942, 27 Jan 1943, PaCC Papers, PHMC, PSA).

42. The explosion resulted from poor ventilation (see the state's investigative report on the disaster in Sen. Leo C. Mundy, Commission Chairman, to Governor Earle, 10 Jan 1939, Volpe Coal Company File, *Times Leader* library; see also Edwin C. Curtis et al., Mine Inspectors, to Hon. M.J. Hartneady, Secretary of Mines., 17 Jun 1938, *ibid*; and Sen. Leo C. Mundy, Commission Chairman, to Governor Earle, 10 Jan 1939, *ibid*.) Most so-called Line companies displayed a much higher concern for safety in the nineteenth and early twentieth centuries than did the lessees of the 1930s and afterwards (see, e.g., Charles P. Kumpas, "Safer Mining at the Hudson Coal Company," *The Miner's Lamp*, 14 [May/Jun 1997], 1, 4). When the large companies began subcontracting and leasing, their inspectors became more concerned with *where* the tenants mined (if they remained within the boundaries of the lease) than *how* they mined.

43. As the chief executive of a successful coal operation, Volpe could afford extended visits to his former home in Montedoro, Sicily. On the eve of a trip in 1920, a large group of well-wishers, including Erie coal mangers and "hundreds of miners," gathered at Scranton's Hotel Casey to wish Volpe a good voyage ("Contract Miner Offers to Quit the Business," *Scranton Times*, 3 Aug 1920).

44. Julian W. Parton, *The Death of a Great Company: Reflections on the Decline and Fall of the Lehigh Coal and Navigation Company*, Easton, PA: Canal History and Technology Press, 1986, 128. As further indication of their relationship, Lewis traveled to northeastern Pennsylvania in 1958 to attend Volpe's funeral.

45. In addition to Fine and Volpe, the other Newport partners were Albert and Lawrence Biscontini. Fine was indicted along with Albert Biscontini and Donald P. Morgan, who was Fine's brother-in-law and tax consultant. The foursome was charged with personal and corporate income-tax evasion. Both Volpe and Lawrence Biscontini had died before the indictments were issued. All of the defendants were acquitted after lengthy trials and arduous jury deliberations. The defense relied on a number of very prestigious character references ("Biscontini and Fine Accused in Tax Case," *Times Leader The Evening News*, 13 Jan 1961; "Fine and Lippi to Face Court Here Thursday," *Scranton Times*, 14 Jan 1961; "Fine and Morgan Acquitted in Tax Case, Ex-Governor Cleared After Eight Ballots," *Sunday Independent*, 24 May 1964; and "Fine Tax Charge Dropped by U.S.," *Wilkes-Barre Record*, 20 Aug 1964). The Glen Alden Coal Corporation sold out to The Blue Coal Corporation in 1966. Journalists Elliot Jaspin and Gil Gaul published five stories on the machinations and corruptions surrounding Blue Coal—including the involvement of organized crime—which appeared between June 12 and 16, 1978, in the *Pottsville Republican*. The series earned a Pulitzer Prize for the writers. On Governor Fine's political career see also Robert Mittrick, "John S. Fine: The Rise and Fall of a Political Boss," unpublished manuscript, Luzerne County Community College, no date.

46. Contract File Index, Contracts 1867–1969, Nos. 274, 335, Box 9, PaCC Papers, PSA.

47. When questioned about the Saporito Coal Company during the Pennsylvania legislature's investigation into the Knox Mine Disaster, Stefano LaTorre disclosed the kickbacks paid to August J. Lippi and other Knox company principals:

Q: Who pushed you out?
SL: Mr. Louis Fabrizio, Mr. Sciandra, and Mr. Lippi.
Q: How did Mr. Lippi push you out?... What was his interest in the Saporito Coal Company?
SL: He was getting six cents a ton.
Q: Was he an officer of the Saporito Coal Company?
SL: No sir.
Q: Did he hold any stock in the Saporito Coal Company?
SL: No sir.
Q: Why did he get six cents a ton?

SL: Well, we formed the company and they didn't talk about that at all. Only me, Fabrizio, and Saporito was the partners. After, they agreed that Mr. Carlo Saporito gets six cents a ton off the price, Mr. Louis Fabrizio gets six cents a ton off the price, Mr. Johnnie Sciandra gets six cents a ton off the price and Mr. Lippi gets six cents a ton off the price.

Q: How about you?

SL: Nothing. I was nobody. ...

(See Stefano LaTorre, testimony, Joint Legislative Committee to Investigate the Knox Mine Disaster, 23 Apr 1959, 1377–85. See also Contract File Index, Contracts 1867–1969, Contract No. 274, Box 9, PaCC Papers, PSA.)

48. Contract File Index, Contracts 1867–1969, No. 335, Box 9, PaCC Papers, PSA. According to the Pennsylvania Crime Commission, Sciandra replaced Volpe as head of the region's organized-crime family in 1933. Sciandra came to Pittston from Buffalo around 1924 to mediate a dispute between mob factions, one of which was headed by Santo Volpe (Pennsylvania Crime Commission, *A Decade of Organized Crime,* 50; William Hastie, audiotaped interview, 31 Jul 1989, NPOLHP). Sciandra died of natural causes in 1949, although he seemed to have lost control of the gang in 1940. He is buried in the Dennison Street Cemetery in Forty Fort, in a section with numerous other alleged crime-family members, including Volpe. More than one Mafia-related internet site mistakenly claims that Sciandra was murdered in 1940.

49. General Contract Index, No. 27. A supplement to the lease, No. 135, was immediately added on 12 Mar 1934, PaCC Papers, PSA.

50. U.S. v. August J. Lippi. U.S. District Court, District of Delaware 190 F. Supp. 604; 47 L.R.R.M. 2537; 42 Lab. Cas. (CCH) P16,792. (2 Feb 1961); U.S. v. August J. Lippi, Criminal A. No. 1269, U.S. District Court, District of Delaware, 190 F. Supp. 604, 47 L.R.R. M. 2537; 42 Lab. Cas. (CCH) P. 16, 792 (19 Apr 1961); U.S. v. George J. Daileda and August J. Lippi, U.S. Court of Appeals Third Circuit, 342 F.2d 218, (9 Mar 1965); August J. Lippi, Plaintiff v. Lester Thomas et al., Defendants, Civil No. 68–336, U. S. District Court for the Middle District of Pennsylvania, 298 F. Supp. 242; 70 L.R.R.M. 3424; 59 Lab. Cas. (CCH) P13, 399 (31 Mar 1969). See also Lippi's testimony in Joint Legislative Committee to Investigate the Knox Mine Disaster (*Final Report*, 27 Jul 1959, Harrisburg, PA: Commonwealth of Pennsylvania, 23 Apr 1959, 14–60); and Dougherty's testimony (*ibid.*, 1391–1444). The three officers of L.U. 8005, at the Knox Coal Company who were convicted of accepting bribes from company officials for "labor peace" were Charles Piasecki, president; Dominick Alaimo, grievance committee member and alleged "soldier" of the Bufalino crime family; and Anthony Argo, committeeman. Argo pleaded guilty and was given a suspended sentence. See U.S. v. Dominick Alaimo, et al., Criminal No. 13225, U.S. District Court, Middle District of Pennsylvania, Scranton, (1959). On the investigations, indictments, and convictions in the Knox case, see Robert P. Wolensky et al., *The Knox Mine Disaster*, Chapter 4.

51. Contract File Index, Pittston Co. & Canceled File, Agreement No. 124, PaCC Papers, PSA.

52. Fabrizio discussed his 19 years of mining experience in testimony before the Joint Legislative Committee to Investigate the Knox Mine Disaster (27 Jul 1959, 932–946, 1351–1375). The Pennsylvania Crime Commission did not name Fabrizio as a member of the region's criminal organization, although he had a history of business relationships with alleged mobsters including Sciandra and LaTorre (Contract File Index, Contracts 1867–1969, Contract Nos. 274, 335, Box 9, PaCC Papers, PSA). Like the other Knox company principals, he was acquitted of illegal mining, conspiracy, and manslaughter in association with the Knox Mine Disaster, but was convicted of income-tax evasion (U.S. v. Louis Fabrizio, Criminal. A. No. 1251, U.S. District Court, Delaware, [19 Apr 1961]; and U.S. v. Knox Coal Company, Robert L. Dougherty, August J. Lippi, Josephine Sciandra and Louis Fabrizio, August J. Lippi Appellant. No. 14803. U.S. Court of Appeals Third Circuit [21 Jun 1965]).

53. "Dougherty Joins Fabrizio in Federal Penitentiary," *Wilkes-Barre Record*, 7 May 1964. Within a year of the Knox disaster, a federal Grand Jury issued indictments against the Peeley and Avon companies, as well as the principal owner, Robert Dougherty, and UMWA officials August J. Lippi, Leonard Statkiewicz and Frank Cardoni—all for corporate-tax evasion and payroll conspiracy amounting to $430,000. The defendants were acquitted ("Peeley Mining Will Pay $2,800 for Overtime; Criminal Counts against Dougherty, Firm Ended," *ibid.*, 21 Jan 1960; "Anniversary of Knox Disaster Recalls Year of Investigations," *Times Leader The Evening News*, 22 Jan 1960), although the companies as corporate entities were found guilty ("Tax Cases Are Joined," *Times Leader The Evening News*, 21 Sept 1963). Dougherty was tried and found guilty of income tax evasion along with other Knox Coal Company principals (U.S. v. Knox Coal Com-

pany, Robert L. Dougherty, August J. Lippi, Josephine Sciandra and Louis Fabrizio, August J. Lippi Appellant. No. 14803. U.S. Court of Appeals Third Circuit, [21 Jun 1965]). For an oral history on the unsafe and hurried working conditions at the Avon Coal Company's Avondale Colliery in Plymouth see John Mikulski, audiotaped interview, 28 July 2001, NPOLHP.

54. Gelso's leases for the No. 14 Colliery were listed as No. 1661, General Contract Index, PaCC Papers, PSA. According to the Crime Commission's *1970 Report*, the brothers "had long been known to federal authorities as close associates of the La Cosa Nostra family run by Russell Bufalino in Luzerne County" (Pennsylvania Crime Commission, *Report on Organized Crime, Harrisburg, PA: Commonwealth of Pennsylvania, 1970,* 59). See also Chapter 7.

55. U.S. v. Philip and Samuel Gelso et al., Criminal No. 13435, U.S. District Court, Middle District of Pennsylvania, Scranton, [1964]. See also Samuel Gelso, unrecorded interview, 19 Aug 2003, NPOLHP. Alex Chamberlain, a mining engineer whose family had been coal operators and leaseholders, said that he appreciated Charles Gelso when he worked as a subcontractor in the 1930s: "Between my father and Bill Jermyn's cousin, whose name I think was Morton Downey, and the Pennsylvania Coal Company's sales branch, Pattison and Bownes, they reopened No. 14 Colliery. They divided the mine up into sectors underground and put individual contractors into either the entire shaft or into a section of the mine. One of the operators was a Charles Gelso, an Italian, who had some very, very interesting concepts. He got rid of the gas problem in the mine workings by taking a churn drill, a water drill, and drilling holes down into the old mine workings ahead of where they were mining, and they vented the gas through them. It made a gassy hell of a mine into a very productive mine. He was one of our [sub]contractors" (Alex Chamberlain, audiotaped interview, 11 Jul 1995, NPOLHP).

56. E. Stewart Milner, audiotaped interview, 3 Aug 1994, NPOLHP. A PaCC employee since 1926, Milner retired as company vice president in 1978 but remained a consultant and keeper of the company's papers and maps at the former corporate headquarters in Dunmore. He donated most of the records to the Pennsylvania State Archives in Harrisburg, and to the Anthracite Heritage Museum in Scranton. However, he either gave or sold a collection of several hundred mine and property maps to a local businessman with alleged criminal connections. The maps provide valuable information related to land values indicating, for example, which properties had been undermined and which had not (see e.g., "Coal Company Unloading Land," *Times Leader The Evening News*, 17 Jul 1978).

57. William A. Hastie, audiotaped interview, 31 Jul 1989, NPOLHP.

58. David Panzitta, audiotaped interview, 14 Mar 1996. See also Joseph Panzitta, audiotaped interview, 25 Apr 2007, NPOLHP.

59. The area's largest producer, Glen Alden Coal Company, leased mines to James Durkin as well as Fred and Palmer Correale whose criminal records were detailed by the Pennsylvania Crime Commission, *A Decade of Organized Crime*, 219, 245–249. LVCC leased to persons with supposed criminal affiliations, such as Louis Pagnotti and James Tedesco, who were among the largest leaseholders. As mentioned, LVCC leased and then sold many of its collieries to Pagnotti. Third-generation coal operator William Butler confided that he sold his company in the middle-eastern field in 1964 mainly because of the growing criminal influence. "That's why I got out," he said (William Butler, audiotaped interview, 25 Jul 1994, NPOLHP).

60. John Kehoe was a coal operator, Pittston political boss, founder of the Pittston *Sunday Dispatch*, and ally of Governor John S. Fine. On Kehoe's involvement with PaCC officials see Lewis Casterline, audiotaped interview, 12 Dec 1988, NPOLHP. Joe Costa described his bootleg alcohol transactions with Kehoe in unrecorded interviews on 4 Jul 1991 and 1 Aug 1992. On Kehoe's life see "Kehoe's Assets Given as $36 in Banks and a Lot; Deposition Taken on His Ability to Pay $549,983 Judgment," *Scranton Tribune*, 14 Dec 1940; "Kehoe Gaining Added Power in County GOP," *Sunday Independent*, 19 Jul 1959; Richard Cosgrove, "Political Leader, Coal Operator, Kehoe Was a Colorful Character," *Sunday Dispatch*, 5 Mar 2000; Anthony DeAngelo, audiotaped interview, 18 Dec 1988, NPOLHP; Richard Cosgrove, audiotaped interview, 13 May 2003, NPOLHP; and John Kehoe, *Autobiography*, Pittston, PA: published by the author, no date.

61. Lewis Casterline, audiotaped interview, 12 Dec 1988, NPOLHP.

62. *Ibid.*

63. After being banned from the UMWA for dual unionism (see Chapter 5), Frank McGarry, the former insurgent president of District 1 and the acknowledged leader of the insurgents at the

No. 6 Colliery, sought reinstatement in the UMWA. In a letter to UMWA Vice President Thomas Kennedy, he wrote: "I have withdrawn all affiliation from the Anthracite Mine Workers Union as Chartered with the State of Pennsylvania and I am not connected in any way with any other dual union movement of Mine Workers that might exist" (Frank McGarry to Thomas Kennedy, 25 Jan 1929, President/District Correspondence, 1920– 1933, Paterno/GST/Z06.02, Box 1, Folder 13, UMWA Papers, PSU). McGarry seems to have faded from the insurgent movement after 1929 and the leadership mantle passed to Thomas Maloney and Rinaldo Cappellini. Also seeking reinstatement with McGarry were Vincent Trettis, Edward Hogan, Joseph Dougher, Walter Harris, Raymond Delaney, and William Murrin. At a hearing on February 26, 1929, all made statements on their own behalf. Some were accused of continuing to engage in dual unionism (John O'Leary, J.R. McCormick, and John Ghizzoni to John L. Lewis, 20 May 1929, International Executive Board Correspondence, AX/B40/ HCLA/01146, Box 6, Folder 32, UMWA Papers, PSU). It is not clear if or when the UMWA reinstated any of the former members.

64. "Cappellini and Lavelle Win Places on Ballot," *Wilkes-Barre Record*, 4 Jun 1929; Douglas K. Monroe, *A Decade of Turmoil: John L. Lewis and the Anthracite Miners, 1926–1936*, unpublished doctoral dissertation, Georgetown University, 1976, 100.

65. "Insurgents May Join Forces," *Wilkes-Barre Record*, 17 Jun 1929; "Boylan Margin 17,350 Votes," *ibid.*, 24 Jun 1929.

66. Staughton Lynd ("*We Are All Leaders*": *The Alternative Unionism of the Early 1930's*, Champaign, IL: University of Illinois Press, 1966, 3) characterized dual or alternative unions as "democratic, deeply rooted in mutual aid among workers in different crafts and work sites, and politically independent."

67. "Miners Would End Special Contracts," *New York Times*, 9 Jul 1930. One of the demands involved the elimination of the fee the company charged workers for compressed air used in jackhammers.

68. *Proceedings of the Reconvened UMWA Tri-District Convention* (Districts 1, 7, 9), 1930, 60–62.

69. "A Decade of Anthracite Peace," *Literary Digest*, 120 (2 Aug 1930), 40.

70. *Ibid.*, 94. The UMWA vice president, Thomas Kennedy, replied that the clause was really nothing new and simply reflected a long-standing policy of the UMWA dating back to the Anthracite Strike Commission of 1902.

71. Proceedings of the Reconvened UMWA Tri-District Convention (Districts 1, 7, 9), 1930, 59, 60.

72. *Ibid.*, 25.

73. *Ibid.*, 164.

74. *Ibid.*, 176.

75. *Ibid.*, 1930, 44, 52.

76. "National Miners' Union Reports Many Meetings; Sessions Held at Pittston, Jessup, Dunmore, Throop and Other Towns in Coal Region," *Scranton Times*, 24 Jun 1930; Walter T. Howard, "Anti-Labor and Anti-Radical Violence in Northeastern Pennsylvania During the Great Depression," 74–100 in Robert Mittrick (ed.), *Proceedings of the Eighth Annual Conference on the History of Northeastern Pennsylvania*, Luzerne County Community College, 6 Oct 1995.

77. John Boylan and Enoch Williams To The Officers and Members of Local Unions in District No. 1, UMWA, 9 Mar 1931, President/District Correspondence, 1920–1933, Paterno/ GST/Z06. 02, Box 1, Folder 6, UMWA Papers, PSU.

78. The *Proceedings of the Twenty-Fourth Consecutive and Ninth Biennial Convention of UMWA District No.1*, 1931, 475–484, 505–523, dealt with vote fraud; see also "Miners Demand a Fight," *Sunday Independent*, 7 Jun 1931; "Miners' Convention Control Fight Looms," *Wilkes-Barre Record*, 20 Jul 1931; and "Administration Gains Convention Control," *ibid.*, 23 Jul 1931.

79. John Boylan et al., To The Officers and Members of District 1, UMWA, 26 Feb 1932, President/District Correspondence, 1920–1933, Paterno/GST/Z06.02, Box 1, Folder 6, UMWA Papers, PSU.

80. "Strike Voted by 44 Locals," *Wilkes-Barre Record*, 12 Mar 1932; "Most Miners Shun Outlaw Coal Strike," *New York Times*, 15 Mar 1932; "Appeal of Union Leaders is Directed to Men's

Loyalty," *Scranton Times*, 17 Mar 1932; "Union Leaders Call on Loyal Miners to Stay at Work," *ibid*., 17 Mar 1932.

81. "Homes Dynamited, Coal Strike Grows," *New York Times*, 19 Mar 1932; Louis Stark, "Insurgents Suffer Mine Strike Check," *ibid*., 22 Mar 1932. See also articles in *The New York Times*, 19 Mar 1932; *Wilkes-Barre Record*, 12 Mar 1932; *Scranton Times*, 17 Mar 1932; and "District Officers Take Steps to Block Strike," *Wilkes-Barre Record*, 14 Mar 1932. Maier Fox (*United We Stand: A History of the United Mine Workers of America, 1890–1990*, Washington, DC: United Mine Workers of America, 1990, 295) wrote that equalization was the key issue in the spring 1932 strike; however, he did not mention the other important precipitating factors, namely, subcontracting and leasing.

82. "Union Leaders Call on Loyal Miners to Stay at Work," *Scranton Times*, 17 Mar 1932.

83. *Ibid.*

84. "Mine Resumes Work, Another to Operate," *Wilkes-Barre Record*, 17 Mar 1932.

85. "Homes Dynamited, Coal Strike Grows," *New York Times*, 19 Mar 1932; Louis Stark, "Insurgents Suffer Mine Strike Check," *ibid*., 22 Mar 1932. See also articles in *The New York Times*, 19 Mar 1932; *Wilkes-Barre Record*, 12 Mar 1932; *Scranton Times*, 17 Mar 1932; and "District Officers Take Steps to Block Strike," *Wilkes-Barre Record*, 14 Mar 1932.

86. "Union Leaders Call on Loyal Miners to Stay at Work," *Scranton Times*, 17 Mar 1932.

87. Telegram, William W. Inglis to John L. Lewis, 31 Mar 1931, President/District Correspondence, 1920–1933, Paterno/GST/Z06.02, Box 1, Folder 25, UMWA Papers, PSU.

88. "Miners Debate Strike 9 Hours, Defer Action," *Wilkes-Barre Record*, 26 Mar 1931.

89. Enoch Williams to John L. Lewis, 6 Apr 1932, Paterno/GST/Z06.02, Box 1, Folder 29, UMWA Papers, PSU.

90. Torma and Maloney were among 28 Glen Alden employees who were fired and barred from future employment at the company, according to Memorandum from Edward Griffith [General Superintendent] to W.L. Davis, Harry Gouldstone, T.R. Gambold, Walter Fahringer, David Girvan, D.J. Joseph, and T.J. Phillips [Colliery Superintendents], 18 Aug 1933, Joseph M. Keating Collection on the UAMP, Nanticoke Historical Society, uncatalogued; see also John Gizzoni to John L. Lewis, 9 Apr 1932, President/District Correspondence, 1920–1933, Paterno/GST/ Z06.02, Box 1, Folder 29, UMWA Papers, PSU; and "Insurgent Miners Vote on Committee Appointees," *Scranton Times*, 8 Aug 1933. Torma had served for more than 10 years as president of the UMWA Local at Glen Alden's Loomis Colliery.

91. "Cappellini Attacks Union Officers at Exeter Meeting," *Scranton Times*, 10 May 1933; "Lax Support of Miners Seen; Interests Neglected by Leaders, Cappellini Tells Jobless," *Wilkes-Barre Record*, 19 May 1933.

92. "Bluff Derided by Insurgents," *Wilkes-Barre Record*, 21 Jul 1933.

93. The Glen Alden Coal Company took over the Lehigh & Wilkes-Barre Coal Company in 1929, and income from the new firm's leasing activities grew precipitously from $100,179 in 1930 to $120,700 in 1931; $97,600 in 1932; $125,300 in 1933; $151,900 in 1934; $162,800 in 1935; $238,291 in 1936; $361,713 in 1937; $441,116 in 1938; $238,165 in 1939; $198,958 in 1940 (Bound unnamed file dealing with balances between 1930 and 1941, Glen Alden Coal Company Papers, Earth Conservancy, Inc., uncatalogued).

94. *Proceedings of the Twenty-Fifth Consecutive and Tenth Biennial Convention of UMWA District No. 1*, 1933, 29.

95. Joe Dougher of Archbald managed to contribute several critical remarks during the first four days of the convention. Maloney finally had him ejected on the fifth day after he admitted, in a heated exchange, that he served as secretary of the local Communist Party. Maloney chastised him following his removal: "I gave him plenty of rope and he hung himself." Cappellini remarked that Dougher was sent by the "Scranton wing of the Communist party" to cause disorder (Walter T. Howard, *Forgotten Radicals: Communists in the Pennsylvania Anthracite, 1919–1950*, Lanham, MD: University Press of America, 2005, 135–137; "Communist is Ejected from Mine Convention," *Wilkes-Barre Record*, 12 Aug 1933). According to communist organizer Steve Nelson, who worked in the anthracite region at this time, Maloney, Cappellini, and Schuster "had been friendly with the NMU [National Miners Union] but were not Party people. ..." He added that the UAMP welcomed the support of the Communist-organized unemploy-

ment councils. Nelson was a leader in the unemployment councils and spoke to UAMP local unions assuring members that the councils would try to prevent the unemployed from working as scabs in the mines (Steve Nelson, *Steve Nelson: American Radical*, 170).

96. The elected UAMP board members were William Martin, Bart Petrini, Anthony Serafini, and John Torma; the committeemen were Sam Loquasto, Joseph Shovlin, James Moleska, William Jones, Willard Saxon, and Clarence Lindy; the auditors were Stanley Ocheski, Joseph Schuster, and Vincent Carey; the Resolutions Committee included Sam Laquasto, Joseph Shovlin, James Moleska, William Jones, William Saxon, and Clarence Lindy. The state officers were Rinaldo Cappellini, president; George Moleski, vice-president; Larry Enderline, secretary-treasurer; and the state board members William Austin, Joseph Gallig, William Saxon, and Stanley Gorgas. Atty. Samuel H. Torchia served as the organization's legal counsel. See "Insurgent Miners Vote on Committee Appointees," *Scranton Times*, 8 Aug 1933; "Insurgents Elect Staff," *Wilkes-Barre Record*, 8 Aug 1933; and "Insurgent Miners Vote on Committee Appointees," *Scranton Times*, 8 Aug 1933.

97. "Insurgents Quit U.M.W.," *Wilkes-Barre Record*, 10 Aug 1933.

98. "Insurgents Elect Staff," *ibid.*, 8 Aug 1933. The National Industrial Recovery Act (NIRA) was enacted on June 16, 1933, but declared unconstitutional by the U.S. Supreme Court on May 27, 1935. Section 7(a) proved especially difficult to enforce and led to considerable labor-management conflict. Congress passed the National Labor Relations Act (NLRA) and President Franklin D. Roosevelt signed it into law on July 5, 1935, which remedied most of the problems with NIRA. John L. Lewis was appointed as a member of the National Labor Relations Board designed by the NLRA to rule on violations of the law.

99. In June 1929 employees at a Lehigh & Wilkes-Barre Coal Company (L&WBCC) colliery in the middle coalfield went on strike over the company's attempt to institute the subcontracting system. According to District 7 president, Michael Hartneady: "... they are trying to put the Pittston [sub]contract system on to us under the L&WBCC. Three weeks ago you can see what happened. They had two miners with eighteen men working before we knew it. The eighteen men were getting days' wages same as you were in the Pittston District. When we discovered it, it didn't take long to get it out of there. It resulted in stopping of twelve hundred men for two days, but if we were not awake it would have stayed. Don't figure Pittston is the only District or the only place for that, because every operator in the district would take it on if he could get it" (Address by President Michael Hartneady of District No. 7, *Proceedings of the Twenty-Third Consecutive and Eighth Biennial Convention of UMWA District No.1*, 17 Jul 1929). On other aspects of labor-management conflict in Districts 7 and 9 see Monroe, *A Decade of Turmoil*, 94–98.

100. John L. Lewis began to support equalization in mid-1933 when it looked like it would be part of an Anthracite Code encouraged by the federal government to boost the industry's recovery. Under the charge of deputy recovery administrator W.H. Davis, the code derived from the National Industrial Recovery Act. However, unlike bituminous, the UMWA and the anthracite operators never agreed on the code's terms and it was never approved ("Equalization Study, Check-Off Voting Proposed for Coal Code on Eve of Strike Conference," *Wilkes-Barre Record*, 9 Jan 1934; "Seeking Recognition, Maloney Says Miners Are Ready for Fight," *ibid.*, 10 Jan 1934; Monroe, *A Decade of Turmoil*, Chapter 6; Randall E. Parker, *The Economics of the Great Depression: A Twenty-first Century Look Back at the Interwar Years*, Northampton, MA: Edward Elgar, 2007, 106).

101. "Insurgents Quit U.M.W.," *Wilkes-Barre Record*, 10 Aug 1933.

102. Lewis's conservative business unionism approach was evident in his book, *The Miners' Fight for the American Standard* (Indianapolis, IN: Bell, 1925, 15) and in statements such as: "The policy of the United Mine Workers of American is neither new nor revolutionary. It does not command the admiration of visionaries and utopians. It ought to have the support of every thinking business man in the United States, because it proposes to allow natural economic laws free play in the production and distribution of coal. ..."

103. Howard, *Forgotten Radicals*, Chapters 6 & 7.

104. Alan Singer ("Communists and Coal Miners: Rank-and-File Organizations in the United Mine Workers of America during the 1920s," *Science and Society*, 55 [1991], 150), concluded that the recurring dual-union problem in anthracite and the larger UMWA was not ultimately caused by Communists or insurgents but by the union's own authority structure. "Where a tendency toward 'dual unionism' did exist," he wrote, "it was not a product of Communist organizing, but a normal feature of the internal UMWA union politics. Entrenched bureaucratic machines

were organized to maintain their control over the union. They refused to run fair union elections or open conventions. Dissident factions were forced to create parallel political structures to challenge these machines. Whether intended or not, these structures had the potential to become dual organizations."

105. Charged and convicted of arson to his own home, Cappellini was sentenced, on February 2, 1931, to serve two to four years in Philadelphia's Eastern State Penitentiary ("Former Chief of Miners Union and Wife Will Face Jury," *Scranton Republican*, 9 Jan 1931). He denied the charges and argued that he had been framed because of his insurgent activities. Realizing the questionable nature of the verdict, Governor Gifford Pinchot pardoned the union leader on March 25, 1932. Following his release, Cappellini rejoined the UAMP movement as a leader and organizer. He was again indicted, in August 1933, for allegedly passing a counterfeit $10 bill, the arrest being carried out on the platform of the inaugural UAMP convention. Cappellini and his colleagues again cried foul, and said his enemies were trying to send him back to prison for parole violations. A grand jury found insufficient evidence to press charges so Cappellini returned to union organizing ("Framed, Says Cappellini, of Money Arrest," *Wilkes-Barre Record*, 14 Aug 1933). For reasons that remain unclear, spouse Marie Cappellini filed for divorce on January 26, 1933, soon after her husband's gubernatorial pardon (*Lunacy, Divorce and Injunction Docket*, 1878–1949, "Persons Against whom Decrees in Divorce Have Been Granted," Prothonatory's Office, Luzerne County Court House, Wilkes-Barre, PA). Five months later, on the eve of a major UAMP strike, Cappellini was again arrested, this time charged with causing a disturbance at Marie's home in Wilkes-Barre. Within days the stunned former union leader was on his way back to state prison for violating parole. He served the remaining 13 months of his term at the Graterford State Penitentiary ("Cappellini Put in Jail: Leader of Striking Miners Arrested on Complaint of Former Wife," *Wilkes-Barre Record*,15 Jan 1934; "Agents Check on Cappellini; State Parole Officers Attend Meeting; Miners' Leader Gives Bail," *ibid*, 15 Jan 1934; "Cappellini Hustled Back into Penitentiary; Arrested on Order of State Parole Board, Union Leader is Stunned by Swift Action," *Scranton Times*, 16 Jan 1934; "Strikers Consider a Coal Peace Plan," *New York Times*, 17 Jan 1934). Sometime after his release in 1934, Rinaldo and Marie reunited as husband and wife. He took a position as manager of an Italian-language newspaper for a brief period. He eventually made peace with John L. Lewis who hired him as an organizer for the Congress of Industrial Organizations (CIO). According to Dubofsky and Van Tine (*John L. Lewis: A Biography*, New York: Quadrangle/New York Times, 1977, 100), Cappellini "later received what had become traditional Lewis treatment for opportunistic union rebels: a well-paid position on the International payroll." Marie thought that Lewis gave her husband a job with the CIO out of fear that he might get involved with another insurgent union movement. She added that many anthracite mineworkers displayed anger toward her husband for having taken the job with Lewis and the CIO. Mr. Cappellini stayed with the CIO until retirement in the 1950s. He returned to northeastern Pennsylvania, where he had always kept a house, and enjoyed a near-celebrity status until his death in 1966 at age 71.

106. Lewis Casterline, audiotaped interview, 12 Dec 1988, NPOLHP.

107. For a detailed study of the UAMP movement see Monroe, *A Decade of Turmoil*, Chapters 6 & 7. It was not a well-known fact that the 1902 Anthracite Strike Commission called for a separate anthracite miners' union, an idea that John Mitchell flatly rejected. See also David Wilcox, "Present Conditions in the Anthracite Coal Industry," *North American Review*, 181 (1905), 216–228; and John Mitchell, *Organized Labor*, Philadelphia, PA: American Book & Bible House, 1903, 268–269.

108. "Railroad Tracks are Dynamited in Mine Outbreak," *The Evening News*, 13 Sept 1933; "Glen Alden's Miners Called Out on Strike," *Wilkes-Barre Record*, 18 Oct 1933.

109. "Two Unions to Battle for Supremacy Today as 50 Mines Call Employees," *ibid.*, 15 Jan 1934. See also "Rump Union Calls Anthracite Strike," *New York Times*, 14 Jan 1934.

110. "Washington Sends Mediator," *ibid.*, 5 Nov 1933. The NLB was later replaced by the National Labor Relations Board after the U.S. Supreme Court declared the NIRA unconstitutional.

111. *Ibid.*

112. "Gorman Rules Against New Union," *Pittston Gazette*, 19 Oct 1934; "Union Renews Threat of Hard Coal Strike," *New York Times*, 20 Oct 1934; "Maloney Makes Strike Threat on Glen Alden," *Wilkes-Barre Record*, 2 Feb 1935; "Bomb Cripples Coal Mine," *New York Times*, 30 Mar 1935; "Five Shot, 21 Injured in New Mine Battle; Scores of Others Beaten in Warfare in Nottingham Strike Near Wilkes-Barre," *ibid.*, 15 May 1935; Stacia Treski, oral history interview

in John Bodnar, *Anthracite People: Family, Unions and Work, 1900–1940*, Harrisburg, PA: PHMC, 1983, 38.

113. "U.A.M. to Act on Court Order This Afternoon," *Wilkes-Barre Record*, 18 Feb 1935; "Jailed Union Heads to Push Court Appeal," *ibid.*, 18 Mar 1935. See also Chester Brozena, audiotaped interview, 3 Dec 1988, NPOLHP. In late March 1934, a bomb shattered a car owned by Judge Valentine but driven by his daughter, Mary. She escaped harm by exiting the vehicle minutes before the device exploded, but a passing newsboy sustained slight injuries. The UAMP condemned the act and offered a $100 reward for information leading to the arrest of the culprits, who, the organization maintained, were not mineworkers but professional criminals ("Girl's Car Bombed in War on Judge," *New York Times*, 29 Mar 29, 1935; Gerald Stout, "Judge Valentine's Car Bombed; Newsboy is Hurt by Explosion in Down Town Area," *Wilkes-Barre Record*, 29 Mar 1935; Lawrence E. Davies, "Hard Coal Strike Battle of Unions; Insurgents Jailed, Charge Operators Control United Mine Workers," *New York Times*, 31 Mar 1935.

114. George Hunter to John L. Lewis, 8 Oct 1931, and Lewis to Hunter, 16 Oct 1931, President/District Correspondence, 1920–1933, Paterno/GST/Z06.02, Box 1, Folder 24, UMWA Papers, PSU.

115. Philip Murray to John L. Lewis, 22 Nov 1930; and John Boylan to John L. Lewis, 2 Dec 1930, President/District Correspondence, 1920–1933, Paterno/GST/Z06.02, Box 1, Folder 19, UMWA Papers, PSU. For another attempt by Gallagher to secure an informal deal at the Underwood Colliery in 1931 see John L. Lewis to Philip Murray, 27 Mar 1931, President/District Correspondence, 1920–1933, Paterno/GST/Z06.02, Box 1, Folder 21, UMWA Papers, PSU). For another attempt at the Butler Colliery in 1932 see "Jobs for 1200 Men to be Provided If Pact Is Approved," *The Evening News*, 19 Oct 1932. See also August J. Lippi's denial that any special rates were offered to the company by the District 1 office (*Proceedings of the Twenty-Fourth Consecutive and Ninth Biennial Convention of UMWA District No.1*, 1931, 336–337).

116. "Dunmore Miners Will Not Join in General Strike," *Scranton Times*, 20 Jun 1930; "Order Strike of 14,000 Mine Workers Today," *Scranton Republican*, 21 Jun 1930; "Leaders Say Men Will Work," *Wilkes-Barre Record*, 21 Jun 1930. Suspected Communist involvement in the equalization strike caused local and state police to harass and arrest left-wing advocates. A series of well-organized and publicized unemployment marches led by Communists no doubt fueled the police reaction (Howard, *Forgotten Radicals*, Chapter 6). John Justin was a garment worker and union organizer, a New York City native, and a Wilkes-Barre resident who helped organize the unemployment marches of the 1930s (see his audiotaped interviews, 28 Jul 1988, 30 May 1996, 25 Mar 1997, NPOLHP).

117. Equalization was again among the main demands agreed upon at the Tri-District UMWA Convention of 1939: "There shall be an equal division of working time, without tolerance, as between operations" (*Proceedings of the UMWA Tri-District Convention* [Districts 1, 7, 9], 1939, 47).

118. "Eight Bullets Strike Pittston Mine Worker," *Scranton Times*, 5 Jan 1931.

119. "Riddled By Bullets While Near His Home," *Trenton Evening News*, 2 Mar 1931.

120. Frank Tomchak, a delegate from L.U. 466, indicated that 4,000 of PaCC's and TPC's approximately 10,000 employees belonged to the UMWA in 1931 (*Proceedings of the Twenty-Fourth Consecutive and Ninth Biennial Convention of UMWA District No.1*, 1931, 173).

121. "Strike Action Awaits Vote of 13 Locals," *Wilkes-Barre Record*, 20 Jul 1931. A letter to John. L Lewis written in Italian by Joe Marianacch of Plains Township on August 13, 1931, complained about subcontractors cheating workers out of pay: "Along with other men I am working with in the headings, we used to be paid by the yard for the bottom rock and for the rock in the middle of the seam of the coal. But now we receive no compensation for this work. I don't accuse the company for this, but the [sub]contractor, superintendent and bosses. At [a] certain day the [sub]contractor comes along with the bosses and engineers and they measure the amount of rock blasted and removed. And this is the end for us. Evidently the Company pays the [subcontractor] for his work, and it seems to us that he and the other members of the gang pocket the money, and the workers are the victims of their greed" (Joe Marianacch to John L. Lewis, 13 Aug 1931, President/District Correspondence, 1920–1933, Paterno/GST/Z06.02, Box 1, Folder 23, UMWA Papers, PSU).

122. "Insurgent Mine Chieftain Maloney Quits Coal Strike," *Scranton Times*, 31 Mar 1932.

123. "Gorman Reveals Some Complainants Hired, Paid by Contractors; Most of 26 Pittston Company Employees' Charges Set for Discrimination," *Wilkes-Barre Record*, 14 Apr 1934.

124. The UMWA reported the income received in dues from local unions at each of the District's coal companies. The area's largest company, Glen Alden, unsurprisingly ranked first, with $438,519. However, TPC ranked second with $109,105, in what surely represented highly inflated figures, possibly to dispel the support the insurgents held at the company's collieries. Hudson ranked third at $78,146, and Lehigh Valley was fourth with $64,599 (*Proceedings of the Twenty-Fifth Consecutive and Tenth Biennial Convention of UMWA District No.1*, 1933).

125. "Rump Meeting is Threatened," *Wilkes-Barre Record*, 14 Jul 1933.

126. "Local Names 14 to Rump Rally," *ibid.*, 2 Aug 1933.

127. *Ibid.* The No. 9 delegates were: Frank Carcivenek, Albert Orlando, Andrew Bauman, Anthony Try, Jacob Cassley, Lee Licata, James Try, George Smith, Thomas Sapolis, Michael Cardone, Samuel Enfileman, Charles Bauman, Michael Kosloski, and James Delmont. The No. 6 Colliery delegates were: Stanley Ocheski, Nick Bonfantem, Joseph Spudis, Joseph Eustice, and Sam Saietta. No delegate names or numbers were listed for the Butler Colliery.

128. "Mine Delegates in Hostile Mood," *Times Leader*, 30 Oct 1933.

139. On the UAMP strikes see Monroe, *A Decade of Turmoil*, Chapter 6; and Bodnar, *Anthracite People.*

130. "Hard Coal Rebels Disband Union; Outlaw Group in Pennsylvania Leaves Control of Field to Lewis," *New York Times*, 27 Oct 1935.

131. On the so-called Good Friday Bombings see Sheldon Spear, *Chapters in Northeastern Pennsylvania History*, Shavertown, PA: Jemags & Co., 1999, Chapter 6. See also Clement Valetta, "'To Battle for Our Ideas': Community Ethic and Anthracite Labor, 1920– 1940," *Pennsylvania History*, 58 (Oct 1991), 311–329. Curran died seven months later, in November 1936.

132. Agreement of May 7, 1936, contained in Anthracite Board of Conciliation, *Award of the Anthracite Coal Strike Commission*, Hazleton, PA, 1939, 64.

133. *Ibid.*, 90–92.

134. In one of the largest property and mineral-rights transfers of the post-World War II era, the Glen Alden Corporation sold all of its Lackawanna County holdings to Robert Y. Moffat on December 12, 1955 ("Assignment of Leases, Glen Alden Corporation to Robert Y. Moffatt, 12 Dec 1955," Glen Alden Coal Company Papers, Earth Conservancy, Inc., uncatalogued). By 1958 the Moffat Coal Company had nine lessees who operated at the Hyde Park Shaft, Continental Colliery, Pyne Colliery, Dodge Colliery, Storrs Colliery, and Baker Colliery. Moffat continued mining and leasing in the Scranton area into the 1970s. The firm's Continental Mine Slope 190, now the site of the popular Lackawanna Coal Mine Tour, produced anthracite until 1966. The sub-leases issued by Moffat were discussed in Garfield Lewis and Harry Welby to August J. Lippi, 26 Sept 1958, UMWA Papers, District 1–Documents, Box ZZ/31, Folder 3, 1958. Moffat, like Pagnotti, was well known for pushing workers toward higher output. After noting that his employees resisted a production speed up, William Graham, the last president of the Penn Anthracite Coal Company, said he learned that Moffat had even higher expectations: "I remember some of the [mineworkers] then came back later and they told me: 'Hey, we didn't know we had it so good. I went to work with Moffat and his mine car is twice ours and I had to load twice the cars.' You know, something like that. So he was practically giving them at least twice what he was doing here" (William Graham, audiotaped interview, 11 Jul 1995, NPOLHP; see also the interview with Moffat employee, Howard Baird, audiotaped interview, 24 Jul 2001, NPOLHP).

135. John Gadamski, audiotaped interview, 12 Dec 1988, NPOLHP; Stanley Roman, audiotaped interview, 12 Dec 1988, NPOLHP; Chester Dunn, audiotaped interview, 4 Aug 1989, NPOLHP; John Mikulski, audiotaped interview, 28 Jul 2001, NPOLHP; John Usefara, audiotaped interview, 28 Jul 2001, NPOLHP.

136. George H. Minor, *The Erie System: The Organization and Corporate History*, published by the Erie Railroad, 1938, 435.

137. Stockholders and Executive Committee Minutes, 1838–1971, Folder: 25 Jan 1880–24 Feb, 1891, Doc No. 1021, PaCC Papers, PSA.

138. Pennsylvania Anthracite Coal Industry Commission, *Report*, 1938.

139. Stockholder and Executive Committee Minutes 1838–1971, Doc. No. 1021, PaCC Papers, PSA.

140. *Ibid.* On TPC's million-dollar losses for 1937 and 1938 see also "Pittston Co. Loss is Cut," *Wilkes-Barre Record*, 20 Apr 1939.

141. Chernow, *The House of Morgan*, 414. Despite heroic measures to save at least part of their business empire, the Van Swerigens lost control of the Alleghany Corporation and most of its holdings, including The Pittston Company (TPC), to the Mid-American Corporation owned by George A. Ball of Muncie Indiana and G.A. Tomlinson of Cleveland (Harwood, *Invisible Giants*, Chapters 22 & 23). Ball then acquired Tomlinson's interests and eventually turned Mid-American over to the George and Frances Ball Foundation for tax purposes. In April 1937, Robert R. Young, Alan P. Kirby, and F.F. Kolbe acquired 43 percent of Alleghany's stock from Mid-American and the Foundation. The investors returned their shares to the Foundation, apparently again for financial reasons, but Kirby and Young repurchased 22 percent of the stock later in 1937 and retained control thereafter (Pennsylvania Anthracite Coal Industry Commission, *Report*, Harrisburg, PA: Commonwealth of Pennsylvania, 1938, 402). Kirby, a native Wilkes-Barrean whose wealth came from the "dime store" empire that his father F.M. Kirby had built with F.W. Woolworth, became president of Alleghany and successfully revived the company. He held the top position until 1961 and served as chairman of Alleghany's board and chief executive officer from 1963 to 1967, and as chairman emeritus from 1967 to 1973 (*Moody's Transportation Manual 1960*, 774–775). Alleghany again rose to Wall Street prominence through a series of takeovers, including the purchase of over one million shares in the New York Central Railroad by 1953. When the railroad suffered financial problems, Alleghany sustained major losses (Chernow, *The House of Morgan*, 511). The precarious fiscal situation forced Alleghany to sell The Pittston Company in 1954. TPC became an independent corporation with 30 subsidiaries, including the Brinks Armored Car division and several coal companies in the bituminous fields (Adele Hast [ed.], "The Pittston Company"). See also "A Loss for Bob Young," *Time* (23 Dec 1957).

142. PaCC, *Annual Report*, 1941, 2.

143. PaCC, *Annual Report*, 1941, *passim*.

144. Royalty Records, 1883–1968, Recapitulation of Tonnage and Coal Royalty Receipts by Lessees for Years 1939–1965, PaCC Papers, PSA.

145. The Glen Alden Coal Company Lease Books (Glen Alden Coal Company Papers, Earth Conservancy, Ashley, PA, uncatalogued) revealed widespread leasing activities, as did the Hudson Coal Company's Lease Books (Hudson Coal Company Papers, Earth Conservancy, Inc., Ashley, PA, uncatalogued). However, unlike PaCC and LVCC, the lease books of GACC and HCC were not complete so a full determination was not possible. The uncatalogued records at the Nanticoke Historical Society Archives, Nanticoke, PA, contain some information on the leasing activities of the Susquehanna Collieries Company.

146. Within six months of the Knox Mine Disaster of January 22, 1959, PaCC's board of directors announced the curtailment of water pumping at all Pittston-area mines (Stockholders and Executive Committee Minutes, Minutes, 1838–1971, Folder: 20 Feb 1953– 15 Dec 1961; Doc. No. 102: 225, PaCC Papers, PSA). Government agencies assumed the responsibility but soon withdrew ("Resume of Meeting Held on Friday, 24 Jul 1959, Relative to the Anthracite Mine Water Situation in the Pittston/Wilkes-Barre Area," Box 1, Folder 1, Joseph T. Kennedy Papers, PSA). As expected, mine pool levels began rising at a rate of three to four feet a day. Although 10 of the 40 mines in the Pittston area were functioning before the disaster, by June 1959 all were closed (Henry A. Dierks, Walter L. Eaton, and Ralph H. Whaite, *Anthracite Mine-Flood Disaster: Breakthrough of Susquehanna River into River Slope Mine, Knox Coal Company, Port Griffith, Pennsylvania*, Washington, DC: U.S. Bureau of Mines, 1960). PaCC began selling its remaining assets—equipment, property, and mineral rights. The dismantling persisted even after the parent company, the Erie Railroad, merged with the Delaware, Lackawanna, & Western Railroad on October 17, 1960. The new Erie-Lackawanna provided freight service to the anthracite region until filing for bankruptcy in June 1972, whereupon its assets were acquired by the government-backed Consolidated Rail Corporation (Conrail) on April 1, 1976. By December 1978, Conrail had liquidated all PaCC's holdings. The company remained in nominal existence until it was finally dissolved in the mid-1980s. Ironically, the last PaCC president, Daniel H. Connelly, served as the chief state mine inspector in January 1959, in the inspection district where the Knox Mine Disaster occurred.

147. *Proceedings of the UMWA Tri-District Convention* (Districts 1, 7, 9), 1939, 7.

148. *Ibid.*, 12. Also at the 1939 District 1 convention, Delegate Bunosky, from L.U. 1159 in Pittston, criticized the tenants for constantly cheating on wages. In language that blurred the distinction between subcontractors and lessees he said: "With regard to leasing coal lands out, we call these people individual contractors. I believe that the negotiating committee should absolutely insist upon doing away with that system. At our colliery we have quite a number of those [sub]contractors, and my local union spends three or four times the amount of money checking on those [sub]contractors in one month as we do at the coal company proper in six months. We have had [sub]contractors up there paying men $1.50 for fourteen hours' work. My committee had collected [back wages] for this one man after we had found out through a lot of persuasion. We went to this man's home and coaxed his daughter, he was a widower, we coaxed his daughter to stand by us and we would collect the money for him. She convinced him that he should [cooperate], and we had collected almost five hundred dollars in back wages for this man. I have affidavits at the hotel and I can show where we collected hundreds of dollars for miners' laborers. Secretary [Thomas] Kennedy says we would like to have a clause in there whereby we hold the coal companies responsible for the wages that these [sub]contractors are paying the men. We have that at our colliery. We insist that our company be responsible for wages that these [sub]contractors are paying. But the idea is that we are continuously in on these [sub]contractors. They are continually trying to put a man in today, a man or two in tomorrow, offering them $2.00 a day, $1.50 a day, and it works along until a committee catches up with them" (Comments by Delegate Bunosky, L.U. 1159, *Proceedings of the UMWA Tri-District Convention* [Districts 1, 7, 9], 1939, 65).

149. Comments by District 7 President Brown, *ibid.*, 73.

150. *Ibid.*, 17, 19.

151. Kennedy was elected Lieutenant Governor of Pennsylvania in 1934 and served one term until January 1939, whereupon he returned to the UMWA.

152. The "Resolution on Leasing" read:

    "WHEREAS, the leasing of collieries, plant units, mines or parcels of coal land, affects the stability of the industry; and

    "WHEREAS, the contracting parties desire to promote the stability of the industry and insure compliance with the general wage agreement;

    "NOW, THEREFORE, BE IT RESOLVED:

    "When any operator proposes to lease a colliery, plant unit, mine or parcels of coal land, and before such lease is executed, he shall notify the President of the District of the United Mine Workers of America in which the property is located. The President of the District shall have the cooperation of the lessor in establishing contact with the proposed lessees and the lessor shall further cooperate and render reasonable assistance to the mineworkers and lessee to have operations under such leases carried on in accordance with the terms and provisions of the wage agreement" ("Resolution In Re: Leasing," Adopted May 26, 1939, *Award of the Anthracite Coal Strike Commission, Subsequent Agreements and Resolutions of Board of Conciliation,* 1 Jul 1953, 179–180).

153. *Ibid.*, 181–182.

154. *Proceedings of the Reconvened UMWA Tri-District Scale Convention* (Districts 1, 7, 9), 1944, 86.

155. Robert H. Zieger, *John L. Lewis: Labor Leader*, Boston, MA: Twayne, 1988, Chapter 3.

156. At least two other researchers examined anthracite leasing systems that operated during the nineteenth century. Clifton K. Yearley, Jr. (*Enterprise and Anthracite: Economics and Democracy in Schuylkill County, 1820–1875*, Baltimore, MD: Johns Hopkins University Press, 1961, Chapter 2) studied a dysfunctional leasing system in the Schuylkill region based on property leases to small, often undercapitalized companies. Without sufficient investments, the firms faced constant financial difficulties and many went out of business. Richard G. Healey researched the mining leases that the large coal corporations secured from landowners between the end of the Civil War and the 1902 strike. Among other things, Healey focused on the relationship between leasing and corporate decision-making under various types of external constraints. For Healey, the key to understanding the industry during the period lay in the relationship between the unintended consequences of leasing coupled with the unpredictability of business cycles (Richard G. Healey, *The Pennsylvania Anthracite Coal Industry, 1860–1902: Economic Cycles, Business Decision-Making and Regional Dynamics*, Scranton, PA: University of Scranton Press, 2008).

Notes to Chapter Seven

1. Raymond E. Murphy and Marion Murphy, "Anthracite Region of Pennsylvania," *Economic Geography*, 14 (1938), 345; E. Willard Miller, "The Anthracite Region of Northeastern Pennsylvania: An Economy in Transition," *Journal of Geography*, 88 (1989), 168; "Gas Competition," *Coal Age*, 65 (Apr 1960), 71–72; "Outlook: Coal vs. Oil, *Coal Age*, 65 (Oct 1960), 72–75.

2. Salvatore Lazzari, Congressional Research Service Report for Congress, "Energy Tax Policy: History and Current Issues," www.fas.org/sgp/crs/misc/RL33578.pdf. Robert Bryce (*Cronies: Oil, the Bushes, and the Rise of Texas, America's Superstate*, New York: Public Affairs, 2004, 48) wrote that "[O]ilmen were getting a tax break that was unprecedented in American business. While other businessmen had to pay taxes on their income regardless of what they sold, the oilmen got special treatment." See also Joseph A. Fox, "Hard Coal's Role in Wartime May Force U.S. to Offer Help," *Philadelphia Evening Star*, 9 Jul 1954.

3. Miller, "The Anthracite Region of Northeastern Pennsylvania," 168.

4. Carter Goodrich et al., *Migration and Economic Opportunity*, Philadelphia, PA: University of Pennsylvania Press, 1936, 430.

5. The 2011 figure is based on the Consumer Price Index value of $3.4 million in 1953, as compared to the same amount in 2011, computed at the website www.measuringworth.com/calculators/uscompare/index.php. The Anthracite Institute's investment figures were procured from the *Proceedings of the UMWA District No. 1 Convention, 1957*, 99. On the Anthracite Institute Laboratories see "Economic Status of Coal-Mining Industry in Last Quarter Century," *Coal Age*, 41 (Feb 1936), 394ff; and Mead, *An Analysis of the Decline of the Anthracite Industry Since 1921*, unpublished doctoral dissertation, University of Pennsylvania, 1933, 353. A good share of the Institute's income likely went for office rent in the Chrysler Building in mid-town Manhattan.

6. Murphy and Murphy, "The Anthracite Region of Northeastern Pennsylvania, 168.

7. The demand for anthracite took a steep decline following the strike of 1925–26, but again the causes were multi-faceted. In addition to the convenience and marketing advantages maintained by other fuels, hard coal's fall-off was partially caused by the railroad companies' shift from steam engines to diesel locomotives. Another decrement came from the deterioration of the anthracite industry itself, which burned a portion of its output. Poor quality control also contributed, for in earlier decades when anthracite had little competition in certain markets, many operators sold "dirty" coal containing slate and other impurities. The fuel's use in steam generation for larger industrial consumers fell precipitously after World War II. High maintenance costs also contributed, for burning hard coal required constant attention to ashes, dust, chimneys, coal orders, and furnace breakdowns; customers who heated with fuel oil and natural gas experienced far fewer problems. The northeastern U.S. constituted anthracite's main home heating market and, when it began to decline, efforts to expand to the Midwest and beyond proved only marginally successful. Finally, the unreliability caused by regular strikes helped erode consumers' confidence. By the 1950s, the switch was on to other fuels even within northeastern Pennsylvania, despite some efforts to expose and ridicule those who changed. In one conspicuous instance, coal operator John Kehoe of Pittston, owner and editor of the *Pittston Dispatch*, published the names of persons and institution that moved from anthracite to other fuels. The military-related demands of World War II buttressed demand for a brief time, but the post-war competition from natural gas and fuel oil further damaged an industry already in retreat. For a statistical portrait of anthracite's final decades see Thomas Dublin and Walter Licht, *The Face of Decline: The Pennsylvania Anthracite Region in the Twentieth Century*, Ithaca, NY: Cornell University Press, 2005, Chapter 4.

8. William M. Steuart, *Special Report: Mines and Quaries*, Washington, DC: U.S. Department of Commerce and Labor, 1902, Chapter 16; Carroll Lawrence Christenson, *Economic Redevelopment in Bituminous Coal: The Special Case of Technological Advance in United States Coal Mines, 1930–1960*, Cambridge, MA: Harvard University Press, 1962, Chapter 4; Harold W. Aurand, *Coalcracker Culture: Work and Values in Pennsylvania Anthracite, 1835–1935*, Selinsgrove, PA: Susquehanna University Press, 2003, 52–58.

9. Chris Tilly and Charles Tilly (*Work Under Capitalism*, Boulder, CO: Westview, 1998, 46) pointed out that geology was not the only factor that allowed the room-and-pillar method to endure: work culture and traditions also played a part. In understanding why room-and-pillar persisted in anthracite even after it had been largely replaced by long-wall mining in bituminous, a number of reasons beyond geology can be considered. First, the anthracite companies were less inclined to invest in the expensive technologies for long-wall mining; second, a readily available

(over)supply of cheap labor discouraged capital investment; third, the workers resisted new technologies that could potentially add to the already serious unemployment problem; and, fourth, the subcontracting and leasing systems increased productivity and stabilized certain aspects of the industry without major technological investments.

10. "Changing Industrial Relations," *Coal Age*, 41 (Oct 1936), 460ff. "Inventory of Machinery and Equipment, Comparison December 20, 1929 and March 15, 1934," Box 1983.177, Folder 14 (b), PaCC Papers, NCMA.

11. "Inventory of Machinery and Equipment, Comparison December 20, 1929 and March 15, 1934," Box 1983.177, Folder 14 (b), PaCC Papers, NCMA.

12. *Ibid.*

13. Harold Barger and Sam H. Schurr, *The Mining Industries, 1899–1939: A Study in Output, Employment and Productivity*, New York: National Bureau of Economic Research, 1944, 185.

14. On coal-processing technologies in anthracite see Edward Pinkowski, "Joseph Batten: Father of the Coal Breaker," *Pennsylvania Magazine of History & Biography*, 73 (Jul 1949), 337–348; and Robert Janosov, "Glen Alden's Huber Breaker: 'A Marvel of Mechanism,'" 103–144 in Lantz Metz (ed.), *Canal History and Technology Proceedings*, Vol. XI, Easton, PA: Canal History and Technology Press, 1992.

15. Stuart Campbell, *Businessmen and Anthracite: Aspects of Change in the Late Nineteenth Century Anthracite Industry*, unpublished doctoral dissertation, University of Delaware, 1978, Chapters 3 & 4.

16. Joe Costa, unrecorded interview, 22 Jan 1999, NPOLHP.

17. The northern field's reserves were originally calculated at 6.55 billion tons when industrial operations commenced in the nineteenth century (George H. Ashley, "Anthracite Reserves," Topographic and Geologic Survey, Progress Report 130, Harrisburg, PA: Commonwealth of Pennsylvania, Department of Internal Affairs, 1945.) In 1922, the reserves were put at 3.27 billion tons (U.S. Bureau of Mines, "Analyses of Pennsylvania Anthracite Coals," Washington, DC: USGPO, 1944, 11). In 1967, the figure was 1.79 billion tons (John J. Schanz, Jr., "Historical Statistics of Pennsylvania's Minerals Industries, 1961–1965," *Bulletin of the Earth and Mineral Sciences Experiment Station*, College of Earth and Mineral Science, Penn State University, 1967, 21). The reserves in all four anthracite fields were estimated at 22.77 billion tons in the early 1800s, and 14 billion tons in 1967, meaning that about one-third of the coal had been mined or lost through mining in over 150 years (*ibid.*).

19. E.W. Miller (*A Geography of Pennsylvania*, University Park, PA: Penn State Press, 1995, 218–219), dismissed the "depletion of reserves" argument as well as the "declining productivity" argument as factors in anthracite's decline. However, he did not consider that, while considerable anthracite reserves were available, they were located at very deep levels, which required large capital investments that the companies did not want to expend. Regarding productivity, while strip mining helped boost output per worker from 2.83 tons a day in 1950 to 7.20 tons in the 1980s, by the 1960s it had little effect because numerous major markets had been lost. Moreover, the productivity gains were tempered to at least some degree by the operating costs associated with federal and state land reclamation and other environmental and safety laws (see Daniel J. Curran, *Dead Laws for Dead Men: The Politics of Federal Coal Mine Health and Safety Legislation*, Pittsburgh University Press, 1994).

19. David Brody, "Market Unionism in America: The Case of Coal," *In Labor's Cause*, New York: Oxford University Press, 1993, 132.

20. George O. Virtue, "The Anthracite Combinations," *Quarterly Journal of Economics* 10 (Apr 1896), 296–323; Clifton K. Yearley, Jr., *Enterprise and Anthracite: Economics and Democracy in Schuylkill County, 1820–1875*, Baltimore, MD: Johns Hopkins University Press, 1961, 158– 164. In 1868, several coal companies reduced output and bolstered prices by supporting a work stoppage sponsored by the Workingmen's Benevolent Association, which called the strike in protest over the coal companies' unwillingness to uphold Pennsylvania's new eight-hour work law (Marvin W. Schlegel, "The Workingmen's Benevolent Association: First Union of Anthracite Miners," *Pennsylvania History*, 10 [1943], 244).

21. Eliot Jones, *The Anthracite Coal Combination in the United States*, Cambridge, MA: Harvard University Press, 1914, Chapter 3 & Appendix, Table 6. The cartel-like efforts by coal companies to manipulate supply and prices were not limited to the U.S. In Britain, operators maintained The Grand Allies between 1710 and 1726 "for the purpose of restricting output and rais-

ing the price. As such it has much in common with other associations of owners, though for a time it was more successful than most" (Michael W. Flinn and David Stoker, *The History of the British Coal Industry*, Vol. 2, Oxford, England: Clarendon, 1984, 40).

22. Priscilla Long, *Where the Sun Never Shines: A History of America's Bloody Coal Industry*, New York: Paragon House, 1989, 123.

23. Flinn and Stoker, *History of the British Coal Industry*, 40.

24. Harold W. Aurand, *From the Molly Maguires to the United Mineworkers: The Social Ecology of an Industrial Union, 1869–1897*, Philadelphia, PA: Temple University Press, 1971, 163.

25. *Ibid.*

26. Ivan A. Given, "The Lewis Era ... 1920–1960 in Coal," *Coal Age*, 65 (Jan 1960), 70.

27. The production percentages allocated to PaCC and its lessees were stated in the Minutes of the Board of Directors' Meeting, 29 Dec 1930, 250–263, Stockholders and Executive Committee Minutes, 1883–1971, 28 Mar 1930—29 Dec 1939, Doc. No. 121, PaCC Papers, PSA.

28. The coal corruptions in Pennsylvania extended to the bituminous industry, as demonstrated in Pennsylvania Crime Commission, *Coal Fraud: Undermining a Vital Resource*, Conshohocken, PA: Commonwealth of Pennsylvania, 1987.

29. *Proceedings of the Twenty-Seventh Consecutive and Twelfth Biennial Convention of UMWA District No. 1*, 1937, 17.

30. Sen. Leo C. Mundy, Commission Chairman, to Governor George Earle, 10 Jan 1939, Volpe Coal Company File, *Times Leader* Archives, Wilkes-Barre, PA. On the Volpe Colliery disaster see also "Blast Inquiry Awaits Mine Ventilation," *Times Leader*, 4 Jun 1938; "Governor Orders Thorough Study Into Mine Safety," *Times Leader*, 7 Jun 1938; "Coroner's Jury to Seal Volpe Blast Verdict," *Times Leader*, 3 Aug, 1938; and Volpe Disaster Investigation Commission, *Report* to M.J. Hartneady, Secretary of the Department of Mines, Commonwealth of Pennsylvania, June 17, 1930, Volpe Coal Company File, *Times Leader* Archives, Wilkes-Barre, PA. On mine safety in the anthracite industry see Alexander Trachtenberg, *The History of Legislation for the Protection of Coal Miners in Pennsylvania, 1824–1915*, New York: International Publishers, 1942; Anthony F.C. Wallace, *St. Clair: A Nineteenth Century Coal Town's Experience with a Disaster-Prone Industry*, New York: Knopf, 1987; and Mark Aldrich, "The Perils of Mining Anthracite: Regulation, Technology and Safety, 1870–1945," *Pennsylvania History*, 64 (1997), 361–383.

31. Unsafe mining through excessive pillar robbing caused significant problems not only for subcontractors and leaseholders. In 1907, about six years before PaCC recommenced subcontracting (see Chapter 2), a coroner's jury found the company liable in the deaths of four mineworkers who were killed at the No. 14 Colliery while robbing pillars. Emphasizing that management had inadequately laid out and supervised the mining plan, the jury criticized the company: "So many pillars had been robbed in the mine that sufficient and safe support had not been left and the roof caved as a consequence." In short, the company violated state mining policy regarding excessive robbing ("Companies Should be Held Responsible for Deaths," *United Mine Workers Journal* [26 Sept 1907], 4). The article's author argued against robbing pillars in any situation unless a company used concrete or some other substance to support the interior structure of the mine: "The danger of mining at the best is bad enough, but the robbing of pillars make the danger all the greater. The robbing of pillars is going on at an extensive scale at present, and men must run the risk to their lives or else get a discharge. It is altogether wrong ... to rob the source of support, of pillars of coal, until concrete or other imperishable supports are substituted. Where deaths result from pillar robbing, the coal corporations should be mulcted to the maximum obtainable sum" (*ibid.*). After 1913, PaCC began issuing subcontracts for pillar removal, no doubt in part because of the problems brought by this incident.

32. The companies and their collieries were listed in W.C. Macquown's *Maps of the Anthracite Coal Fields of Northeastern Pennsylvania*, Pittsburgh, PA: National Coal Publications, 1942 & 1952.

33. Mining engineer [no name], Glen Alden Coal Company, to Edward Griffith, vice president and general manager, Glen Alden Coal Company, Memorandum, 20 Apr 1935, box labeled "Wilkes-Barre Railroad Traction Co., File No. 175–182," Glen Alden Coal Company Papers, Earth Conservancy, Inc., uncatalogued.

34. See the case, Plymouth Coal Co. v. Pennsylvania, 232 U.S. 531 (1914), No. 102, Argued January 15, 1914, Decided February 24, 1914, 232. In the 1891 Anthracite Mining Laws of Pennsyl-

vania, Article 3, Section 10 stated: "It shall be obligatory on the owners of adjoining coal properties to leave, or cause to be left, a pillar of coal in each seam or vein of coal worked by them, along the line of adjoining property, of such width, that taken in connection with the pillar to be left by the adjoining property owner, will be a sufficient barrier for the safety of the employes [*sic*] of either mine in case the other should be abandoned and allowed to fill with water; such width of pillar to be determined by the engineers of the adjoining property owners together with the inspector of the district in which the mine is situated, and the surveys of the face of the workings along such pillar shall be made in duplicate and must practically agree. A copy of such duplicate surveys, certified to, must be filed with the owners of the adjoining properties and with the inspector of the district in which the mine or property is situated" (Pennsylvania Department of Mines and Mineral Industries, *Annual Report: Anthracite Division*, Harrisburg, PA: Commonwealth of Pennsylvania, 1901, xv). Subsequent regulations stipulated 100 feet of solid coal as a barrier pillar's thickness. S.J. Phillips, who served as a state mine inspector and secretary of the state Water Hazards Commission, summarized the law regarding barrier pillars: "The Mining Laws are violated when an operator removes or weakens a barrier-pillar legally established for the purpose of preventing the flow of water from one mine to another. The law is also violated when an operator carries on mining operations dangerously close to a body of water, either in a mine or on the surface" (from a paper presented to the Anthracite Section of the American Institute of Mining and Metallurgical Engineers, 29 Jun 1935, Anthracite Heritage Museum, Scranton, uncatalogued, p. 7). See also Solomon H. Ash, "Barrier Pillars in Wyoming Basin, Northern Field," Bulletin 583, U.S. Bureau of Mines, Washington, DC: USGPO, 1954.

35. Agreement between PaCC and LVCC, 18 Feb 1925, LVCC Lease Books, Vol. 12, Lease No. 29, LVCC Papers, Pagnotti Enterprises Archives, Wilkes-Barre, PA. The companies agreed to the mutual robbing of the barrier pillar: "And Whereas the said Clark Vein barrier-pillar is no longer required to be left for the protection of the mines of the parties hereto," it can be mined "... and [if] the coal in said pillar is not easily accessible from workings of the LVCC, but is contiguous to the workings of the PaCC," then PaCC shall mine it (Edward Griffith, vice president & general manager, Glen Alden Coal Company to William Inglis, president of Glen Alden Coal Company, memorandum, 14 Apr 1936, box labeled "Wilkes-Barre Railroad Traction Co., File No. 175–182," Glen Alden Papers, Earth Conservancy, Inc., uncatalogued, p. 2).

36. The previous source also divulged a conversation between Griffith and TPC president H.J. Connolly regarding the barriers in an apparent mutual understanding for robbing: "Connelly ... informed me that the two boreholes between Heidelberg and No. 9 are 4 inches in diameter each and not plugged. He also stated that when the water reached elevation 535 feet, their base, it would flow around the anticlinal and into Central through Heidelberg where the barriers have been removed. This, of course, would make a direct connection between the present body of water and Central and Old Forge when it reaches 535 feet, Pittston Co. base" (Edward Griffith, vice president & general manager, Glen Alden Coal Company, to William Inglis, president, Glen Alden Coal Company, 14 Apr 1936, box labeled "Wilkes-Barre Railroad Traction Co., File No. 175–182," Glen Alden Coal Company Papers, Earth Conservancy, Inc., uncatalogued, 3).

37. "Governor Will Plea to Continue Mine Pumps," *Times Leader Evening News*, 27 Jul 1959.

38. In 1954, Solomon Ash conducted a study for the U.S. Bureau of Mines that examined each of the barrier pillars in Luzerne County. He found that many had been "robbed, weakened, or destroyed." After concluding that, "their stability was made uncertain," he added: "This is evident in many leased properties. ..." (Solomon H. Ash,"Barrier Pillars in Wyoming Basin, Northern Field," Bulletin 583, U.S. Bureau of Mines, Washington, DC: USGPO, 1954. See also S.J. Phillips, "Water Hazards in the Lackawanna Valley"). Numerous mineworkers provided oral history evidence of barrier-pillar robbing. Al Kanarr worked as a miner for PaCC, the Knox Coal Company, the Ferretti Coal Company, and other firms, and witnessed the robbing firsthand: "Sure they'd steal," he said. "They'd chip away and you'd get two companies chipping at the same barrier from both sides and soon you'd have no more barrier or a weak one" (Al Kanarr, audiotaped interview, 27 Oct 1988, NPOLHP). William A. Hastie described the robbing at his employer, the Knox Coal Company: "For a time during the ensuing period, several crews worked the downriver perimeter of the mine, robbing the barrier pillar of the Dial Rock mine. The Dial Rock workings lie under the Borough of Wyoming and extended out under the river. ... Now coal companies were required to leave barrier pillars between mines. Each company must leave a hundred feet of coal in each vein, thus providing a barrier of two hundred feet of solid coal. Some of the larger companies honored the barrier pillars, but many others, particularly those like the Knox, companies that were mining other people's coal, raked the hell out of the barriers" (William A. Hastie, audiotaped interview, 31 Jul 1989, NPOLHP). Anthony

Waitkevich discussed the specific procedure of barrier-pillar robbing while he was an employee of the Knox Coal Company (Anthony Waitkevich, audiotaped interview, 22 Jan 1988, NPOLHP). For other commentaries on the subject see George Gushanas, audiotaped interview, 13 Jan 1994, NPOLHP; Thomas Supey, Jr., audiotaped interview, 7 Jul 1999, NPOLHP; and Joseph Stonionis, audiotaped interview, 12 Sept 2001, NPOLHP.

39. William Loftus, "Total Shutdown Looks Possible; State, U.S. Chiefs Act," *Scranton Times*, 23 Jan 1959.

40. Ash, "Barrier Pillars in Wyoming Basin, Northern Field, 249.

41. The well-known Quecreek mining disaster of 2002, in western Pennsylvania's Lincoln Township, Somerset County, occurred when workers at the Black Wolf Coal Company broke into the adjoining Harrison No. 2 mine, formerly owned by the Saxman Coal and Coke Company. An evidently weakened barrier pillar held back some 50 million gallons of water between the mines. The Quecreek men were apparently not mining illegally because investigations revealed that an inaccurate mine map caused the disaster. Of the 18 workers in the Quecreek pit, nine escaped quickly, leaving nine others trapped for 72 hours. The daring and successful rescue effort captured the nation's and the world's attention. On the Quecreek disaster see Jeff Goodell, *Our Story: 77 Hours That Tested Our Friendship and Our Faith*, New York: Hyperion, 2002.

42. The billions of gallons of water that coursed underground from the Knox disaster did not affect the Scranton area's mines because the coal to the northeast, in Lackawanna County, is situated in a separate basin or "bowl" of anthracite.

43. The indicted individuals were Dominick Alaimo, Anthony Argo, and Charles Piasecki (L.U. 8005 officials at the Knox Coal Company; accepting bribes); Albert Biscontini (co-owner, Newport Excavating Co.; personal and corporate income tax evasion); Frank Cardoni (assistant to August J. Lippi and District 1 board member; personal and corporate income tax evasion); William Dombrowski (secretary, L.U. 7519; accepting bribes); Robert Dougherty (co-owner, Knox Coal Company; conspiracy to violate mining laws; involuntary manslaughter; personal and corporate income tax evasion); Louis Fabrizio (co-owner, Knox Coal Company; conspiracy to violate mining laws, involuntary manslaughter, personal income tax evasion); former Governor John Fine (co-owner, Newport Excavating Co.; personal and corporate income tax evasion); Ralph Fries (Engineering Department, PaCC; conspiracy to violate mining laws, involuntary manslaughter); Robert Groves (superintendent, Knox Coal Company; involuntary manslaughter); Philip Gelso (co-owner, No. 14 Coal Company; bribing union officials); Sam Gelso (co-owner, No. 14 Coal Company; bribing union officials); Thomas Larkin (Umpire, Anthracite Conciliation Board; personal income tax evasion); August Lippi (President, UMWA District 1 & co-owner, Knox Coal Company; conspiracy to violate mining laws, involuntary manslaughter, personal income tax evasion, bank fraud); Donald Morgan (brother-in-law and tax advisor to Gov. John S. Fine at the Newport Excavating Co.; personal and corporate income tax evasion); William Receski (assistant foreman, Knox Coal Company; involuntary manslaughter); Fritz Renner (Engineering Department, PaCC; conspiracy to violate mining laws; involuntary manslaughter); John Salvo (committeeman, L.U. 7519; accepting bribes); Josephine Sciandra (co-owner, Knox Coal Company; personal and corporate income tax evasion); John Shipula (president L.U. 7519; accepting bribes); Leonard Statkewicz (board member, UMWA District 1; personal and corporate income tax evasion). The indicted companies were the Avon Coal Company, the Newport Excavating Company, the Knox Coal Company, and the Peeley Coal Company. Although several of the accused were convicted, upon appeal many cases were overturned and only the following persons were ultimately convicted: Alaimo, Argo, Dombrowski, Dougherty, Fabrizio, Philip Gelso, Sam Gelso, Lippi, Piasecki, Salvo, Sciandra, and Shipula. They received sentences ranging from suspended to five years in prison. Three companies were convicted: Knox, Avon, and Peeley.

44. Allegations of illicit labor-management dealings in the northern field surfaced more than two years before the Knox mine disaster. In 1957, U.S. Senator Barry Goldwater furnished the names of five District 1 officials and one coal company executive to Robert F. Kennedy, chief counsel to the Select Committee on Improper Activities in the Labor or Management Field, chaired by Senator John McClellan. "I am recommending to the chairman of the committee that an investigation be conducted in Luzerne County," said Goldwater, who served on the McClellan committee. The senator charged company and union officials with collusion involving payoffs in exchange for labor peace. He further asserted that certain unnamed UMWA officials secured tenancy agreements or held other interests in a coal company. Rumors circulated that the allegations referred to the Glen Alden Coal Company, the largest producer in the northern field. Goldwater based the charges on over two dozen letters and complaints he had

received from mineworkers and interested citizens. Lippi joined with Thomas Kennedy and Glen Alden Coal Company president Francis O. Case in denying the charges. Said Case: "I don't go in for collusion ... I don't condone it. If we knew of people involved with kickbacks, they would be fired." The *Scranton Times* reached Lippi in New York City where he was attending a union conference. "I'd say no," he replied when asked about the connivance. "I consider it sufficient to state that there is adequate machinery in our [labor-management] contract to assure individual members the right to raise any grievance they have, either real or imaginary. I have always made it a policy to process all grievances brought to my attention before the Anthracite Board of Conciliation, a labor-management body whose record over more than a half-century certainly is above reproach. As a matter of fact, the 1956 Anthracite Wage Agreement negotiated by the anthracite operators and the UMWA provides specifically for the protection of wage rates not only with the major producers but with each and every small operator who might be mining under lease from the 'Line' companies" ("Names Not Disclosed by Senator Goldwater," *Scranton Times*, 4 May 1957). Federal agents entered the northern field and spent several months investigating the allegations. They uncovered little wrongdoing and produced no indictments because they could not garner sufficient evidence to support the claims in the letters ("Collusion is Alleged by Republican Senator, Luzerne Will be Center," *Scranton Times*, 3 May 1957).

45. Bootlegging—the illegal mining, processing, and selling of coal—represented another illicit aspect of the business during the 1930s. Thousands of unemployed workers without mineral rights' access took coal in small mines called *dogholes*, typically located on property owned by the large companies. In 1936–1937, the bootleg trade amounted to 2.5 million tons or five percent of total output. By 1939, an estimated 2,500 illegitimate mines produced approximately 3.5 million tons. The participants sold the run-of-mine coal in a thriving underground market of processors, shippers, sellers, and buyers. At its peak, the sector employed between 13,000 and 20,000 men and boys. Police were reluctant to make arrests and the judges were hesitant to prosecute violators because of the anthracite regions' economic plight. The black market industry thrived mainly in the southern field where veins outcropped close to the surface, but it also found many practitioners the middle and northern fields. On bootleg mining see Pennsylvania Anthracite Coal Industry Commission, "Bootlegging or Illegal Mining of Anthracite Coal In Pennsylvania, A Census and Survey of the Facts," 39–146 in *Report of the Pennsylvania Anthracite Coal Industry Commission*, Harrisburg, PA: Commonwealth of Pennsylvania, 1938, 43; and Barger and Schurr, *The Mining Industries*, 181. On bootlegging in the northern-field borough of Plymouth see John Usefara, audiotaped interview, 28 Jul 2001, and Chester Brozena, audiotaped interview, 3 Dec 1988; in the borough of Swoyersville see John Wolinsky, unrecorded interview, NPOLHP, 29 Jul 2001, and John Piazza, unrecorded interview 19 May 1994; in Moosic see Charles Orloski, audiotaped interview, 9 Jun 1998, NPOLHP; and in the borough of Jessup see Secundo Sabastianelli, audiotaped interview, 29 Jan 1999. For an overview of the bootleg system in the 1930s see "Coal Bootlegging, An Octopus," *Literary Digest*; Louis Adamic, "The Great Bootleg Coal Industry," *The Nation* (9 Jan 1935), 46–49; Pennsylvania Anthracite Coal Industry Commission, *Report*, 5; Goodrich et al., *Migration and Economic Opportunity*, 429; and William Gustafson, "Bootleg Mining In Pennsylvania," 1–24 in Robert Mittrick (ed.), *Proceedings of the Ninth Annual Conference on the History of Northeastern Pennsylvania*, Nanticoke, PA: Luzerne County Community College, 1997.

46. Mark Aldrich, "The Perils of Mining Anthracite: Regulation, Technology and Safety, 1870–1945," *Pennsylvania History*, 64 (1997), 361–383. Strip mining represented another of the operators' responses to economic decline, especially in the southern field. Douglas K. Monroe (*A Decade of Turmoil: John L. Lewis and the Anthracite Miners, 1926–1936*, unpublished doctoral dissertation, Georgetown University, 1976, 273), noted that strip mining was as popular in the southern field during the 1930s as subcontract mining was in northern field. See also Dublin and Licht, *The Face of Decline*, 23, 52, 61.

47. Pacifico "Joe" Stella, audiotaped interview, 1 Nov 1988, NPOLHP.

48. Alex Chamberlain, audiotaped interview, 11 Jul 1995, NPOLHP.

49. Charles Volpe, audiotaped interview, 22 Jan 1998, NPOLHP.

50. Pacifico "Joe" Stella, audiotaped interview, 1 Nov 1988, NPOLHP.

51. *Ibid.*

52. Joe Costa, unrecorded interview, 20 Jun 1992, NPOLHP.

53. *Ibid.*, unrecorded interview, 1 Aug 1992, NPOLHP.

54. *Ibid.*, unrecorded interview, 22 Jul 1995, NPOLHP.

55. *Ibid.*, unrecorded interview, 1 Aug 1992, NPOLHP.

56. George DeGerolamo, *Autobiography*, Pittston, PA: Published by the author, 1980.

57. William Graham, Penn Anthracite Colliery Company's last president, discussed the company's history, including its tenancy structure, in an audiotaped interview, 11 Jul 1995, NPOLHP. The company's other collieries in the 1930s and 1940s included the Ontario in Peckville, Riverside in Archbald, Raymond in Archbald, Johnson in Dickson City, Harry Taylor in Scranton, and Rushbrook in Archbald—all in Lackawanna County.

58. Oliver L. Davis, chief inspector, to John H. Harvey, general superintendent, Penn Anthracite Collieries Company, 10 Feb 1936, General File on Leases, Penn Anthracite Collieries Company Papers, uncatalogued.

59. Memorandum, Herbert Thomas Contract, 1 Sept 1936, No author, General File on Leasing, Penn Anthracite Collieries Company Papers, uncatalogued.

60. T.S. Shoemaker, chief mining engineer, to Charles Dorrance, president, 9 Jul 1936, General Leasing File, Penn Anthracite Collieries Company Papers, uncatalogued. President Dorrance, who also served on the Board of Directors of the Wyoming National Bank of Wilkes-Barre and on the boards of various other companies, came from one the Wyoming Valley's most prestigious families within the so-called Anthracite Aristocracy (Edward J. Davies, *The Anthracite Aristocracy: Leadership and Social Change in the Hard Coal Regions of Northeastern Pennsylvania, 1800–1930*, DeKalb, IL: Northern Illinois University Press, 1985). For further information on Penn Anthracite Collieries Company see *Report of the Anthracite Coal Industry Commission*, 1938, 413–414.

61. "Memorandum," Herbert Thomas Contract, 1 Sept 1936, No author, General File on Leasing, Penn Anthracite Collieries Company Papers, uncatalogued.

62. Joe Costa, unrecorded interviews, 24 Mar 1995 and 22 Jul 1995. As indicated in Chapter 5, Costa also served as one of two financial secretaries for L.U.1703 at the No. 6 Colliery in January 1928, soon after the insurgents took control. Because of his ties to organized crime during this period, exemplified by his job "stealing coal" for a subcontractor as discussed in Chapter 6, he may have been acting as an infiltrator or "double agent" on behalf of his criminally related bosses.

63. *Proceedings of the UMWA Tri-District Convention*, 1939, 51. While inadequate record keeping undoubtedly occurred, so did calculated theft, as evidenced by Costa's undertakings.

64. Joe Costa, unrecorded interview, 24 Mar 1995. It was a common practice for ordinary citizens—often youth charged with the task by parents—to steal coal from railroad cars after they had been loaded for the trip to market, and from culm, or coal waste, banks, which contained some anthracite. By the 1930s, the thievery had become fairly well organized ("Coal Stealing Highly Organized, Say Police," *Wilkes-Barre Record*, 12 July 1933).

65. George DeGerolamo, *Autobiography*, 1980, 40–41.

66. The Pennsylvania Crime Commission did not mention Lippi as a member or associate of the region's crime family, but he clearly had business dealings with alleged gang members John Sciandra, Stefano LaTorre, and Nick Alaimo. (On Alaimo see his unrecorded interview, 8 Aug 1989, NPOLHP). Lippi's relationship with alleged crime boss John Sciandra dated back to their younger days as mine laborers for PaCC, said Edwin M. Kosik, Lippi's attorney in one of his income tax evasion cases ("Union Leader Given 3 Years, Fined $5,000," *Times Leader Evening News*, 20 Jan 1964). Lippi's friendships with Fabrizio, LaTorre, and Volpe also dated back decades. Further indication of Lippi's involvement with alleged crime figures can be inferred from PaCC's simultaneous sale in October 1940 of three adjoining properties to Santo Volpe, John Sciandra, and August Lippi; the board of directors unanimously passed one motion for the sale of the three lots on October 23, 1940 (Stockholders and Executive Committee Minutes, Minutes, 1838–1971, Minutes from 19 Jan 1940–16 Oct 1945; Doc. No. 1021, PaCC Papers, PSA). On corruption within the International UMWA around this time see Joseph E. Finley, *The Corrupt Kingdom: The Rise and Fall of the United Mine Workers*, New York: Simon and Schuster, 1972.

67. Dublin and Licht, *The Face of Decline*, Chapters 6 & 7.

68. See, e.g., Joseph Kopcza, audiotaped interview, 21 Dec 1988, NPOLHP; Tony Waitkevich, audiotaped interview, 8 Aug 1989, NPOLHP; Lewis Casterline, audiotaped interview, 25 May

1990, NPOLHP; Charles Volpe, audiotaped interview, 22 Jan 1998, NPOLHP; and John Usefara, audiotaped Interview, 28 Jul 2001, NPOLHP.

69. The workers' grievances can be found in the many *Reports of the Anthracite Board of Conciliation*, included in the papers of the ABC located at Indiana University of Pennsylvania. Grievances by PaCC employees were being submitted as late as the 1950s (see, e.g., Vol. XXXIII, Grievances from 3 Feb 1955 to 21 Jun 1957).

70. PaCC used tenancy to curb strikes as part of an effort to enhance workplace control. The company's earliest leases did not hold tenants liable for royalty payments or coal deliveries during work stoppages. But beginning in 1940, a new clause appeared in certain leases: "This agreement of lease shall not be construed to permit or allow the Lessee to terminate his mining operations or to give up possession of the demised lands on account of strikes, general suspensions, lockouts or any other suspension or suspensions of operations or prolonged idle time at any colliery of the Lessor where the coal to be mined hereunder is for the time being to be delivered. *On the contrary, the Lessee will resume mining operations under this agreement of lease and resume delivery of coal to the Lessor after any suspension or suspensions or idle times upon written notice from the Mining Engineer of the Lessor directing work to resume.*" (emphasis added) The consequences were harsh if the lessee failed to resume mining as directed, for such inaction "*shall be warrant and sufficient cause of the cancellation of this agreement of lease and the termination of all the Lessee's rights and privileges herein granted without liability to the Lessor.*" (emphasis added) (The wording is from the Edward J. Quinn lease, 13 Nov 1940, Contract File Index, Contracts 1867–1969, Contract No. 326, Box 9, PaCC Papers, PSA). Therefore, PaCC expected lessees to settle or, if necessary, break strikes in order to keep the coal flowing. After a thorough search of hundreds of PaCC's leasing documents, the Quinn lease appeared to be the earliest with the anti-strike clause. Leases to the Knox Coal Company contained a similar clause.

71. Frederick W. Taylor, *The Principles of Scientific Management*, New York: Harper & Row, 1911, 48–49. See also Frederick W. Taylor, *Scientific Management: Comprising Shop Management, the Principles of Scientific Management, Testimony Before the Special House Committee*, New York: Harper & Row, 1947; David Montgomery, *Workers' Control in America: Studies in the History of Work, Technology, and Labor Struggles*, New York: Cambridge University Press, 1979; and Daniel Nelson, *Frederick Winslow Taylor and the Rise of Scientific Management*, Madison, WI: University of Wisconsin Press, 1980.

72. Taylor, *Principles of Scientific Management*, 83, original emphasis.

73. H. Martyn Chance, *Second Geological Survey of Pennsylvania, 1883: Report on the Mining Methods and Appliances used in the Anthracite Coal Fields*, Harrisburg, PA: Board of Commissioners for the Second Geological Survey, Washington, DC: USGPO, 1883, 346.

74. Carter Goodrich, *The Miner's Freedom: A Study of the Working Life in a Changing Industry*, New York: Arno, 1977 [1925], 59. As a production manager from the Pittsburgh Coal Company put it: "Therefore, the fundamental characteristic of scientific management is control—based upon facts—by the use of scientifically determined standards" (Jerome C. White, "Scientific Management and the Coal Industry," *Coal Age*, 41 [Oct 1936], 424–425).

75. *Ibid.*, 177.

76. Tilly and Tilly, *Work Under Capitalism*, 51.

77. As Tilly and Tilly (*ibid.*, 48) suggested, the lag in technological development in American coal mining was due not only to investment policies and geology but also labor organization and supply. Regarding the latter, David Montgomery (*The Fall of the House of Labor: The Workplace, the State, and American Labor Activism, 1865–1925*, New York: Cambridge University Press, 1987, 334) pointed out that, as late as 1930, bituminous mining was still very labor intensive, with contract miners and their laborers constituting 75 percent of the workforce. This figure was probably comparable in anthracite, although the number of workers employed by subcontractors was surely higher in hard coal than in soft coal.

78. The authors found no data that would have allowed an assessment of the tenant companies' precise impact on overall productivity. However, oral histories as well as the information presented in these three tables support the proposition that tenancy elevated output to at least some degree. See also notes 80 and 81 below.

79. The argument was often stated or implied by coal operators, lease holders, and managers as exemplified by Alex Chamberlain, audiotaped interview, 11 Jul 1995, NPOLHP; David Panzitta, audiotaped interview, 14 Mar 1996, NPOLHP; James Tedesco, audiotaped interview, 8 Oct

1996, NPOLHP; David Randall, audiotaped interview, 7 Jun 1998, NPOLHP; and Joseph Panzitta, audiotaped interview, 25 Apr 2007, NPOLHP.

80. Morgan Bird and his brother, Samuel, operated the Bird Mining Company in the 1940s and 1950s. The company secured leases from the Glen Alden Coal Company, the Kingston Coal Company, and others to take coal from deep mines and strip mines in Plymouth and Wyoming boroughs and Hanover Township. They ran mines that were union in name only. John Oshirak and John Usefara worked for the company and described it as an essentially non-union firm that eliminated the stint by requiring each worker to produce seven cars per shift (John Oshirak and John Usefara, audiotaped interview, 2 Aug 2001, NPOLHP).

81. For example, as discussed in Chapter 6, when one of PaCC's largest tenants, Louis Pagnotti, took a lease he typically circumvented the stint by increasing the size of the car and insisting that workers put in a full 7-hour shift. Pagnotti was also renowned for lowering wage rates and having his men take coal in off-limit places (George DeGerolamo, *Autobiography*, 34, 40–1; William A. Hastie, audiotaped interview, 31 Jul 1989, NPOLHP; Joe Costa, unrecorded interview, 1 Aug 1992, NPOLHP; Hubert Amos, audiotaped interview, 12 Jan 1995, NPOLHP). Regarding bonuses, the Knox Coal Company was one of a number of lessees that instituted a bonus plan that paid workers an extra amount for each car produced over the stint. According to the company's part owner and first president, Robert L. Dougherty, in the late 1940s, Knox instituted a bonus-pay system whereby workers received an escalating amount for every car produced over the stint of three cars per man. As Dougherty put it: "The tenth car on that job paid 25 cents extra, the eleventh car on that job paid 50 cents extra, up to a point where if they load enough coal maybe the last car of the day that they loaded over the basic car rate would be maybe $3.00 a car extra on the pyramiding system. ... The men earned between $30 and $50 a day—not a week, a day" (Robert L. Dougherty, testimony before the Joint Legislative Committee to Investigate the Knox Mine Disaster, 23 Apr 1959, 1423–25). He said it was not uncommon for a Knox work team of three men to produce 25 to 35 cars per shift. However, Knox employee William A. Hastie said the bonus was much lower than Dougherty stated: 35 cents for one extra car and 10 cents for every car thereafter. Moreover, said Hastie, while Knox disaster survivor Nick Lucas headed the crew that produced the River Slope mine's greatest yield for one shift, 45 cars, rarely did a crew produce over 20. Hastie added that the incentive system encouraged output at the expense of safety.

82. On the Blue Coal Corporation see also: In re Blue Coal Corporation, Debtor [and] In re Glen Nan, Inc., Debtor, 47 B.R. 754 (Bankruptcy, M.D. Pa., 1985); U.S. v. Tabor Court Realty Corp., et al., 803 F. 2d 1288 (3d. Cir., 1986); and Janosov, "Glen Alden's Huber Breaker.

83. On Durkin's organized-crime affiliations see Pennsylvania Crime Commission, *A Decade of Organized Crime*, Harrisburg. PA: Commonwealth of Pennsylvania, 1980, 219, 245–249.

84. David Brody, "Market Unionism in America: The Case of Coal," *In Labor's Cause*, New York: Oxford University Press, 1993, 133.

85. To be sure, anthracite was not the only American mining industry to employ subcontractors. By 1902, they were actually most prevalent in gold and silver mining while coal had the second largest number (William M. Steuart, *Special Report: Mines and Quarries*, Chapter 10).

86. *Proceedings of UMWA District No. 1 Convention, 1957*, 19.

87. John Mikulski, audiotaped interview, 28 Jul 2001 NPOLHP.

88. Regarding local government leases, in 1964 the Luzerne County Tax Claim Bureau secured the mineral rights to a mine as part of a tax case. The county then leased the pit to a small operator, the Glyon Coal Company, for $.40 per ton royalty. The lessee further agreed to pay property taxes on the land during the course of the mining ("40c-A-Ton Royalty Offered by Company; County OKs Mining Proposal," *Times Leader Evening News*, 18 Oct 1964).

89. The UMWA's participation in leasing came after the Lehigh Coal & Navigation Company (LC&N)—historically the Lehigh region's largest producer—shut down operations in the early 1950s. An investment consortium headed by James H. Pierce of Scranton, described by former LC&N president W. Julian Parton as "a long-time confidant of John L. Lewis," established the Coaldale Mining Company and secured a lease for the LC&N's Coaldale Colliery in November 1954 (W. Julian Parton, *The Death of a Great Company: Reflections on the Decline and Fall of the Lehigh Coal and Navigation Company*, Easton, PA: Canal History and Technology Press, 1986, 116–130). Another investor in the Coaldale concern was former alleged crime boss Santo Volpe. In his history of LC&N, Parton wrote that the leases to the Coaldale were made possible by the financial backing of the UMWA's International office in Washington: "Pierce had a

longstanding relationship with John L. Lewis. It was widely reported that he had financial backing of the UMWA through the Riggs National Bank in Washington, D.C., of $2,5000,000 and eventually up to $8,000,000" (*ibid.*, 116). In May 1955, the Coaldale Coal Company bought the Panther Valley Coal Company, another LC&N firm, which had been experiencing labor unrest. In 1957, the Panther Valley company terminated the LC&N lease when "apparently its UMWA 'shadow' supporters decided that the losses were too great for this company and that it could not continue its financial support." The Coaldale Mining Company "was to continue operations and was expected to produce all the anthracite which could be sold from the company properties in 1958" (*ibid.*, 128). The authors are indebted to Lantz Metz and Thomas Dublin for bringing these leases to our attention.

90. The Biscontini Coal Company went into business in 1946 with a strip mine in Glen Lyon, Luzerne County, under a lease from Glen Alden. The firm also operated a deep mine at Wanamie, Newport Township, Luzerne County, also under a lease from Glen Alden, and remained in business until 1956. In another operation, father Lorenzo and son Albert Biscontini founded the Newport Excavating Company in 1950 and began strip mining in Newport Township, Luzerne County, again with a Glen Alden lease. John S. Fine and Santo Volpe were partners in Newport Excavating Company, which had a lease on the Wanamie mine in Newport Township from Glen Alden. The firm remained in business until 1956. Information on Biscontini's and Newport's holdings can found at www.northernfieldinfo.com.

91. Attorney Harold Rosenn was another highly respected community member who secured a lease. A partner in Wilkes-Barre's premiere law firm of Rosenn, Jenkins and Greenwald, he took a PaCC lease on September 24, 1963: "PaCC leases to H. Rosenn 17.45 acres of Top Checker Vein coal, part of the Zenas Barnum (lots 41 and 42) Tract, in Duryea Boro. Luz. Co." The lease was amended on June 22, 1964, reducing the royalty from 75 cents per gross ton, run-of-mine coal, to 60 cents, effective May 1, 1964. Rosenn released the property on December 14, 1964 (Agreement, No. 14416, PaCC Lease Books, PaCC Papers, PSA).

92. Albert O. Hirschman, *Exit, Voice and Loyalty: Responses to Decline in Firms, Organizations and States*, Cambridge, MA: Harvard University Press, 1970.

93. In 1938 and 1939, at least two UMWA local unions joined the Progressive Miners Union (PMU) of the American Federation of Labor (AFL). In an intra-union dispute, Lewis seceded from the rival AFL and joined with Philip Murray and others to create the Congress of Industrial Organizations (CIO) in 1935. In June 1939, over 450 out of 500 men at the Alden Colliery in Newport Township signed cards to join the PMU. The men also tried to gain control of a bank account containing $4,000 that had been deposited by the now defunct UAMP. The UMWA used the courts to block access to the money and joined with the companies to thwart the latest dual-union effort. The PMU also succeeded in organizing the Cameron Colliery of the Stevens Coal Company in the Schuylkill region. See "AFL Miners Restrained; Judge Grants Injunction to UMWA to Prevent Seizure of $4,000," *Wilkes-Barre Record*, 12 Apr 1939; "Alden Workers Ballot To Go Back to U.M.W.," *Wilkes-Barre Record*, 26 May 1939; "Declares Alden Is Dismantled; Hermansen Asks for Labor Board Election Among Mine Workers," *Wilkes-Barre Record*, 5 Jun 1939. On the formation of the CIO see Robert Zieger, *The CIO: 1935–1955*, Chapel Hill, NC: University of North Carolina Press, 1995.

Notes to Chapter Eight

1. Victor R. Greene, *The Slavic Community on Strike: Immigrant Labor in Pennsylvania Anthracite*, South Bend, IN: University of Notre Dame Press, 1968. See also Victor R. Greene, "A Study in Slavs, Strikes, and Unions: The Anthracite Strike of 1897," *Pennsylvania History*, 31 (1964), 199–215.

2. Greene, *The Slavic Community on Strike*, 210.

3. According to Commons: "The [union] defeat at this time is ascribed with unanimity to the presence of the cheap labor of southern [and eastern] Europe, which could not be controlled or organized according to the methods then used. The operators were able to play ... one nationality against another nationality" (Quoted in Greene, *The Slavic Community on Strike*, 80).

4. *Ibid.*, 87. Fr. John Wolansky was a distant relative of the first author.

5. Harold W. Aurand, *From the Molly Maguires to the United Mineworkers: The Social Ecology of an Industrial Union, 1869–1897*, Philadelphia, PA: Temple University Press, 1971, 121.

6. Greene, *The Slavic Community on Strike*, 214.

7. Aurand, *From the Molly Maguires ...*, 122. Peter Roberts (*The Anthracite Coal Industry: A Study of the Economic Conditions and Relations of the Co-operative Forces in the Development of the Anthracite Coal Industry of Pennsylvania*, New York: Macmillan, 1901, 171) observed that the Slavs' UMWA involvement contravened and undermined the coal operators' intentions when they recruited the eastern Europeans as mineworkers after 1875: "It was the anthracite coal operators who first brought Sclavs [*sic*] to the coal fields, to break the power of Anglo-Saxon labor, but these foreigners have proved capable of forming labor organizations that are more compact and united than any which ever existed among various English-speaking nationalities, who first constituted these communities."

8. Edna Bonacich, "A Theory of Ethnic Antagonism: The Split Labor Market," *American Sociological Review*, 37 (Oct 1972), 547–559. See also William Julius Wilson, *The Declining Significance of Race*, Chicago, IL: University of Chicago Press, 1980.

9. Peter Roberts, *Anthracite Coal Communities: A Study of the Demography, the Social, Educational and Moral Life of the Anthracite Regions*, New York: Macmillan, 1904, 13.

10. Robert Cornell, *The Anthracite Coal Strike of 1902*, Washington, DC: Catholic University Press, 1957.

11. Greene, *The Slavic Community on Strike*, 173. See also John Bodnar, "Socialization and Adaptation: Immigrant Families in Scranton, 1880–1900," *Pennsylvania History*, 43 (1976), 154–172; and John Bodnar, "Immigration and Modernization: The Case of Slavic Peasants in Industrial America," *Journal of Social History*, 14 (1976), 45–65.

12. However, at least some Italians from an unnamed local union were critical of Mitchell as evidenced by a resolution offered at the UMWA convention in Indianapolis in January 1905: "The resolution came from a local [union] made up principally of Italians and declared that Mitchell had acted the coward by staying away from the places of danger when he went west and that he had deserted the miners who were on strike at the most critical period in the struggle for independence" ("Declared Mitchell Acted the Coward," *Scranton Republican*, 24 Jan 1905). The resolution was voted down.

13. Michael A. Barendse, *Social Expectations and Perception: The Case of the Slavic Anthracite Workers*, University Park, PA: Penn State University Press, 1981, 99. See also Michael A. Barendse, "American Perceptions Concerning Slavic Immigrants in the Pennsylvania Anthracite Fields, 1880–1910: Some Comments on the Sociology of Knowledge," *Ethnicity*, 8 (1981), 96–105. For examples of scholarship embodying prejudices against immigrants see George O. Virtue, "The Anthracite Mine Laborers," 728–774 in Carroll D. Wright and Oren W. Weaver (eds.), *Bulletin of the Department of Labor*, Washington, DC: USGPO, Nov 1897; Frank J. Warne, "The Anthracite Coal Strike," *Annals of the American Academy of Political and Social Science*, 17 (1901), 15–52; Frank J. Warne, *The Slav Invasion and the Mineworkers*, Philadelphia, PA: Lippincott, 1904; Roberts, *The Anthracite Coal Industry*; and Roberts, *Anthracite Coal Communities*. For a literary example see Henry Edward Rood, "A Pennsylvania Colliery Village: A Polyglot Community," *Century Magazine*, 55 (Apr 1898), 809–821.

14. For example, Roberts' jeremiad against the eastern and southern European immigrants included tracts regarding their harmful effects on the intellectual and moral state of the anthracite regions: "The character of the population of this area has perceptibly deteriorated in the last thirty years. A selection has been affected in a retrogressive sense. The physical strength of the accretions of the last quarter of a century may favorably compare with that of any previous period, but their intellectual and moral qualities are decidedly lower." As an indicator of moral depravity, Roberts chastised the newcomers (which he grouped together as "Sclavs") for drinking, partying, and otherwise disregarding the true meaning of Sunday: "The Sclavs attend church but they do not observe the Sabbath. They buy, drink, dance, sing etc. on Sunday without scruple." He quoted a fellow man of the cloth who visited the Slavic neighborhoods of Mahanoy City on the Lord's Day: "It was terrible; saloons full blast; singing and dancing and drinking everywhere; it was Sodom and Gomorrah revived; the judgment of God, sir, will fall upon us" (Roberts, *Anthracite Coal Communities*, 53; see also his Chapter 8, "Men at the Bar"). Although Roberts had hope for the "improvement" of the younger generation, his overall assessment was derisive: "Sclavs are ignorant, clannish, unclean, suspicious of strangers, revengeful and brutal. ... They are dirty, under the influence of drink [, and] they soon turn to the unstable nature of their barbarous ancestors" (*ibid.*, 40).

15. On the immigrants' contribution Roberts wrote: "It is conceded by men intimate with the situation throughout the coal fields that its [UMWA's] universality was more due to the Sclav than to any other nationality" (*ibid.*, 172).

16. *Ibid.*, 191. Carter Goodrich (*The Miner's Freedom: A Study of the Working Life in a Changing Industry*, New York: Arno, 1977 [1925], 109) agreed that the Slavs and other immigrant mineworkers were essential to the unionization drive. He cited a Dillingham Commission report, which found that more than 80 percent of the British-born immigrants to the American coal fields had mining experience, while only about 10 percent of the southern and eastern Europeans had previously mined. Yet, "the union that was so largely founded by British miners has passed on its traditions and policies [to the immigrants] with very little change in spite of its changing membership."

17. Barendse, *Social Expectations and Perception*, 102.

18. Dillingham Commission, *Reports of the Immigration Commission: Dictionary of Races and Peoples*, Washington, DC: USGPO, 1911, 84. By 1920, about 78,000 or 53 percent of the anthracite mineworkers in all three fields were foreign born, with more than half from Poland and other Slavic countries, and 9,600 or 6.5 percent, from Italy. One-third of all non-English-speaking immigrant mineworkers were illiterate (Marie L. Obenauer, "Living Conditions in the Anthracite Region and Composition of the Mining Population," 527–571 in Annex to *The General Report of the U.S. Coal Commission*, Washington, DC: USGPO, 1925, 535; Table 4). Harold W. Aurand (*Coalcracker Culture: Work and Values in Pennsylvania Anthracite, 1835–1935*, Selinsgrove, PA: Susquehanna University Press, 2003, 23) stated that, in 1912, over 11,500 of the 18,500 workers in the Delaware Lackawanna & Western's coal department were either Slavs or Italians. Dublin and Licht (*The Face of Decline*, 48) reported that, in 1920, only 4 percent of the Italian workers at the Lehigh Coal & Navigation Company (in the Lehigh Region) were employed as contract miners, while for Slavic workers the figure was 25 percent. Overall, the Italians held lower-skilled occupations and had lower social mobility than did immigrants from eastern and central Europe.

19. The percentage breakdown of foreign-born workers in anthracite in 1917 was Polish 13 percent, Austrian 8 percent, Italian 8 percent, Lithuanian 7 percent, Russian 7 percent, Slovenian 6 percent, Irish 3 percent, Welsh 2 percent, and English 2 percent. Most of the Austrians were actually Ukrainians and Slovaks, but for geo-political reasons related to conquest and boundaries, they resided in Austria before emigrating to the U.S. See Pennsylvania Department of Mines and Mineral Industries. *Annual Report: Anthracite Division*, Harrisburg, PA: Commonwealth of Pennsylvania, 1919.

20. Italian immigration to Philadelphia in southeastern Pennsylvania began much earlier than the mass migration to the U.S. and to the anthracite regions in the late nineteenth century. See Richard N. Juliani, *Building Little Italy: Philadelphia's Italians before Mass Migration*, University Park, PA: Penn State Press, 1998.

21. Richard D. Grifo and Anthony F. Noto, *Italian Presence in Pennsylvania*, University Park, PA: Pennsylvania Historical Associations, 1990, 8.

22. Anthony F.C. Wallace, *St. Clair: A Nineteenth-Century Coal Town's Experience with a Disaster-Prone Industry*, New York: Knopf, 1987, 338.

23. Priscilla Long, *Where the Sun Never Shines: A History of America's Bloody Coal Industry*, New York: Paragon, 1989, 114. See Greene (*The Slavic Community on Strike*, 61, 75) on the Slavic groups who were brought to anthracite as strikebreakers.

24. Grifo and Noto, *Italian Presence in Pennsylvania*, 8.

25. "Hazleton Italians Will Hold Out," *Wilkes-Barre Record*, 14 Dec 1887; "Italians to Aid the Strikers," *ibid.*, 15 Dec 1887.

26. "Hazleton Italians Will Hold Out," *ibid.*, 14 Dec 1887.

28. "Italians Out on Strike: Exciting Scenes in Archbald on Saturday," *ibid.*, 7 Dec 1896; see also Perry K. Blatz, *Democratic Miners: Work and Labor Relations in the Anthracite Coal Industry, 1875–1925*, Albany, NY: SUNY Press, 1994, 64.

28. "Italians Out on Strike," *Wilkes-Barre Record*, 7 Dec 1896.

29. "Affairs at Archbald Mine," *ibid.*, 8 Dec 1896.

30. "Italians Out on Strike," *ibid.*, 7 Dec 1896.

31. *Ibid.*

32. *Ibid.*

33. *Ibid.*

34. "Affairs at Archbald Mine," *ibid.*, 8 Dec 1896.

35. *Ibid.*

36. *Ibid.*

37. "All is Quiet in Archbald," *ibid.*, 9 Dec 1896.

38. On the Lattimer Massacre see Edward Pinkowski, *Lattimer Massacre*, Philadelphia, PA: Sunshine, 1950; Greene, *The Slavic Community on Strike*, Chapter 7; Aurand, *From the Molly Maguires* ..., 139–142; George A. Turner, "The Lattimer Massacre and its Sources," *Slovakia*, 27 (1977), 9–43; George Turner, "Ethnic Responses to the Lattimer Massacre," 126–153 in David L. Salay (ed.), *Hard Coal, Hard Times: Ethnicity and Labor in the Anthracite Region*, Scranton, PA: Anthracite Museum Press, 1984; Donald L. Miller and Richard E. Sharpless, *The Kingdom of Coal: Work, Enterprise, and Ethnic Communities in the Mine Fields*, Philadelphia, PA: University of Pennsylvania Press, 1985, Chapter 7; Kenneth C. Wolensky, "The Lattimer Massacre," Historic Pennsylvania Leaflet, Harrisburg, PA: PHMC, 1997; Kenneth C. Wolensky, "Freedom to Assemble and the Lattimer Massacre of 1897," Historical Society of Pennsylvania, http://173.203.96.155/node/2915, and George A. Turner, "The Lattimer Massacre: A Perspective from the Ethnic Community," in Harold W. Aurand, (guest ed.), *Pennsylvania History*, Special Issue: The Lattimer Massacre—1897, 69 (2002), 11–30.

39. *Wilkes-Barre Record*, 15 Sept 1897.

40. "Frenzied Strikers," *Wilkes-Barre Times*, 2 Sept 1897; "Marching Miners," *ibid.*, 3 Sept 1897; "In the Strike Region," *ibid.*, 4 Sept 1897; "The Hazleton Strike," *ibid.*, 4 Sept 1897.

41. Alyssa Murphy, "The Martyred Miners of Lattimer," http://www.pabook.libraries.psu.edu/palitmap/Lattimer.html. Burska likely represented the Slavs.

42. "In the Strike Region," *Wilkes-Barre Times*, 4 Sept 1897.

43. *Ibid.*

44. *Ibid.*

45. *Ibid.* Several authors, including Aurand (*From the Molly Maguires* ..., 137), cited newspaper reports indicating that the strike did not have "any recognized leader." Clearly, however, the leader of this particular strike was an Italian. He was probably not formally recognized because he was an ordinary mineworker and not affiliated with the UMWA.

46. *Ibid.*

47. *Ibid.*

48. The authors are indebted to Prof. George A. Turner for consultation on the Lattimer Massacre, including the status of the deputies on whom he has conducted detailed research.

49. "The Hazleton Strike," *Wilkes-Barre Times*, 4 Sept 1897.

50. Pinkowski, *Lattimer Massacre*, 5.

51. Turner, "Ethnic Responses to the Lattimer Massacre," 126.

52. "Bloodshed!" *Wilkes-Barre Times*, 10 Sept 1897.

53. On the Twin Shaft disaster see Ellis W. Roberts, *The Breaker Whistle Blows: Mining Disasters and Labor Leaders in the Anthracite Region*, Scranton, PA: Anthracite Museum Press, 1984, Chapter 4. On the Avondale disaster see Robert P. Wolensky and Joseph M. Keating, *Tragedy at Avondale: The Causes, Consequences, and Legacy of the Pennsylvania Anthracite Coal Industry's Most Deadly Mining Disaster, September 6, 1869*, Easton, PA: Canal Museum and Technology Press, 2008.

54. Quoted in Turner, "Ethnic Responses to the Lattimer Massacre," 136.

55. Three of the men wounded at Lattimer may have been Italian, based on surnames: John Contra (listed as Hungarian by *Daily Sentinel* [Hazleton] 11 Sept 1897), Joseph Meci (listed as Hungarian in *Philadelphia Inquirer*, 12 Sept 1897), and John Fora (listed as Hungarian, *Daily Sentinel* [Hazleton] 11 Sept 1897). Three other wounded men whose nationalities were not officially established were Constantine Manoulso, John Postia, and Anthony Mitscula (the names were listed in Henry Hoyt, United States Assistant Attorney-General, to William Day, Secretary of State, 8 Apr 1899, in *Papers Relating to the Foreign Relations of the United States, with the Annual Message of the President Transmitted to Congress December 6, 1898*, Washington: U.S. Govern-

ment Printing Office, 1910, 82–87, cited in Turner, "The Lattimer Massacre: A Perspective from the Ethnic Community").

56. Greene, *The Slavic Community on Strike*, 140–141.

57. *Ibid*. See also Aurand, *From the Molly Maguires* . . .; Miller and Sharpless, *The Kingdom of Coal*; Blatz, *Democratic Miners*; and Dublin and Licht, *The Face of Decline*.

58. Howard Zinn ("Massacres of History," *The Progressive*, 62 [Aug 1998], 17) noted that slaughters such as those at Lattimer had received very little attention in history books even as other contemporary non-labor events were given much greater notice: "When, the following year, the press set out to create a national excitement over the mysterious sinking of the battleship Maine in Havana harbor, a machinists' journal pointed to the Lattimer Massacre, saying that the deaths of workers resulted in no such uproar. It pointed out that 'the carnival of carnage that takes place every day, month, and year in the realm of industry, the thousands of useful lives that are annually sacrificed to the Moloch of greed . . . brings forth no shout for vengeance and reparation.'" See also Paul Shackel, Michael Roller, and Kristin Sullivan, "Historical Amnesia and the Lattimer Massacre," *News: A Publication of the Society of Applied Anthropology*, 2011, http://sfaanews.sfaa.net/2011/05/01/historical-amnesia-and-the-lattimer-massacre.

59. Aurand (guest ed.), *Pennsylvania History*, Special Issue: The Lattimer Massacre. However, Michael Novak's fictionalized account of the massacre did include references to, and circumstances that involved, Italians (*The Guns of Lattimer*, New York: Basic Books, 1978).

60. The first author attended the events to commemorate the Lattimer Massacre's 100th anniversary, held in Lattimer and also at PHMC's Eckley Miners' Village museum near Hazleton.

61. In unrecorded interviews on 5 Apr 2003 and 21 Apr 2010, Prof. Greene reflected on his research and admitted that, in retrospect, the title of his book was somewhat inaccurate because, although Lithuanians and Hungarians are not Slavs, he included them because they were eastern Europeans. He added that he should have devoted greater attention to the Italians because they were also supportive of the early anthracite unions.

62. Peter Berger, *Invitation to Sociology: A Humanistic Perspective*, Garden City, NY: Doubleday, 1963, 57. See also Norman K. Denzin, *Interpretive Biography*, Thousand Oaks, CA: Sage Publications, 1989; David R. Maines and Jeffrey Ulmer, "The Relevance of Narrative for Interactionist Thought," *Studies in Symbolic Interaction*, 14 (1993), 109–124; and Norman K. Denzin and Yvonna S. Lincoln (eds.), *Collecting and Interpreting Qualitative Materials*, Thousand Oaks, CA: Sage, 1998.

63. Raphael Samuels, *Island Stories: Unraveling Britain*, London: Verso, 1999, 204.

64. Roberts, *The Anthracite Coal Industry*, 1901, 205.

65. *Ibid*., 206.

66. *Ibid*., 205–206.

67. *Ibid*., 189.

68. *Ibid*., 205. Roberts presumably wanted to include the Welsh and Irish in the "Anglo-Saxon" category because they were important elements in the English-speaking workforce even though they were neither Angles nor Saxons, but of Celtic ancestry. See also note 71.

69. Rev. Peter Roberts, Testimony, 12 Feb 1903, in *Proceedings of the Anthracite Mine Strike Commission*, 1902–03, Scranton, PA: Scranton Tribune, 1903. Roberts' statement that the immigrants lacked respect for "personal rights" may have, in fact, reflected his strong concern for property rights. As such, he seemed to have had a double standard. That is, he was willing to criticize the immigrants for calling strikes (an expression of personal rights), but he rarely castigated the coal companies for wage violations, excessive docking, and other unfair practices (because they controlled property rights). Similarly, he reprimanded the workers for their strike militancy but not the Coal & Iron Police who used forceful methods under the direction of the coal operators. The authors are indebted to Megan Hastie for these insights.

70. Barendse, "American Perceptions Concerning Slavic Immigrants," 102, 103.

71. Though the Welsh are not Anglo-Saxons and had historically been in conflict with the English in Britain, they sought approval and acceptance by the Anthracite Aristocracy and other members of the Anglo establishment in hard-coal country. Their Protestant religion, English language (although many spoke Welsh), and Republican politics facilitated their acceptance. As prominent members of the anthracite work force, the Welsh came into conflict with Italians

and Slavs, who were mainly Catholic in religion and Democratic in politics (William A. Gudelunas, Jr. and William G. Shade, *Before the Molly Maguires: The Emergence of the Ethno-Religious Factor in the Politics of the Lower Anthracite Region, 1844–1872*, New York: Arno, 1976). As shown later, the conflict with newcomers carried over into a prominent Welsh membership in the Ku Klux Klan of northeastern Pennsylvania.

72. Virtue, "The Anthracite Mine Laborers," 753.

73. Roberts, *Anthracite Coal Communities*, 33.

74. Caroline Golab, *Immigrant Destinations*, Philadelphia, PA: Temple University Press, 1977, 61, 109.

75. Virtue, "The Anthracite Mine Laborers," 752.

76. Marcella Bencivenni, *Italian Immigrant Radical Culture: The Idealism of the Sovversivi in the United States, 1890–1940*, New York: New York University Press, 2011, 25.

77. PaCC employee records from 1892 showed only a small number of Italian mineworkers, based on last name identification (Mining Records, Local Breakers—Workers and Wages Folder, Lackawanna Historical Society, uncatalogued). The breakdown by ethnicity was:
    - Ewen Breaker—143 total employees, 15 Italians including Mike Binaco ("dumping"– $1.45/day wage), Pasquale Britardo (plate man– $1.45/day), Jno Morinzo (plate man– $1.25/day); the others were slate pickers for $1.00/day: Nick Verio, Joseph Coshetta, Jno Kallotto, Mike Veendura, Jno Barrette, Joe Vennella, Jno Locato, Frank Leon, Joe Dluuno, Geo. Kallotto, Frank Pissillo, and Mike Frank.
    - No. 14 Breaker—112 total employees, no Italians; mostly Irish with some Slavic, English, Welsh and Scottish.
    - No. 1 Breaker—100 total employees, no Italians; mostly Irish with some other groups.
    - Barnum Breaker—100 total employees, two who may have been Italian, Jno Vano (general work–$1.45/day), and Joe Vossie (loading culm–$1.45/day).
    - No. 1 Breaker—55 total employees, no Italians, mostly Irish.
    - No. 6. Breaker—91 total employees, no Italians, mostly Irish.
    - Old Forge Breaker—100 total employees, including one Italian: John Kossicci (slate picker– $1.00/day).
    - No. 8 Breaker—88 employees, no Italians, mostly Irish and Slavs.
    - Central Breaker—101 employees, no Italians.
    - No. 5 Breaker—97 employees, no Italians.
    - Gipsy Grove Breaker—53 employees, no Italians.

78. On Italian mineworkers in other countries see Robert F. Foerster, *The Italian Emigration of Our Times*, Cambridge, MA: Harvard University Press, 1924 (138–39, France; 155–58, Germany; 206, Belgium); Stephen Catterall and Keith Gildart, "Outsiders: Trade Union Responses to Polish and Italian Coal Miners in Two British Coalfields, 1945–54," 164–176 in Stefan Berger, Andy Croll, and Norman LaPorte (eds.), *Toward a Comparative History of Coalfield Societies*, Burlington, VT: Ashgate, 2005; Philip Mosley, "Screen Images of Italian Immigrants to the Belgian Coal Mines," unpublished manuscript, Penn State University-Worthington Scranton, 2005; and Bencivenni, *Italian Immigrant Radical Culture*. Italians constituted the majority of mineworkers at the Le Bois du Cazier mine in Marcinelle, Belgium, where, on August 8, 1956, an underground fire took 262 lives; 136 of the victims were Italian. Italian mineworkers were also employed in the Cape Breton, Canada, mining industry (Sam Migliore and A. Evo DiPierro, *Italian Lives, Cape Breton Memories*, Sydney, Nova Scotia: Cape Breton University Press, 1999).

79. Quoted in Aurand, *Coalcracker Culture*, 36.

80. Quoted in Long, *Where the Sun Never Shines*, 132.

81. Quoted in Matthew Frye Jacobson, *Whiteness of a Different Color, European Immigrants and the Alchemy of Race*, Cambridge, MA: Harvard University Press, 1998, 56.

82. Matthew Stanley Kemp, *Boss Tom: The Annals of an Anthracite Mining Village*, Akron, OH: Saalfield, 1904, 296, 72–73. Both Italians and Slavs were stereotyped in Kemp's book. He was

much more sympathetic to the Irish mineworkers whom he saw as poor but honest. See also Peter Goin and Elizabeth Raymond, "Living in Anthracite: Mining Landscape and Sense of Place in Wyoming Valley, Pennsylvania," *The Public Historian*, 23 (Spring 2001), 29–45.

83. Arno Dosch, "Just Wops," *Everybody's Magazine*, 25 (Nov 1911), 579. The manager expanded his definition of Wops to include "Dagos, Niggers, and Hungarians—the fellows that did the work. They don't know anything, and they don't count."

84. Foerster, *The Italian Immigration of Our Times*, 504. The main slur directed against Italians has been Wop, a word whose origin remains unclear. It may have begun in U.S. immigration where it could have meant "without papers." Another possibility is that it derived from the southern Italian dialect term *guappo*, meaning swagger. After documenting the extensive migration from Italy, Foerster remarked: "The Greater Italy is an empire—but a proleteriate [*sic*] empire. It bestrides the world like a Colossus—but a Colossus arrayed in rags" (504).

85. Dillingham Commission, "Immigrants in Industries," in *Reports of the Immigration Commission*, Vol. 16, Part 19, Part 2, 697.

86. W. Jett Lauck, "*Copper Mining and Smelting; Iron Ore Mining; Anthracite Coal Mining; Oil Refining*," in Dillingham Commission, *Reports of the United States Immigration Commission*, Vol. 16, 686.

87. Clive Webb, "The Lynching of Sicilian Immigrants in the American South, 1886–1910," *American Nineteenth Century History*, 3 (Spring 2002), 45–76. See also Jacobson, *Whiteness of a Different Color*, 56–62; and Marco Rimanelli and Sherl L. Postman, *The 1891 New Orleans Lynching and U.S.-Italian Relations: A Look Back*, New York: Lang, 1991.

88. Turner, "Ethnic Responses to the Lattimer Massacre," 137.

89. http://images.library.wisc.edu/FRUS/EFacs/1896/reference/frus.frus1896.i0021.pdf.

90. "Lynching of Persons of Italian Origin in Tallulah, Louisiana," Count Vinci to Mr. Hay, 22 Jul 1899, in *Papers Relating to Foreign Relations of the United States*, Washington, DC: USGPO, 1901, 440.

91. "An Act of Simple Justice," *New York Times*, 4 Feb 1896; Senator Augusto Pierantoni, "Italian Feeling on American Lynching," *The Independent*, 55 (1905), 2040–2042; Conrad L. Woodall, "The Italian Massacre of Walsenburg, Colorado, March 1895," 297–317 in *Italian Ethnics—Their Languages, Literature, and Lives, Proceedings of the 20th Annual Conference of the American Italian Historical Association*, Chicago, Illinois, November 11–13, 1987.

92. Herbert G. Gutman, "The Buena Vista Affair, 1874–1875," *Pennsylvania Magazine of History and Biography*, 88 (Jul 1964), 251–292.

93. "The Right to Work: The Story of Non-Striking Miners," *McClures Magazine*, 20 (Nov 1902–Apr 1903), 332–236. Whether Castelli and Kiblotti were murdered because they were strikebreakers or were seen as *Italian* strikebreakers was not clear. The Maltby Colliery workers were on the lookout for a group of scabs who were scheduled to take over the mine that morning. Because Castelli shouted out when the assault began, the assailants may have detected his accent and believed that he was not just an imported worker but an Italian imported worker. However, the dense fog was a mitigating factor in the murders as was the fact that the victims were carrying hunting guns, which may have contributed to the rush-to-judgment attack by the picketers. The pair had set out on a small-game hunting trip to secure food for their families during the strike.

94. Stanley Feldstein and Lawrence Costello (eds.), *The Ordeal of Assimilation: A Documentary History of the White Working Class, 1830s to the 1970s*, Garden City, NY: Anchor-Doubleday, 1974, 186.

95. Cristogianni Borsella, *On Persecution, Identity, and Activism: Aspects of the Italian-American Experience from the Late 19th Century to Today*, Boston, MA: Dante University Press, 2005. Certainly, the Chinese, Japanese, Jewish, Mexican and other Latino groups can also make a claim for mistreatment in this regard.

96. William J. Connell and Fred Gardaphe (eds.), *Anti-Italianism: Essays on Prejudice*, New York: Palgrave Macmillan, 2010.

97. A Zogby opinion poll in 2000 found that 70 percent of the respondents associated the word "Italian" with "crime" (Kenneth A. Ciongoli and Jay Parini, *Passage to Liberty: The Story of Italian Immigration and the Rebirth of America*, New York: Regan, 2002, 30). Hundreds of persecutory acts against Italians in the U.S. have been documented between the 1860s and early 2000s

(see e.g., Borsella, *On Persecution, Identity, and Activism*, 52–129). Branded as "enemy aliens" during World War II, over 600,000 Italian-born persons faced property confiscations and travel restrictions, and were forced to carry identification cards. About 10,000 Italians on the West Coast were dispatched from their homes and a few hundred were placed in military camps in a manner similar to, but not as extensive nor as harsh as, the treatment of Japanese Americans (Paula Branca-Santos, "Injustice Ignored: The Internment of Italian-Americans during World War II," *Pace International Law Review*, 13 [2001], 151–182; James Brooke, "After Silence, Italians Recall the Internment: An Official Apology Is Sought from the U.S.," *New York Times*, 11 Aug 1997; Lawrence DiStasi [ed.], *Una Storia Segreta: The Secret History of Italian American Evacuation and Internment during World War II*, Berkeley, CA: Heyday, 2004). A documentary exhibit on the Italian internment, *Una Storia Segreta*, has been shown in over 50 American museums and at other sites since 1994. On the Japanese internment see Alan Austin, *From Concentration Camp to Campus: Japanese American Students and World War II*, Urbana, IL: University of Illinois Press, 2007.

98. Jacobson, *Whiteness of a Different Color*, Chapter 2.

99. Scelsa v. City University of New York, 1994, No. 1063, Docket 95–7975, Argued 4 Dec 1995–25 Jan 1996, United States District Court for the Southern District of New York. See also Joseph V. Scelsa, "Italian Americans and Civil Rights: A Case Study at the City University of New York," Italian American Review 3 (12 Oct 1994), 21–32; Joseph V. Scelsa, "Affirmative Action for Italian Americans: The City University of New York Story," Chapter 7 in Connell and Gardaphe (eds.), *Anti-Italianism*.

100. Stefano Luconi, "Mafia-Related Prejudice and the Rise of Italian Americans in the United States," *Patterns of Prejudice*, 33 (1999), 43–57.

101. Jennifer Guglielmo and Salvatore Salerno, *Are Italians White? How Race Is Made in America*, New York: Routledge, 2003. The "whiteness" issue was not confined to Italians. See, for example, Noel Ignatiev, *How the Irish Became White*, New York: Routledge, 1995; Karen Brodkin Sacks, "How Did Jews Become White Folks?" 78–102 in Steven Gregory and Roger Sanjek (eds.), *Race*, New Brunswick, NJ: Rutgers University Press, 1994; and Karen Brodkin, *How Jews Became White Folks: And What That Says About Race in America*, New Brunswick, NJ: Rutgers University Press, 2000.

102. Eduardo Bonilla-Silva, *Racism without Racists: Color-Blind Racism and Racial Inequality in Contemporary America*, Landham, MD: Rowman & Littlefield, 2010, 73. For other works within the whiteness literature see David R. Roedinger, *Working Toward Whiteness: How America's Immigrants Became White, the Strange Journey from Ellis Island to the Suburbs*, New York: Basic, 2005; David R. Roediger, *The Wages of Whiteness: Race and the Making of the American Working Class*, New York: Verso, 1999 [1991]; Jacobson, *Whiteness of a Different Color*; Theodore W. Allen, *The Invention of the White Race (Vol. 1), Racial Oppression and Social Control*, London: Verso, 1994; Theodore W. Allen, *The Invention of the White Race (Vol. 2), Origin of Racial Oppression in Anglo-America*, London, Verso, 1997; Karyn D. McKinney, *Being White: Stories of Race and Racism*, New York: Routledge, 2005; and Phylis Cancilla Martinelli, *Undermining Race: Ethnic Identities in Arizona Copper Camps, 1880–1920*, Tuscon, AZ: University of Arizona Press, 2009.

103. Maria Laurino, *Were You Always Italian?*, New York: Norton, 2000, 37.

104. Azar Nafisi, *Reading Lolita in Tehran*, New York: Random House, 2003.

105. Salvatore J. LaGumina, "Prejudice and Discrimination: The Italian-American Experience Yesterday and Today," 108–115 in Connell and Gardaphe (eds.), *Anti-Italianism*; Jerome Krase, "Shark Tale—'Puzza da Cap': An Attempt at Ethnic Activism," 137–150 in *ibid.*

106. "Italian Americans Slam 'Jersey Shore,'" (4 Dec 2009) UPI online, http://www.upi.com/Entertainment_News/TV/2009/12/04/Italian-Americans-slam-Jersey-Shore/UPI-85771259960979/; Lisa Wade, "What Exactly Is Wrong with Jersey Shore?" *Sociological Images* on-line (27 Dec 2009), http://thesocietypages.org/socimages/2009/12/27/what-exactly-is-problematic-about-jersey-shore/.

107. E.A. Ross, *The Old World in the New*, New York: The Century Co., 1914, original emphasis. See also by the same author, "Racial Consequences of Immigration," *Century Magazine*, 87 (Feb 1914), 615–622. In *The Old World in the New* Ross wrote: "Among the foreign-born, the Italians rank lowest in adhesion to trade unions, lowest in ability to speak English, lowest in proportion to naturalized citizens after ten years' residence, lowest in proportion of children in school, and highest in proportion of children at work. Taking into account the innumerable 'birds of passage' without family or future in this country, it would be safe to say that half, perhaps two-thirds

of our Italian immigrants are *under* America, not *of* it. Far from being borne along our onward life, they drift round and round in a 'Little Italy' eddy, or lie motionless in some industrial pocket or crevice at the bottom of the national current." Ross had negative evaluations of the Slavs as well, commenting that a physician once told him that they "are immune to certain kinds of dirt," and "can stand what would kill a white man" (Ross, *The Old World in the New*, 291).

108. Aristide Zolberg, *A Nation by Design: Immigration Policy in the Fashioning of America*, Cambridge, MA: Harvard University Press, 2008.

109. Edward Banfield, *The Moral Basis of a Backward Society*, New York: Free Press, 1958, 84.

110. Filippo Sabetti, *The Search for Good Government: Understanding the Paradox of Italian Democracy*, Montreal, Canada: McGill-Queen's University Press, 2000, Chapter 8.

111. For a thorough review of 50 years of criticism and reanalysis of Banfield's case study see Emanuele Ferragina, "The Never-Ending Debate about *The Moral Basis of a Backward Society*: Banfield and 'Amoral Familism'," JASO-Online (formerly the *Journal of the Anthropological Society of Oxford*), 1 (Winter 2009), 141–160. See also the sociological criticism of cultural interpretations in ethnic research by Stephen Steinberg, *The Ethnic Myth: Race, Ethnicity, and Class in America*, Boston, MA: *Beacon Press*, 1989.

112. Nathan Glazer and Daniel Patrick Moynihan, *Beyond the Melting Pot: The Negroes, Puerto Ricans, Jews, Italians, and Irish of New York City*, Cambridge, MA: Harvard University Press, 1963; Robert Putnam, *Making Democracy Work: Civic Traditions in Modern Italy*, Princeton, NJ: Princeton University Press, 1992. Even noted Italian-American scholar Rudolph J. Vecoli essentially reinforced Banfield's culturally biased thesis in "Contadini in Chicago: A Critique of the Uprooted," *Journal of American History*, 51 (Dec 1964), 404–17. Banfield's view is still very much alive as indicated by its inclusion in an article in *The Economist* dealing with Italy's current financial crisis ("The Ins and Outs: Italians Are Deeply Anti-meritocratic," *The Economist*, 9 Jun 2011).

113. Philip Jenkins, *Hoods and Shirts: The Extreme Right in Pennsylvania*, 1925–1950, Chapel Hill, NC: University of North Carolina Press, 1997, 73. Chapter 3 of Jenkins' book, "The White Giant," deals with the KKK in the coal fields. The Loucks reference was to a study of the KKK by Emerson H. Loucks, *The Ku Klux Klan in Pennsylvania: A Study in Nativism*, Harrisburg, PA: Telegraph Press, 1936.

114. William Fitzhugh Brundage, *Lynching in the New South: Georgia and Virginia, 1880–1930*, Urbana, IL: University of Illinois Press, 1993; Jonathan Markovitz, *Legacies of Lynching: Racial Violence and Memory*, Minneapolis, MN: University of Minnesota Press, 2004; Peter Vellon, "'Between White Men and Negroes': The Perception of Southern Italian Immigrants through the Lens of Italian Lynchings," 23–32 in Connell and Gardaphe (eds.), *Anti-Italianism*.

115. "Hooded Mob Burn Crosses in This Valley," *The Evening News*, 27 Dec 1924. Several persons who contributed interviews to the Northeastern Pennsylvania Oral and Life History Project (NPOLHP) witnessed KKK activities in the 1920s and 1930s. Claire Hart Cummings, who resided in Forty Fort (a town once known for anti-Catholic sentiment), in Luzerne County, recalled one incident from her youth: "In 1926, we moved over to Yeager Avenue in Forty Fort. We were one of the first Catholics to live in Forty Fort. … I can remember standing in my parents' bedroom which was on the front of the house and watching them burn a cross in the empty lot across the street. I forget who I was with just the other day that said they had the same experience. They were Catholics and they had the same experience at Forty Fort about the same time. … Oh, the parade, yes, in the middle-to-late twenties. Sonny Gibbons, who's now dead, God rest his soul, and Bob Weaver, who's also now dead, were my buddies. We were just little kids and we grew up on Yeager Avenue, all of us. It was announced, and I don't remember how, that a Ku Klux Klan parade was going to go down Wyoming Avenue from Forty Fort to Wyoming, you know in that direction. I don't know how far they went. And we were told not to go anywhere near it. Well, that was like a red flag for us, you know! So the three of us hid behind the Methodist Church. Oh, and my sister was also with us. Then we got bolder as the parade went on and we crept out, and then we were finally standing on the curb watching them. There was a man down the street from us, whose name I won't mention, who always wore his shoes too long, and they'd turn up at the toes. All the kids used to laugh about it. So we spotted him [in the parade]. I think he was on a horse. Well, anyway, he was in the hood and the bit. I was friends with his children and he had a brown leather chair in his den. [He lived] only a few houses from where I grew up. [The next day] I walked in and I said, 'Oh, hi. I saw you in the parade yesterday.' Well, that was the worst thing I could've done. He said, 'What do you mean you saw me in the parade?' 'See how you've got the same shoes on; see how they turn up at the toes.

We saw you.' (laughs) I was never allowed in the house again. (laughs)" (Claire Hart Cummings, audiotaped interview, 15 Jan 1995). Other persons who provided oral history accounts of KKK activities in northern field towns were Chester Brozena (3 Dec 1988), Helen Pinkowski (7 Dec 1988), Lewis Casterline (12 Dec 1988), and, in a joint interview, Ethel Bankovitch and Ann Major (5 Jun 1993).

116. Oral history evidence produced but a few instances of KKK-initiated acts directed against Italians or any other specific ethnic group. In one case, Lewis Casterline recalled that when local Italian entrepreneur Andrew J. Sordoni ran for public office in 1926, "We had the Ku Klux Klan against us," and "We had one of the rottenest private detectives in Hazleton [who] was going around saying that Sordoni was going to be representing the Pope of Rome and all that stuff" (Lewis Casterline, audiotaped interview, 12 Dec 1988, NPOLHP). Sordoni won the election by receiving votes from across the ethnic spectrum.

117. Jenkins, *Hoods and Shirts*, 73.

118. *Ibid.*, 69.

119. William A. Hastie recalled the activities of the KKK in the Pittston area in the 1930s, and said that recruiters tried to endear themselves with local Protestants by entering local churches on Sunday and walking down the center isle to present the church with money they had collected. They tried to appeal to the anti-Catholic and anti-union sentiments, said Hastie, because "anti-labor elements were always trying to splinter the labor movement along ethnic and religious differences." He further recalled that two Klansmen who lived in West Pittston were known as such because their children boasted about their fathers' membership. However, most community members had little respect for them because of their bigotry. Richard Cronin, an Irish Catholic who served as vice president and executive secretary of the Greater Wilkes-Barre Chamber of Commerce from the 1950s to the 1970s, discussed the KKK and other forms of anti-Catholicism among local business leaders in the Wilkes-Barre area. See his audiotaped interview, 24 May 1983, NPOLHP.

120. Jenkins, *Hoods and Shirts*, 75.

121. *Ibid.*, 79, 80.

122. *Ibid.*, 84–88. See also "Paul M. Winter in Police Cell," *Wilkes-Barre Record*, 17 Jul 1931.

123. George A. Theodorson and Achilles G. Theodorson (*A Modern Dictionary of Sociology*, New York: Crowell, 1969, 87) defined organized crime as "crime committed by members of a formal organization devoted to activities that are in violation of the law. Such criminal organizations have a division of labor with certain roles filled by skilled specialists, a hierarchy of status and authority, their own system of norms, and strict organizational loyalty and discipline. These organizations also have arrangements with members of the local police and sometimes with certain influential community leaders." Joseph Albini (*The American Mafia: Genesis of a Legend*, New York: Appleton-Century-Crofts, 1971) categorized the KKK as one of four types of organized crime, namely, political-social. The other types were mercenary (predatory/ theft), in-group (gangs), and syndicated (function as businesses offering goods and services). Researchers have also included threats and violence as essential characteristics of organized criminal activity.

124. As Jenkins pointed out, the Italian Blackshirt groups who supported Fascist dictator Benito Mussolini were prominent in Pennsylvania during the 1930s and 1940s. They took pride in *Il Duce's* accomplishments, which "buoyed the morale of Italian Americans, who perceived themselves as victims of a dual assault from the Ku Klux Klan and the anti-immigration campaign" (Jenkins, *Hoods and Shirts*, 92). In historian David L. Salvaterra's assessment: "Within the Italian enclaves there was broad community support and approval of the fascists' activities" (David L. Salvaterra, Book Review of Philip Jenkins, *Hoods and Shirts* (1997), at H-Net, July 1997, www.h-net.org/reviews/showrev.php?id=1140). In Pennsylvania, the Blackshirts were most active in Philadelphia but also had followers in anthracite towns such as Scranton and Pittston, although the latter city also had a weekly antifascist newspaper, *Il Paese*, that began publishing in 1938 (Salvatore J. LaGumina, *The Italian American Experience: An Encyclopedia*, New York: Garland 2000, 513–514). Antifascist Italians likely had a hand in the bombing, in April 1928, of *La Voce Italiana*, a fascist newspaper published in Scranton ("Office of '*La Voce Italiana*' Wrecked by Bombers Late Last Night," *Scranton Times*, 21 Apr 1928). On the Italian community's support for Fascism in San Francisco see Dino Cinel, *From Italy to San Francisco: The Immigrant Experience*, Stanford, CA: Stanford University Press, 1982, 247–261. On the antifascist movement in the U.S. see Fraser M. Ottanelli, "'If Fascism Comes to American We Will Push It Back into the Ocean': Italian-American Antifascism in the 1920s and 1930s," 178–95 in

Donna R. Gabaccia and Fraser M. Ottanelli (eds.), *Italian Workers of the World: Labor Migration and the Formation of Multiethnic States*, Urbana, IL: University of Illinois Press, 2005.

125. John A. Agnew, *Place and Politics in Modern Italy*, Chicago, IL: University of Chicago Press, 2002; Nelson Moe, *The View from Vesuvius: Italian Culture and the Southern Question*, Berkeley, CA: University of California Press, 2002; Roedinger, *Working Toward Whiteness*, 111–114. The entries in the 1911 Dillingham Commission's *Dictionary of Races or Peoples* appear to have had considerable staying power. The dictionary cited Italian sociologist Alfredo Niceforo's characterization of southerners as "excitable, impulsive, highly imaginative, impracticable," with "little adaptability to highly organized society." The northerners, on the other hand, were "patient, practical and ... capable of great progress in political and social organizations." Niceforo was a student of Italian criminal sociologist Cesare Lombroso whose early writings stated that Sicilians possessed a violent nature as a result of racial degeneracy. See the Dillingham Commission, *Dictionary of Races or Peoples*, 82.

126. For other acts of prejudice and discrimination perpetuated by Italians and Italian-Americans see Roedinger, *Working Toward Whiteness*, 111–113.

127. On the Italian immigration and assimilation experience see Irvin L. Child, *Italian or American? The Second Generation in Conflict*, New Haven, CT: Yale University Press, 1943; Michael Lalli, "The Italian-American Family: Assimilation and Change, 1900–1965," *The Family Coordinator*, 18 (Jan 1969), 44–48; Feldstein and Costello (eds.), *The Ordeal of Assimilation*; Andrew Greeley, "The Transmission of Cultural Heritages: The Case of the Irish and the Italians," 209–235 in Nathan Glazer and Daniel P. Moynihan (eds.), *Ethnicity: Theory and Experience*, Cambridge, MA: Harvard University Press, 1975; Rudolph J. Vecoli, "The Coming of Age of Italian Americans: 1945–1974," *Ethnicity* 5 (1978), 119–147; James A. Crispino, The Assimilation of Ethnic Groups: The Italian Case, New York: Center for Migration Studies, 1980; Charles D. Ferroni, *Italians in Cleveland: A Study in Assimilation*, New York: Arno, 1980; Luciano John Iorizzo, *Italian Immigration and the Impact of the Padrone System*, New York: Arno, 1980 [1966]; Herbert Gans, *Urban Villagers: Group and Class in the Life of Italian-Americans*, New York: Free Press, 1982 [1962]; Humbert S. Nelli, *From Immigrants to Ethnics: The Italian Americans*, New York: Oxford University Press, 1983; John Bodnar, *The Transplanted: A History of Immigration in Urban America*, Bloomington, IN: Indiana University Press, 1985; Stanley Lieberson, *From Many Strands: Ethnic and Racial Groups in Contemporary America*, New York: Russell Sage Foundation, 1988; Mary Waters, *Ethnic Options: Choosing Identities in America*, Berkeley, CA: University of California Press, 1990; William F. Whyte, *Street Corner Society: The Social Structure of an Italian Slum*, Chicago, IL: University of Chicago Press, 1993 [1943]; Jerome Krase, "Between Columbus and Cuomo: The Italian-American Experience in America," *Italian American Review* 6 (Spring-Summer 1997), 29–44; Alejandro Portes (ed.), *The Economic Sociology of Immigration: Essays on Networks, Ethnicity, and Entrepreneurship*, New York: Russell Sage Foundation, 1995; Salvatore J. LaGumina, *Wop: A Documentary History of Anti-Italian Discrimination*, Toronto: Guernica, 1999 [1973]; Richard N. Juliani, *Priest, Parish, and People: Saving the Faith in Philadelphia's "Little Italy,"* South Bend, IN: University of Notre Dame Press, 2007; Paolo A. Giordano and Anthony Julian Tamburri (eds.), *Italian-Americans in the Third Millennium*, New York: Bordighera, 2009.

128. Milton Gordon, *Assimilation in American Life: The Role of Race, Religion, and National Origins*, New York: Oxford University Press, 1964; Gabaccia and Ottanelli (eds.), *Italian Workers of the World.*

129. Roedinger (*Working Toward Whiteness*, Chapter 4) observed that the ethnic revival was part of a white reaction to the progress made by blacks and other minorities during the Civil Rights movement of the 1950s and 1960s.

130. Linda Brandi Cateura, *Growing Up Italian: How Being Brought Up as an Italian-American Helped Shape the Characters, Lives, and Fortunes of Twenty-four Celebrated Americans*, New York: William Morrow, 1987. Among the two dozen successful Americans who contributed oral history memoirs to the volume were Mario Cuomo, Joseph Cardinal Bernardin, Gay Talese, Francis Coppola, Yogi Berra, John Ciardi, Ken Auletta, Rudolph Giuliani, Tony Bennett, and Geraldine Ferraro.

131. See, e.g., Frederico and Stephen Moramarco, *Italian Pride: 101 Reasons to Be Proud You're Italian*, New York: Citadel/Kensington, 2000; A. Kenneth Ciongoli and Jay Parini, *Passage to Liberty*; Antonio Santi, *The Book of Italian Wisdom*, New York: Kensington, 2003; LaGumina et al., (eds.), *The Italian-American Experience*; Stephanie Longo, *Italians of Northeastern Pennsylvania*, Chicago, IL: Arcadia, 2004; and Steve Puleo, *The Boston Italians: A Story of Pride, Perseverance,*

and Paesani from the Years of the Great Immigration to the Present Day, Boston, MA: Beacon Press, 2007.

132. Acclaimed U.S. citizens with Sicilian ancestry include Frank Capra, Chick Corea, Jon Bon Jovi, Joe DiMaggio, Dean Martin, Natalie Merchant, Liza Minnelli, Al Pacino, Ferdinand Pecora, Mario Puzo, Martin Scorsese, Frank Sinatra, Bruce Springsteen, Jack Valenti, and Philip Zimbardo. See, e.g., LaGumina et al., (eds.), *The Italian-American Experience*.

133. G.M. Tuoni and G. Brogelli, *Attività Italiane in California*, published by the authors, 1929. Special thanks to Robert B. Enright Jr. for bringing this book to our attention and for his observations on assimilation. Antonio Martin Cogliandro was Prof. Enright's maternal grandfather. San Francisco had been the immigrant destinations of several successful Italian-Americans including Andrea Sbarboro, A.P. Giannini, and Marco J. Fontana, who were called "the architects of the San Francisco Italian economy." Sbarboro began as an importer who founded the Italian-American Bank and the Italian Swiss Colony enterprise. Fontana, who envisioned a wide market for California's agricultural products, built the highly successful Del Monte Corporation. Giannini established the Bank of Italy, later renamed the Bank of America (Cinel, *From Italy to San Francisco*, 233–40). Giuseppe Paulo DiMaggio (Joe DiMaggio) was another well-known San Franciscan, as were mayors Angelo Joseph Rossi, Joseph Alioto, and George Moscone, as well as U.S. Rep. Nancy Pelosi. See also Deanna Paoli Gumina, *The Italians of San Francisco, 1850–1930*, New York: Center for Migration Studies, 1978. On radicalism among the city's Italians see Paola A. Sensi-Isolani, "Italian Radicals and Union Activists in San Francisco, 1900–1920," 189–204 in Philip V. Cannistraro and Gerald Meyer (eds.), *The Lost World of Italian-American Radicalism*, Westport, CT: Praeger, 2003.

134. Edward F. Hanlon, *The Wyoming Valley: An American Portrait*, Woodland Hills, CA: Windsor Publications, 1983, 123.

135. Joan Saverino, "'Domani Ci Zappa': Italian Immigration and Ethnicity in Pennsylvania," *Pennsylvania Folklife*, 44 (Autumn 1995), 2–22.

136. As best determined, among the dozens and perhaps hundreds of legitimate and successful lessee coal operators in the northern field between 1930 and 1960 were Dominick Panzitta and his sons, Carlo Sebastianelli and his sons, Pasquale Adonizio, Bruno and Nello Ferretti, Gerard Damiani, Rocco D'Angelo, Louis Sarf, Ross Loiacono, Enrico Matteucci, Samuel Insalaco, Santo Sperazzo, Angelo Albanesi, Daniel Constantino, John Gigliello, Raniero Giombetti, John Gigliello, Raymond Popple, Louis Coccodrilli, Rudolph Rosetti, and Louis, Harry, and Benedict Bernardi.

137. Rep. Nancy Pelosi was first elected to the House of Representatives in 1987 and served as the first female Speaker of the House between 2007 and 2010. Her father, Thomas D'Alesandro, Jr., had served as a U.S. House representative from Maryland as well as the mayor of Baltimore. On Pelosi's biography see the *Biographical Directory of the U.S. Congress*, http://bio guide.congress.gov/scripts/biodisplay.pl?index=P000197.

138. Regarding lower crime rates, "Italians in the United States" (*Literary Digest*, 93 [23 Apr 1927], 30–32), reported that Italians had fewer cases of pauperism and alcohol-related problems, lower illiteracy rates, and less criminality than most other contemporary immigrant groups. Seventy-five years later in 2002, Ciongoli and Parini (*Passage to Liberty*, 30) cited a University of Chicago study, which found that Italian-Americans had among the lowest incarceration rates of any ethnic group. The authors also reported that, of the 25 million Italian-Americans, fewer than 5,000 (or 0.0002 percent of the group's population) have been members of organized crime. Moreover, between 1952 and 2002, of the 469 people on the FBI's most-wanted list, only 4.9 percent were Italian-American, which is less than one-half their proportion of the U.S. population. Italian-Americans have had among the lowest divorce and unemployment rates, and near the highest percentage of two-parent families, elderly family members who live at home, and families that regularly eat together.

139. See, e.g., Pellegrino D'Acierno, "Cinema Paradiso: The Italian Presence in American Cinema," 563–689 in Pellegrino D'Acierno, *The Italian-American Heritage: A Companion to Literature and Arts*, New York: Routledge-Garland Reference Library of the Humanities, 1999, 559–579. The "Godfather Industry" refers to the three exceedingly successful Godfather films, directed by Francis Ford Coppola in collaboration with author Mario Puzo, as well as the plethora of mobster pictures that have continued to arrive in theaters. The genre dates to the 1910s and 1920s. One of the most popular television shows in the late 1950s and early 1960s was *The Untouchables*, based upon the diaries of FBI agent Elliot Ness, which dealt with the Italian criminal gangs of Chicago during the 1930s (the show was much less successful when it was revived in 1993).

Italian-American director Martin Scorsese has been chastised in some circles for portraying Italians as gangsters in films such as *Casino* and *Goodfellas*, although his more recent productions have focused on Irish organized criminals—*The Gangs of New York*, *The Departed*, and *The Irishman* (forthcoming). On Scorsese see Robert Casillo, *Gangster Priest: The Italian American Cinema of Martin Scorsese*, Toronto: University of Toronto Press, 2006; and Ellis Cashmore, *Martin Scorsese's America*, Malden, MA: Polity, 2009. On the negative images of Italian-Americans in the media see Mary Jo Bona and Anthony J. Tamburri (eds.), *Through the Looking Glass: Italian & Italian/American Images in the Media, Selected Essays from the 27th Annual Conference of the American Italian Historical Association*, Nov 10–12, 1994, Staten Island, NY: American Italian Historical Association, 1996. On Irish-American hoodlums see T.J. English, *Paddy Whacked: The Untold Story of the Irish-American Gangster*, New York: HarperCollins, 2005.

140. Allen Rucker, *The Sopranos: A Family History*, New York: New American Library, 2001.

141. Oscar Wilde spoke to the American fascination with the criminal-as-hero phenomenon as he passed the Missouri hometown of the bank robber and outlaw Jesse James during a lecture tour in 1882: "Americans are certainly great hero-worshippers and always take their heroes from the criminal classes" (Alvin Redman, *The Wit and Humor of Oscar Wilde*, New York: Dover, 1959, 123).

142. Peter Maas, *The Valachi Papers*, New York: Bantam, 1969; Vincent Teresa (with Thomas C. Renner), *My Life in the Mafia*, New York: Doubleday, 1973. On the Apalachin crime meeting see Stephen Fox, *Blood and Power: Organized Crime in Twentieth Century America*, NY: Penguin, 1989, 326–327, 380. See also Maryclair Dale, "Judge Weighs Scope of Mob History at Philly Trial," *Philadelphia Inquirer*, 9 Aug 2012.

143. Robert J. Kelly, *The Upperworld and the Underworld: Case Studies of Racketeering and Business Infiltrations in the United States*, New York: Kluwer Academic/Plenum, 1999, 157. See also Joint Legislative Committee on Government Operations, on the Gangland Meeting in Apalachin, New York, *Interim Report*, Albany, NY: State of New York, 25 Jun 1958.

144. Kelly, *ibid*. Interestingly and appropriately, Kelly included not only the Mafia in his analysis but also the Molly Maguires, an Irish immigrant gang that operated in the lower anthracite fields in the 1860s and 1870s. On the Molly Maguires see Walter J. Coleman, *The Molly Maguire Riots*, Richmond, VA: Garrett & Massey, 1936, 23–24; Kevin Kenny, *Making Sense of the Molly Maguires*, New York: Oxford University Press, 1998; and Wolensky and Keating, *Tragedy at Avondale*, Chapter 4.

145. Alba, *Italian Americans*, 37. See also Peter Gottschalk, "Stages of Financial Crime by Business Organizations," *Journal of Financial Crime*, 15 (2008), 38–48. Documentary filmmaker Ken Burns, in his three-part PBS series *Prohibition* (2011), included the fact that, despite a ban on alcohol sales, President Warren G. Harding had a local bootlegger come to the White House on a weekly basis to restock the liquor cabinet.

146. Michael Woodiwiss, *Organized Crime and American Power*, Toronto: University of Toronto Press, 2001, 11.

147. *Ibid*., 13.

148. Michael Woodiwiss, *Gangster Capitalism: The United States and the Global Rise of Organized Crime*, New York: Carroll & Graff, 2005, 4.

149. Marshall B. Clinard and Peter C. Yeager, *Corporate Crime*, New York: Free Press, 2006 [1980]; Marshall B. Clinard, *Corporate Corruption: The Abuse of Power*, New York: Praeger, 1990.

150. Al Gedicks, *The New Resource Wars: Native and Environmental Struggles Against Multinational Corporations*, Boston, MA: South End, 1993; Al Gedicks, *Resource Rebels: Native Challenges to Mining and Oil Corporations*, Brooklyn, NY: South End, 2001; Al Gedicks, *Dirty Gold: Indigenous Alliances to End Global Resource Colonialism*, Brooklyn, NY: South End, 2009. See also Robert P. Wolensky, "A Review of Al Gedicks' *The New Resource Wars:* An Organized Crime Approach to Multinational Mining and Energy Corporations," Panel presentation, Meetings of the Wisconsin Sociological Association, La Crosse, WI, 1995.

151. See the U.S. Department of Justice website: http://www.justice.gov/civil/cases/tobacco2/index.htm.

152. "Stringfellows: Rupert Murdoch and News Corporation," *The Economist*, 28 Apr 2012; the article is a review of Watson's co-authored book (with Martin Hickman), *Dial M for Murdoch: News Corporation and the Corruption of Britain*, London: Alan Lane, 2012.

153. Matt Taibbi, "The Scam Wall Street Learned from the Mafia," *Rolling Stone*, 5 Jul 2012; see also his book on American financial corruption in the highest places, *Griftopia: Bubble Machines, Vampire Squids, and the Long Con That Is Breaking America*, New York: Spiegel & Grau, 2010.

154. Yves Smith, *ECONned: How Unenlightened Self-Interest Undermined Democracy and Corrupted Capitalism*, New York: Palgrave Macmillan, 2010, 4.

155. J. Davitt McAteer et al., *Upper Big Branch, The April 5, 2010, Explosion: A Failure of Basic Coal Mine Safety Practices*, Report to the Governor of West Virginia by the Governor's Independent Investigation Panel, May, 2011; Sabrina Tavernise, "Report Faults Mine Owner for Explosion That Killed 29," *New York Times*, 16 May 2011; "Mine Probe Faults Massey," *Wall Street Journal*, 20 May 2011. ABC News referred to Blankenship as "notorious" (see abcnews. go.com/ Blotter/massey-boss-mining-coal/story?id=15095868).

156. E. Stewart Milner, audiotaped interview, 3 Aug 1994, NPOLHP.

157. A heightened interest in the northeastern Pennsylvania crime family has been prompted by three recent items: (1) the fraud, bribery, and corruption charges against two crooked Luzerne County judges, Mark A. Ciavarella Jr. and Michael T. Conahan (the latter allegedly mob-connected), who were convicted of racketeering and conspiracy in the privatization of the county juvenile detention center, and given long prison sentences in 2011 (see William Encenbarger, *Kids for Cash: Two Judges, Thousands of Children, and a $2.6 Million Kickback Scheme*, New York: The New Press, 2012); (2) the publication of Charles Brandt's book, *I Heard You Paint Houses: Frank "The Irishman" Sheeran and the Inside Story of the Mafia, the Teamsters, and the Last Ride of Jimmy Hoffa* (Hanover, NH: Steerforth, 2005), which examined local crime boss Russell Bufalino's role in the murder of Teamsters union president James R. Hoffa, who had some questionable investments in northeastern Pennsylvania; and (3) the publication of Thomas Hunt and Michael A. Tona's investigative article, "Men of Montedoro," *Informer*, (Apr 2011, informer-journal.blogspot.com), which detailed the history of the region's crime family. See also Dave Janoski, "The Bufalino File: An Exclusive Look Inside the Massive FBI Paper Trail on Northeastern Pa.'s Most Notorious Mobster," *Citizens Voice*, 17 Jul 2011.

158. Lewis Casterline, audiotaped interview, 16 Jan 1970, interviewed by Clement Valetta and Mary Barrett, King's College Oral History Collections, Corgan Library, Wilkes-Barre, PA. On oral history as related to Italian-Americans see Luisa Del Giudice (ed.), *Oral History, Oral Culture, and Italian Americans*, New York: Palgrave Macmillan, 2009.

159. Cannistraro and Meyer (eds.), *The Lost World of Italian-American Radicalism*; Donna R. Gabaccia, *Militants and Migrants: Rural Sicilians Become American Workers*, New Brunswick, NJ: Rutgers University Press, 1988. On the comparative-internationalist method in Irish research see Kenny, *Making Sense of the Molly Maguires*, and Kevin Kenny (ed.), *New Directions in Irish-American History*, Madison, WI: University of Wisconsin Press, 2003.

160. Bruno Cartosio, "Sicilian Radicals in Two Worlds," 117–128 in Debouzy (ed.), *In the Shadow of the Statue of Liberty: Immigrants, Workers, and Citizens in the American Republic, 1880-1920*, Urbana, IL: University of Illinois Press, 1992. See also Donna Gabaccia, "Neither Padrone Slaves nor Primitive Rebels: Sicilians on Two Continents," 95–117 in Dirk Hoerder (ed.), *"Struggle a Hard Battle": Essays on Working-Class Immigrants*, DeKalb, IL: University of Northern Illinois Press, 1986.

161. Sicilian peasant revolts occurred in 1772, 1779, 1820–1821, 1848, 1860, 1866, 1892–94, 1901–1904, 1717–1921, and 1945–50 (Gabaccia, *Militants and Migrants*, 27; Eric J. Hobsbawm, *Primitive Rebels: Studies in Archaic Forms of Social Movement in the 19th and 20th Centuries*, New York: Norton, 1959).

162. Gabaccia, *Militants and Migrants*, 27–30.

163. Kenny, *Making Sense of the Molly Maguires*, 8. Kenny argued that the murders, beatings, and acts of arson perpetrated by the Molly Maguires during the 1860s and 1870s were considered necessary and right only when they fostered collective goals in the face of an overpowering company and its officials. In the eyes of members, therefore, such actions rested on a moral foundation of retributive justice. Another motivation for the insurgents was *distributive justice*, or a concern with fairness in the distribution of scarce resources and rewards such as wages. See also James Konow, "Which Is the Fairest One of All?: A Positive Analysis of Justice Theories," *Journal of Economic Literature*, 41 (2003), 1188–1239.

164. Gabaccia, *Militants and Migrants*, 29.

165. *Ibid.*, 68.

166. *Ibid.*, 151; see also Bencivenni, *Italian Immigrant Radical Culture*, 23.

167. *Ibid.*, 173.

168. Gabaccia, *Militants and Migrants*, 174.

169. *Ibid.*, Chapter 2. "The labor question" has been concerned with the relationship between capital and labor. See, e.g., Wendell Phillips, *The Labor Question*, Boston, MA: Lee & Shepard, 1884; Steve Fraser, "The 'Labor Question'," 55–84 in Steve Fraser and Gary Gerstle (eds.), *The Rise and Fall of the New Deal Order, 1930–1980*, Princeton, NJ: Princeton University Press, 1989; Rosanne Currarino, *The Labor Question in America: Economic Democracy in the Gilded Age*, Urbana, IL: University of Illinois Press, 2011.

170. Elizabetta Vezzosi, "Radical Ethnic Brokers: Immigrant Socialist Leaders in the United States Between Ethnic Community and the Larger Society," 121–138 in Gabaccia and Ottanelli (eds.), *Italian Workers of the World*, 126.

171. Vezzosi, "Radical Ethnic Brokers," 126. See also Rudolf Vecoli, "Free Country: The American Republic Viewed by the Italian Left, 1880–1920," 23–44 in Debouzy (ed.), *In the Shadow of the Statue of Liberty*.

172. Rudolph Vecoli, "The Italian Immigrants in the United States' Labor Movement from 1880 to 1920," in Bruno Bezza (ed.), *Gli italiani fouri de'Italia: Gli emigrati Italiana nei Moveimenti Operai dei paesi d'adpzione*, 1880–1940," Milan, Italy: Franco Angeli, 1983, 279.

173. In 1927, mineworkers in the Pittston area established the UMWA District 1 Sacco-Vanzetti Committee. It included representatives from 40 UMWA local unions within the northern field, as well as several Italian-American clubs and organizations. The group held parades and rallies. Rinaldo Cappellini spoke at one gathering and declared that the condemned men were being persecuted because of their involvement with the labor movement (he did not mention their Italian ethnicity as a factor). Local activists Pat Toohy, Frank Vrataric, William Brennan, Andred Pascucci, Albert F. Fagin, and Walter Harris were among the leaders of the committee (Howard, *Forgotten Radicals*, 35–37). For a bibliography on the vast Sacco and Vanzetti literature see Jerry Kaplan, "Sacco and Vanzetti Bibliography," http://sacco andvanzetti.org/sacco vanzetti_bibliography.pdf. See also Paul Avrich, *Sacco and Vanzetti: The Anarchist Background*, Princeton, NJ: Princeton University Press, 1991.

174. Kenneth C. Wolensky, Nicole H. Wolensky, and Robert P. Wolensky, *Fighting for the Union Label: The International Ladies' Garment Workers' in Pennsylvania*, University Park, PA: Penn State Press, 2002.

175. For example, as discussed in Chapter 5, Italian strikers could have been among those involved in the dynamiting of James Joyce's business and home, the bombing of numerous other buildings and homes, as well as the murder of Samuel Spachia. Italian immigrant Sam Bonita, the local union president at the No. 6 Colliery, was convicted of manslaughter in the shooting death of District 1 union official (and alleged organized crime affiliate) Frank Agati.

176. On the UMWA's radical element, labor historian John Laslett wrote: "… the radicals in the UMWA were very largely the Anglo-Saxon element, not the Italians, or Slavs or other eastern Europeans who by now [1920] constituted more than half of the mining labor force." Laslett warned against "the danger of assuming that most of the militants in the labor movement were foreign-born" (John H.M. Laslett, *Labor and the Left: A Study of Socialist and Radical Influences in the American Labor Movement, 1881–1924*, New York: Basic Books, 1970, 230). Although no firm data were available, it seems safe to conclude that northern anthracite field radicals in 1920 were fairly evenly divided between the "old" and the "new" immigrants groups.

177. Dozens and perhaps hundreds of letters to Lewis and other UMWA officials, written in the 1940s and 1950s, can be found in various files of the President/District Correspondence, 1920–1933, GST/2/06.02, and President/District Correspondence, D.1 1954–1960, UMWA Papers, HCLA, PSU.

178. The labor wars in anthracite during the first third of the twentieth century were reminiscent of the many workingmen's uprisings in late eighteenth century Britain in that both were based upon demands for human rights and justice. In discussing the protests, E.P. Thompson observed that the British uprisings were based not on mere radical outbursts, but on "legitimizing notion" of rights: "Here is something unusual—pitmen, keelmen, cloth-dressers, cutlers: not only the weavers and labourers of Wapping and Spitafields, whose rowdy and colourful demonstrations had often come out in support of Wilkes, but working men in villages and towns over the whole country claiming *general* rights for themselves. It was this—and not the French terror—which

threw the propertied classes into terror" (E.P. Thompson, *The Making of the English Working Class*, New York: Pantheon, 1964, 105, original emphasis).

179. Joint Legislative Committee to Investigate the Knox Mine Disaster, *Final Report*, Harrisburg, PA: Commonwealth of Pennsylvania, 27 Jul 1959.

180. On other "labor martyrs" see, e.g., Gibbs M. Smith, *Joe Hill*, Salt Lake City, UT: University of Utah Press, 1969; and Vito Marcantonio, *Labor's Martyrs (Haymarket, 1887; Sacco and Vanzetti, 1927)*, Project Gutenberg e-Book, 2004, http://www.gutenberg.org/files/11009/11009-h/11009-h.htm

181. On Italians in the IWW see Salvatore Salerno, "No God, No Master: Italian Anarchists and the Industrial Workers of the World," 172–187 in Cannistraro and Meyer (eds.), *The Lost World of Italian American Radicalism*.

182. Edwin Fenton, "Italians in the Labor Movement," *Pennsylvania History*, 26 (Apr 1959), 143.

183. Four of the many contemporary Italian Americans who have continued the tradition of activism on behalf of human rights and social and economic justice are civil rights organizer Fr. James Groppi; human rights and free-speech proponent Mario Savio; urban social development advocate Fr. Geno Baroni (from Acosta, Pennsylvania); and public official Joseph A. Califano, Jr. On Groppi see Patrick Jones, "'Not a Color but An Attitude': Fr. James Groppi and Black Power Politics in Milwaukee," 259–81 in Jeanne Theoharris and Komozi Woodard (eds.), *Groundwork: Local Black Freedom Movements*, New York: New York University Press, 2005; Patrick Jones, *The Selma of the North: Civil Rights Insurgency in Milwaukee*, Cambridge, MA: Harvard University Press, 2009; and Jackie Di Salvo, "Father James E. Groppi (1930– 1985): The Militant Humility of a Civil Rights Activist," 229–244 in Cannistraro and Meyer (eds.), *The Lost World of Italian-American Radicalism*. On Baroni see Lawrence O'Rourke, *Geno: The Life and Mission of Geno Baroni*, Mahwah, NJ: Paulist, 1991. On Califano see his memoir *Inside—A Public and Private Life*, New York: Public Affairs, 2004; and *The Triumph and Tragedy of Lyndon Johnson*, College Station, TX: Texas A&M University Press, 2000. On Savio see Gil Fagiani, "Mario Savio: Resurrecting an Italian American Radical," 245–250 in Cannistraro and Meyer (eds.), *The Lost World of Italian-American Radicalism*; and Gabaccia, *Militants and Migrants*. In Sicily, attorneys Giovanni Falcone and Paolo Borsellino were assassinated in Palermo in 1992 after undertaking dangerous but ultimately successful legal actions against over 350 Mafia figures. The city's airport was named as a memorial in their honor. In a eulogy at his best friend Falcone's funeral, shortly before his own death, Borsellino said: "The fight against Mafia, which is the first problem to solve in our unfortunate and beautiful land, must be not only a cold repressive action, but a moral and cultural movement, involving everyone, especially younger generations, the most fit to feel the beauty of the fresh taste of freedom that sweeps away the foulness of moral compromise, of indifference, of contiguity and, hence, of complicity" (Paolo Borsellino, "State Funeral of Giovanni Falcone," 25 May 1992, www. studentpulse.com/articles /292/3/italian-politics-and-the-sicilian-mafia-an-account-from-1983-to-present). See also "60 Years of Heroes: Giovanni Falcone and Paolo Borsellino," *Time* (13 Nov 2006); and Paulo Borsellino, "Prepared Remarks for Assistant Attorney General Lanny Breuer at the Falcone Commemoration Ceremony," Palermo, Italy, 23 May 2010, www.justice.gov/criminal/pr/speeches-testimony/2010/05-23-10aag-palermo-remarks.pdf.

184. E.P. Thompson, "History from Below," *Times Literary Supplement*, 7 Apr 1966, 279–280; Melvyn Dubofsky, "The 'New' Labor History: Achievements and Failures," *Reviews in American History*, 5 (Jun 1977), 249–254; Howard Zinn, *A People's History of the United States*, New York: Harper Perennial, 1980; David Brody, "Reconciling the Old Labor History and the New," *Pacific Historical Review*, 72 (Feb 1993), 111–126; Michael Rogin, "How the Working Class Saved Capitalism: The New Labor History and The Devil and Miss Jones," *Journal of American History*, 89 (Jun 2002), 87–114.

185. Sara Lawrence-Lightfoot and Jessica Hoffman Davis, *The Art and Science of Portraiture*, San Francisco, CA: Josey-Bass, 1997, 9; Norman K. Denzin, "The New Ethnography," *Journal of Contemporary Ethnography*, 27 (1998), 405–415.

186. William Grimes, "Rudolph J. Vecoli, Scholar of Immigration, Is Dead at 81," *New York Times*, 23 Jun 2008.

187. The film *Black Fury* (Warner Brothers, 1935), directed by Michael Curtiz and staring Paul Muni, was a contemporary representation of the "labor wars" in Pennsylvania mining (probably in the bituminous fields). It included a strong ethnic component. "Underworld" organized crime was not part of the story although a crooked "upperworld" detective agency, which pur-

posely despoiled labor-management relations, was. The film proposed a conservative union leadership as labor's answer to its problems. See Rogin, "How the Working Class Saved Capitalism," 93–95.

188. Mary Jo Bona, *By the Breath of Their Mouths: Narratives of Resistance in Italian America*, Albany, NY: SUNY Press, 2010, 3.

189. Rudolph J. Vecoli, "Are Italian Americans Just White Folks?" 3–17 in Bona and Tamburri (eds.), *Through the Looking Glass*.

190. We acknowledge that the "mineworker as hero" has been forcefully reexamined especially by feminist writers who have critiqued the persistent patriarchy and the subservient and often exploitative situation of women in mining communities (see, e.g., Raphael Samuels, *Island Stories: Unraveling Britain*, London: Verso, 1999, 153–171). We also acknowledge that the anthracite mineworkers and their labor movements never had a lasting transformative political effect at the local or state level, nor was there any intention of their doing so. The election of Thomas Kennedy as lieutenant governor of Pennsylvania in 1933 was an historic moment for mineworkers and other unionists. Along with Governor George H. Earl, Kennedy became part of Pennsylvania's "Little New Deal." Yet, the labor policies promoted by the New Deal in Washington, Pennsylvania, and elsewhere focused on so-called business unionism and, therefore, incorporated few if any of the social democratic reforms achieved by western European labor unions. On the New Deal in Pennsylvania see Kenneth C. Wolensky, "An Activist Government in Harrisburg: Governor George H. Earl III and Pennsylvania's 'Little New Deal'," *Pennsylvania Heritage*, 34 (Winter 2008), 14–23; and Dublin and Licht, *The Face of Decline*, Chapter 3. Although women in the anthracite regions have often played front-line roles in labor movements, we found little evidence of their participation in the labor wars at the Erie Coal Companies. In Chapter 3 we discussed a group of women who joined a strike against the Erie Coal Companies in 1910, and in this chapter we reported on a group of Italian women who stormed the Squire of Archbald and his hired deputies during a subcontracting-related strike in 1896 at an Archbald Coal Company mine in Lackawanna County. Furthermore, we know that women were active in the strikes and protests initiated by the UAMP in the 1930s, as John Bodnar has shown in his book, *Anthracite People: Families, Unions, and Work, 1900–1940* (Harrisburg, PA: PHMC, 1983). However, while we can speculate about newspaper coverage and the role of women especially in Italian families, we are frankly uncertain as to why there was no historical record of women-initiated marches or demonstrations during the many strikes at PaCC and HC&I.

191. Daniel Taylor, *Tell Me a Story: The Life-Shaping Power of Stories*, St. Paul, MN: Bog Walk, 2001, 73.

192. Ronald J. Berger and Richard Quinney, *Storytelling Sociology: Narrative as Social Inquiry*, Boulder, CO: Lynne Reinner, 2005, 5.

REFERENCES

I. *Books, Articles, Dissertations, Academic Proceedings, Theses, and Unpublished Monographs*

Abern, Martin. International Labor Defense Activities, 1 Jan–1 Jul 1928, www.marxists.org/history/usa/parties/cpusa/1928/ildactabern.htm

Agnew, John A. *Place and Politics in Modern Italy*, Chicago, IL: University of Chicago Press, 2002.

Alba, Richard D. *Italian Americans: Into the Twilight of Ethnicity*, Englewood Cliffs, NJ: Prentice-Hall, 1985.

Albini, Joseph. *The American Mafia: Genesis of a Legend*, New York: Appleton-Century-Crofts, 1971.

Aldrich, Mark. "The Perils of Mining Anthracite: Regulation, Technology and Safety, 1870– 1945," *Pennsylvania History*, 64 (1997), 361–83.

———. *Safety First: Technology, Labor, and Business in the Building of American Work Safety, 1870–1934*, Baltimore, MD: Johns Hopkins University Press, 1997.

Allen, Frederick Lewis. *The Lords of Creation*, New York: Quadrangle, 1966 [1935].

Allen, Theodore W. *The Invention of the White Race (Vol. 1), Racial Oppression and Social Control*, London: Verso, 1994.

———. *The Invention of the White Race (Vol. 2), Origin of Racial Oppression in Anglo-America*, London, Verso, 1997.

Amsden, Jon, and Stephen Brier. "Coal Miners on Strike: The Transformation of Strike Demands and the Formation of a National Union," *Journal of Interdisciplinary History*, 8 (Spring 1977), 583–616.

Andrews, E. Benjamin. *History of the United States*, Vol. V, New York, Scribner's, 1912.

Appalachian Land Ownership Task Force. *Who Owns Appalachia: Landownership and Its Impact*, Lexington, KY: University Press of Kentucky, 1983.

Ashton, T.S., and Joseph Sykes. *The Coal Industry of the Eighteenth Century*, Manchester, England: Manchester University Press, 1964 [1929].

Aurand, Harold W. *From the Molly Maguires to the United Mineworkers: The Social Ecology of an Industrial Union, 1869–1897*, Philadelphia, PA: Temple University Press, 1971.

———. "Mine Safety and Social Control in the Anthracite Industry," *Pennsylvania History*, 52 (1985), 227-41.

———. *Anthracite Heritage Museum and Scranton Iron Furnaces, Pennsylvania Trail of History*, Mechanicsburg, PA: Stackpole, 2002.

———, guest ed. *Pennsylvania History*, Special Issue: The Lattimer Massacre—1897, 69 (2002).

———. *Coalcracker Culture: Work and Values in Pennsylvania Anthracite, 1835–1935*, Selinsgrove, PA: Susquehanna University Press, 2003.

Austin, Alan. *From Concentration Camp to Campus: Japanese American Students and World War II*, Urbana, IL: University of Illinois Press, 2007.

Avrich, Paul. *Sacco and Vanzetti: The Anarchist Background*, Princeton, NJ: Princeton University Press, 1991.

Bakerman, Theodore. *Anthracite Coal: A Study in Advanced Industrial Decline*, New York: Arno, 1979 [1956].

Banfield, Edward. *The Moral Basis of a Backward Society*, New York: Free Press, 1958.

Barendse, Michael A. "American Perceptions Concerning Slavic Immigrants in the Pennsylvania Anthracite Fields, 1880–1910: Some Comments on the Sociology of Knowledge," *Ethnicity*, 8 (1981), 96–105.

———. *Social Expectations and Perception: The Case of the Slavic Anthracite Workers*, University Park, PA: Penn State Press, 1981.

Barger, Harold, and Sam H. Schurr. *The Mining Industries, 1899–1939: A Study in Output, Employment and Productivity*, New York: National Bureau of Economic Research, 1944.

Baumol, W.J., A.S. Blinder, and E.N. Wolff. *Downsizing in America: Reality, Causes and Consequences*, New York: Russell Sage Foundation, 2003.

Bayard, Charles J. "The 1927–1928 Colorado Coal Strike," *Pacific Historical Review*, 32 (Aug 1963), 235–50.

Becker, Thea Gallo. *Images of America: Cleveland, 1796–1929*, Chicago, IL: Arcadia, 2004.

Beers, Paul. *Pennsylvania Politics, Today and Yesterday*, University Park, PA: Penn State Press, 1980.

Bencivenni, Marcella. *Italian Immigrant Radical Culture: The Idealism of the Sovversivi in the United States, 1890–1940*, New York: New York University Press, 2011.

Benson, Ronald M. "Commentary: The Family Economy and Labor Protest in Industrial America and the Coal and Iron Police in Anthracite Country," 120–25 in David L. Salay (ed.), *Hard Coal, Hard Times: Ethnicity and Labor in the Anthracite Region*, Scranton, PA: Anthracite Museum Press, 1984.

Berger, Peter. *Invitation to Sociology: A Humanistic Perspective*, Garden City, NY: Doubleday, 1963.

Berger, Ronald J., and Richard Quinney. *Storytelling Sociology: Narrative as Social Inquiry*, Boulder, CO: Lynne Reinner, 2005.

Bernstein, Irving. *The Lean Years: The History of the American Worker, 1920–1933*, Boston, MA: Houghton Mifflin, 1960.

Blatz, Perry K. "Local Leadership and Local Militancy: The Nanticoke Strike of 1899 and the Roots of Unionization in the Northern Anthracite Fields," *Pennsylvania History*, 58 (1991), 278–97.

———. *Democratic Miners: Work and Labor Relations in the Anthracite Coal Industry, 1875–1925*, Albany, NY: SUNY Press, 1994.

———. "Workplace Militancy and Unionization: The UMWA and the Anthracite Miners, 1890–1912," Chapter 3 in John H.M. Laslett (ed.), *The United Mine Workers of America: A Model of Industrial Solidarity?*, University Park, PA: Penn State Press & Penn State University Libraries, 1996.

———. "Titanic Struggles, 1873–1917," 83–148 in Howard Harris and Perry K. Blatz (eds.), *Keystone of Democracy: A History of Pennsylvania Workers*, Harrisburg, PA: PHMC, 1999.

Blatz, Perry K., and Robert P. Wolensky. "Rinaldo Cappellini, the Knox Mine Disaster, and the Decline of the Anthracite Coal Industry," 154–57 in Howard Harris and Perry Blatz (eds.), *Keystone of Democracy: A History of Pennsylvania Workers*, Harrisburg, PA: PHMC, 1999.

Bluestone, Barry, and Bennett Harrison. *The De-Industrialization of America: Plant Closings, Community Abandonment, and the Dismantling of Basic Industry*, New York: Basic, 1982.

Boal, William M. "New Estimates of Paid-up Membership in the United Mineworkers, 1920–1929, by State and Province," *Labor History*, 47 (2006), 537–46.

Bodnar, John. "Immigration and Modernization: The Case of Slavic Peasants in Industrial America," *Journal of Social History*, 14 (1976), 45–65.

———. "Socialization and Adaptation: Immigrant Families in Scranton, 1880–1900," *Pennsylvania History*, 43 (1976), 154–72.

———. *Anthracite People: Families, Unions and Work, 1900–1940*, Harrisburg, PA: PHMC, 1983.

———. "Commentary: The Family Economy and Labor Protest in Industrial America: Hard Coal Miners in the 1930s," 78–99 in David L. Salay (ed.), *Hard Coal, Hard Times: Ethnicity and Labor in the Anthracite Region*, Scranton, PA: Anthracite Museum Press, 1984.

———. *The Transplanted: A History of Immigration in Urban America*, Bloomington, IN: Indiana University Press, 1985.

Bogen, Jules I. *The Anthracite Railroads*, New York: Ronald, 1927.

Bohl, Don Lee. *Responsible Reductions in Force: An American Management Association Research Report on Downsizing and Outplacement*, New York: American Management Association, 1987.

Bona, Mary Jo. *By the Breath of Their Mouths: Narratives of Resistance in Italian America*, Albany, NY: SUNY Press, 2010.

Bona, Mary Jo, and Anthony J. Tamburri, eds. *Through the Looking Glass: Italian and Italian/American Images in the Media, Selected Essays from the 27th Annual Conference of the American Italian Historical Association, November 10–12, 1994*, Staten Island, NY: American Italian Historical Association, 1996.

Bonacich, Edna. "A Theory of Ethnic Antagonism: The Split Labor Market," *American Sociological Review*, 37 (Oct 1972), 547–59.

Bonilla-Silva, Eduardo. *Racism Without Racists: Color-Blind Racism and Racial Inequality in Contemporary America*, Landham, MD: Rowman & Littlefield, 2010.

Borisenkova, Anna., "Narrative Foundations of Knowing: Towards a New Perspective in the Sociology of Knowledge," *Sociological Research Online*, 14(5)17, www.socresonline.org.uk/14/5/17.html.

Borsella, Cristogianni. *On Persecution, Identity, and Activism: Aspects of the Italian-American Experience from the Late 19th Century to Today*, Boston, MA: Dante University Press, 2005.

Bradsby, Henry C., ed. *History of Luzerne County Pennsylvania*, Chicago, IL: Nelson, 1893.

Branca-Santos, Paula. "Injustice Ignored: The Internment of Italian-Americans during World War II," *Pace International Law Review*, 13 (2001), 151–82.

Brandt, Charles. *I Heard You Paint Houses: Frank "The Irishman" Sheeran and the Inside Story of the Mafia, the Teamsters, and the Last Ride of Jimmy Hoffa*, Hanover, NH: Steerforth, 2005.

Briggs, John W. *An Italian Passage: Immigrants to Three American Cities*, New Haven, CT: Yale University Press, 1978.

Brody, David. "Market Unionism in America: The Case of Coal," 131–74 in David Brody (ed.), *In Labor's Cause*, New York: Oxford University Press, 1993.

———. "Reconciling the Old Labor History and the New," *Pacific Historical Review*, 72 (Feb 1993), 111–26.

Brophy, John (John O.P. Hall, ed.). *A Miner's Life*, Madison, WI: University of Wisconsin Press, 1964.

Brundage, William Fitzhugh. *Lynching in the New South: Georgia and Virginia, 1880–1930*, Urbana, IL: University of Illinois Press, 1993.

Bryce, Robert. *Cronies: Oil, the Bushes, and the Rise of Texas, America's Superstate*, New York: Public Affairs, 2004.

Busbey, T. Addison, ed. *The Biographical Directory of Railway Officials of America*, Chicago, IL: Railway Age, 1901.

Bussacco, James. *Pittston's Coal Mining Era*, Pittston, PA: published by the author, 1995.

Bussel, Michael R. *From Harvard to the Ranks of Labor: Powers Hapgood and the American Working Class*, University Park, PA: Penn State Press, 1999.

Califano, Joseph A. Jr. *The Triumph and Tragedy of Lyndon Johnson*, College Station, TX: Texas A&M University Press, 2000.

———. *Inside–A Public and Private Life*, New York: Public Affairs, 2004.

Campbell, Stuart. *Businessmen and Anthracite: Aspects of Change in the Late Nineteenth Century Anthracite Industry*, unpublished doctoral dissertation, University of Delaware, 1978.

Cannistraro, Philip V., and Gerald Meyer, eds. *The Lost World of Italian-American Radicalism*, Westport, CT: Praeger, 2003.

Cannon, James P. "The Story of Alex Howat," *The Liberator*, 4 (Apr 1921), 25–26.

Carlton, Frank Tracy. *The Industrial Situation*, New York: Fleming H. Revell, 1914.

Carnes, Cecil. *John L. Lewis: Leader of Labor*, New York: Robert Speller, 1936.

Cartosio, Bruno. "Sicilian Radicals in Two Worlds," 117–28 in Marianne Debouzy (ed.), *In the Shadow of the Statue of Liberty: Immigrants, Workers, and Citizens in the American Republic, 1880–1920*, Urbana, IL: University of Illinois Press, 1992.

Casey, Hon. John. *Miners' Strike in the Bituminous Coal Fields*, Washington, DC: USGPO, 8 Feb 1928.

Cashmore, Ellis. *Martin Scorsese's America*, Malden, MA: Polity, 2009.

Casillo, Robert. *Gangster Priest: The Italian American Cinema of Martin Scorsese*, Toronto: University of Toronto Press, 2006.

Catterall, Stephen, and Keith Gildart. "Outsiders: Trade Union Responses to Polish and Italian Coal Miners in Two British Coalfields, 1945–54," 164–176 in Stefan Berger, Andy Croll, and Norman LaPorte (eds.), *Toward a Comparative History of Coalfield Societies*, Burlington, VT: Ashgate, 2005.

Cateura, Linda Brandi. *Growing Up Italian: How Being Brought Up as an Italian-American Helped Shape the Characters, Lives, and Fortunes of Twenty-four Celebrated Americans*, New York: William Morrow, 1987.

Chandler, Alfred D. "Anthracite Coal and the Beginnings of the Industrial Revolution in the United States," *Business History Review*, 46 (1972), 141–81.

Chernow, Ron. *The House of Morgan: An American Banking Dynasty and the Rise of Modern Finance*, New York: Simon & Schuster-Touchstone, 1990.

Child, Irvin L. *Italian or American? The Second Generation in Conflict*, New Haven, CT: Yale University Press, 1943.

Church, Roy, and Quentin Outram, *Strikes and Solidarity: Coalfield Conflict in Britain, 1889–1966*, New York: Cambridge University Press, 1998.

Cinel, Dino, *From Italy to San Francisco: The Immigrant Experience*, Stanford, CA: Stanford University Press, 1982.

Ciongoli, A. Kenneth, and Jay Parini. *Passage to Liberty: The Story of Italian Immigration and the Rebirth of America*, New York: Regan, 2002.

Clark, Dennis J. *The Irish in Philadelphia: Ten Generations of Urban Experience*, Philadelphia, PA: Temple University Press, 1973.

Clark J.A., ed. *The Wyoming Valley, Upper Waters of the Susquehanna, and the Lackawanna Coal-Region*, Scranton, PA: J.A. Clark, 1875.

Clark, Norman H. *Mill Town: A Social History of Everett, Washington*, Seattle, WA: University of Washington Press, 1970.

Clawson, Dan. *Bureaucracy and the Labor Process: The Transformation of U.S. Industry, 1860–1920*, New York: Monthly Review, 1980.

Clinard, Marshall B. *Corporate Corruption: The Abuse of Power*, New York: Praeger, 1990.

Clinard, Marshall B., and Peter C. Yeager. *Corporate Crime*, New York: Free Press, 2006 [1980].

Colatrella, Steven. *Workers of the World: African and Asian Migrants in Italy in the 1990s*, Trenton, NJ: Africa World Press, 2001.

Colby, Frank Moore, et al. *The New International Year Book*, New York: Dodd & Mead, 1914.

Coleman, McAlister. *Men and Coal*, New York: Farrar & Rinehart, 1943.

Coleman, Walter J. *The Molly Maguire Riots*, Richmond, VA: Garrett & Massey, 1936.

Conlin, Joseph R. *A Study of the Industrial Workers of the World before World War I*, unpublished doctoral dissertation, University of Wisconsin, 1966.

Connell, William J., and Fred Gardaphe, eds. *Anti-Italianism: Essays on Prejudice*, New York: Palgrave Macmillan, 2010.

Connolly, Mary Theresa. *'The Gravity': History of the Pennsylvania Coal Company Railroad, 1850–1885*, Olyphant, PA: Barrett, 1972.

Cooke, Edward F., and Edward G. Janosik. *Pennsylvania Politics*, New York: Holt, Rinehart & Winston, 1965.

Corbin, David A. *Life, Work, and Rebellion in the Coal Fields: The Southern West Virginia Miners, 1880–1922*, Champaign, IL: University of Illinois Press, 1981.

Corlsen, Carl. *Buried Black Treasure: The Story of Pennsylvania Anthracite*, Dansville, NY: F.A. Owen, 1954.

Cornell, Robert. *The Anthracite Coal Strike of 1902*, Washington, DC: Catholic University Press, 1957.

Couch, Stephen R. "The Coal and Iron Police in Anthracite Country," 100–19 in David L. Salay (ed.), *Hard Coal, Hard Times: Ethnicity and Labor in the Anthracite Region*, Scranton, PA: Anthracite Museum Press, 1984.

Crispino, James A. *The Assimilation of Ethnic Groups: The Italian Case*, New York: Center for Migration Studies, 1980.

Critchley, David. *The Origin of Organized Crime in America: The New York City Mafia, 1891–1931*, New York: Routledge, 2009.

Curran, Daniel J. *Dead Laws for Dead Men: The Politics of Federal Coal Mine Health and Safety Legislation*, University of Pittsburgh Press, 1993.

Currarino, Rosanne. *The Labor Question in America: Economic Democracy in the Gilded Age*, Urbana, IL: University of Illinois Press, 2011.

D'Acierno, Pellegrino. *The Italian-American Heritage: A Companion to Literature and Arts*, New York: Routledge-Garland Reference Library of the Humanities, 1999.

Dacus, J.A. *Annals of the Great Strikes in the United States*, Chicago, IL: L.T. Palmer, 1877.

Daggett, Stuart. *Railroad Reorganization*, Washington, DC: Beard, 1999 [1908].

D'Angelo, Pascal. *Son of Italy*, New York: Arno, 1975 [1924].

Davies, Edward J. *The Anthracite Aristocracy: Leadership and Social Change in the Hard Coal Regions of Northeastern Pennsylvania, 1800–1930*, DeKalb, IL: Northern Illinois University Press, 1985.

DeGerolamo, George. *Autobiography*, Pittston, PA: published by the author, 1980.

Del Giudice, Luisa, ed. *Oral History, Oral Culture, and Italian Americans*, New York: Palgrave Macmillan, 2009.

Denzin, Norman K. *Interpretive Biography*, Thousand Oaks, CA: Sage, 1989.

———. "The New Ethnography," *Journal of Contemporary Ethnography*, 27 (1998), 405–15.

Denzin, Norman K., and Yvonna S. Lincoln, eds. *Collecting and Interpreting Qualitative Materials*, Thousand Oaks, CA: Sage, 1998.

Dickie, John. *Cosa Nostra: The History of the Sicilian Mafia*, New York: Palgrave Macmillan, 2004.

DiSalvo, Jackie. "Father James E. Groppi (1930–1985): The Militant Humility of a Civil Rights Activist," 229–43 in Philip V. Cannistraro and Gerald Meyer (eds.), *The Lost World of Italian-American Radicalism*, Westport, CT: Praeger, 2003.

DiStasi, Lawrence, ed. *Una Storia Segreta: The Secret History of Italian American Evacuation and Internment during World War II*, Berkeley, CA: Heyday, 2004.

Dix, Keith. *What's a Coal Miner to Do? The Mechanization of Coal Mining*, Pittsburgh, PA: University of Pittsburgh Press, 1988.

Douglas, Norman. *Old Calabria*, New York, Cosimo, 2007 [1915].

Douglass, Dave. "The Durham Pitman," 226–36 in Raphael Samuel (ed.), *Miners, Quarrymen and Saltworkers*, London: Routledge & Kegan Paul, 1977.

Dublin, Thomas. "The Equalization of Work: An Alternative Vision of Industrial Capitalism in the Anthracite Region of Pennsylvania in the 1930s," 81–98 in Lance E. Metz (ed.), *Canal History and Technology Proceedings* Vol. XIII, Easton, PA: Canal History and Technology Press, 1994.

Dublin, Thomas, and Walter Licht. *The Face of Decline: The Pennsylvania Anthracite Region in the Twentieth Century*, Ithaca, NY: Cornell University Press, 2005.

Dubofsky, Melvyn. *We Shall Be All: A History of the Industrial Workers of the World*, New York: Quadrangle/New York Times, 1969.

———. "The 'New' Labor History: Achievements and Failures," *Reviews in American History*, 5 (Jun 1977), 249–54.

Dubofsky, Melvyn, and Warren Van Tine. *John L. Lewis: A Biography*, New York: Quadrangle/New York Times, 1977.

Dury, Horace B. "Wages in the Coal Industry as Compared with Wages in other Industries (With Special Reference to the Anthracite Situation)," *Annals of the American Academy of Political and Social Science*, 111 (1924), 314–43.

English, T.J. *Paddy Whacked: The Untold Story of the Irish-American Gangster*, New York: HarperCollins, 2005.

Encenbarger, William. *Kids for Cash: Two Judges, Thousands of Children, and a $2.6 Million Kickback Scheme*, New York: The New Press, 2012

Feldstein, Stanley, and Lawrence Costello, eds. *The Ordeal of Assimilation: A Documentary History of the White Working Class, 1830s to the 1970s*, Garden City, NY: Anchor-Doubleday, 1974.

Fenton, Edwin. "Italians in the Labor Movement," *Pennsylvania History*, 26 (Apr 1959), 143.

Ferragina, Emanuele. "The Never-Ending Debate about *The Moral Basis of a Backward Society: Banfield and 'Amoral Familism'*," JASO-Online (formerly *Journal of the Anthropological Society of Oxford*), 1 (Winter 2009), 141–60.

Ferroni, Charles D. *Italians in Cleveland: A Study in Assimilation*, New York: Arno, 1980.

Filippelli, Ronald L. *The Schuylkill Navigation Company and Its Role in the Development of the Anthracite Coal Trade and Schuylkill County, 1815–1845*, unpublished M.A. thesis, Penn State University, 1966.

Finley, Joseph E. *The Corrupt Kingdom: The Rise and Fall of the United Mine Workers*, New York: Simon and Schuster, 1972.

Flinn, Michael W., and David Stoker. *The History of the British Coal Industry*, Vol. 2, Oxford, England: Clarendon, 1984.

Flynn, Elizabeth Gurley. *The Rebel Girl–An Autobiography, My First Life (1906–1926)*, New York: International Publishers, 1973 [1955].

Flynn, William J. *The Barrel Mystery*, New York: James A. McCann, 1919.

Foerster, Robert F. *The Italian Emigration of Our Times*, Cambridge, MA: Harvard University Press, 1924.

Folsom, Burton W. Jr. *Urban Capitalists: Entrepreneurs and City Growth in Pennsylvania's Lackawanna and Lehigh Regions, 1800–1920*, Baltimore, MD: Johns Hopkins University Press, 1981.

Foner, Philip S. *The History of Labor in the United States: The Policies and Practices of the American Federation of Labor, 1900–1909*, New York: International Publishers, 1981 [1964].

———. *History of the Labor Movement in the United States, Vol. 4: The Industrial Workers of the World 1905–1917*, International Publishers, 1972.

———. *Women and the American Labor Movement*, New York: Free Press, 1982.

———. *History of the Labor Movement in the United States, The T.U.E.L., 1925–1929*, Vol. 10, New York: International Publishers, 1994.

Fox, Maier. *United We Stand: A History of the United Mine Workers of America, 1890–1990*, Washington, DC: United Mine Workers of America, 1990.

Fox, Stephen. *Blood and Power: Organized Crime in Twentieth Century America*, New York: Penguin, 1989.

Fraser, Steve. "The 'Labor Question'," 55–84 in Steve Fraser and Gary Gerstle (eds.), *The Rise and Fall of the New Deal Order, 1930–1980*, Princeton, NJ: Princeton University Press, 1989.

Frisch, Michael. *A Shared Authority: Essays of the Craft and Meaning of Oral History*, Albany, NY: SUNY Press, 1990.

Gabaccia, Donna R. "Neither Padrone Slaves nor Primitive Rebels: Sicilians on Two Continents," 95–117 in Dirk Hoerder (ed.), *"Struggle a Hard Battle": Essays on Working-Class Immigrants*, DeKalb, IL: University of Northern Illinois Press, 1986.

———. *Militants and Migrants: Rural Sicilians Become American Workers*, New Brunswick, NJ: Rutgers University Press, 1988.

Gabaccia, Donna R., and Fraser M. Ottanelli, eds. *Italian Workers of the World: Labor Migration and the Formation of Multiethnic States*, Urbana, IL: University of Illinois Press, 2005.

Gallagher, Rev. John P. *A Century of History: The Diocese of Scranton, 1868–1968*, Scranton, PA: The Diocese of Scranton, 1968.

Gans, Herbert. *Urban Villagers: Group and Class in the Life of Italian-Americans*, New York: Free Press, 1982 [1962].

Gedicks, Al. *The New Resource Wars: Native and Environmental Struggles Against Multinational Corporations*, Boston, MA: South End, 1993.

———. *Resource Rebels: Native Challenges to Mining and Oil Corporations*, Brooklyn, NY: South End, 2001.

———. *Dirty Gold: Indigenous Alliances to End Global Resource Colonialism*, Brooklyn, NY: South End, 2009.

Giordano, Paolo A., and Anthony Julian Tamburri, eds. *Italian-Americans in the Third Millennium*, New York: Bordighera, 2009.

Glazer, Nathan, and Daniel Patrick Moynihan. *Beyond the Melting Pot: The Negroes, Puerto Ricans,* Jews, Italians, *and Irish of New York City*, Cambridge, MA: Harvard University Press, 1963.

Gluck, Elsie. *John Mitchell, Miner: Labor's Bargain with the Gilded Age*, New York: John Day, 1929.

Goin, Peter, and Elizabeth Raymond. "Living in Anthracite: Mining Landscape and Sense of Place in Wyoming Valley, Pennsylvania," *The Public Historian*, 23 (Spring 2001), 29–45.

Golab, Caroline. *Immigrant Destinations*, Philadelphia, PA: Temple University Press, 1977.

Gooch, Glenn. "Pennsylvania Gas & Water Company: A History," unpublished manuscript, Pennsylvania Gas & Water Company, Wilkes-Barre, PA, 1976.

Goodell, Jeff. *Our Story: 77 Hours That Tested Our Friendship and Our Faith*, New York: Hyperion, 2002.

Goodrich, Carter. *The Miner's Freedom: A Study of the Working Life in a Changing Industry*, New York: Arno, 1977 [1925].

———. "Contract Labor," 342–44 in Edwin R.A. Seligman (ed.), *Encyclopaedia of the Social Sciences*, Vol. 4, New York: Macmillan, 1931.

Goodrich, Carter, et al. *Migration and Economic Opportunity*, Philadelphia, PA: University of Pennsylvania Press, 1936.

Gordon, David M., Richard Edwards, and Michael Reich. *Segmented Work, Divided Workers: The Historical Transformation of Labor in the United States*, New York: Cambridge University Press, 1982.

Gordon, Milton. *Assimilation in American Life: The Role of Race, Religion, and National Origins*, New York, Oxford University Press, 1964.

Gordon, Robert. "Custom and Consequence: Early Nineteenth-Century Origins of the Environmental and Social Costs of Mining Anthracite," 240–77 in Judith A. McGraw (ed.), *Early American Technology: Making and Doing Things From the Colonial Era to 1850*, Chapel Hill, NC: University of North Carolina Press, 1994.

Gottschalk, Peter. "Stages of Financial Crime by Business Organizations," *Journal of Financial Crime*, 15 (2008), 38–48.

Gowaskie, Joe. "John Mitchell and the Anthracite Mine Workers: Leadership Conservatism and Rank-and-File Militancy," *Labor History*, 27 (1985), 54–83.

Green, James R. "Tying the Knot of Solidarity: The Pittston Strike of 1989–1990," Chapter 21 in John H.M. Laslett (ed.), *The United Mine Workers of America: A Model of Industrial Solidarity?*, University Park, PA: Penn State Press & Penn State University Libraries, 1996.

Greene, Victor R. "A Study in Slavs, Strikes, and Unions: The Anthracite Strike of 1897," *Pennsylvania History*, 31 (1964), 199–215.

———. *The Slavic Community on Strike: Immigrant Labor in Pennsylvania Anthracite*, South Bend, IN: University of Notre Dame Press, 1968.

Greenhouse, Steven. *The Big Squeeze: Tough Times for American Workers*, New York: Alfred A. Knopf, 2008.

Greeley, Andrew. "The Transmission of Cultural Heritages: The Case of the Irish and the Italians," 209–35 in Nathan Glazer and Daniel P. Moynihan (eds.), *Ethnicity: Theory and Experience*, Cambridge, MA: Harvard University Press, 1975.

Grifo, Richard D., and Anthony F. Noto. *Italian Presence in Pennsylvania*, University Park, PA: Pennsylvania Historical Association, 1990.

Gudelunas, William A. Jr., and William G. Shade. *Before the Molly Maguires: The Emergence of the Ethno-Religious Factor in the Politics of the Lower Anthracite Region, 1844–1872*, New York: Arno Press, 1976.

Guglielmo, Jennifer, and Salvatore Salerno, eds. *Are Italians White?: How Race Is Made in America*, New York: Routledge, 2003.

Gumina, Deanna Paoli. *The Italians of San Francisco, 1850–1930*, New York: Center for Migration Studies, 1978.

Gustafson, William. "Bootleg Mining in Pennsylvania," 1–24 in Robert Mittrick (ed.), *Proceedings of the Ninth Annual Conference on the History of Northeastern Pennsylvania*, Nanticoke, PA: Luzerne County Community College, 1997.

Gutman, Herbert G. "The Buena Vista Affair, 1874–1875," *Pennsylvania Magazine of History and Biography*, 88 (Jul 1964), 251–92.

———. "Work, Culture, and Society in Industrializing America, 1815–1919," *American Historical Review*, 78 (Jun 1973), 531–58.

Hanlon, Edward F. *The Wyoming Valley: An American Portrait*, Woodland Hills, CA: Windsor, 1983.

Harrison, Royden, ed. *Independent Collier: The Coal Miner as Archetypal Proletarian Reconsidered*, New York: St. Martin's, 1978.

Harvey, Oscar J., and Ernest G. Smith. *A History of Wilkes-Barre*, Vol. 4, Wilkes-Barre, PA: Smith Bennett, 1929.

Harvey, Oscar J., and Harrison G. Smith. *A History of Wilkes-Barre*, Vol. 5, Wilkes-Barre, PA: Smith Bennett, 1930.

Harwood, Herbert H. Jr. *Invisible Giants: The Empires of Cleveland's Van Sweringen Brothers*, Bloomington, IN: Indiana University Press, 2003.

Hast, Adele, ed. "The Pittston Company," 180–82 in *International Directory of Company Histories*, Vol. 4, Chicago, IL: St. James, 1991.

Hays, Samuel P. *The Response to Industrialism, 1885–1924*, Chicago, IL: University of Chicago Press, 1957.

Healey, Richard G. *The Pennsylvania Anthracite Coal Industry, 1860–1902: Economic Cycles, Business Decision-Making and Regional Dynamics*, Scranton, PA: University of Scranton Press, 2008.

Hine, Lewis. "The High Cost of Child Labor," *The Child Labor Bulletin*, 3 (1915), 63–67.

Hirschman, Albert O. *Exit, Voice and Loyalty: Responses to Decline in Firms, Organizations and States*, Cambridge, MA: Harvard University Press, 1970.

History of Luzerne, Lackawanna, and Wyoming Counties, New York: Munsell, 1880.

Hobsbawm, Eric J. *Primitive Rebels: Studies in Archaic Forms of Social Movement in the 19th and 20th Centuries*, New York: Norton, 1959.

———. *Industry and Empire*, New York: Pantheon, 1968.

Hoffman, Frederick L. "Problems of Labor and Life in Anthracite Coal Mining," *Engineering and Mining Journal*, 74 (1902), 675–76.

Hollister, Horace. *History of the Lackawanna Valley*, Philadelphia, PA: Lippincott, 1885.

Hostetter, G.L., and T.Q. Beesley. "20th Century Crime," 49–58 in Gus Tyler (ed.), *Organized Crime in America*, Ann Arbor, MI: University of Michigan Press, 1962 [1933].

Howard, Walter T. "Anti-Labor and Anti-Radical Violence in Northeastern Pennsylvania During the Great Depression," 74–100 in Robert Mittrick (ed.), *Proceedings of the Eighth Annual Conference on the History of Northeastern Pennsylvania*, Luzerne County Community College, 1995.

———. "The International Labor Defense and Antiradical Repression in The Pennsylvania Anthracite, 1928–1929," *Carver*, 15 (Spring 1998), 41–46.

———. "The National Miners Union: Communists and Miners in the Pennsylvania Anthracite, 1928–1931," *Pennsylvania Magazine of History and Biography*, 125 (Jan/Apr 2001) 104–05.

———, ed. *Anthracite Reds: A Documentary History of Communists in Northeastern Pennsylvania during the 1920s*, Vol. I, New York: iUniverse, 2004.

———. *Forgotten Radicals: Communists in the Pennsylvania Anthracite, 1919–1950*, Landham, MD: University Press of America, 2005.

Hudson Coal Company. *The Story of Anthracite*, New York: Hudson Coal Company, 1932.

Hudson, Samuel. *Pennsylvania and its Public Men*, Philadelphia, PA: published by the author, 1909.

Hunt, Edward E., F.G. Tryon, and Joseph H. Willits. *What the Coal Commission Found: An Authoritative Summary by the Staff*, Baltimore, MD: Williams & Wilkins, 1925.

Ignatiev, Noel. *How the Irish Became White*, New York: Routledge, 1995.

Iorizzo, Luciano John. *Italian Immigration and the Impact of the Padrone System*, New York: Arno, 1980 [1966].

Jacobson, Matthew Frye. *Whiteness of a Different Color: European Immigrants and the Alchemy of Race*, Cambridge, MA: Harvard University Press, 1998.

Janosov, Robert A. "Glen Alden's Huber Breaker: 'A Marvel of Mechanism'," 103–44 in Lantz Metz (ed.), *Canal History and Technology Proceedings* Vol. XI, Easton, PA: Canal History and Technology Press, 1992.

Jenkins, Philip. *Hoods and Shirts: The Extreme Right in Pennsylvania, 1925–1950*, Chapel Hill, NC: University of North Carolina Press, 1997.

Jones, Eliot. *The Anthracite Coal Combination in the United States*, Cambridge, MA: Harvard University Press, 1914.

Jones, Patrick. "'Not a Color But an Attitude': Fr. James Groppi and Black Power Politics in Milwaukee," 259–81 in Jeanne Theoharris and Komozi Woodard (eds.), *Groundwork: Local Black Freedom Movements*, New York: NYU Press, 2005.

———. *The Selma of the North: Civil Rights Insurgency in Milwaukee*, Cambridge, MA: Harvard University Press, 2009.

Josephson, Matthew. *The Robber Barons*, New York: Harcourt, Brace & World, 1962 [1934].

Joy, Agnes. "John Mitchell and Religious Leaders of the Period 1900–1910," unpublished M.A. thesis, Catholic University of America, 1954.

Juliani, Richard N. *Priest, Parish, and People: Saving the Faith in Philadelphia's "Little Italy,"* South Bend, IN: University of Notre Dame Press, 2007.

Kanarek, Harold. "Disaster for Hard Coal: The Anthracite Strike of 1925–1926," *Labor History*, 15 (1974), 44–62.

———. "The Pennsylvania Anthracite Strike of 1922," *Pennsylvania Magazine of History and Biography*, 99 (1975), 207–25.

Kehoe, John. *Autobiography*, Pittston, PA: published by the author, no date.

Keil, Thomas. *On Strike! Capital Cities and the Wilkes-Barre Newspaper Unions*, Tuscaloosa, AL: University of Alabama Press, 1988.

Kelly, Robert J. *The Upperworld and the Underworld: Case Studies of Racketeering and Business Infiltrations in the United States*, New York: Kluwer Academic/Plenum, 1999.

Kemp, Matthew Stanley. *Boss Tom: The Annals of an Anthracite Mining Village*, Akron, OH: Saalfield, 1904.

Kenny, Kevin. *Making Sense of the Molly Maguires*, New York: Oxford University Press, 1998.

———, ed. *New Directions in Irish-American History*, Madison, WI: University of Wisconsin Press, 2003.

Knies, Michael. "The Pennsylvania Coal Company: New Insights from the James Archbald Papers, 1852–1853: 'We Are Now at War'," 45–76 in Lance Metz (ed.), *Canal History and Technology Proceedings* Vol. XXIII, Easton, PA: Canal History and Technology Press, 2004.

Korson, George. *Minstrels of the Mine Patch: Songs and Stories of the Anthracite Industry*, Philadelphia, PA: University of Pennsylvania Press, 1938.

———. *Black Land: The Way of Life in the Coal Fields*, Evanston, IL: Row & Petersen, 1941.

Krase, Jerome. "Between Columbus and Cuomo: The Italian-American Experience in America," *Italian American Review* 6 (Spring–Summer, 1997), 29–44.

———. "Shark Tale–'Puzza da Cap': An Attempt at Ethnic Activism," 137–50 in William J. Connell and Fred Gardaphe (eds.), *Anti-Italianism: Essays on Prejudice*, New York: Palgrave Macmillan, 2010.

LaGumina, Salvatore J. *Wop: A Documentary History of Anti-Italian Discrimination*, Toronto: Guernica, 1999 [1973].

———. "Prejudice and Discrimination: The Italian-American Experience Yesterday and Today," 108–15 in William J. Connell and Fred Gardaphe (eds.), *Anti-Italianism: Essays on Prejudice*, New York: Palgrave Macmillan, 2010.

LaGumina, Salvatore J. et al., eds. *The Italian American Experience: An Encyclopedia*, New York: Garland, 2000.

Lalli, Michael. "The Italian-American Family: Assimilation and Change, 1900–1965," *The Family Coordinator*, 18 (Jan 1969), 44–48.

Larkin, Thomas E. "History of Labor Relations in the Anthracite Industry," 205–18 in *Transactions of the Fifth Annual Anthracite Conference of Lehigh University*, Bethlehem, PA, 8–9 May 1947.

Laslett, John H.M. *Labor and the Left: A Study of Socialist and Radical Influences in the American Labor Movement, 1881–1924*, New York: Basic Books, 1970.

———, ed. *The United Mine Workers of America: A Model of Industrial Solidarity?*, University Park, PA: Penn State Press & Penn State University Libraries, 1996.

La Sorte, Michael. *La Merica: Images of Italian Greenhorn Experience*, Philadelphia, PA: Temple University Press, 1985.

Lauck, W. Jett, and Edgar Sydenstricker. *Conditions of Labor in American Industries*, New York: Arno, 1917.

Laurino, Maria. *Were You Always Italian?* New York: Norton, 2000.

Lawrence-Lightfoot, Sara, and Jessica Hoffman Davis. *The Art and Science of Portraiture*, San Francisco, CA: Josey-Bass, 1997.

Lewis, John L. *The Miners' Fight for the American Standard*, Indianapolis, IN: Bell, 1925.

Lieberson, Stanley. *From Many Strands: Ethnic and Racial Groups in Contemporary America*, New York: Russell Sage Foundation, 1988.

Long, Priscilla. *Where the Sun Never Shines: A History of America's Bloody Coal Industry*, New York: Paragon, 1989.

Longo, Stephanie. *Italians of Northeastern Pennsylvania*, Chicago, IL: Arcadia, 2004.

Loucks, Emerson H. *The Ku Klux Klan in Pennsylvania: A Study in Nativism*, Harrisburg, PA: Telegraph Press, 1936.

Lovejoy, Owen R. "The Extent of Child Labor in the Anthracite Coal Industry," Annals of the American Academy of Political and Social Science, 29 (Jan 1907), 35–49.

Luconi, Stefano. "'Mobsters' at the Polls: The Mafia Stereotype of the Media and Italian-American Voters in Philadelphia in the Early 1950s," 38–50 in Mary Jo Bona and Anthony J. Tamburri (eds.), *Through the Looking Glass: Italian & Italian/American Images in the Media, Selected Essays from the 27th Annual Conference of the American Italian Historical Association, November 10–12, 1994*, Staten Island, NY: American Italian Historical Association, 1996.

———. "Mafia-Related Prejudice and the Rise of Italian Americans in the United States," *Patterns of Prejudice*, 33 (1999), 43–57.

Lupo, Salvatore. *History of the Mafia*, New York: Columbia University Press, 2009.

Lynch, Patrick M. *Pennsylvania Anthracite: A Forgotten IWW Venture, 1906–1916*, unpublished M.A. thesis, Bloomsburg State College, 1974.

Lynd, Staughton. *"We Are All Leaders": The Alternative Unionism of the Early 1930's*, Champaign, IL: University of Illinois Press, 1966.

Maas, Peter. *The Valachi Papers*, New York: Bantam, 1969.

MacGregor, Morris J. *Steadfast in the Faith: The Life of Patrick Cardinal O'Boyle*, Washington, DC: Catholic University of America Press, 2006.

Macquown, W.C. *Maps of Anthracite Coal Fields of Northeastern Pennsylvania*, Pittsburgh, PA: National Coal Publications, 1942, 1952.

Maines, David R., and Jeffrey Ulmer. "The Relevance of Narrative for Interactionist Thought," *Studies in Symbolic Interaction*, 14 (1993), 109–24.

Maran, Arnold George D. *Mafia—Inside the Dark Heart: The Rise and Fall of the Sicilian Mafia*, New York: Thomas Dunne, 2010.

Marcantonio, Vito. *Labor's Martyrs (Haymarket, 1887; Sacco and Vanzetti, 1927)*, Project Gutenberg EBook, 2004, www.gutenberg.org/files/11009/11009-h/11009-h.htm.

Markovitz, Jonathan. *Legacies of Lynching: Racial Violence and Memory*, Minneapolis, MN: University of Minnesota Press, 2004.

Martinelli, Phylis Cancilla. *Undermining Race: Ethnic Identities in Arizona Copper Camps, 1880– 1920*, Tucson, AZ: University of Arizona Press, 2009.

Mattes, Philip V. "The Mine-Cave Struggle," 368–83 in Thomas Murphy (ed.), *Jubilee History of Lackawanna County, Pennsylvania*, Topeka, KS: Historical Publishing, 1928.

———. "The Mine-Cave Problem," and "Kohler-Fowler Acts," 15–28 in *Tales of Scranton*, Scranton, PA: Published by the author, no date.

McKinney, Karyn D. *Being White: Stories of Race and Racism*, New York: Routledge, 2005.

Mead, Richard R. *An Analysis of the Decline of the Anthracite Industry Since 1921*, unpublished doctoral dissertation, University of Pennsylvania, 1933.

Merrick, Sr. Mary Annunciata, R.S.M. *A Case in Practical Democracy: Settlement of the Anthracite Coal Strike of 1902*, unpublished doctoral dissertation, University of Notre Dame, 1942.

Migliore, Sam, and A. Evo DiPierro. *Italian Lives, Cape Breton Memories*, Sydney, Nova Scotia, Canada: Cape Breton University Press, 1999.

Miller, Donald L., and Richard E. Sharpless. *The Kingdom of Coal: Work, Enterprise, and Ethnic Communities in the Mine Fields*, Philadelphia, PA: University of Pennsylvania Press, 1985.

Miller, E. Willard. "The Anthracite Region of Northeastern Pennsylvania: An Economy in Transition," *Journal of Geography*, 88 (1989), 167–71.

———, ed. *A Geography of Pennsylvania*, University Park, PA: Penn State Press, 1995.

Miller, Sally M. *The Radical Immigrant*, New York: Twayne Publishers, 1974.

Minor, George H. *The Erie System: The Organization and Corporate History*, New York: Erie Railroad, 1938.

Mitchell, John. *Organized Labor*, Philadelphia, PA: American Book & Bible House, 1903.

Mittrick, Robert. "John S. Fine: The Rise and Fall of a Political Boss," unpublished manuscript, Luzerne County Community College, Nanticoke, PA, no date.

Moe, Nelson. *The View from Vesuvius: Italian Culture and the Southern Question*, Berkeley, CA: University of California Press, 2002.

Monroe, Douglas K. *A Decade of Turmoil: John L. Lewis and the Anthracite Miners, 1926–1936*, unpublished doctoral dissertation, Georgetown University, 1976.

Montgomery, David. *Workers' Control in America: Studies in the History of Work, Technology, and Labor Struggles*, New York: Cambridge University Press, 1979.

———. *The Fall of the House of Labor: The Workplace, the State, and American Labor Activism, 1865–1925*, New York: Cambridge University Press, 1987.

Moramarco, Frederico and Stephen. *Italian Pride: 101 Reasons to Be Proud You're Italian*, New York: Citadel/Kensington, 2000.

Mosley, Philip. "Screen Images of Italian Immigrants to the Belgian Coal Mines," unpublished manuscript, Penn State University–Worthington Scranton, 2005.

Mott, Edward H. *Between the Ocean and the Lakes: The Story of Erie*, New York: Ticher, 1908.

Murphy, Alyssa. "The Martyred Miners of Lattimer," www.pabook.libraries.psu.edu/palitmap/Lattimer.html.

Murphy, Raymond E., and Marion Murphy. "Anthracite Region of Pennsylvania," *Economic Geography*, 14 (1938), 338–48.

Murray, Alan. *Holding the Line: A Narrative History of Australian Coal Miners and their Union in the 1980s*, Sydney, Australia: Coal Forestry Mining & Energy Union, 2009.

Nafisi, Azar. *Reading Lolita in Tehran: A Memoir in Books*, New York: Random House, 2003.

Nearing, Scott. *Anthracite: An Instance of Natural Resource Monopoly*, Freeport, NY: Books for Libraries, 1915.

Nelli, Humbert S. "The Italian Padrone System in the United States," *Labor History*, 5 (1964), 153–67.

———. *The Business of Crime*, New York: Oxford University Press, 1976.

———. *From Immigrants to Ethnics: The Italian Americans*, New York: Oxford University Press, 1983.

Nelson, Daniel. *Frederick Winslow Taylor and the Rise of Scientific Management*, Madison, WI: University of Wisconsin Press, 1980.

Nelson, Steve, James R. Barrett, and Rob Ruck. *Steve Nelson: American Radical*, Pittsburgh, PA: University of Pittsburgh Press, 1981.

Novak, Michael. *The Guns of Lattimer*, New York: Basic, 1978.

O'Rourke, Lawrence M. *Geno: The Life and Mission of Geno Baroni*, Mahwah, NJ: Paulist, 1991.

Ottanelli, Fraser M. "'If Fascism Comes to American We Will Push It Back into the Ocean': Italian-American Antifascism in the 1920s and 1930s," 178–95 in Donna R. Gabaccia and Fraser M. Ottanelli (eds.), *Italian Workers of the World: Labor Migration and the Formation of Multiethnic States*, Urbana, IL: University of Illinois Press, 2005.

Parker, Randall E. *The Economics of the Great Depression: A Twenty-first Century Look Back at the Interwar Years*, Northhampton, MA: Edward Elgar, 2007.

Parton, W. Julian. *The Death of a Great Company: Reflections on the Decline and Fall of the Lehigh Coal and Navigation Company*, Easton, PA: Canal History and Technology Press, 1986.

Patton, Spiro. "Delaware & Hudson Company vs. Pennsylvania Coal Company During the 1850s," 7–24 in Lance E. Metz (ed.), *Canal History and Technology Proceedings* Vol. XI, Easton, PA: Canal History and Technology Press, 1992.

Peck, Gunther. "Reinventing Free Labor: Immigrant Padrones and Contract Labor in North America, 1885–1925," *Journal of American History*, 83 (1996), 848–71.

Perlman, Selig. *Upheaval and Reorganization (Since 1876)*, Part 4, Vol. 2, in John R. Commons et al., *History of Labour in the United States*, New York: Macmillan, 1918.

———. *A History of Trade Unionism in the United States*, New York: Macmillan, 1922.

Phelan, Craig. *Divided Loyalties: The Public and Private Life of Labor Leader John Mitchell*, Albany, NY: SUNY Press, 1994.

———. *Grand Master Workman: Terrence Powderly and the Knights of Labor*, Westport, CT: Greenwood, 2000.

Phillips, S.J. "Water Hazards in the Lackawanna Valley," Paper presented to the Anthracite Section of the American Institute of Mining and Metallurgical Engineers, 29 Jun 1935, Anthracite Heritage Museum, Scranton, PA, uncatalogued.

Phillips, Wendell. *The Labor Question*, Boston, MA: Lee & Shepard, 1884.

Pierantoni, Senator Augusto. "Italian Feeling on American Lynching," *The Independent*, 55 (1905), 2040–42.

Pinchot, Gifford. "Wages, Margins and Anthracite Prices," *Annals of the American Academy of Political and Social Science*, 111 (1924), 61–81.

Pinkowski, Edward. "Joseph Batten: Father of the Coal Breaker," *Pennsylvania Magazine of History & Biography*, 73 (Jul 1949), 337–48.

———. *Lattimer Massacre*, Philadelphia, PA: Sunshine, 1950.

Pitkin, Thomas M. and Francesco Cordasco. *The Black Hand: A Chapter in Ethnic Crime*, Totowa, NJ: Littlefield & Adams, 1977.

Polletta, Francesca, Pang Ching, et al. "The Sociology of Storytelling," *Annual Review of Sociology*, 37 (2011), 109–30.

Portelli, Allessandro. *The Death of Luigi Trastulli and Other Stories: Form and Meaning in Oral History*, Albany, NY: SUNY Press, 1991.

———. *The Battle of Valle Giulia: Oral History and the Art of Dialogue*, Madison, WI: University of Wisconsin Press, 1997.

———. *They Say in Harlan County: An Oral History*, New York: Oxford University Press, 2010.

Portes, Alejandro, ed. *The Economic Sociology of Immigration: Essays on Networks, Ethnicity, and Entrepreneurship*, New York: Russell Sage Foundation, 1995.

Powell, H. Benjamin. *Philadelphia's First Fuel Crisis: Jacob Cist and the Developing Market for Pennsylvania Anthracite*, University Park, PA: Penn State Press, 1978.

Puleo, Steve. *The Boston Italians: A Story of Pride, Perseverance, and Paesani from the Years of the Great Immigration to the Present Day*, Boston, MA: Beacon Press, 2007.

Putnam, Robert. *Making Democracy Work: Civic Traditions in Modern Italy*, Princeton, NJ: Princeton University Press, 1992.

Redman, Alvin. *The Wit and Humor of Oscar Wilde*, New York: Dover, 1959.

Riessman, Catherine Kohler. *Narrative Methods for the Human Sciences*, Thousand Oaks, CA: Sage, 2008.

Rhone, Rosamond D. "Anthracite Coal Mines and Mining," *American Monthly* Review of Reviews, 26 (1902), 54–63.

Roberts, Ellis W. A *History of Land Subsidence and its Consequences Caused by the Mining of Anthracite Coal in Luzerne County, Pennsylvania*, unpublished doctoral dissertation, New York University, 1948.

———. *The Breaker Whistle Blows: Mining Disasters and Labor Leaders in the Anthracite Region*, Scranton, PA: Anthracite Museum Press, 1984.

Roberts, Peter. *The Anthracite Coal Industry: A Study of the Economic Conditions and Relations of the Co-operative Forces in the Development of the Anthracite Coal Industry of Pennsylvania*, New York: Macmillan, 1901.

———. *Anthracite Coal Communities: A Study of the Demography, the Social, Educational and Moral Life of the Anthracite Regions*, New York: Macmillan, 1904.

Rochester, Anna. *Labor and Coal*, New York: International Publishers, 1931.

Roedinger, David R. *The Wages of Whiteness: Race and the Making of the American Working Class*, New York: Verso, 1999 [1991].

———. *Working Toward Whiteness: How America's Immigrants Became White, the Strange Journey from Ellis Island to the Suburbs*, New York: Basic, 2005.

Rimanelli, Marco, and Sheryl L. Postman, *The 1891 New Orleans Lynching and U.S.-Italian Relations: A Look Back*, New York: Lang, 1991.

Rosner, David, and Gerald Markowitz, eds. *Dying for Work: Workers' Safety and Health in Twentieth Century America*, Bloomington, IN: Indiana University Press, 1989.

Ross, E.A. *The Old World in the New*, New York: Century, 1914.

Roy, Andrew. *The Coal Mines: Containing a Description of the Various Systems of Working and Ventilating Mines*, Cleveland, OH: Robinson & Savage, 1876.

———. A History of the Coal Miners of the United States, Columbus, OH: Trauger, 1903.

Rucker, Allen. *The Sopranos: A Family History*, New York: New American Library, 2001.

Sabetti, Filippo. *The Search for Good Government: Understanding the Paradox of Italian Democracy*, Montreal, Canada: McGill-Queen's University Press, 2000.

Sacks, Karen Bodkin. "How Did Jews Become White Folks," 78–102 in Steven Gregory and Roger Sanjek (eds.), *Race*, New Brunswick, NJ: Rutgers University Press, 1994.

Salerno, Salvatore. *Red November, Black November: Culture and Community in the Industrial Workers of the World*, Albany, NY: SUNY Press, 1989.

———. "*I Telitti Della Razza Bianca* (Crimes of the White Race): Italian Anarchists' Racial Discourse as Crime," 11–23 in Jennifer Guglielmo and Salvatore Salerno (eds.), *Are Italians White? How Race Is Made in America*, New York: Routledge, 2003.

———. "No God, No Master: Italian Anarchists and the Industrial Workers of the World," 172–87 in Philip V. Cannistraro and Gerald Meyer (eds.), *The Lost World of Italian American Radicalism*, Westport, CT: Praeger, 2003.

Salvaterra, David L. Book Review of Philip Jenkins, *Hoods and Shirts* (1997), H-Net, Jul 1997, www.h-net.org/reviews/showrev.php?id=1140.

Samuels, Raphael. *Island Stories: Unraveling Britain*, London: Verso, 1999.

Savage, Lon. *Thunder In the Mountains: The West Virginia Mine War, 1920–21*, Pittsburgh, PA: University of Pittsburgh Press, 1990.

Saverino, Joan. "'Domani Ci Zappa': Italian Immigration and Ethnicity in Pennsylvania, *Pennsylvania Folklife*, 44 (Autumn 1995), 2–22.

Saward, Frederick. *The Coal Trade: The Year Book of the Coal and Coke Industry*, New York: Published by the author, 1889.

Scelsa, Joseph V. "Italian Americans and Civil Rights: A Case Study at the City University of New York," Italian American Review 3 (12 Oct 1994), 21–32.

———. "Affirmative Action for Italian Americans: The City University of New York Story," Chapter 7 in William J. Connell and Fred Gardaphe (eds.), *Anti-Italianism: Essays on Prejudice*, New York: Palgrave Macmillan, 2010.

Schlegel, Marvin W. "The Workingmen's Benevolent Association: First Union of Anthracite Miners," *Pennsylvania History*, 10 (1943), 243–67.

Schnell, George A. "Anthracite's Impact on Northeastern Pennsylvania's Population and Environment, 1880–1995," *Journal of the Pennsylvania Academy of Science*, 72 (1998), 22–8.

Shackel, Paul, Michael Roller, and Kristin Sullivan. "Historical Amnesia and the Lattimer Massacre," *News: A Publication of the Society of Applied Anthropology*, 2011, http://sfaanews.sfaa.net/2011/05/01/historical-amnesia-and-the-lattimer-massacre/.

Shogan, Robert. *The Battle of Blair Mountain: The Story of America's Largest Labor Uprising*, Boulder, CO: Westview, 2004.

Singer, Alan. "Communists and Coal Miners: Rank-and-File Organizations in the United Mine Workers of America during the 1920s," *Science and Society*, 55 (1991), 132–57.

Smith, Gibbs M. *Joe Hill*, Salt Lake City, UT: University of Utah Press, 1969.

Smith, Walker. *The Everett Massacre*, Chicago, IL: IWW Publishing Bureau, 1918.

Smith, Yves. *ECONned: How Unenlightened Self-Interest Undermined Democracy and Corrupted Capitalism*, New York: Palgrave Macmillan, 2010.

Sondern, Frederick Jr. *Brotherhood of Evil: The Mafia*, New York: Farrar, Straus & Cudahy, 1959.

Spear, Sheldon. *Chapters in Northeastern Pennsylvania History*, Shavertown, PA: Jemags, 1999.

Stanislaus, Richard. "The Role of the Italian Immigrant in the Anthracite Coal Industry of Northeastern Pennsylvania During the 1920s," unpublished manuscript, 1997.

Staudohar, Paul D., and Holly E. Brown, eds. *Deindustrialization and Plant Closure*, Lexington, MA: D.C. Heath, 1987.

Steers, Edward. "The Pennsylvania Coal Company's Gravity Railroad," 155–228 in Lance E. Metz (ed.), *Canal History and Technology Proceedings* Vol. I, Easton, PA: Canal History and Technology Press, 1982.

Steinberg, Stephen. *The Ethnic Myth: Race, Ethnicity, and Class in America*, Boston, MA: Beacon Press, 1989.

Stevenson, George E. *Reflections of An Anthracite Engineer*, Published by the author, 1931.

Storch, Randi. *Red Chicago: American Communism at Its Grassroots, 1928–35*, Urbana, IL: University of Illinois Press, 2009.

Sydenstricker, Edgar. "The Settlement of Disputes under Agreements in the Anthracite Industry," *The Journal of Political Economy*, 24 (Mar 1916), 254–83.

Taibbi, Matt. *Griftopia: Bubble Machines, Vampire Squids, and the Long Con That Is Breaking America*, New York: Spiegel & Grau, 2010.

Tarrow, Sydney G. *Power in Movement: Social Movements and Contentious Politics*, New York: Cambridge University Press, 2011.

Taylor, A.J. "The Sub-Contract System in the British Coal Industry," 215–35 in L.S. Pressnell (ed.), *Studies in the Industrial Revolution*, London: University of London-Athlone, 1960.

Taylor, Daniel. *Tell Me a Story: The Life-Shaping Power of Stories*, St. Paul, MN: Bog Walk, 2001.

Taylor, Frederick W. *The Principles of Scientific Management*, New York: Harper & Row, 1911.

———. *Scientific Management: Comprising Shop Management, the Principles of Scientific Management, Testimony before the Special House Committee*, New York: Harper & Row, 1947.

Teresa, Vincent, with Thomas C. Renner. *My Life in the Mafia*, New York: Doubleday, 1973.

Theodorson, George A., and Achilles G. Theodorson. *A Modern Dictionary of Sociology*, New York: Crowell, 1969.

Thompson, E.P. *The Making of the British Working Class*, Harmondsworth, England: Penguin, 1980 [1963].

______. "History from Below," *Times Literary Supplement*, 7 Apr 1966, 279–80.

Thompson, Fred. *The I.W.W.: Its First Seventy Years (1905–1975)*, Chicago, IL: Industrial Workers of the World, 1976 [1955].

Thompson, Paul R. *The Voice of the Past: Oral History*, New York: Oxford University Press, 1988.

Thrush, Paul W. et al., eds. *A Dictionary of Mining, Mineral, and Related Terms*, Washington, DC: USGPO, 1968.

Tilly, Charles. *From Mobilization to Revolution*, Reading, MA: Addison-Wesley, 1978.

———. "War Making and State Making as Organized Crime," 169–86 in Peter Evans, Dietrich Rueschemeyer, and Theda Skocpol (eds.), *Bringing the State Back in*, Cambridge, England: Cambridge University Press, 1985.

Tilly, Chris, and Charles Tilly. *Work Under Capitalism*, Boulder, CO: Westview, 1998.

Trachtenberg, Alexander. *The History of Legislation for the Protection of Coal Miners in Pennsylvania, 1824–1915*, New York: International Publishers, 1942.

Trotta, Liberato. *Biography of Louis Pagnotti*, Translated by Michael Panzitta, unpublished manuscript, 1981.

Tuoni, G.M., and G. Brogelli. *Attività Italiane in California*, published by the authors, 1929.

Turner, George A. "The Lattimer Massacre and its Sources," *Slovakia*, 27 (1977), 9–43.

———. "Ethnic Responses to the Lattimer Massacre," 126–53 in David L. Salay (ed.), *Hard Coal, Hard Times: Ethnicity and Labor in the Anthracite Region*, Scranton, PA: Anthracite Museum Press, 1984.

———. "The Lattimer Massacre: A Perspective from the Ethnic Community," *Pennsylvania History*, Special Issue: The Lattimer Massacre (1897), Harold W. Aurand (guest ed.), 69 (2002), 11–30.

Valetta, Clement. "'To Battle for Our Ideas': Community Ethic and Anthracite Labor, 1920– 1940," *Pennsylvania History*, 58 (Oct 1991), 311–29.

Varbero, Richard A. "Workers in City and County: The South Italian Experience in Philadelphia, 1900–1950," 16–32 in Richard N. Juliani, and Philip V. Cannistraro (eds.), *Italian Americans: The Search for a Usable Past, Proceedings of the 19th Annual Conference of the American Italian Historical Association*, Staten Island, NY: Italian Historical Association, 1989.

Vecoli, Rudolph J. "Contadini in Chicago: A Critique of the Uprooted," *Journal of American History*, 51 (Dec 1964), 404–17.

———. "The Coming of Age of Italian Americans: 1945–1974," *Ethnicity*, 5 (1978), 119–47.

———. "Free Country: The American Republic Viewed by the Italian Left, 1880–1920," 23–44 in Marianne Debouzy (ed.), *In the Shadow of the Statue of Liberty: Immigrants, Workers, and Citizens in the American Republic, 1880–1920*, Urbana, IL: University of Illinois Press, 1992.

———. "Are Italian Americans Just White Folks?" 3–17 in Mary Jo Bona and Anthony Julian Tamburri (eds.), *Through the Looking Glass: Italian & Italian/American Images in the Media*, Staten Island, NY: American Italian American Association, 1996.

———. "The Italian Immigrants in the United States' Labor Movement from 1880 to 1920," in Bruno Bezza (ed.), *Gli italiani fuori d'Italia: gli emigrati italiani nei movimenti operai dei paesi d'adozione, 1880–1940*, Milan, Italy: Franco Angeli, 1983.

Vellon, Peter. "'Between White Men and Negroes': The Perception of Southern Italian Immigrants through the Lens of Italian Lynchings," 23–32 in William J. Connell and Fred Gardaphe (eds.), *Anti-Italianism: Essays on Prejudice*, New York: Palgrave Macmillan, 2010.

Vezzosi, Elizabetta. "Radical Ethnic Brokers: Immigrant Socialist Leaders in the United States Between Ethnic Community and the Larger Society," 121–138 in Donna R. Gabaccia and Fraser M. Ottanelli (eds.), *Italian Workers of the World: Labor Migration and the Formation of Multiethnic States*, Urbana, IL: University of Illinois Press, 2005.

Virtue, George O. "The Anthracite Combinations," *Quarterly Journal of Economics*, 10 (Apr 1896), 296–323.

———. "The Anthracite Mine Laborers," 728–44 in Carroll D. Wright and Oren W. Weaver (eds.), *Bulletin of the Department of Labor*, Washington, DC: USGPO, Nov 1897.

Voskuil, Walter H. *Minerals in Modern Industry*, New York: Wiley, 1930.

Wallace, Anthony F.C. *St. Clair: A Nineteenth-Century Coal Town's Experience with a Disaster-Prone Industry*, New York: Knopf, 1987.

Wallerstein, Immanuel. *World-Systems Analysis: An Introduction*, Durham, NC: Duke University Press, 2004.

Warne, Frank J. "The Anthracite Coal Strike," *Annals of the American Academy of Political and Social Science*, 17 (1901), 15–52.

———. "The Effects of Unionism upon the Mine Workers," *Annals of the American Academy of Political and Social Science*, 21 (Jan 1903), 20–35.

———. *The Slav Invasion and the Mineworkers*, Philadelphia, PA: Lippincott, 1904.

Washington, Booker T. "Child Labor and the Sulfur Mines," Chapter 11 in Booker T. Washington, *The Man Farthest Down: A Record of Observations and Study in Europe*, Garden City, NY: Doubleday & Page, 1913.

Waters, Mary. *Ethnic Options: Choosing Identities in America*, Berkeley, CA: University of California Press, 1990.

Watson, Tom, and Martin Hickman. *Dial M for Murdoch: News Corporation and the Corruption of Britain*, London: Alan Lane, 2012.

Webb, Clive. "The Lynching of Sicilian Immigrants in the American South, 1886–1910," *American Nineteenth Century History*, 3 (Spring 2002), 45–76.

Whyte, William F. *Street Corner Society: The Social Structure of an Italian Slum*, Chicago, IL: University of Chicago Press, 1993 [1943].

Wiebe, Robert H. "The Anthracite Strike of 1902: A Record of Confusion," *Mississippi Valley Historical Review*, 48 (1961), 229–51.

Wilcox, David. "Present Conditions in the Anthracite Coal Industry," *North American Review*, 181 (1905), 216–28.

Wilson, William Julius. *The Declining Significance of Race*, Chicago, IL: University of Chicago Press, 1980.

Wolensky, Kenneth C. "Freedom to Assemble and the Lattimer Massacre of 1897," Historical Society of Pennsylvania, www.hsp.org/node/2915.

———. *The Lattimer Massacre,* Historic Pennsylvania Leaflet No. 15. Harrisburg, PA: PHMC, 1997.

———. "An Activist Government in Harrisburg: Governor George H. Earl III and Pennsylvania's 'Little New Deal'," *Pennsylvania Heritage* 34 (Winter 2008), 14–23.

Wolensky, Kenneth C., Nicole H. Wolensky, and Robert P. Wolensky. *Fighting for the Union Label: The International Ladies' Garment Workers' in Pennsylvania*, University Park, PA: Penn State Press, 2002.

Wolensky, Robert P. "Power Structure and Group Mobilization Following Disaster: A Case Study," *Social Science Quarterly*, 64 (Mar 1983), 96–110.

———. "POWER: Collective Action and the Anthracite Region Water Crisis," 230–61 in S.R. Couch and S. Kroll-Smith (eds.), *Communities at Risk: Collective Responses to Technological Hazards*, New York: Peter Lang, 1991.

———. "Review of Al Gedicks' *The New Resource Wars*: An Organized Crime Perspective on Multi-National Mining and Energy Companies," panel presentation, Meetings of the Wisconsin Sociological Association, La Crosse, WI, 1995.

———. "The Contracting-Leasing System in the Northern Anthracite Field," 67–94 in Lance Metz (ed.), *The Great Strike: Perspectives on the 1902 Anthracite Coal Strike*, Easton, PA: Canal History and Technology Press, 2002.

———. "Working Class Heroes: Rinaldo Cappellini and the Anthracite Mineworkers," Chapter 18 in Ronald Berger and Richard Quinney (eds.), *Storytelling Sociology: Narrative as Social Inquiry*, Boulder, CO: Lynne Rienner, 2004.

Wolensky, Robert P., Thomas J. Baldino, and John V. Hepp. "Remaking Municipal Government? Charter Reform in Wilkes-Barre, Pennsylvania, 1968–2001," *Pennsylvania History* 73 (Autumn 2006), 446–79.

Wolensky, Robert P., and William A. Hastie. "The Pennsylvania Coal Company and the Anthracite Subcontracting System: Italians, Wildcatters, and the Industrial Workers of the World, 1905–1916," 95–112 in Lance E. Metz (ed.), *Canal History and Technology Proceedings*, Vol. XXX, Easton, PA: Canal History and Technology Press, 2011.

Wolensky, Robert P., and Joseph M. Keating. *Tragedy at Avondale: The Causes, Consequences, and Legacy of the Pennsylvania Anthracite Coal Industry's Most Deadly Mining Disaster, September 6, 1869*, Easton PA: Canal History and Technology Press, 2008.

Wolensky, Robert P., and Edward J. Miller. "The Everyday Versus the Disaster Role of Local Officials: Citizen and Official Definitions," *Urban Affairs Quarterly* 16 (1981), 483–504.

Wolensky, Robert P., Kenneth C. Wolensky, and Nicole H. Wolensky. *The Knox Mine Disaster: The Final Years of the Northern Anthracite Industry and the Effort to Rebuild a Regional Economy*, Harrisburg, PA: PHMC, 1999.

Wolensky, Robert P., Nicole H. Wolensky, and Kenneth C. Wolensky. *Voices of the Knox Mine Disaster: Stories, Remembrances, and Reflections on the Anthracite Coal Industry's Last Major Catastrophe, January 22, 1959*, Harrisburg, PA: PHMC, 2005.

Woodall, Conrad L. "The Italian Massacre of Walsenburg, Colorado, March 1895," 297–317 in *Italian Ethnics—Their Languages, Literature, and Lives, Proceedings of the 20th Annual Conference of the American Italian Historical Association*, Chicago, IL, 1987.

Woodiwiss, Michael. *Organized Crime and American Power*, Toronto: University of Toronto Press, 2001.

———. *Gangster Capitalism: The United States and the Global Rise of Organized Crime*, New York: Carroll & Graff, 2005.

Wright, Carroll D. "Padrones in Chicago," 691–727 in *Italians in Chicago: A Social and Economic Study*, Washington, DC: USGPO, 1897.

———. *The Italians in Chicago: A Social and Economic Study*, Washington, DC: USGPO, 1897.

Yearley, Clifton K. Jr. *Enterprise and Anthracite: Economics and Democracy in Schuylkill County, 1820–1875*, Baltimore, MD: Johns Hopkins University Press, 1961.

Zieger, Robert H. "Pinchot and Coolidge: The Politics of the 1923 Anthracite Crisis," *Mississippi Valley Historical Review*, 52 (1967), 566–81.

———. "Pennsylvania Coal and Politics: The Anthracite Strike of 1925–26," *Pennsylvania Magazine of History and Biography* 93 (1969), 244–62.

———. *John L. Lewis: Labor Leader*, Boston, MA: Twayne, 1988.

———. *The CIO: 1935–1955*, Chapel Hill, NC: University of North Carolina Press, 1995.

Zinn, Howard. *A People's History of the United States*, New York: Harper Perennial, 1980.

———. "Massacres of History," *The Progressive*, 62 (Aug 1998), 17.

Zolberg, Aristide. *A Nation by Design: Immigration Policy in the Fashioning of America*, Cambridge, MA: Harvard University Press, 2008.

II. *Reports, Labor-Management Awards, Government Documents, Hearings, Organizational Proceedings, Directories, Manuals, Yearbooks, Corporate Documents.*

Ashley, George H. "Anthracite Reserves," Topographic and Geologic Survey, Progress Report 130, Harrisburg, PA: Commonwealth of Pennsylvania, Department of Internal Affairs, 1945.

Anthracite Board of Conciliation. *Award of the Anthracite Coal Strike Commission, Subsequent Agreements, and Resolutions of the Board of Conciliation*, Hazleton, PA: Anthracite Board of Conciliation, 1916, 1923, 1939, 1953.

Anthracite Board of Conciliation. Grievance No. 1126, Vol. 11, 82, 1920, Pennsylvania Coal Co. vs. Employees, Contained in Digest of the Decisions of the Anthracite Board of Conciliation, Wage Agreements, Carl A. Peterson Papers, HCLA, Penn State University, Box 1, Folder 3.

Anthracite Board of Conciliation. *Report* (Grievances and Actions), Grievance No. 2035, Hazleton, PA: Anthracite Board of Conciliation, 1926, 16, 30–48.

Anthracite Bureau of Information. *The Anthracite Strike of 1925–1926*, Philadelphia, PA: The Anthracite Bureau of Information, 1926.

Anthracite Coal Commission. *Report, Findings, and Award of the Anthracite Coal Commission of 1920*, Washington, DC: USGPO, 1920.

The Anthracite Coal Controversy, Demands of 1906, Scranton Public Library Donation Boxes, Anthracite Heritage Museum, uncatalogued information sheets.

Anthracite Coal Strike Commission. *Report to the President on the Anthracite Coal Strike of May–October 1902*, Washington, DC: USGPO, 1903.

Ash, Solomon H. "Barrier Pillars in Wyoming Basin, Northern Field," Bulletin 583, U.S. Bureau of Mines, Washington, DC: USGPO, 1954.

Biographical Directory of the U.S. Congress, http://bioguide.congress.gov/scripts/biodisplay.pl?index=P000197.

Chance, H. Martyn. *Second Geological Survey of Pennsylvania, 1883: Report on the Mining Methods and Appliances used in the Anthracite Coal Fields*, Harrisburg, PA: Board of Commissioners for the Second Geological Survey, Washington, DC: USGPO, 1883.

Dierks, Henry A., Walter L. Eaton, and Ralph H. Whaite. *Anthracite Mine-Flood Disaster: Breakthrough of Susquehanna River into River Slope Mine, Knox Coal Company, Port Griffith, Pennsylvania*, Washington, DC: U.S. Bureau of Mines, 1960.

Dillingham Commission. *Reports of the United States Immigration Commission*, 41 Volumes, Washington, DC: USGPO, 1911.

Ferry, Neal J. "Minority Report," in *Report, Findings, and Award of the Anthracite Coal Commission of 1920*, Washington, DC: USGPO, 1920, 312–23.

Joint Legislative Committee on Government Operations, on the Gangland Meeting in Apalachin, New York. *Interim Report*, Albany, NY: State of New York, 25 Jun 1958.

Joint Legislative Committee to Investigate the Knox Mine Disaster. *Final Report*, Harrisburg, PA: Commonwealth of Pennsylvania, 27 Jul 1959.

Lauck, W. Jett. "Copper Mining and Smelting; Iron Ore mining; Anthracite Coal Mining; Oil Refining," Dillingham Commission, *Report of the United States Immigration Commission*, Vol. 16, Washington, DC: USGPO, 1911, 581–740.

Lazzari, Salvatore. Congressional Research Service Report for Congress, "Energy Tax Policy: History and Current Issues," www.fas.org/sgp/crs/misc/RL33578.pdf.

McClellan Hearings. *Hearings Before the Select Committee on Improper Activities in the Labor or Management Field*, Senator John L. McClellan, Chairman, Eighty-fifth & Eighty-sixth Congress, Washington, DC: USGPO, 1958 & 1959.

"The Miners' Demands," *The Anthracite Strike of 1922*, Philadelphia, PA: The Anthracite Bureau of Information, 1923.

Moody's Transportation Manual, New York: Moody, 1960, 1980.

Naison, Mark, and Maurice Isserman, eds. *Research Collections in American Radicalism*, Department of Justice, Investigative Files, Part II, The Communist Party, University Publications of America, Section 5, Previously Restricted Materials, 1921–1930, 90 pages; http://library. lexisnexis.com/ksc_assets/catalog/10836.pdf.

Obenauer, Marie L. "Living Conditions in the Anthracite Region and Composition of the Mining Population," 527–71 in *Annex to The General Report of the U.S. Coal Commission*, Washington, DC: USGPO, 1925.

Papers Relating to Foreign Relations of the United States. Washington, DC: USGPO, 1901.

Pennsylvania Anthracite Coal Industry Commission. "Bootlegging or Illegal Mining of Anthracite Coal in Pennsylvania, A Census and Survey of the Facts," 39–146 in *Report of the Pennsylvania Anthracite Coal Industry Commission*, Harrisburg, PA: Commonwealth of Pennsylvania, 1938.

Pennsylvania Anthracite Coal Industry Commission. *Report of the Pennsylvania Anthracite Coal Industry Commission*, Harrisburg, PA: Commonwealth of Pennsylvania, 1938.

Pennsylvania Bureau of Mines. *Report of the Bureau of Mines*, Department of Internal Affairs of Pennsylvania, Harrisburg, PA, 1898, 1899.

Pennsylvania Coal Company. *Annual Report*, 1941.

"The Pennsylvania Coal Company." *Scranton Directory*, 1867.

Pennsylvania Crime Commission. *A Decade of Organized Crime*, Harrisburg, PA: Commonwealth of Pennsylvania, 1980.

———. *Report on Organized Crime*, Harrisburg, PA: Commonwealth of Pennsylvania, 1970, 1983, 1984.

———. *Coal Fraud: Undermining a Vital Resource*, Conshohocken, PA: Commonwealth of Pennsylvania, 1987.

Pennsylvania Department of Mines. *Anthracite Mining Laws of Pennsylvania*, Harrisburg, PA: Pennsylvania Department of Mines, 1948.

Pennsylvania Department of Mines and Mineral Industries. *Annual Report: Anthracite Division* (aka *The Inspectors' Reports*), Harrisburg, PA: Commonwealth of Pennsylvania, 1901, 1916, 1919, 1939, 1940, 1941, 1942, 1944, 1949, 1950, 1954, 1959, 1961, 1962, 1963.

Proceedings of the UMWA Tri-District Convention (Districts 1, 7,& 9), 1902, 1916, 1923, 1930, 1939.

The Pennsylvania Manual, Harrisburg, PA: Bureau of Publications, 1955.

Proceedings of the UMWA District No. 1 Convention, 1901, 1921, 1939, 1957.

Proceedings of the Joint Convention of UMWA Districts Nos. 1, 7, and 9, 1902.

Proceedings of the Anthracite Mine Strike Commission, 1902–03. Scranton, PA: Scranton Tribune, 1903.

Proceedings of the Twelfth Annual Convention of UMWA District No. 1, 1910.

Proceedings of the Fourteenth Annual Convention of UMWA District No. 9, 1913.

Proceedings of the Nineteenth Successive and Fourth Biennial Convention of UMWA District No. 1, 1921.

Proceedings of the Twenty-Third Consecutive and Eighth Biennial Convention of UMWA District No.1, 1929.

Proceedings of the Reconvened UMWA Tri-District Convention (Districts 1, 7, & 9), 1930, 1944.

Proceedings of the Twenty-Fourth Consecutive and Ninth Biennial Convention of UMWA District No.1, 1931.

Proceedings of the Twenty-Fifth Consecutive and Tenth Biennial Convention of UMWA District No.1, 1933.

Proceedings of the Twenty-Seventh Consecutive and Twelfth Biennial Convention of UMWA District No. 1, 1937.

Proceedings of the Reconvened UMWA Tri-District Scale Convention, 1944.

"Report of the Anthracite Coal Strike Commission of 1902," Reprinted with Subsequent Agreements and the Majority and Minority Reports of the U.S. Anthracite Commission of 1920, Philadelphia, PA: Anthracite Bureau of Information, October 1920.

Schanz, John J. Jr. "Historical Statistics of Pennsylvania's Mineral Industries, 1961–1965," *Bulletin of the Earth and Mineral Sciences Experiment Station*, College of Earth and Mineral Science, Penn State University, 1967.

Stump, Herman, J.H. Senner, and Edward F. McSweeney. *Report of the Immigration Investigation Commission to the Honorable Secretary of the Treasury*, Washington, DC: USGPO, 1895.

United Mine Workers of America. *Minutes of the Twelfth Annual Convention of the United Mine Workers of America*, Indianapolis: Hollenback, 1929.

U.S. Bureau of the Census. *Historical Statistics of the United States: Colonial Times to 1970*, Washington, DC: USGPO, 1975.

U.S. Bureau of the Census. Thirteenth Census of the U.S., 1910.

U.S. Bureau of Mines. "Analyses of Pennsylvania Anthracite Coals," Washington, DC: USGPO, 1944.

U.S. Coal Commission. *General Report*, Washington, DC: USGPO, 1923.

U.S. Coal Commission. *Report*, Part II–Anthracite, Washington, DC: USGPO, 1925.

U.S. Geological Survey. *Mineral Resources of the United States, Part II–Nonmetals*, Washington, DC: USGPO, 1910, 1913, 1917.

U.S. Treasury Department, Bureau of Narcotics. *Mafia: The Government's Secret File on Organized Crime*, Foreword by Sam Giancanna, New York: Harper Collins, 2007.

Volpe Disaster Investigation Commission. *Report* to M.J. Hartneady, Secretary of the Department of Mines, Commonwealth of Pennsylvania, June 17, 1930, Volpe Coal Company File, *Times Leader* Archives, Wilkes-Barre, PA.

Warriner, Samuel D. "*Reply of the Operators [to the Miners' Demands]*," in *The Anthracite Strike of 1925–1926*, Philadelphia, PA: The Anthracite Bureau of Information, 1926.

Washington Coal Company. Minutes of the meeting of the directors of the Washington Coal Co. at Carbondale, Pennsylvania, 25 Jan 1848, 1989.343, folder 1, NCMA.

Wyoming Coal Association. Agreement between the Delaware & Hudson Canal Co. and the Wyoming Coal Association, 31 Aug 1847, 1989.343, folder 1, NCMA.

Wyoming Coal Association and Washington Coal Association. Resolution of the Managers of the Wyoming Coal Association for the transfer of real estate to the Pennsylvania Coal Company, 8 Nov 1848, 1989.343, folder 1, NCMA.

III. *Magazines, Almanacs, Reference Books, Industry Publications, Websites, Newsletters, and Religious Documents.*

Adamic, Louis. "The Great Bootleg Coal Industry," *The Nation* (9 Jan 1935), 46–49.

"A Decade of Anthracite Peace," *Literary Digest*, 120 (2 Aug 1930), 40.

"Allegheny [*sic*] Corporation," *Time* (11 Feb 1929).

"Anthracite," *Time* (12 Mar 1928).

"Anthracite Boosters; Unique Club is Organized in Philadelphia," *United Mine Workers Journal*, 39 (1 Sept 1928), 8.

"At Wilkes-Barre," *Time* (30 Jul 1923).

Borsellino, Paulo. "Prepared Remarks for Assistant Attorney General Lanny Breuer at the Falcone Commemoration Ceremony," Palermo, Italy, 23 May 2010, www.justice.gov/criminal/pr/speeches-testimony/2010/05-23-10aag-palermo-remarks.pdf.

"Can Anthracite Come Back?" *Business Week* (13 Feb 1954), 182–84.

"Changing Industrial Relations," *Coal Age*, 41 (Oct 1936), 460ff.

"Coal and Coke News," *Coal Age*, 6 (19 Sept 1914), 482.

"Coal Bootlegging: An Octopus," *Literary Digest*, 126 (12 Dec 1936), 40.

The Colliery Engineer, 25 (Aug 1914–Jul 1915), 391.

"Companies Should be Held Responsible for Deaths," *United Mine Workers Journal* (26 Sept 1907), 4.

"Contract Mining System Abolished at Underwood Colliery in District No. 1," *United Mine Workers Journal*, 39 (15 Dec 1928), 5.

"District No. 1 Executive Board Expels Group Identified in Forming Dual Organization," *United Mine Workers Journal*, 39 (1 Oct 1928), 5.

Dosch, Arno. "Just Wops," *Everybody's Magazine*, 25 (Nov 1911), 579–89.

"Dual Unions Roundly Denounced by Murray and Kennedy at John Mitchell Memorial Meeting," *United Mine Workers Journal*, 39 (15 Nov 1928), 10.

"Economic Status of Coal-Mining Industry in Last Quarter Century," *Coal Age*, 41 (Feb 1936), 394ff.

"Elected President; John J. Boylan Becomes Head of District No. 1," *United Mine Workers Journal*, 39 (1 Aug 1928), 7.

"Empire for Sale," *Time* (23 Sept 1935).

Engdahl, Louis. "The Yellow Streak in Coal," *The Liberator*, 9 (Sept 1923), 23–25.

"Five-Year Agreement Ends Longest Strike in History of Anthracite Industry," *Coal Age*, 29 (Feb 1926), 268–70.

"Frank Agati Slain by Gunmen in Offices of Union at WB," *United Mine Workers Journal*, 39 (1 Mar 1928), 17.

"Gas Competition," *Coal Age*, 65 (Apr 1960), 71–72.

Given, Ivan A. "The Lewis Era ... 1920–1960 in Coal," *Coal Age*, 65 (Jan 1960), 66–71.

"Great Industry; Anthracite Mining Highly Important in Pennsylvania," *United Mine Workers Journal*, 39 (1 Oct 1928), 17.

Grimes, William. "Rudolph J. Vecoli, Scholar of Immigration, Is Dead at 81," *New York Times*, 23 Jun 2008.

Hapgood, Powers, and Mary Donovan. "Murdered Miners," *The Nation* (14 Mar 1928), 293–94.

Hard, Anne. "Anthracite Country," *The Nation*, 121 (25 Nov 1925), 2–3.

"Hard-Coal Operators Prepared for Finish Fight in War against Future Strikes," *Coal Age*, 28 (Nov 1925), 737.

Heller, Bernard. "The Anthracite Industry," *The Nation*, 128 (27 Feb 1929), 255–56.

Hunt, Thomas and Michael A. Tona. "Men of Montodoro," *Informer* (Apr 2011), informer-journal.blogspot.com.

"Increased Production and Efficiency Urged in Address by Lewis," *The Record American* (2 Oct 1953).

"Industrial Workers of the World Join Union," *Coal Age*, 9 (4 Mar 1916), 419.

"The Ins and Outs: Italians are Deeply Anti-meritocratic," *The Economist* (9 Jun 2011).

"Italians in the United States," *Literary Digest*, 93 (23 Apr 1927), 30–2.

Kaplan, Jerry. "Sacco and Vanzetti Bibliography," saccoandvanzetti.org/saccovanzetti_bibliography.pdf.

Kumpas, Charles P. "Safer Mining at the Hudson Coal Company," *The Miner's Lamp*, 14 (May/Jun 1997), 1 & 4.

———. "The Peoples Coal Company and the Law," *The Miner's Lamp*, 16 (Nov/Dec 1999), 1 & 4.

"A Loss for Bob Young," *Time* (23 Dec 1957).

"Mr. Cappellini," *Time* (9 Jul 1923).

"New 'Baloney' Organization in Anthracite Region Will Seek Assistance from Anti-Union Employers," *United Mine Workers Journal*, 39 (15 Nov 1928), 7.

The New York Society for the Prevention of Cruelty to Children. *The Italian Padrone Case: U.S. Against Antonio Giovanni Ancarola*, New York: Styles & Cash, 1880.

"Official Letter to Local Unions in District No. 1," *United Mine Workers Journal*, 39 (15 Jun 1928), 7.

"Outlook: Coal Vs. Oil," *Coal Age*, 65 (Oct 1960), 72–75.

"Pennsylvania Anthracite: Pittston," *Coal Age*, 9 (29 Apr 1916), 775.

"The Pittston Story," *Mechanization*, 21 (Oct 1957), 60–94.

Pope Leo XIII, Rerum Novarum, *Rights and Duties of Capital and Labour: On the Conditions of Labour*, New York: Paulist, 1939.

"President Boylan Files Complaint against Contract Mining with Board," *United Mine Workers Journal*, 39 (15 Sept 1928), 10.

"The Right to Work: The Story of Non-Striking Miners," *McClures Magazine*, 20 (Nov 1902–Apr 1903), 332–336.

"Robert DeNiro Says Martin Scorcese's 'The Irishman' Still Happening," www.slashfilm.com/robert-de-niro-martin-scorsese-irishman-happening/.

Roller, Anne H. "Wilkes-Barre, An Anthracite Town," *The Survey*, 55 (Feb 1926), 534–7.

Rood, Henry Edward. "A Pennsylvania Colliery Village: A Polyglot Community," *Century Magazine*, 55 (Apr 1898), 809–21.

Ross, E.A. "Racial Consequences of Immigration," *Century Magazine*, 87 (Feb 1914), 615–22.

Selekman, Ben. "Miners and Murder: What Lies Back of the Labor Feud in Anthracite," *Survey Graphic*, 60 (1 May 1928), 151ff.

Serrin, William. "Historians See Lessons for Present in Fate of Dead Anthracite Coal Towns," *New York Times*, 29 Dec 1985.

"60 Years of Heroes: Giovanni Falcone and Paolo Borsellino," *Time* (13 Nov 2006).

Taibbi, Matt. "The Scam Wall Street Learned from the Mafia," *Rolling Stone*, 5 (Jul 2012), www.rollingstone.com/politics/news/the-scam-wall-street-learned-from-the-mafia-20120620.

"Two More Slain; Local Union Officers Meet Death in Pennsylvania," *United Mine Workers Journal*, 39 (15 Mar 1928), 16.

United States Conference of Catholic Bishops. *Economic Justice For All: Pastoral Letter on Catholic Social Teaching and the U.S. Economy*, Washington, D.C.: USCCB, 1986.

Van Winkle, Marybeth. "The Other Side of the Story," *The Searcher*, 13 (2010), 6.

White, Jerome C. "Scientific Management and the Coal Industry," *Coal Age*, 41 (Oct 1936), 424–25.

Wilkes-Barre Record Almanac, Wilkes-Barre, PA: Wilkes-Barre Publishing Co., 1928, 1931, 1932, 1934, 1952.

"Will Not Arbitrate," *Coal Age*, 28 (Oct 1925), 451–52.

www.adbusters.org.

www.everyculture.com

www.fbartcreations.com.

www.gutenberg.org.

www.h-net.org.

www.hsp.org.

www.justice.gov.

www.latinamericanstudies.org

www.marxists.org.

www.measuringworth.com.

www.miningartifacts.org.

www.nanticokecity

www.northernfield.info/.
www.oldforgecoalmine.com.
www.old-time.com/sights.
www.pabook.libraries.psu.edu.
www.rollingstone.com.
www.socresonline.org.uk.
www.studentpulse.com.
www.usembassy.it.
www.vatican.va.
www.wabash.edu.

IV. *Newspapers Cited*

Albany Evening Journal
Anthracite Tri-District News (Hazleton, PA)
The Coal Digger (Pittsburgh, PA)
Citizens Voice (Wilkes-Barre, PA)
Daily Sentinel (Hazleton, PA)
Daily Worker (New York, NY)
Duluth News (Duluth, MN)
Evening Herald (Shenandoah, PA)
The Evening News (Wilkes-Barre, PA)
Gazette Telegraph (Denver, CO)
Harrisburg Patriot
Idaho Statesman (Boise, ID)
News Tribune (Duluth, MN)
The New York Times
New-York Tribune
Philadelphia Evening Star
Philadelphia Inquirer
Philadelphia Ledger
Pittston Dispatch
Pittston Gazette
Plain Dealer (Cleveland, OH)
Daily Republican (Pottsville, PA)
The Record American (Boston, MA)
Scranton Republican
Scranton Times
Scranton Tribune
Standard-Speaker (Hazleton, PA)
Sunday Independent (Wilkes-Barre, PA)
Times Leader (Wilkes-Barre, PA)
Times Leader Evening News (Wilkes-Barre, PA)
Trenton Evening Times

Trenton Sunday Times Advertiser

Wilkes-Barre Record

Wilkes-Barre Times

V. *Archives and Special Collections*

Albertson Learning Resource Center. University of Wisconsin-Stevens Point, Newspaper Microfilms, Stevens Point, WI.

Albright Memorial Public Library. Newspaper Microfilms, Scranton, PA.

Anthracite Board of Conciliation Collection. MG 108, Indiana University of Pennsylvania Archives, Indiana, PA.

Citizens' Voice. Newspaper Archives. Wilkes-Barre, PA.

Curilla, George Jr. Collection. AC 97.1, Anthracite Heritage Museum, Scranton, PA.

Curran, Rev. John J. Papers. University of Notre Dame Archives, Notre Dame, IN.

Department of Justice Investigative Files. Industrial Workers of the World, Mark Naison and Maurice Isserman (eds.), No. 753.PT1, Reel 8, Reel 11, Andruss Library, Bloomsburg University.

Glen Alden Coal Company Papers. Earth Conservancy, Inc., Ashley, PA.

Golda Meir Library, University of Wisconsin-Milwaukee. Newspaper Microfilms, Milwaukee, WI.

Hillside Coal & Iron Company Papers. MG 282, Pennsylvania State Archives, PHMC, Harrisburg, PA.

Hudson Coal Company Collection. Historical Collections and Labor Archives, Penn State University, University Park, PA.

Hudson Coal Company Papers. Earth Conservancy, Inc., Ashley, PA.

Keating, Joseph M. Collection on the United Anthracite Miners of Pennsylvania. Nanticoke Historical Society, Nanticoke, PA.

Kennedy, Joseph T. Papers. MG 191, Pennsylvania State Archives, PHMC, Harrisburg, PA.

Labor History Archives. Wisconsin Historical Society, Madison, WI.

Lackawanna Historical Society Archives. Scranton, PA.

Lehigh & Wilkes-Barre Coal Company Papers. Earth Conservancy, Inc., Ashley, PA.

Lehigh Valley Coal Company Papers. Pagnotti Enterprises, Inc., Wilkes-Barre, PA.

Luzerne County Historical Society Archives. Wilkes-Barre, PA.

New York, Susquehanna & Western Coal Company Papers. MG 282, Pennsylvania State Archives, PHMC, Harrisburg, PA.

Osterhaut Memorial Library. Newspaper Microfilms, Wilkes-Barre, PA.

Penn Anthracite Collieries Company Papers. Private collection of William Graham, now at the Anthracite Heritage Museum, Scranton, PA.

Pennsylvania Coal Company Papers. E. Stewart Milner Collection, Anthracite Heritage Museum, Scranton, PA.

Pennsylvania Coal Company Papers. MG 282, Pennsylvania State Archives, PHMC, Harrisburg, PA.

Pennsylvania Coal Company Papers. National Canal Museum Archives, Easton, PA.

Peterson, Carl A. Papers. Historical Collections and Labor Archives, Penn State University, University Park, PA.

Scranton Times. Newspaper Archives. Scranton, PA.

Times Leader. Newspaper Archives. Wilkes-Barre, PA.

United Mine Workers of America Papers. Historical Collections and Labor Archives, Penn State University, University Park, PA.

Weinberg Memorial Library, University of Scranton. Newspaper Microfilms, Scranton, PA.

VI. *Court Cases, Legal References, and Related Literature*

Brandes, Evan B. "Legal Theory and Property Jurisprudence of Oliver Wendell Holmes, Jr. and Louis D. Brandeis: An Analysis of Pennsylvania Coal Company v. Mahon," *Creighton Law Review*, 38 (2005) 1179.

Commonwealth v. Erico, Appellant, 1922, 508–511, Superior Court of Pennsylvania, May and Oct Terms, (1921), Jan Term (1922).

Commonwealth of Pennsylvania v. Ralph Melissari, Court of Quarter Sessions of Luzerne County, Apr Session (1928).

Commonwealth v. Puntario, Appellant, 1922, 501–08, Superior Court of Pennsylvania, May and Oct Terms (1921), Jan Term (1922).

Friedman, Lawrence M. "A Search for Seizure: Pennsylvania Coal Co. v. Mahon in Context," *Law and History Review*, 4 (Spring 1986), 1–22.

In re Blue Coal Corporation, Debtor [and] In re Glen Nan, Inc., Debtor, 47 B.R. 754 (Bankruptcy, M.D. PA., 1985).

Kens, Paul. *Locher v. New York*, Lawrence, KS: University Press of Kansas, 1990.

Keystone Bituminous Coal Association, et al., Petitioners v Nicholas DeBenedictis. (1987) 480 US 470, 94 L Ed 2d 472, 107 S Ct 1232.

Kohler Act, 27 May 1921, P.L. 1198, Section 1.

Lippi, August J. Plaintiff v. Lester Thomas et al., Defendants, Civil No. 68–336, U. S. District Court for the Middle District of Pennsylvania, 298 F. Supp. 242; 70 L.R.R.M. 3424; 59 Lab. Cas. (CCH) P13, 399 (31 Mar 1969).

Mahon v. Pa. Coal Co., Pennsylvania Supreme Court, 118 A. 491 (Pa. 1922).

McMichael, Susan Manges. "Mahon Revisited: *Keystone Bituminous Coal Ass'n v. DeBenedictis*, 480 U.S. 470 (1987)," *Natural Resources Journal*, 29 (1989) 1067–77.

Pennsylvania Coal Company v. Mahon, 260 U.S. 393 (1922).

Plymouth Coal Co. v. Pennsylvania, 232 U.S. 531 (1914).

Scelsa v. City University of New York, 1994, No. 1063, Docket 95–7975, Argued 4 Dec 1995–25 Jan 1996, United States District Court for the Southern District of New York.

U.S. v. August J. Lippi, Criminal A. No. 1269, U.S. District Court, District of Delaware, 190 F. Supp. 604, 47 L.R.R. M. 2537; 42 Lab. Cas. (CCH) P. 16, 792 (19 Apr 1961).

U.S. v. August J. Lippi. U. S. District Court, District of Delaware, 190 F. Supp. 604, 47 L.R.R.M. 2537; 42 Lab. Cas, (CCH) P16,792 (2 Feb 1961).

U.S. v. Dominick Alaimo, et al., Criminal No. 13225, U.S. District Court, Middle District of Pennsylvania, Scranton (1959).

U.S. v. George J. Daileda and August J. Lippi, U.S. Court of Appeals Third Circuit, 342 F.2d 218 (9 Mar 1965).

U.S. v. Knox Coal Company, Robert L. Dougherty, August J. Lippi, Josephine Sciandra and Louis Fabrizio, August J. Lippi Appellant. No. 14803. U.S. Court of Appeals Third Circuit (21 Jun 1965).

U.S. v. Louis Fabrizio, Criminal. A. No. 1251, U.S. District Court, District of Delaware (19 Apr 1961).

U.S. v. Philip and Samuel Gelso et al., Criminal No. 13435, U.S. District Court, Middle District of Pennsylvania, Scranton (1964).

U.S. v. Tabor Court Realty Corp., et al., 803 F. 2d 1288 (3d. Cir., 1986).

VII. *Oral History Collections*

King's College Oral History Collections. Corgan Library, King's College, Wilkes-Barre, PA.

Northeastern Pennsylvania Oral and Life History Project. Center for the Small City, University of Wisconsin-Stevens Point.

VIII. *Audiotaped and Unrecorded Interviews. (Unless otherwise indicated, Robert P. Wolensky conducted the interview and the Center for the Small City, University of Wisconsin-Stevens Point, is the repository.)*

Adonizio, Charles Jr. Audiotaped interview, 14 Mar 1996, NPOLHP.

Alaimo, Nick. Unrecorded interview, 8 Aug 1989, NPOLHP.

Amos, Hubert. Audiotaped interview, 12 Jan 1995, NPOLHP.

Baird, Howard. Audiotaped interview, 24 Jul 2001, NPOLHP.

Bankovitch, Ethel, and Ann Major. Audiotaped joint interview, 5 Jun 1993, NPOLHP.

Brozena, Chester. Audiotaped interview, 3 Dec 1988, NPOLHP.

Butler, William. Audiotaped interview, 25 Jul 1994, NPOLHP.

Cappellini, Marie. Audiotaped interviews, 8 Dec 1988, 29 Dec 1988, NPOLHP.

Casterline, Lewis (Angelo Valentino Casterlani). Audiotaped interview, 16 Jan 1970, interviewed by Clement Valetta and Mary Barrett, King's College Oral History Collections, Wilkes-Barre, PA.

Casterline, Lewis (Angelo Valentino Casterlani). Audiotaped interviews, 12 Dec 1988, 25 May 1990, NPOLHP.

Chamberlain, Alex. Audiotaped interview, 11 Jul 1995, NPOLHP.

Cornell, Fr. Robert. Unrecorded interview, 3 Feb 1996, NPOLHP.

Cosgrove, Richard. Audiotaped interview, 13 May 2003, NPOLHP.

Costa, Joe (Costanzio Lopes), Audiotaped interview, 14 Aug 1985, NPOLHP.

———. Unrecorded interviews, 4 Jul 1991, 20 Jun 1992, 1 Aug 1992, 11 Oct 1993, 24 Mar 1995, 22 Jul 1995, 22 Jan 1999, NPOLHP.

Costanza, James. Unrecorded interview, 22 Jun 1999, NPOLHP.

Cronin, Richard. Audiotaped interview, 24 May 1983, NPOLHP.

Cummings, Claire Hart. Audiotaped interview, 15 Jan 1995, NPOLHP.

DeAngelo, Anthony. Audiotaped interview, 18 Dec 1988, NPOLHP.

Dorish, John. Audiotaped interview, 31 Jul 1988, NPOLHP.

Dunn, Chester. Audiotaped interview, 4 Aug 1989, NPOLHP.

Gadamski, John. Audiotaped interview, 12 Dec 1988, NPOLHP.

Gelso, Samuel. Unrecorded interview, 19 Aug 2003, NPOLHP.

Graham, William. Audiotaped interview, 11 Jul 1995, NPOLHP.

Greene, Victor. Unrecorded interviews, 5 Apr 2003, 21 Apr 2010, NPOLHP.

Gushanas, George. Audiotaped interview, 13 Jan 1994, NPOLHP.

Hastie, William A. Sr. Audiotaped interviews, 31 Jul 1989, 28 Jun 1990, 2 Jul 1990, 23 May 1991, 13 Mar 1996, 18 Mar 1999; videotaped interview, 15 May 2003, NPOLHP.

Hastie, William A. Sr. Unrecorded interviews, 18 Jan 2011, 12 Jan 2012, 24 Apr 2012, NPOLHP.

Justin, John. Audiotaped interviews, 28 Jul 1988, 30 May 1996, 25 Mar 1997, NPOLHP.

Kanaar, Al. Audiotaped interview, 27 Oct 1988, NPOLHP.

Kopcza, Joseph. Audiotaped interview, 21 Dec 1988, NPOLHP.

Kupstas, Agnes. Audiotaped interview, 4 Dec 1988, NPOLHP.

Lazar, Peter. Audiotaped interview, 25 Jun 1990, NPOLHP.

Mikulski, John. Audiotaped interview, 28 Jul 2001, NPOLHP.

Milner, E. Stewart. Audiotaped interview, 3 Aug 1994, NPOLHP.

Mullay, Sam. Audiotaped interview, 22 Jan 2005, NPOLHP.
Musto, Raphael. Audiotaped interview, 30 Jul 2001, NPOLHP.
Orloski, Charles. Audiotaped interview, 9 Jun 1998, NPOLHP.
Oshirak, John and John Usefara. Audiotaped interview, 2 Aug 2001, NPOLHP.
Panzitta, David. Audiotaped interview, 14 Mar 1996, NPOLHP.
Panzitta, Joseph. Audiotaped interview, 25 Apr 2007, NPOLHP.
Piazza, John. Unrecorded interview, 19 May 1995, NPOLHP.
Piccillo, Joseph. Unrecorded interview, 6 Jul 1999, NPOLHP.
Pinkowski, Helen. Audiotaped interview, 7 Dec 1988, NPOLHP.
Randall, David. Audiotaped interview, 7 Jun 1998, NPOLHP.
Roman, Stanley. Audiotaped interview, 12 Dec 1988, NPOLHP.
Rosenn, Judge Max. Unrecorded interview, 18 Jan 1995, NPOLHP.
Sabastianelli, Secundo. Audiotaped interview, 29 Jan 1999, NPOLHP.
Shupnik, Fred. Audiotaped interview, 11 Jan 1995, NPOLHP.
Siracuse, Angelo. Audiotaped interviews, 19 Jul 1989, 6 Oct 1989, NPOLHP.
Stanton, Earl. Audiotaped interview, 15 Mar 1999, NPOLHP.
Stella, Pacifico "Joe". Audiotaped interview, 1 Nov 1988, NPOLHP.
Stonionis, Joseph. Audiotaped interview, 12 Sept 2001, NPOLHP.
Supey, Thomas Jr. Audiotaped interview, 7 Jul 1999, NPOLHP.
Tedesco, James. Audiotaped interview, 8 Oct 1996, NPOLHP.
Usefara, John. Audiotaped interview, 28 Jul 2001, NPOLHP.
Volpe, Charles. Audiotaped interview, 22 Jan 1998, NPOLHP.
Voystock, Phil. Audiotaped interview, 19 Nov 2002, NPOLHP.
Waitkevich, Anthony. Audiotaped interview, 8 Aug 1981, NPOLHP.
Wolinsky, John. Unrecorded interview, 29 Jul 2001, NPOLHP.

APPENDIX I

A Glossary of Anthracite-Related Terms, Organizations, and Events*

(Note: A capitalized word in an entry is itself defined within the Glossary)

Alleghany Corporation—A corporation created in 1929 by the Van Sweringen Brothers of Cleveland, Ohio, to serve as the holding company for their expanding business empire, which, after 1924, included the ERIE RAILROAD and its subsidiaries, HC&I, NYS&WCC, and PaCC. Alleghany faced bankruptcy in the 1930s and went through a series of ownership and organizational changes beginning in the mid-1930s. Alan P. Kirby, a native Wilkes-Barrean, became a partner in the late 1930s and held the presidency until 1961; he also served as chairman of the board and chief executive officer from 1963 to 1967, and as chairman emeritus from 1967 to 1973.

Anarchism, Anarchist(s)—Anarchism is a political philosophy that is critical of state, corporate, or other hierarchical or bureaucratic control. Anarchists advocate a stateless society based on non-hierarchical voluntary associations controlled through direct democracy by workers and community members. Many members of the INDUSTRIAL WORKERS OF THE WORLD, which organized ANTHRACITE MINEWORKERS in the 1910s, were anarchists. See SYNDICALISM.

Anthracite—Geologists have classified coal into three general categories: lignite, BITUMINOUS, and anthracite, with grades in between. Also known as HARD COAL because of its difficulty in ignition, it is almost pure carbon and burns more consistently and efficiently than bituminous or SOFT COAL. Anthracite produces less waste and smoke than bituminous. A ten-county area in northeastern Pennsylvania contained up to 95 percent of the northern hemisphere's anthracite deposits.

Anthracite Board of Conciliation (ABC)—A GRIEVANCE hearing and adjudicating body created by the ANTHRACITE STRIKE COMMISSION in 1903 as part of the settlement of the STRIKE OF 1902. The body consisted of the presidents of UMWA anthracite Districts 1, 7 and 9, plus three management members selected by the coal companies, and one appointed umpire. Its purpose was to adjudicate workers' grievances and thereby reduce labor-management animosity and avoid strikes. MINEWORKERS generally saw the rulings of the tribunal as favoring management. The ABC remained in existence from 1903 to 1962. See LABOR-MANAGEMENT RELATIONS.

Anthracite Code—Envisioned as a set of rules and regulations negotiated and agreed upon by labor and management as part of the National Industrial Recovery Act of 1933, it was intended to facilitate industrial progress and economic recovery during the Great Depression. However, unlike the BITUMINOUS code, the anthracite principals could not agree upon the terms and the Anthracite Code was never adopted.

Anthracite Coal Fields—Four geographic areas or fields in northeastern Pennsylvania contained the ANTHRACITE deposits: the SOUTHERN COAL FIELD headquartered at Pottsville; the WESTERN-MIDDLE COAL FIELD between Mahanoy City and Shamokin; the EASTERN-MIDDLE COAL FIELD in the Minersville and Hazleton area; and the NORTHERN COAL FIELD in the Wilkes-Barre/WYOMING VALLEY and Scranton/ LACKAWANNA VALLEY areas. The fields covered parts of ten counties that once contained up to 95 percent of the northern hemisphere's anthracite deposits. See ANTHRACITE REGIONS.

Anthracite Industry—Comprised all of the organizations and persons involved in the mining, production, distribution, and sales of ANTHRACITE. It included workers, unions, coal companies, railroads, canal operators, trade associations, marketing and sales firms, etc. Anthracite employment peaked in 1913 at over 180,000 men and boys while production peaked in 1917 at over 100 million tons. The industry was situated in a ten-county area of northeastern Pennsylvania, which contained up to 95 percent of the northern hemisphere's anthracite deposits.

Anthracite Institute—Established in 1929 by the coal companies to pursue technological innovation and enhance consumer markets; the Manhattan-based agency in turn created the ANTHRACITE LABORATORIES to develop new technologies (including furnaces) and approve and certify existing ones. During four decades of operation, the Institute did little to stem the movement away from ANTHRACITE to other fossil fuels.

Anthracite Laboratories—See ANTHRACITE INSTITUTE.

Anthracite Miners Union of Pennsylvania (AMUP)—A short-lived labor organization founded in 1929 by Frank McGarry, Rinaldo F. Cappellini, and numerous other INSURGENTS in DISTRICT 1 as an alternative to the UMWA. The DUAL UNION received a state charter but faded within months. It was the predecessor of the UNITED ANTHRACITE MINERS OF PENNSYLVANIA (UAMP), an insurgent organization formed in 1933 and disbanded in 1935. The UMWA expelled several persons associated with the AMUP.

Anthracite Mineworkers—A term referring to all men and boys employed in the ANTHRACITE INDUSTRY.

Anthracite Regions—The COAL OPERATORS created three anthracite regions for marketing purposes: Schuylkill, Lehigh, and Wyoming. The SCHUYLKILL ANTHRACITE REGION encompassed most of the SOUTHERN COAL FIELD and all of the WESTERN-MIDDLE COAL FIELD; the LEHIGH ANTHRACITE REGION covered the EASTERN-MIDDLE COAL FIELD and the eastern tip of the SOUTHERN COAL FIELD; the WYOMING ANTHRACITE REGION included the NORTHERN COAL FIELD from above Scranton to below Wilkes-Barre.

Anthracite Strike Commission of 1902–1903—Appointed by President Theodore Roosevelt to settle the STRIKE OF 1902, the seven-member body heard testimony from 558 witnesses, including workers, managers, community leaders, and others over a three-month period. The body produced a final report that issued several binding measures, which included an average ten percent wage increase for MINEWORKERS and the creation of the ANTHRACITE BOARD OF CONCILIATION, among others. The INDIVIDUAL CONTRACT, which was a main issue for employees of THE ERIE COAL COMPANIES, the Hudson Coal Company, and the DL&W Railroad's Coal Division, was among the UMWA's demands, and the Strike Commission heard testimony about its harmful consequences on workers. However, the matter was not seriously considered because Strike Commission members considered the issue beyond their charge.

Apalachin Organized Crime Meeting—A national meeting of top organized-crime bosses and their lieutenants and on November 14, 1957, in the small New York state town of Apalachin. The gathering was organized by Joseph Barbara, the alleged crime boss of northeastern Pennsylvania from 1950 to 1959. Police raided the meeting and arrested several individuals, although many escaped. Indictments and court cases followed. The meeting advanced the federal government's drive against organized crime. Others present from northeastern Pennsylvania were Russell Bufalino, Dominick Alaimo, James A. Osticco, and Angelo Sciandra. Alaimo and Sciandra were affiliated with the KNOX COAL COMPANY and also held interests in the regional ladies' garment manufacturing industry.

Battle of Archbald—On December 5, 1896, a group of Italian immigrant LABORERS staged a WILDCAT STRIKE over what they saw as an unfair wage change by their employer, the Forest

Mining Company, at a mine in Archbald, Lackawanna County. Violence marked the day-long confrontation between the workers on one side, and mine bosses, the Squire of Archbald, and the hired deputies of the Barrington & McSweeney Detective Agency on the other. The STRIKE was broken and the company dismissed the workers and expelled them from company housing. It was one of the earliest labor protests by Italian immigrant workers in the NORTHERN ANTHRACITE FIELD.

Barrier Pillar(s)—Pennsylvania's state mining law required that large, solid blocks of unmined coal, called barrier pillars, be left intact in a mine to separate adjacent collieries' workings. The pillars were required to be 100 feet in width. Their purpose was to insure that a flood, fire, or explosion in one mine would not spread to its neighbor(s). See ROBBING THE BARRIER.

Bituminous—A type of coal ranking below ANTHRACITE but above lignite in energy value. It is also known as SOFT COAL because it lacks the hardness of anthracite, can flake and split more easily, and can be readily ground for easy combustion. It contains between 15 and 50 percent volatile matter and is "younger" and of lower heat value than anthracite. Bituminous is the most abundant coal in the world and in the U.S., which has deposits in 35 states. Processing does not require an elaborate BREAKER, in contrast to anthracite.

Black Hand (*Mano Nero*)—A term referring to localized Italian gangs in the late nineteenth and early twentieth centuries who engaged in extortion and protection rackets. Their primary targets were Italian individuals (including MINEWORKERS) and small-business owners, and they threatened violence or murder as the consequences for noncompliance. While the gangs were not generally associated with early Sicilian MAFIA gangs, Sicilians did form Black Hand groups along with Calabrians and other Italian immigrants. One Black Hand group had its base in the Browntown section of Pittston, and its threats, crimes, arrests, and trials received wide newspaper coverage in the early 1900s. The Black Hand gangs generally declined during the Prohibition era and their leaders had little if any impact in the succeeding criminal networks. Santo Volpe was alleged to have been the leader of organized crime—Black Hand and Mafia—in northeastern Pennsylvania between 1908 and 1933. He exerted considerable influence over the SUBCONTRACTING SYSTEM and the LEASING SYSTEM at PaCC–HC&I in the 1920s and 1930s, as well as UMWA affairs in DISTRICT 1 in the 1930s. To counteract the Pittston Black Hand, law-abiding Italian citizens in northeastern Pennsylvania formed the WHITE HAND SOCIETY in the early twentieth century.

Blackleg—Another term for STRIKEBREAKER or SCAB, it has been more commonly used in British industry and in nineteenth-century American industry.

Blacklist, Blacklisting—A term referring to the informal exclusion of a worker from a particular place of employment (and sometime from an entire industry) because she/he had committed some egregious violation as defined by management.

Blasting Powder—Before the introduction of dynamite in the ANTHRACITE INDUSTRY during the 1880s, MINERS used blasting powder, or black powder, to assist in mining at the COAL FACE.

Bootleg Mining, Bootlegging Coal—The illegal mining, processing, and selling of coal mainly by unemployed MINEWORKERS, especially during the national economic depression of the 1930s. In 1936–1937, the bootleg trade amounted to some 2.5 million tons or five percent of total output. By 1939, an estimated 2,500 illegitimate ANTHRACITE mines produced approximately 3.5 million tons. The bootleggers sold the coal in a thriving underground market of processors, shippers, sellers, and buyers. At its peak, the sector employed between 13,000 and 20,000 men and boys. Police were reluctant to make arrests and the courts were hesitant to prosecute violators because of the industry's dire economic circumstances. The black market sector thrived mainly in the SOUTHERN COAL FIELD where veins outcropped close to the surface, but it also found practitioners the MIDDLE and NORTHERN COAL FIELDS.

Boss System—Another term for the PADRONE SYSTEM, "*padrone*" meaning boss in Italian.

Boycott—As used here, it was another term for STRIKE, SHUTDOWN, and WORK SUSPENSION, where workers closed a mine to express GRIEVANCES and demand redress from employers. See WILDCAT STRIKE.

Breaker—An ANTHRACITE coal-processing and preparation plant, it was the main structure on the COLLIERY grounds. Although initially using relatively simple technology, the breaker grew to house the extensive machinery used for crushing, sorting, sizing, "cleaning," weighing, and washing the coal. The structures were made of wood until the 1910s when steel and concrete came into use. See BREAKERMAN, BREAKER BOY.

Breakerman, Breakermen—A MINEWORKER who toiled in the BREAKER on some aspect of processing coal (crushing, "cleaning," sizing, washing, loading, weighing, etc.).

Breaker Boy(s)—A young lad between six and twelve years of age who picked WASTE out of coal during its processing in the BREAKER. They worked for long hours at extremely low wages and were often mistreated by the bosses. Mine-working families needed their wages and the coal companies needed their labor.

Burgess—An elective position in Pennsylvania municipalities analogous to mayor.

Business Unionism—A practical, non-ideological, and cooperative approach to LABOR-MANAGEMENT RELATIONS. Fostered by Samuel Gompers of the AF of L in the late nineteenth century, it was also supported by labor leaders such as John Mitchell and John L. Lewis, presidents of the UMWA. It came to represent the orientation of most U.S. unions whereby wages, benefits, work rules, and seniority rights became the main concerns, as opposed to policy-related matters such as investment decisions, the right to issue contracts, and production methods, which were left to management.

Butty System—The name given to a SUBCONTRACTING SYSTEM in the British coal-mining industry in the nineteenth century. See CHARTER MASTER SYSTEM.

Campbell Act (Act No. 139)—Formally termed The Alien Tax Act, it went into effect on July 1, 1897, after passage by the Pennsylvania Legislature and the signature of Governor Daniel H. Hastings. The statute imposed a three-cents-per-day tax on coal companies for each adult immigrant employed in their mines. The companies and the immigrant workers opposed the measure while the UMWA initially supported it as a bulwark against cheap foreign labor. The statute figured prominently in the protests preceding the LATTIMER MASSACRE of September 1897. Within two months following its implementation, the U.S. Court of Appeals for the Third Circuit in Pittsburgh declared it unconstitutional as a violation of the equal protection clause of the fourteenth amendment. In his opinion on the decision, Judge Marcus Wilson Acheson wrote that the law discriminated against new immigrants. The UMWA also opposed it by this time.

Capitalism—An economic system based on private ownership of the means of production. Other key components include private property, the profit motive, free competition, and a pricing mechanism built on supply and demand. Beginning in the eighteenth century with the writings of Adam Smith and others, among capitalism's historical variants have been mercantilism, free-market capitalism, state capitalism, social capitalism, and corporate capitalism. Adherents of SYNDICALISM, COMMUNISM, and SOCIALISM—including some ANTHRACITE MINEWORKERS, and former Sicilian SULFUR WORKERS—generally opposed capitalism. See SOCIALISM, STATE SOCIALISM, SYNDICALISM.

Car(s) of Coal—MINERS and LABORERS in the NORTHERN COAL FIELD were most commonly paid by the PIECE or the number of cars of coal produced. By the turn of the nineteenth century, the WORK CULTURE stipulated that each worker produce no more than two cars per shift. Therefore, a crew of one miner and two laborers was expected to harvest six cars of coal; however, the actual number varied by the size of the car and other factors. The COAL CARS varied in size from two to five tons each, depending on the company. The coal companies' desire for more than two cars led to extensive labor-management conflict before and after the STRIKE OF 1902, and also to broader use of SUBCONTRACTING and LEASING, especially at PaCC and HC&I. Pagnotti Company mines were known to have used among the largest cars in the industry, five tons each.

Car Runner—A MINEWORKER who collected fully loaded COAL CARS from CHAMBERS after a shift had been completed and pushed them by hand into the GANGWAY. MULE DRIVERS would later hitch the trip of cars to their mules and transport them to the SHAFT, SLOPE, or CONVEYOR where the RUN-OF-MINE COAL was sent to the surface for processing. See MULE DRIVERS

Card of Recommendation—A practice whereby a MINEWORKER dismissed at one PaCC or HC&I COLLIERY remained unemployed until his former bosses sent his employment card or record to the main office where his discharge was entered and the card returned to the former workplace. The process often took weeks and the workers saw the policy as both a punishment and a form of BLACKLISTING.

Cave-in—See SUBSIDENCE.

Chamber—Also referred to as a ROOM, it was the part of a COAL VEIN from which a miner and his laborer(s) removed coal. The forward-most position in the chamber is called the COAL FACE. Each chamber was connected to the main GANGWAY with rail tracks so that the RUN-OF-MINE COAL could be taken to the surface for processing in the BREAKER. See ROOM-AND-PILLAR METHOD.

Charter Master System—The name given to a SUBCONTRACTING SYSTEM used in the British mining industry during the eighteenth and nineteenth centuries. See BUTTY SYSTEM.

Check-Docking Boss—The final ruling of the ANTHRACITE STRIKE COMMISSION in 1903 gave members in LOCAL UNIONS the option to elect and pay a fellow workman to watch over or check the decisions of the company's DOCKING BOSS. The ruling applied to COLLIERIES that paid workers by the PIECE PAYMENT METHOD or the number of COAL CARS produced each shift, as opposed to those that paid by the WEIGHT PAYMENT METHOD; workers at the latter collieries had the option to elect and pay a CHECK-WEIGHMAN. In practice, the holders of these positions were not always allowed to assume their duties because the company denied them access to the colliery.

Check-Off—MINEWORKERS' pay deductions for union dues collected by the coal companies and forwarded to the UMWA. After years of negotiations, the UMWA finally achieved a partial check-off through the labor-management contract of 1930.

Check-Weighman—The final ruling of the ANTHRACITE STRIKE COMMISSION in 1903 gave members of LOCAL UNIONS the option to elect and pay a fellow worker to watch over or check the decisions of the company's WEIGHMAN. The ruling applied to COLLIERIES that paid workers by the WEIGHT METHOD OF PAY, as opposed to those that paid the PIECE METHOD OF PAY or the number of cars produced; workers at the latter collieries had the option to elect and pay a CHECK-DOCKING BOSS. In practice, the holders of these positions were not always allowed to assume their duties because the company denied them access to the colliery. Alex Campbell was one such person who, as the elected check-weighman of L.U. 1703 at PaCC's No. 6 Colliery, was prevented by the company from assuming his post.

Closed Shop—A union-only workplace where all employees must belong to the recognized labor organization representing the workers. The major ANTHRACITE producers became closed-shop operators in 1920 when, as part of labor-management contract negotiations, the OPERATORS finally recognized the UMWA as the legitimate bargaining agent of the MINEWORKERS.

Coal and Iron Police—A constabulary force created and sanctioned by the Pennsylvania General Assembly in 1866. Its members were employed and paid not by the state but by coal, iron, railroad, and other corporations and used to break strikes, protect property, and otherwise reinforce management's intentions. It essentially functioned as the companies' private police force. Following the STRIKE OF 1902 and the resulting violence, the state legislature voted to create a statewide public police force. The Pennsylvania State Police, established in May 1905, were under the direct supervision of the state government. However, the Coal and Iron Police continued under private auspices into the late 1920s and early 1930s. Governor Gifford Pinchot finally moved against private police commissions in 1931, thereby effectively terminating the organization.

Coal Car(s)—A wooden container on metal wheels used to transport RUN-OF-MINE COAL out of a mine. There were many different types and sizes but a typical car held from two to five tons. The coal companies required a sizeable TOPPING on each car or else the MINEWORKERS' wages were DOCKED. The workers' pay could also be docked for too much WASTE in a car, as judged by the DOCKING BOSS, or for an insufficient weight, as judged by the WEIGHMAN. See CAR OF COAL.

Coal Face—The PLACE in a mine at the head or front of a CHAMBER where the workers removed the coal.

Coal Vein—See VEIN OF COAL.

Collier—A general term for a MINEWORKER, referring to any person employed in or around a COLLIERY, above or below ground. See MINER, LABORER, BREAKER BOY, DIGGER, HEWER, NIPPER.

Collier-Serf System—In 1606, the Scottish Parliament enacted a law that bonded all colliers and their families to their places of work, irrespective of ownership. Scholars and others have come to regard the system as a form of slavery. Attempts at escape were severely punished as breeches of contract that deprived the landlord of the colliers' labor. The contract was usually established at a child's baptism when, for a small fee, the father-collier pledged the life-long labor of the child—male or female—to the landowner or "laird" on whose property the collier mined; this act was called *arling*. Entire families participated in the coal extraction and haulage in Scotland's mines, including women and very young children. The collier-serf law was overturned in 1799, but women continued to work underground until the mid-nineteenth century, and male children of legal age did so for many decades beyond. The 1606 law likewise was overturned a few decades later.

Colliery, Collieries—The entire coal-production plant, including the mines, BREAKER, and all the related buildings.

Communism, Communist(s)—An economic, political, and social ideology that seeks to create a SOCIALIST economy with workers collectively owning and controlling the means of production, as well as a stateless social order and classless society, among other things. Many variants have been fostered including Marxist Communism, ANARCHIST Communism, Eurocommunism, Leninism, Stalinism, and Maoism. Some MINEWORKERS in the NORTHERN COAL FIELD subscribed to Communism during the 1920s and 1930s. See COMMUNIST PARTY.

Communist Party—The Communist Party of the USA (CPUSA) sought to organize ANTHRACITE MINEWORKERS during the 1920s and 1930s. Estimates put CPUSA membership at about 50 persons in the NORTHERN COAL FIELD during the late 1920s, and over 200 by the mid-1930s. The SAVE-THE-UNION COMMITTEE had CPUSA affiliation, as did the NATIONAL MINERS UNION. The CPUSA never gained a significant following in the ANTHRACITE COAL FIELDS, nor did it have any significant influence on LABOR-MANAGEMENT RELATIONS.

Company Miner—In contrast to the CONTRACT MINER, company miners did not typically work at the COAL FACE. They were paid by the day to undertake various projects such as clearing CAVE-INS, making repairs, and securing the main GANGWAYS.

Contract Miner—Not to be confused with SUBCONTRACTOR, a contract miner worked at the COAL FACE in one CHAMBER at a time with one or two LABORERS. The term derived from the British industry where skilled miners were required to sign INDIVIDUAL CONTRACTS in the spring, binding them for one year's work at a particular mine. Some contract miners in the NORTHERN ANTHRACITE FIELD became subcontractors when they secured MINERAL-RIGHTS agreements from the coal companies.

Contracting System—See SUBCONTRACTING SYSTEM.

Control—See WORKPLACE CONTROL.

Conveyor—One of a number of mining technologies that gained wider use in ANTHRACITE during the 1920s. Intended to speed the removal of RUN-OF-MINE COAL from a mine, it consisted of a rubber belt connected to an electric drive motor onto which the LABORER(S) shoveled or dumped coal. The raw product went directly into empty COAL CARS or, in some cases, directly to the surface into the BREAKER or into waiting railroad cars for shipment to a breaker at another site. MINEWORKERS often resisted such new technologies because of their potential negative effects on employment and WORKPLACE CONTROL.

Culm—The slate, rock, and other WASTE that results from the processing of coal in the BREAKER. See CULM BANKS.

Culm Banks—A pile of CULM or coal WASTE that can assume mountain-like proportions. They have been among the most visible legacies of the deep mining era in ANTHRACITE. Many have disappeared in recent years after having been reprocessed to extract additional combustible materials for use in modern power plants.

Dago—A pejorative term or slur directed at persons of Italian heritage. Similar insults include "wop," "gringo," "carcamano," "badola," "cincali," and "macaroni."

Davis Act (1913)—A Pennsylvania state law intended to limit pillar removal from within a mine in order to protect the surface from SUBSIDENCE or CAVE-INS. It was never effectively implemented. See FOWLER ACT

Dead Work—Work conducted by a mining crew that does not involve extraction, conveyance, or processing of coal (i.e., production), but which is necessary to facilitate production and insure safety. Dead work includes tasks such as removing rock from a CHAMBER and propping the ROOF.

De-industrialization—A term made popular in the 1980s to describe the loss of manufacturing and related blue-collar jobs in the U.S., it also described the situation in the ANTHRACITE INDUSTRY especially after 1926, when numerous mines began closing in a trend that continued into the 1960s and 1970s.

Development Work—The opening of new sections of a mine so that workers can access virgin VEINS OF COAL, it involved the removal of rock and coal and in the twentieth century was typically undertaken by SUBCONTRACTORS working with SPECIAL CONTRACTS. It proved controversial when subcontractors harvested a greater number of CARS OF COAL than the work-culture standard of two cars per man.

Digger—A vernacular term used mainly in the nineteenth century for MINER or LABORER. See COLLIER, HEWER, MINEWORKER.

Dillingham Commission—Officially termed The United States Immigration Commission but popularly known as the Dillingham Commission, it was headed by Senator William Paul Dillingham of Vermont. This joint House-Senate committee was formed in 1907 to investigate the growing national concern over immigration. The commission's final report in 1911, issued in 41 volumes, concluded that southern and eastern European immigrants posed a serious threat to the nation and their numbers should be significantly reduced. The findings led to the immigration restriction measures of the 1920s, including the Emergency Quota Act of 1921, the Immigration Act of 1924, and the National Origins Formula of 1929, which greatly diminished the number of immigrants from southern and eastern Europe and Asia, while favoring those from Britain and northern Europe.

Dissident(s)—In this volume, a term used to describe militant MINEWORKERS who, during the 1910s, 1920s, and 1930s, were dissatisfied with TENANCY, wage rates, work rules, and other matters, as well as the ineffectiveness of the UMWA to address their GRIEVANCES. Many were employed by PaCC and HC&I where they called themselves INSURGENTS. The INSURGENT MOVEMENT spread beyond THE ERIE COAL COMPANIES in the late 1920s and early 1930s to include the all of DISTRICT 1 and the NORTHERN ANTHRACITE FIELD. The movement led to the formation of the UAMP in 1933. The term has not been used to describe other labor militants, including COMMUNISTS and SOCIALISTS, who were also active in District 1. See *GLI INSORTI*.

Distributive Justice—A type of justice that emphasizes equity in the dissemination of outcomes in a society or other organization. It focuses on fairness and justice in answering the question: who gets what, where, when, and why? See RETRIBUTIVE JUSTICE, JACQUERIE.

Districts, Anthracite—The UMWA designated three anthracite districts corresponding to the three ANTHRACITE REGIONS. DISTRICT 1 encompassed the WYOMING ANTHRACITE REGION; District 7 the LEHIGH ANTHRACITE REGION; and District 9 the SCHUYLKILL ANTHRACITE REGION.

District 1—One of three UMWA districts covering ANTHRACITE MINEWORKERS, it included the entire WYOMING ANTHRACITE REGION or NORTHERN ANTHRACITE FIELD. With over 60,000 workers in the early 1930s, it had the largest membership of the three ANTHRACITE DISTRICTS, and one of the largest in the UMWA. Criminal corruption, some of it instigated by ORGANIZED CRIME, began infecting the District administration in the 1920s and culminated with President August J. Lippi's illegal ownership of the Knox Coal Company in 1950.

Docked, Dockage, Docking—Reduction in a MINER's wages by the DOCKING BOSS for an insufficient TOPPING on a CAR OF COAL, inadequate weight, or excessive WASTE. Docking led to numerous conflicts as workers charged the DOCKING BOSS with cheating them out of income by docking excessively. The ANTHRACITE STRIKE COMMISSION's 1903 final ruling gave LOCAL UNION members the option to elect and pay a CHECK-DOCKING BOSS to help insure fair treatment. In practice, coal company management often prevented the check-docking boss from assuming his duties by denying him access to the COLLIERY.

Docking Boss—In mines that paid by the PIECE PAYMENT METHOD (as opposed to WEIGHT PAYMENT METHOD), coal companies employed a person with this title to examine the cars of RUN-OF-MINE COAL coming out of a mine. He looked for excessive WASTE or an insufficient TOPPING, and "docked" or reduced a miner's wages accordingly. Workers regularly charged that the docking boss reduced wages excessively and unfairly. The final ruling of the ANTHRACITE STRIKE COMMISSION in 1903 gave members of LOCAL UNIONS the option to elect and pay a CHECK-DOCKING BOSS to work alongside the DOCKING BOSS and, at least in principle, ensure against excessive and unfair DOCKAGE. In practice, coal company management often prevented the check-docking bosses from assuming their duties by denying them access to the COLLIERY. See WEIGHMAN and CHECK-WEIGHMAN.

Door Tender(s), Door Boy(s)—Young boys who sat in the dark and opened and closed the wooden air-lock doors in a mine so that CARS OF COAL, equipment, and workers could pass.

The doors had to be kept closed to insure the flow of air currents. Also known as NIPPERS, they represented an example of child labor in mining.

Downsizing—A term that came into use in the 1990s, it describes the practice of companies laying off or terminating employees in order to streamline operations, cut costs, and boost efficiency. It aptly characterizes the decisions made by the coal corporations in the 1920s and 1930s as the ANTHRCITE INDUSTRY began to decline.

Driver Boy(s)—A teenage boy who led a mule in hauling empty CARS OF COAL via rail to MINERS and LABORERS at the COAL FACE and later hauling the loaded cars to the SHAFT where they were sent to the BREAKER for coal processing. See MULE DRIVER. MULE TENDER.

Dual Union, Dual Unionism—A term used by the leadership of an established union to characterize the formation of, or participation in, another, alternative union within the same company or industry. The second union is usually the result of worker dissatisfaction with the established organization. Dual unionism, therefore, creates a situation where two labor organizations compete to represent the same workers. Within DISTRICT 1, the established UMWA faced dual-union threats from the INDUSTRIAL WORKERS OF THE WORLD between 1907–1916; the ANTHRACITE MINERS UNION OF PENNSYLVANIA in 1928; the COMMUNIST-affiliated NATIONAL MINERS UNION between 1928 and 1931; the UNITED ANTHRACITE MINERS OF PENNSYLVANIA between 1933 and 1935; and the PROGRESSIVE MINERS UNION in 1938–1939, which was an affiliate of the American Federation of Labor. The UMWA successfully defeated each threat to its position as the sole representative of the ANTHRACITE workers, often relying on the assistance of the coal companies, local and state police, the courts, community leaders, the natinoal government, and ORGANIZED CRIME.

Due Bill—A wage receipt from the company to a MINEWORKER stating the amount of income due for work. The due bill listed hours and pay rates minus deductions for supplies, materials, and other charges.

Eastern-Middle Anthracite Field—The smallest of the four ANTHRACITE COAL FIELDS, most of its territory lies in southern Luzerne County and includes small areas of Schuylkill and Carbon counties. The field is 26 miles long and 10 miles wide and consists of a series of small trough-like basins with coal veins running in an east to west direction. Hazleton in Luzerne County is the main city and other towns include West Hazleton, Lattimer, Harwood, Harley, Eckley, and Audenreid. Although the Lehigh & Wilkes-Barre Coal Company was a main producer, the field had numerous smaller, family-owned COAL OPERATORS including the Van Winkles, Pardees, Markles, Coxes, and Fells. Anthracite is still strip mined in the field by the Jeddo-Highland Coal Co., a subsidiary of Pagnotti Enterprises, Inc. See NORTHERN ANTHRACITE FIELD, SOUTHERN ANTHRACITE FIELD, WESTERN-MIDDLE ANTHRACITE FIELD.

Eastern-Middle Coal Field—See EASTERN-MIDDLE ANTHRACITE FIELD.

Easy Coal—An informal term for highly accessible, easy-to-mine coal. In a clear act of corruption, bosses often dispensed easy coal to favored employees or to employees who offered monetary KICKBACKS (which some bosses required). Mining in such coal usually allowed workers to toil in safer conditions and meet production quotas relatively quickly.

Equalization—The principle of sharing scarce available work among MINEWORKERS within the COLLIERIES of a particular coal company. During the second half of the 1920s and the first half of the 1930s, coal companies closed or cut production at their less efficient collieries and kept the more efficient operations running. The policy led to thousands of unemployed and underemployed men who argued for a job-sharing program to provide some income for all workers. The idea derived from, and was an expression of, the communitarian values within the workers' ethnic and working-class cultures, while also being a practical response to mass unemployment. With few exceptions, the companies were concerned only with the efficiency and, therefore, resisted the idea. President John L. Lewis and the UMWA were not enthusiastic about the plan and did not push for it. The never-approved ANTHRACITE CODE proposed in 1933 by the National Recovery Administration would have reduced the work week from 48 to 40 hours, thereby providing more jobs for workers. The 40-hour week under the WAGNER ACT OF 1935 was intended to have the same effect. While workers at certain collieries in the SCHUYLKILL ANTHRACITE REGION did gain a viable equalization plan, and while PaCC did implement a short-lived plan at the Underwood Colliery, equalization was not realized to any large extent in the anthracite fields.

The Erie Coal Companies—Three coal companies—HC&I, PaCC, and NYS&WCC—were controlled by the ERIE RAILROAD by 1901 and became known as the Erie Coal Companies. At the time, PaCC operated ten collieries while HC&I had four, all of them in the NORTHERN ANTHRACITE FIELD. Their combined workforces totaled about 12,000 men and boys. NYS&WCC mined none of its own coal but contracted for the finished product from INDEPENDENT COAL COMPANIES and initially sold the coal under its own name. HC&I and PaCC were the leaders in issuing SUBCONTRACTS during the 1890s and resumed the practice between 1913 and 1920, and again after 1925. Numerous STRIKES occurred when workers resisted the SUBCONTRACTING SYSTEM. The Van Sweringen brothers gained control of the Erie Railroad and its subsidiaries in 1924. Michael Gallagher was appointed president of the coal companies in 1928 and soon became chairman of the board. The Van Sweringens formed THE ALLEGHANY CORPORATION in 1929 as a holding company for railroad interests, including Erie and its subsidiaries. They created THE PITTSTON COMPANY (TPC) as a subsidiary of PaCC in 1929, whose main purpose was to issue and manage coal LEASES. The three Erie companies turned over all of their coal properties (except PaCC's Ewen Colliery) to TPC in a major expansion of the LEASING SYSTEM. LABOR-MANAGEMENT RELATIONS deteriorated and conflict spread from HC&I and PaCC collieries to others throughout DISTRICT 1. In an effort to gain WORKPLACE CONTROL and boost efficiency and profits, the Erie Coal Companies became the leaders in issuing TENANCY agreements to independent operators. Hundreds of the subcontracts and leases were issued to companies owned by persons with alleged ORGANIZED CRIME connections in a prime example of UPPERWORLD cooperation with the UNDERWORLD.

Erie Railroad —With roots in the New York and Erie Rail Road, which formed in 1832, the Erie went through numerous transformations, including bankruptcies, during the nineteenth century. Following a bankruptcy in 1859, it reorganized as the Erie Railway in 1861. The scandalous "Erie Wars," fought between Cornelius Vanderbilt and Jay Gould, followed in the 1860s. Early in the decade, Erie contracted with PaCC to haul anthracite from Lackawaxen, PA, to New York City. Erie took control of another anthracite firm, HC&I, in 1871 and, like other major anthracite railroads, became a coal producer. A reorganized Erie resumed control of HC&I in 1895 at a cost of $1 million, a deal financed by J.P Morgan banks. After decades of sometimes difficult relations, PaCC moved to break its dependency on Erie by establishing an alternative rail line. Erie's president, Eben B. Thomas, fought the idea and, in 1901, convinced Morgan interests to take over PaCC at a cost $32 million ($874,000,000 in 2011 dollars). When PaCC's production was added to the output from HC&I and the coal contracts held by another minor Erie-owned ANTHRACITE company, the New York Susquehanna & Western Coal Company (NYS&WCC), Erie accounted for nearly six million tons in 1902. As subsidiaries, PaCC, HC&I, and NYS&WCC became known as THE ERIE COAL COMPANIES. PaCC and HC&I became the leaders in fostering the INDIVIDUAL CONTRACT and TENANCY SYSTEMS in the NORTHERN ANTHRACITE FIELD; many of the agreements went to alleged organized criminals.

Face—See COAL FACE.

Fasci—Translated as "bundle of sticks," this Italian term refers to the peasants' and workers' organizations formed mainly by SOCIALISTS in Sicily during the early 1890s. At their peak, the groups had more than 300,000 members from diverse occupational categories, including share croppers, day laborers, industrial workers, and SULFUR MINERS and LABORERS. The *Fasci* led a wave of *JACQUERIE*-like STRIKES in dozens of towns between 1892 and 1894. They protested the terribly brutal and exploitative working conditions in the sulfur mines, the owners of which used SUBCONTRACTORS and PADRONES often with ORGANIZED CRIME affiliations. The failure of the organizations and their WORK SUSPENSIONS prompted many Sicilians to emigrate to the U.S. and other countries.

Fines—A term used for the smallest sizes of mined coal. For many years, the companies had surface workers discard the "fines" in CULM BANKS, refusing to count them as part of a miner's production and pay. Beginning in the 1920s, markets developed for the "fines" but the companies delayed before agreeing, through union negotiations, to pay workers for the product.

Fire Boss—A company official who entered a mine before a work shift began and used a safety lamp to check for methane gas buildup in a VEIN OF COAL that could result in an explosion.

First Mining—The removal of VIRGIN COAL from a mine, which, in ANTHRACITE, involved the ROOM-AND-PILLAR METHOD. See SECOND MINING, THIRD MINING.

Foran Law—Passed in 1885 by the U.S. Congress in response to the abuses associated with the PADRONE SYSTEM, it was designed to prevent the immigration of persons who had signed work contracts before entering the U.S.

Fowler Act (P.L. 1198, 1921)—A Pennsylvania law designed to protect the public from damage caused by mine SUBSIDENCE. It stipulated that, regardless of provisos in deeds and other documents releasing coal companies from liability in certain instances, any company mining so as to cause subsidence had committed a criminal act. It established the Anthracite Mine Cave Commission to oversee the implementation of the law. See DAVIS ACT, KOHLER ACT, PENN COAL CASE, PENNSYLVANIA COAL COMPANY V. MAHON (1922).

Gangway—A term in ANTHRACITE referring to a main underground haulage road. After HEADING MEN had quarried a tunnel, other workers would lay track in the gangway by means of which loaded and empty CARS OF COAL could then travel to and from the SHAFT or the SLOPE for transmission to the BREAKER on the surface.

General Grievance Committees—The labor-management agreement of 1912 allowed union members to elect GRIEVANCE COMMITTEES at each COLLIERY. The three-person bodies were charged with working to adjust employee GRIEVANCES with company officials before an employee filed a formal grievance with the ANTHRACITE BOARD OF CONCILIATION. Later, General Grievance Committees were established at each coal company consisting of representatives from each individual COLLIERY's grievance committee. As the General Grievance Committees grew in power, they called WILDCAT STRIKES and otherwise challenged the authority of the coal companies as well as the UMWA's Anthracite District administrations. In 1928, the committees from all of DISTRICT 1's collieries gathered in weekly meetings in anticipation of the special RUMP convention organized by INSURGENTS. Because their power and independence represented such a threat to the companies and to the UMWA, the committees were greatly weakened by the labor-management agreement of 1930.

General Strike—A term for an industry-wide SHUTDOWN called by a labor union.

Gli Insorti—An Italian term meaning "the INSURGENTS," which referred to a direct-action, anti-organizational Italian ANARCHIST group active in Paterson, New Jersey, and elsewhere during the 1910s and early 1920s. There was no evidence that the self-imposed insurgent label attached to DISSIDENTS at THE ERIE COAL COMPANIES beginning in the 1910s (many of whom were Italian immigrants) and continuing into the 1920s and 1930s derived from this group. The IWW was active in Paterson and when WOBBLIES started organizing PaCC and HC&I workers in the 1910s, they might have brought the term to the Pittston area.

Gravity Railroad, PaCC—Using gravity to propel large COAL CARS downhill and steam engines to pull them up hills, it was built by PaCC and opened in 1850 to transport processed coal from mines in the Pittston and Dunmore areas to Hawley, PA, from where it was taken toward New York and other eastern markets on the Delaware & Hudson Canal. "The Gravity" declined in use after 1865 when canal shipment was eclipsed by overland transport by the Erie Railway. It finally closed in 1885, to be superseded by PaCC's Erie & Wyoming Valley Railroad. John B. Smith served as The Gravity's general superintendent between 1850 and 1886.

Grievance(s)—A complaint filed by a MINEWORKER for a perceived management violation or abuse in the workplace. With the settlement of the STRIKE OF 1902, the federal government established the ANTHRACITE BOARD OF CONCILIATION to review and adjudicate grievances.

Grievance Committees—The labor-management agreement of 1912 allowed union members to elect these three-person bodies at each COLLIERY. They were charged with adjusting employee GRIEVANCES with company officials in advance of a formal grievance-filing with the ANTHRACITE BOARD OF CONCILIATION. Later, representatives of the individual COLLIERY committees sent representatives to a GENERAL GRIEVANCE COMMITTEE created at each coal company. The 1930 labor-management agreement weakened the committees, which had become the most powerful local representative bodies in anthracite and contributed to the labor uprising of the early 1930s that culminated in the formation of the UAMP.

Hard Coal—Another term for ANTHRACITE, it refers to the fuel's difficulty in ignition due to a high kindling point, high carbon content, and other physical properties as compared to BITUMINOUS or SOFT COAL.

Haulers—A BITUMINOUS term for a MINEWORKERS who took the coal from the mine to the surface for processing using mules or, later, electric locomotives referred to as MOTORS. In ANTHRACITE, haulers were referred to as MULE DRIVERS or DRIVER BOYS in earlier years, and MOTORMEN later when the mines became electrified and "motors" replaced mules.

Heading—A tunnel driven by HEADING MEN to access VIRGIN COAL.

Heading Men—MINEWORKERS who drive HEADINGS to access VIRGIN COAL.

Hewer—A term used mainly in the early British coal industry in reference to a MINER or a LABORER who hewed, or chopped, the coal by hand from the FACE. See COLLIER, DIGGER, MINEWORKER.

Hillside Coal & Iron Co. (HC&I)—An ANTHRACITE-mining company established in 1867, it was purchased by the ERIE RAILROAD in 1871 and re-purchased by a reorganized Erie in 1895. Three coal companies—HC&I, PaCC, and NYS&WCC—shared the same management after 1901 and became known as THE ERIE COAL COMPANIES. HC&I and PaCC vigorously expanded TENANCY while resisting unionization. They remained the largest non-union anthracite OPERATORS until the UMWA organized the companies' workforces following a major STRIKE in 1920. HC&I operated four anthracite collieries in 1895: the Butler and Consolidated in Luzerne County, the Erie in Lackawanna County, and the Forest City in Lackawanna, Susquehanna, and Wayne counties. Its output of 934,306 tons in 1899 made it one of the smallest LINE COAL COMPANIES. It enjoyed among the highest profits in the industry.

Independent Operator(s), Independent Coal Company(ies)—An incorporated coal-producing firm not owned or controlled by a railroad. See LINE COAL COMPANIES.

Individual Contract—With roots in the British mining industry, it was typically a legally binding agreement between a coal company and a miner, originally used to bind the miner to a PIT for a specific period of time. It was in effect in ANTHRACITE for most of the nineteenth century. It formed the basis of a variant called the SUBCONTRACTING SYSTEM late in the century. MINEWORKERS rebelled against the individual contract in anthracite and elsewhere, preferring a collectively bargained contract with the OPERATORS. See CONTRACT MINER, SPECIAL CONTRACT, TENANCY.

Industrial Workers of the World (IWW)—A labor union founded in Chicago in 1905 also known as the WOBBLIES (a term of unknown derivation). The group began organizing the ANTHRACITE FIELDS in 1907, receiving its strongest support in the NORTHERN COAL FIELD among Italian and other workers at PaCC and HC&I collieries, and at the Greenwood and Jermyn Collieries of the D&H. The union had up to 5,000 members by the failed STRIKE OF 1916, after which it was repressed and withdrew from anthracite. Elimination of the INDIVIDUAL CONTRACT and the SUBCONTRACTING SYSTEM was the IWW's top priority.

Insurgent(s)—Activist MINEWORKERS at PaCC and HC&I adopted the moniker in the 1920s; it was later applied it to militants at other NORTHERN ANTHRACITE FIELD companies in the 1930s as the rebellion spread. The militants wanted to increase employment and improve wages and working conditions, which included eliminating the INDIVIDUAL CONTRACT and TENANCY. The movement challenged the coal companies and also the UMWA by calling numerous WILDCAT STRIKES. The insurgents ran candidates for DISTRICT 1 elections during the 1920s and early 1930s. The effort culminated in the formation of the UAMP in 1933. The term did not apply to other DISSIDENTS in the field such as the COMMUNISTS, who were involved in the SAVE-THE-UNION COMMITTEE and the NATIONAL MINERS UNION. The term may have been taken from an Italian ANARCHIST group called *GLI INSURGENTI* (*The Insurgents*), which was a direct-action, anti-organizational radical group active in Paterson, New Jersey, and elsewhere during the 1910s and early 1920s.

Insurgent Movement—A labor movement led by INSURGENTS in the NORTHERN ANTHRACITE FIELD, it began at THE ERIE COAL COMPANIES soon after the settlement of the STRIKE OF 1902 and remained strong into the late 1920s, when it spread to all of DISTRICT 1 and culminated in the formation of the UAMP in 1933.

Jacquerie—The term for the Sicilian peasant and worker revolts that occurred once or twice each generation between 1772 and 1950. Revenge against the wealthy and powerful provided the underlying motivation for the uprisings (RETRIBUTIVE JUSTICE); however, they were also intended to redress rampant social inequalities and injustices (DISTRIBUTIVE JUSTICE). Approximately 86 percent of the island's towns witnessed *jacquerie*-like STRIKES after 1860, including Sicily's 14 sulfur-mining towns, which experienced the uprisings more often than other types of settlements. Failed strikes in the 1890s and in 1900 caused many SULFUR MINEWORKRS to immigrant to the U.S., some undoubtedly to Pennsylvania's ANTHRACITE COAL FIELDS and THE ERIE COAL COMPANIES.

Jalopy—A term for a type of chain-driven MECHANICAL LOADER used to move RUN-OF-MINE COAL from the FACE and into a COAL CAR. It consisted of metal "pans" or troughs arranged in seven-foot lengths into which LABORERS shoveled raw coal. The chain-driven mechanism within the pans pulled the coal into an awaiting coal car.

Kickback(s)—A term for a form of bribery or graft in which there is collusion between parties such that one person makes offers a monetary sum while another person takes the payment and grants some favor. In the ANTHRACITE INDUSTRY, kickbacks typically involved a monetary amount paid by an employee to a boss for a specific job or for a special PLACE in a mine.

Bosses often demanded the kickbacks, especially during tight labor markets, even though the practice had been outlawed by the Pennsylvania General Assembly in 1897.

Knights of Labor—The first successful national labor union in the U.S., it formed in 1869 as a secret fraternal order called the Noble and Holy Order of the Knights of Labor under the leadership Uriah Smith Stephens and James L. Wright of Philadelphia. After removing "Noble and Holy Order" from its name—in part as a response to objections by the Catholic Church—the Knights transformed into a more traditional labor union. It rejected SOCIALISM and other forms of RADICALISM while pursuing the eight-hour day, an end to child labor, growth of cooperatives, and other conservative goals. Thousands of MINEWORKERS joined and supported the union during the 1870s and 1880s. Grand Master Workman Terence Powderly, of Scranton, led the group between 1879 and 1893 when it became one of the largest labor organizations in the U.S. It began to decline in the late 1880s. The Knights organized women and other workers across skills but did little to counteract racial segregation in local unions, especially in the South. The Miners' National Trade Assembly 135 of the Knights of Labor participated in the formation of the UMWA in 1890.

Knox Coal Company—Known as the most infamous of PaCC's many LEASEHOLDERS, the company secured its first TENANCY agreement in 1941. The owners were alleged crime boss John Sciandra as well as Louis Fabrizio, Robert Dougherty, and UMWA DISTRICT 1 President August J. Lippi. The company illegally mined under the Susquehanna River, causing the epoch-ending KNOX MINE DISASTER. See ORGANIZED CRIME.

Knox Mine Disaster—On January 22, 1959, 12 men died in the NORTHERN COAL FIELD at the KNOX COAL COMPANY's River Slope Mine in Port Griffith, Jenkins Township, near Pittston, when the Susquehanna River broke into the mine. Owned and operated by alleged organized criminals and criminally associated persons, including UMWA DISTRICT 1 President August J. Lippi, the company mined illegally under the river causing a major flood that expedited the end of deep mining in the Pittston/Wilkes-Barre area as well as the lower section of the WYOMING VALLEY. Sixty-nine workers escaped the deluge, many taking up to seven hours. Three of the owners, along with three local union officers, were jailed as a result of the corruption surrounding the disaster.

Kohler Act (P.L. 1198, 1921)—A Pennsylvania law designed to protect the life, health, and safety of the public from SUBSIDENCE. The act made it unlawful for a company to mine coal so "as to cause the caving-in, collapse, or subsidence" of homes, public buildings, private buildings, streets, bridges, and other structures. It was declared unconstitutional by the U.S. Supreme Court in PENNSYLVANIA COAL COMPANY V. MAHON (1922), also known as the PENN COAL CASE. See DAVIS ACT, FOWLER ACT.

La Cosa Nostra—A controversial but widely accepted term referring to a unified Sicilian ORGANIZED CRIME syndicate. Despite wide public use since the 1950s, especially after the testimony of hoodlum Joe Valachi in the 1960s, many criminologists now see organized crime—Sicilian and otherwise—as decentralized with numerous different territorial and ethnic groups. See BLACK HAND, MAFIA.

Labor-Management Relations—The conflicts and negotiations between workers and managers over wage rates, work rules, DOCKING policies, SUBCONTRACTING, LEASING, contract talks, safety, and related matters.

Laborer—An unskilled MINEWORKER who toiled above or below ground. Most laborers worked under the direction of a CONTRACT MINER in harvesting coal, but some (called company laborers because they worked directly for the company) were employed on various surface and subsurface tasks. By tradition, laborers were paid by their supervising miner at a rate of one-third of a crew's gross earnings for the laborer(s) and two-thirds for the miner. Also by tradition, one miner worked in one CHAMBER with one laborer who assisted with tasks such as carrying props and supplies, setting TIMBERS, loading CARS OF COAL, and TOPPING the COAL CARS. By the 1920s, pressure for increased production led to one miner working with two, three, or more laborers. SUBCONTRACTORS incurred the ire of MINERS and laborers because they typically gained access to more than one CHAMBER and employed numerous laborers in what amounted to a production speed-up that often devoted less attention to safety.

Lackawanna Valley—Part of the Scranton/Wilkes-Barre Metropolitan Statistical Area (population 563,631 in 2010) in northeastern Pennsylvania, it is situated in Lackawanna County. It extends from Carbondale on the north to Old Forge on the south for a length of about 40 miles, and from the Moosic Mountain on the northeast and West Mountain on the southwest at a width of about six miles. The valley's largest city is Scranton (population 76,089 in 2010) and its main river is the Lackawanna. The area contained one of the richest ANTHRACITE deposits in the world. Coal mining in the valley ended in the 1960s.

Lattimer Massacre—August and early September of 1897 witnessed a wave of STRIKES and marches over wages, working conditions, the CAMPBELL ACT, company stores, and other workers' concerns in the EASTERN-MIDDLE ANTHRACITE FIELD around Hazleton. On September 10, 1897, a group of 400 unarmed men marched six miles from Harwood (a town with mainly Slavic residents and workers) to Lattimer (a town with mostly Italian residents and workers). The goal was to close down the one remaining Pardee Coal Company COLLIERY that had not been shuttered by the recent demonstrations. As the protestors approached Lattimer, deputies under the command of Luzerne County Sheriff James Martin fired on them. Nineteen Slavic and Hungarian workers were killed, while 32 others were wounded. Martin and 19 deputies were charged and tried for murder in the death of one victim, Mike Cheslak. In what has been called one of the great failures of American justice, they were all found not guilty. Lattimer stands as the largest American labor-related massacre of the nineteenth century.

Lease(s), Coal—See LEASEHOLDER, LEASING SYSTEM, LESSEE, TENANCY, TENANT.

Leaseholder(s)—A term referring to an INDEPENDENT COAL COMPANY (i.e., not affiliated with a railroad) that secured a COAL LEASE from a larger MINERAL RIGHTS-controlling company that typically was, or had been, a LINE COAL COMPANY. The lease allowed the TENANT to mine in a specific VEIN(S) OF COAL. See LEASING SYSTEM, TENANCY.

Leasing—See LEASE, LEASING SYSTEM, LEASEHOLDER, LESSEE, TENANCY, TENANT.

Leasing System—A form of TENANCY whereby a large MINERAL RIGHTS-controlling and, typically, railroad-owned or LINE COAL COMPANY let sections of its mines or even entire collieries to smaller, INDEPENDENT COAL COMPANIES (i.e., not affiliated with a railroad). Although PaCC had issued COAL LEASES as early as 1909, the company greatly expanded the practice in the 1930s, especially following the defeat of the UAMP in 1935. As with SUBCONTRACTING, management used leasing to "discipline" and control the workforce, boost productivity, and bolster profits. In 1929, PaCC established THE PITTSTION COMPANY (TPC) as a subsidiary to lease PaCC and HC&I collieries and mines, and collect royalties from LEASEHOLDERS. TPC failed to make royalty payments and, in 1938, PaCC dissolved the relationship and resumed control of its leasing system. By 1940, PaCC had leased out virtually all of its mineral rights. Other large companies began issuing leases which facilitated the transformation of the ANTHRACITE INDUSTRY over time from a highly concentrated sector dominated by a small number of large, railroad-owned companies to a very decentralized form with numerous leaseholders. The leasing system helped management gain WORKPLACE CONTROL, circumvent the militancy of organized labor, overcome STINTS, and undercut the collectively bargained contract between the companies and the UMWA. Alleged members of ORGANIZED CRIME gained many of the leases at PaCC, HC&I, and other companies. See LESSEE COMPANY, LESSOR COMPANY, TENANCY.

Lessee—See LESSEE COMPANY.

Lessee Company—Small to mid-sized INDEPENDENT COAL COMPANIES that did not own MINERAL RIGHTS and, therefore, had to LEASE the rights from a LESSOR COMPANY that did. The COAL LEASES were usually for SECOND or THIRD MININGS but occasionally for a FIRST MINING. The lessee usually exerted greater WORKPLACE CONTROL than did the lessor and brought greater "discipline" and control, lower costs, and higher production. Many engaged in corrupt mining and labor practices and some had ORGANIZED CRIME affiliations. The workers engaged in regular STRIKES and other protests against many lessees. See LEASING SYSTEM, LEASEHOLDER, LESSEE, SUBCONTRACTING SYSTEM, TENANCY.

Lessor Company—Typically, a large former LINE COAL COMPANY that owned or otherwise controlled MINERAL RIGHTS. The lessor company LEASED coal properties to a LESSEE COMPANY that mined the coal and delivered it to the lessor for processing. Through often illegal or unethical means, the lessee exerted greater WORKPLACE CONTROL than did the lessor and brought greater discipline, lower costs, and higher production. Labor militancy and STINTS were among the reasons the lessor companies instituted the LEASING SYSTEM. See LEASEHOLDER, LESSEE, SUBCONTRACTING SYSTEM, TENANCY.

Lessor—See LESSOR COMPANY.

Lehigh Anthracite Region—One of three ANTHRACITE REGIONS fashioned by the COAL OPERATORS for marketing purposes. It encompassed the WESTERN-MIDDLE and EASTERN-MIDDLE COAL FIELDS, which covered parts of Carbon, Columbia, Luzerne, Northumberland, and Schuylkill counties. Main cities include Ashland, Hazleton, Mahanoy City, Mount Carmel, and Shamokin. The Philadelphia and Reading Coal & Iron Company and the Susquehanna Coal Co. were the most prominent operators, but there were several smaller, family-owned COAL OPERATORS especially in the eastern-middle field, including the Van Winkles, Pardees, Markles, Coxes, and Fells.

Line Coal Company(ies)—An ANTHRACITE coal company owned or controlled by one of the main railroad lines that transported the coal to market. By the early 1900s, the six major railroads owned or controlled virtually all of the large coal companies in all three ANTHRACITE COAL FIELDS. The railroads were, in turn, owned or controlled mainly by J.P. Morgan's financial interests in New York City. See INDEPENDENT COAL COMPANIES.

Local Union(s)—An official organization representing a labor union at an individual COLLIERY, or in some cases two or more collieries. Colliery workers elect officers to represent them in local unions. The UMWA had hundreds of local unions in the ANTHRACITE COAL FIELDS.

Local Union 1703—The UMWA local union representing employees of PaCC's No. 6 COLLIERY in Jenkins Township and HC&I's Butler Colliery in Pittston Township. It stood out as one of the most militant locals in the NORTHERN ANTHRACITE FIELD during the 1920s and 1930s. Leaders included Rinaldo F. Cappellini between1920 and 1923; Alexander Campbell between 1920 and 1925 and again between1927 and his assassination in February, 1928; and Frank McGarry between 1928 and 1930. It underwent a significant reorganization in January 1928, when the previously elected officers—all them SUBCONTRACTORS—were deposed by INSURGENT members who elected new leaders. Four current or former L.U. 1703 officers were subsequently murdered in January and February of 1928 as part of "The Feud at No. 6," which occurred during a protracted STRIKE over the SUBCONTRACTING SYSTEM. Alleged ORGANIZED CRIMINALS, as well as union members, were involved in the killings. Alleged crime boss Santo Volpe and his Volpe Coal Company gained control over the No. 6 and Butler Collieries in 1937 and 1938, respectively, under the LEASING SYSTEM.

Lochner v. New York—A landmark U.S. Supreme Court decision in 1905 that held that the due process clause of the Fourteenth Amendment guaranteed the "liberty of contract" for businesses. The case involved a New York law that limited the number of hours a baker could work each day to ten, and the number of hours per week to 60. By a 5–4 vote, the Court rejected the argument that the law was necessary to protect the health of bakers and said it actually regulated the terms of employment such that it presented an "unreasonable, unnecessary and arbitrary interference with the right and liberty of the individual to contract." The decision had the effect of solidifying business control of the workplace and hindering the passage of progressive labor laws. The Lochner Era ended in the late 1930s when the high court began taking a more restrictive view toward business activity.

Lockout—A management maneuver used in anticipation of a STRIKE, during an impasse on contract negotiations, or to break a strike whereby employees are prevented from entering the workplace.

Machine Mining—The use of modern mining technologies that cut and/or loaded the RUN-OF-MINE COAL. Companies began using machines during the latter part of the nineteenth century and expanded their deployment in the twentieth century. ANTHRACITE generally lagged behind BITUMINOUS in implementing most new machines, in some cases because geology made them inapplicable.

Mafia—A Sicilian term for the ORGANIZED CRIME gangs (termed Mafias) that operated within a local or regional territory. The groups originated in Sicily during the nineteenth century. There was no specific plan by the gangs to establish "branches" in the U.S. The 13 years of alcohol prohibition in the U.S. (1920–1933) brought a major expansion of their power and wealth. They began cooperating more closely over territory and markets, and attempted to arbitrate disputes through a "commission." The Pittston crime "family" has been called the second oldest in the U.S., allegedly founded in 1908 by Santo Volpe and Stefano LaTorre, SULFUR MINERS who had emigrated from Montedoro, Sicily, at the turn of the century. They began working at PaCC early in the twentieth century. They and other gang members eventually became SUBCONRACTORS and LESSESS of PaCC and HC&I as well as factory owners and subcontractors in the region's garment and other industries. Charles Consagra, Russell Bufalino, Louis Fabrizio, and August J. Lippi were other organized crime-affiliated persons who, at some point, were employed by THE ERIE COAL COMPANIES. Oral-history evidence suggested that a small number of mine-working emigrants from Sicily entered the U.S. as so-called "Mafia bond slaves," whose passage had been paid by organized criminals, or by mob-contracted PADRONES, and who worked for gangsters in ANTHRACITE mines until they had paid off their bond. See BLACK HAND, LA COSA NOSTRA.

Make-Water—A term for the natural and incessant seepage of groundwater into the mines.

McClellan Committee—Formally known as the United States Senate Select Committee on Improper Activities in Labor and Management, the committee was established by the U.S. Senate on January 30, 1957, under chairman Sen. John L. McClelland of Arkansas. Attorney Robert F. Kennedy served as the chief counsel. The committee was charged with investigating organized criminal activities in LABOR-MANAGEMENT RELATIONS. However, the UNDERWORLD biases of McClelland, Kennedy, and other committee members caused it to conduct hearings and commission investigations into unions only. An UPPERWORLD perspective on the problem was not pursued. The committee disbanded after submitting a final report on March 31, 1960. Its investigations led to the passage of the Labor-Management Reporting and Disclosure Act, also known as the Landrum-Griffin Act, on September 14, 1959.

Mechanical Loader—A technology brought into ANTHRACITE as a result of the labor-management agreement of 1916, it allowed for more efficient loading of RUN-OF MINE coal into CARS OF COAL. Some loaders such as the SHAKER CHUTE vibrated the coal into a car, while the SCRAPER LOADER used a chain-driven mechanical arm to pull the coal into an awaiting car. See JALOPY, SHAKER CHUTE.

Middle Coal Fields—See EASTERN-MIDDLE COAL FIELD, WESTERN-MIDDLE COAL FIELD, LEHIGH ANTHRACITE REGION.

Miner—While the term has often been applied to the broader category of underground MINE-WORKERS, it accurately refers to a specific occupational category. To be a miner, a man had to have the skill, experience, and authority to plan and direct the coal harvest. In 1889, the Commonwealth of Pennsylvania required the certification of miners, and the law was strengthened in 1897 to require a two-year apprenticeship followed by a written examination. A state-certified miner was said to have gained his "mining papers" or an official certificate. As the highest-skilled person in a mine, only the certified miner was allowed to fire dynamite charges and perform certain other specific duties. The miner was responsible for the safety of his crew. A miner typically headed a work crew of one to two LABORERS and received two-thirds of the pay while the laborer(s) received one-third. For historic reasons associated with having to sign an INDIVIDUAL CONTRACT with the owners during the nineteenth century, they were referred to as CONTRACT MINERS (not to be confused with SUBCONTRACTORS). A WORK CULTURE norm in the NORTHERN ANTHRACITE FIELD held that the miner and his crew produce no more than two CARS OF COAL per man per day. The great majority of miners in the Northern Anthracite Field were paid by car of coal but some were paid by the ton.

Mineral Rights—In legal parlance, real-estate property contracts designate two types of property rights: surface rights and subsurface (or mineral) rights. Typically, the property owner possesses both rights and may sell or lease the subsurface to another party for mining or drilling. In the mid-1800s, PaCC owned large tracts of land, controlling both surface and subsurface rights. The company often sold the surface rights to individuals while retaining the mineral rights. The deeds often exempted the company from liability for any surface damage caused by mining, including CAVE-INS of houses and other buildings. In the late 1800s, the large ANTHRACITE coal companies that owned or otherwise leased most of the coal-laden properties in the NORTHERN COAL FIELD began granting mineral-rights access through SUBCONTRACTS to MINERS and other entrepreneurs. In the 1910s, PaCC and HC&I began LEASING mineral rights to small incorporated firms. By the late 1930s, most other large coal companies throughout the field began issuing COAL LEASES to hundreds of newly formed INDEPENDENT COAL COMPANIES. A new order of production and distribution developed and, along with it, significant labor-management conflict over the policies and practices of the LEASEHOLDERS, including many who were affiliated with organized crime.

Miner's Freedom—A term used by economist Carter Goodrich and others to describe the relatively unsupervised nature of underground mine work. Because of the vastness of coal mines, bosses rarely visited and, when they did, they did not linger. In conjunction with his high skill level, the MINER viewed himself as a craftsman who required independence from management in order to do his job properly. Managers at THE ERIE COAL COMPANIES and other coal companies viewed the miner's freedom as a form of indiscipline and inefficiency. Management's move to establish the SUBCONTRACTING and LEASING systems were part of a concerted effort to establish WORKPLACE CONTROL and eliminate the miner's freedom by imposing the factory model of hierarchy and work organization within the mines. Miners and LABORERS opposed and resisted the TENANCY SYSTEMS because they infringed upon underground "freedom," circumvented the STINT, encouraged corruption, and brought other untoward consequences.

Miners' and Laborers' Amalgamated Association—A labor union active in the ANTHRACITE COAL FIELDS in the 1880s, it was dominated by British and German MINEWORKERS. It attempted to cooperate with the other union active at this time, the KNIGHTS OF LABOR. The unions called a GENERAL STRIKE on September 10, 1887, which proved unsuccessful mainly because of divisions among the MINEWORKERS in the ANTHRACITE REGIONS. Another strike commenced in February 1888 and terminated unsuccessfully on March 4, 1888. Both the Knights and the Amalgamated soon faded from anthracite. See STRIKES OF 1877–1888.

Mineworker—A general term referring to a worker from any occupational category involved in harvesting, hauling, or processing coal, or participating in ancillary activities whether they were in the BREAKER or elsewhere aboveground, or within the mine underground. MINERS, LABORERS, SUBCONTRACTORS, electricians, timbermen, pump operators, motormen, engineers, MULE DRIVERS, STABLEMEN, NIPPERS, BREAKERMEN, carpenters, and other occupational categories were all considered mineworkers.

Mining Machine(s)—See MACHINE MINING, MECHANICAL LOADER, JALOPY.

Mining Papers—A vernacular term for the official state certification for MINERS. Beginning in 1889, Pennsylvania law required miners to pass a written examination and secure certification.

Motor—A slang term for an electric underground locomotive used to transport empty COAL CARS into a mine and loaded cars out. See MOTORMAN.

Motorman, Motormen—The driver of an electric underground locomotive used to transport empty CARS OF COAL into a mine and loaded cars out. See MOTOR.

Mule Driver—Typically, a teenage boy who led a mule in delivering empty COAL CARS via rail to MINEWORKERS at the COAL FACE and later hauling the loaded coal cars to the SHAFT for transport to the BREAKER for coal processing. See DRIVER BOY.

Mule Tender—The person in charge of leading a mule to take empty CARS OF COAL via rail to MINEWORKERS at the COAL FACE and later hauling the loaded cars to the SHAFT where they were sent to the BREAKER for coal processing. He may or may not have been a teenage boy as with DRIVER BOYS. See MULE DRIVERS.

National Miners Union (NMU)—A communist-affiliated organization that operated in ANTHRACITE between 1929 and 1932, it grew out of the SAVE-THE-UNION COMMITTEE. Despite concerted efforts to recruit members from the INSURGENT anthracite movement, the NMU could not gain a substantial membership mainly because of the workers' desire to pursue an independent activist course. See COMMUNIST PARTY.

New York, Susquehanna & Western Railway (NYS&WRR)—Formed in 1881 through the merger of six smaller railroad companies, it came under the control of the ERIE RAILROAD in 1898. It owned the New York, Susquehanna & Western Coal Company (NYS&WCC).

New York, Susquehanna & Western Coal Company (NYS&WCC)—Owned by the NYS&WRR, it was one of THE ERIE COAL COMPANIES, along with PaCC and HC&I. The firm mined none of its own coal but contracted for the finished product from INDEPENDENT COAL COMPANIES, which it initially sold under its own name.

Nipper—See DOOR TENDER.

Northern Anthracite Field—The northernmost of the four ANTHRACITE COAL FIELDS, it is 55 miles long and has a maximum width of about seven miles. It extended from Forest City in Lackawanna County to Mocanaqua in Luzerne County, and included small areas of Wayne and Susquehanna counties, also in the northeast. Main cities include Carbondale, Nanticoke, Pittston, Scranton, and Wilkes-Barre. It was coterminous with the WYOMING ANTHRACITE REGION. Major producers were D&H, THE ERIE COAL COMPANIES, Glen Alden, LVCC, and Susquehanna Collieries Company. See EASTERN-MIDDLE ANTHRACITE FIELD, SOUTHERN ANTHRACITE FIELD, WESTERN-MIDDLE ANTHRACITE FIELD.

Northern Coal Field—See NORTHERN ANTHRACITE FIELD.

Operator(s), Coal Operators—A general term for any incorporated coal company, whether it owned or leased MINERAL RIGHTS.

Organized Crime—A general term for criminal gangs who systematically violated the law in pursuit of profit and other gain. Along with their illegal manipulations, the gangs used threats and violence to attain goals. From the earliest colonial times to the westward expansion, they were typically led by persons of British (including Irish) extraction. By the late nineteenth and early twentieth centuries, the dominant gangs were of Irish, Sicilian, Jewish, and mixed ethnicities. Terms such as BLACK HAND (late nineteenth and early twentieth centuries), "the syndicate" (late 1920s and 1930s), MAFIA and LA COSA NOSTRA (especially after 1950), have referred to national crime organizations dominated by Sicilians but often including Irish, Jewish, and other groups. In recent decades the term has become a generic reference to gangs controlled by Chinese, Russian, Columbian, Puerto Rican, African American, Mexican, and other ethnic groups. These UNDERWORLD entities almost invariably work with persons and institutions from the UPPERWORLD to attain their ends. Among the ANTHRACITE OPERATORS, THE ERIE COAL COMPANIES had the greatest number of business dealings with organized-crime-affiliated persons and companies mainly through SUBCONTRACTING and LEASING, although other MINERAL RIGHTS-controlling operators also participated in such TENANCY schemes. Alleged organized criminals owned some of the largest anthracite producers after 1935, including the VOLPE COAL COMPANY and, in the 1950s, the KNOX COAL COMPANY.

Padrone System, Padrone(s)—Also known as the BOSS SYSTEM (*Padrone* meaning "boss" in Italian), it grew out of the drive to exploit low-cost contract labor in several industrializing countries during the second half of the nineteenth century. Although used in numerous countries, it was most widespread in Italy, including in the island of Sicily's SULFUR MINING industry. *Padrones* served as brokers who operated between bonded or contracted laborers, and corporations and public agencies in need of workers. In the U.S., the *Padrone* often paid the transportation costs of his contracted laborers in return for payments from a contracting business or agency. *Padrones* were well known to have cheated and otherwise abused their charges. See CONTRACTING SYSTEM, FORAN LAW, SUBCONTRACTING SYSTEM.

Penn Coal Case—See PENNSYLVANIA COAL COMPANY V. MAHON.

Pennsylvania Coal Company (PaCC)—Chartered in 1838 by the Commonwealth of Pennsylvania, it remained an INDEPENDENT COAL COMPANY until the ERIE RAILROAD gained control of its stock in 1901. PaCC, along with the HC&I and the NYS&WCC, were known as THE ERIE COAL COMPANIES because the Erie Railroad held the controlling interest in them. After 1901, the Erie Coal Companies operated under the same management. During the early twentieth century, PaCC paid the lowest wages in the ANTHRACITE INDUSTRY, which helped yield the highest profits. Together PaCC and HC&I employed 12,000 workers during the 1910s and 1920s, approximately 10,000 of them at PaCC. These two companies also employed the largest percentage of Italian immigrant workers in the ANTHRACITE INDUSTRY who constituted up to one-third of the workforce. The Erie Coal Companies remained the largest non-union operation in the industry until the workers joined the UMWA following the STRIKE OF 1920. Beginning in the 1890s, and after a hiatus between 1903 and 1913, PaCC and HC&I fostered the SUBCONTRACTING SYSTEM. In the 1930s, the companies were in the forefront in developing and expanding the LEASING SYSTEM. They also led the way in subcontracting and leasing mines and collieries to alleged ORGANIZED CRIME figures. When the Van Sweringen brothers purchased the Erie Railroad in 1924 and thereby gained control of the coal companies, they greatly encouraged TENANCY under President Michael Gallagher.

Pennsylvania Coal Company v. Mahon 260 U.S. 393 (1922)—A U.S. Supreme Court decision also known as the PENN COAL CASE, it concerned the constitutionality of the FOWLER ACT (1921) and KOHLER ACT (1921), which were Pennsylvania laws that held coal companies liable for surface damage from mine SUBSIDENCE or CAVE-INS. PaCC was the first company to challenge the legislation in suits that led to this decision. By a 5–4 margin, the high court overturned the lower courts and ruled the law was unconstitutional, meaning that the companies were not responsible for surface damage. Judge Oliver Wendell Homes wrote the majority opinion and Judge Louis Brandeis wrote for the minority. In the aftermath, the companies expanded ROBBING THE PILLARS, which produced greater subsidence, by issuing

COAL LEASES for SECOND and THIRD MININGS. The NORTHERN COAL FIELD experienced widespread subsidence and surface damage in the years and decades following. See ROOM-AND-PILLAR METHOD.

Petty Boss—A MINEWORKER who secured a SUBCONTRACT from a mineral-rights-controlling coal company and, unlike a PETTY OPERATOR, remained unincorporated. He hired a labor crew of a few to several dozen men who assisted in harvesting coal in one or more CHAMBERS.

Petty Operator—As the principal owner in a small, incorporated INDEPENDENT COAL COMPANY, he typically worked with SUBCONTRACTS and/or COAL LEASES obtained from a MINERAL-RIGHTS-controlling company and conducted SECOND MININGS or THIRD MININGS. See PETTY BOSS

Picket Lines—Protest queues in front of, or surrounding, a workplace that are established by striking workers who carry placards containing words or phrases regarding the STRIKE. The line is intended to prevent workers, suppliers, and others from entering the workplace, thereby closing the facility. Crossing a picket line can lead to a violent confrontation.

Piece, Piece Payment Method—Wage payments to MINEWORKERS based on the CARS OF COAL produced. See CHECK-DOCKING BOSS, DOCKING BOSS, WEIGHT PAYMENT METHOD.

Pillar—A solid block or wall of coal left behind by MINEWORKERS in a FIRST MINING under the ROOM-AND-PILLAR METHOD of removing coal. SECOND and THIRD MININGS involved ROBBING THE PILLARS, a potentially dangerous undertaking since the pillars supported the structural integrity of a mine, as well as the surface.

Pit—Another term for a coal mine, more commonly used in British and Australian mining.

The Pittston Company (TPC)—Created in 1930 by the Van Sweringen brothers as a subsidiary of PaCC, its main purposes was to lease PaCC and HC&I collieries and coal mines to INDEPENDENT COAL COMPANIES. At this time, PaCC was a subsidiary of the ERIE RAILROAD which, in turn, had become a subsidiary of the Van Sweringen brothers' ALLEGHANY CORPORATION. Under President Michael Gallagher, who was president of PaCC, Alleghany, and TPC, the company faced financial difficulties during the mid-1930s and could not maintain the agreed-upon royalty payments to PaCC. The PaCC–TPC relationship was dissolved in 1938, when TPC went bankrupt. PaCC resumed control of its collieries and mines and continued leasing. TPC reorganized and remained a subsidiary of the Alleghany Corporation until 1954 when new owners sold it off. TPC then continued as an independent diversified enterprise, gaining ownership of the Brinks Armored Car Division, bituminous coal companies, and up to 30 other subsidiaries.

Place—A slang term used by MINEWORKERS to refer to their specific work location in a CHAMBER underground. See ROOM.

Pools, Coal—To deal with a constant overproduction problem, as early as the 1850s and continuing on an irregular basis into the 1930s, ANTHRACITE companies organized to establish coal-production "pools" that set quotas on the tonnage going to market. The pools were seldom effective and were usually short-lived because the participants constantly overshot their limits, resulting in more coal going to market than planned, thus depressing prices.

Progressive Miners Union—A DUAL UNION threat to the UMWA, this AFL affiliate entered the NORTHERN COAL FIELD in 1938–1939 and gained adherents at the Alden COLLIERY in Newport Township, Luzerne County. It fought to gain control of the treasury of its predecessor, the UAMP, but the UMWA blocked the move and the organization soon faded.

Prop—A shaved wooden post used to support the ROOF of a mine. The workers sized, set, and shimmed the posts, which were essential to mine safety. They were stored in the prop yard on the COLLIERY grounds until LABORERS hauled them underground.

Putting-Out System—An early term for industrial SUBCONTRACTING.

Radical, Radicalism—Terms originally used in reference to adherents of various class-based ideologies on the Left during the nineteenth and twentieth centuries, including ANARCHISM, COMMUNISM, SOCIALISM, and SYNDICALISM. The terms derive from the Latin word *radix*, meaning root. Marxist-based radical theories beginning in the 1850s took the view that CAPITALISM has fundamental—root level—tendencies to concentrate economic power and wealth through exploitation of workers and nature, and that such problems cannot be addressed fully or adequately by reform or through cooperation between intrinsically adversarial classes—capitalists and workers. The terms have also been used more generally in reference to any person or group that militantly challenges the status quo. As such the INSURGENTS discussed in this volume were not RADICALS in the Marxist sense, but in the sense of challenging the status quo of labor and workplace relations.

Rank-and-File—A term used in this volume to refer to the larger body of MINEWORKERS and/or union members.

Rate Sheet—A listing of the pay rates negotiated by organized labor and management for each job in and around a mine.

Red-Baiting—The tactic of accusing certain individuals of having COMMUNIST or other RADICAL affiliation, or holding such views, or of blaming specific problems or unrest on them. UMWA President John L. Lewis, as well as DISTRICT 1 President Rinaldo F. Cappellini, used the tactic during the labor upheavals of the 1920s, during the Red Scare period, and Lewis continued the practice in the 1930s and afterwards. In the process, the underlying causes of substantive labor-management problems were often avoided or ignored.

Robbing the Barrier(s)—The usually illegal removal of the 100-foot thick BARRIER PILLARS of solid coal left in the ground to separate the workings of adjoining collieries so that a flood or fire or explosion in one would not spread to its neighbor.

Robbing the Pillar(s)—The removal of the blocks or pillars of coal left behind from a FIRST MINNG under the ROOM-AND-PILLAR METHOD. It was usually a legal undertaking designed to extract all of the coal from a mine; however, following the PENN COAL CASE of 1922, companies undertook the practice without regard to, or liability for, surface SUBSIDENCE, which often occurred. By the late 1910s, pillar robbing became one of the main sources of coal taken by SUBCONTRACTORS.

Rock Contractor—Usually a certified CONTRACT MINER who functioned as an independent entrepreneur, he secured an INDIVIDUAL CONTRACT or a SPECIAL CONTRACT from a coal company to quarry a tunnel or a SLOPE from the surface through sold rock to access a VEIN OF COAL within a mine, or to connect two COAL VEINS within a mine

Rock Tunnel—A tunnel quarried through solid rock that, when completed, allowed MINEWORKERS to access coal in unmined veins. Rock tunnels were typically dug under INDIVIDUAL CONTRACTS or SPECIAL CONTRACTS issued by a coal company to SUBCONTRACTORS.

Roof—A vernacular term for the top or "ceiling" rock in a VEIN OF COAL. It is almost always supported by PROPS lest it collapse and injure or kill the workers. Roof falls have been the most common cause of death in ANTHRACITE mines. See ROOM-AND-PILLAR MINING.

Room—See CHAMBER, PLACE.

Room-and-Pillar Mining—A coal mine can be visualized as an underground multi-story building, having an elevator (SHAFT) going down and stopping at each floor (VEINS OF COAL).

The floors are separated by rock strata of varying thicknesses. In ANTHRACITE, to get at the coal, ROCK CONTRACTORS, who were a type of SUBCONTRACTOR, conducted DEVELOPMENT WORK, which involved quarrying through solid rock either from the surface or within a mine. COMPANY MINERS then laid track to create a GANGWAY, or main haulage road. In a FIRST MINING, MINEWORKERS entered the gangway and began harvesting coal in rectangular ROOMS or CHAMBERS off the gangway. The chambers were usually 12 to 16 feet wide, but could be up to 60 feet wide if the vein was close to the surface and the ROOF was solid. A vein's pitch determined the length of the room but 300 feet was typical. The MINER and LABORER(S) used PROPS to support the ROOF and constantly checked its solidity lest the roof collapse and injure or kill the workers. Solid blocks of coal (PILLARS) up to 300 feet by 25 feet by 60 feet were left behind on either side of the room. On the other side of a pillar, further up the gangway, another mining crew would begin creating another chamber in the same vein, and so it went to the end of the vein or to the end of the company's property. The deeper the vein of coal, the thicker the pillar required to support the internal structure of the mine. For example, a mine 120 feet deep might have pillars only 25 by 60 feet thick but a mine 600 feet could need a pillar 50 feet by 120 feet. In a SECOND MINING, workers engaged in ROBBING THE PILLARS.

Royalty—A sum of money paid by a LESSEE to a LESSOR, usually on a per-ton basis, as part of a lease agreement that granted MINERAL RIGHTS access to the lessee.

Rump—An adjective describing an unsanctioned or unofficial gathering or group that challenges some form of established authority. See RUMP CONVENTION, RUMPERS.

Rump Convention—Between May 21 and June 1, 1928, the INSURGENT faction within UMWA DISTRICT 1 convened a special, unauthorized convention that, among other things, led to the "impeachment" of President Rinaldo F. Cappellini and the other sitting UMWA officers, and elected an alternative slate who still maintained their UMWA affiliation. Neither the established national or District 1 leadership of the UMWA recognized the convention or the officers. See RUMPERS.

Rumpers—A slang term for INSURGENTS or DISSIDENTS who challenged the authority and legitimacy of the UMWA District 1. See RUMP, RUMP CONVENTION. The UAMP convened rump gatherings in the early 1930s with the same consequence.

Run-of-Mine Coal—A term for the raw, unprocessed hewn or mined coal that is transported out of the mine in COAL CARS to the BREAKER for processing.

Sacco and Vanzetti Case—Italian immigrant ANARCHSTS Nicola Sacco and Bartolomeo Vanzetti were arrested for a murder conducted by a gang of bandits during a payroll robbery in South Braintree, Massachusetts, in 1920. Despite weak evidence against them, they were tried, convicted, and sentenced to death. They were executed in 1927 despite a worldwide protest from many quarters. In 1927, prior to the executions, MINEWORKERS in the Pittston area established the UMWA-DISTRICT 1 Sacco-Vanzetti Committee. It included representatives from 40 UMWA local unions within DISTRICT 1, as well as several Italian-American clubs and organizations. The committee convened parades and rallies. Rinaldo F. Cappellini, president of District 1, spoke at one gathering and declared that the condemned men were persecuted because of their involvement with the labor movement. Local members of the CPUSA also participated in the committee. The case inspired several literary works including *Justice Denied in Massachusetts* (1927) by Edna St. Vincent Millay; *Firehead* (1929), a narrative poem by Lola Ridge; *Boston: A Documentary Novel of the Sacco-Vanzetti Case* (1928), by Upton Sinclair (1928); and *Winterset* (1935), a play by Maxwell Anderson.

Save-the-Union Committee—Formed at a convention of RADICAL activists, its purpose was to challenge President John L. Lewis and the UMWA as the only representative of the nation's MINEWORKERS. The committee unsuccessfully challenged Lewis and his administration in the national UMWA election of 1926, with central Pennsylvania bituminous leader John

Brophy as the presidential candidate and PaCC employee William J. Brennan as the vice-presidential aspirant. The committee called short-lived STRIKES at BITUMINOUS mines in western Pennsylvania and later yielded to the NATIONAL MINERS UNION, which continued the fight against the UMWA.

Scab(s)—A slang term for a STRIKEBREAKER or BLACKLEG.

Scraper Loader—A type of MECHANICAL LOADER designed to facilitate the loading of RUN-OF-MINE COAL within a mine. It functioned to drag the coal toward an empty COAL CAR and scoop it into the car. The devices were either electric or pneumatic.

Schuylkill Anthracite Region—One of three ANTHRACITE REGIONS fashioned by the COAL OPERATORS for marketing purposes. It encompassed the SOUTHERN COAL FIELD, which was the largest of the anthracite fields and the first to produce coal in large quantities in the mid-nineteenth century. Major cities included Coaldale, Jim Thorpe (Mauch Chunk), Lansford, Lykens, Minersville, Nesquehoning, Tamaqua, and Tower City. The field had numerous small operators but was eventually dominated by the Philadelphia & Reading Coal and Iron Co. See LEHIGH ANTHRACITE REGION, WYOMING ANTHRACITE REGION.

Seam(s) of Coal—Another term for a VEIN OF COAL.

Second Mining—The removal of most or all of the coal in the PILLARS remaining after a FIRST MINING. The major coal companies engaged in the practice during the latter part of the nineteenth century and, after 1900, it became the realm of SUBCONTRACTORS and, in the 1930s, LEASEHOLDERS. An often dangerous but usually legal undertaking, it could result in mine accidents and SUBSIDENCE. See FIRST MINING, LEASING SYSTEM, ROOM-AND-PILLAR METHOD, SUBCONTRACTING SYSTEM, THIRD MINING.

Shaft—A vertical entry into a mine that passes through COAL VEINS. In ANTHRACITE mining, shafts sometimes extended to a depth of 1,000 feet but the typical depth was between 200 to 500 feet. The shaft housed a steam-driven (and later an electric) "cage" or elevator that transported men, mules, equipment, and material into and out of the mine, as well as CARS OF COAL between the mine and the BREAKER on the surface. See ROOM-AND-PILLAR MINING, SLOPE.

Shaker Chutes—A slang term for a MINING MACHINE that consisted of a trough (or chute) into which LABORERS shoveled raw coal, and that advanced the coal into an awaiting mine car by means of continuous and vigorous vibration.

Short-Weighing—MINEWORKERS often complained that the companies short-weighted them by deducting excessively for WASTE (rock, slate, "dust") from the CARS OF COAL they sent to the surface. The companies hired a WEIGHMAN and/or a DOCKING BOSS to scrutinize for waste. The workers eventually secured the option of paying for a CHECK-WEIGHMAN or a CHECK DOCKING BOSS (depending upon the pay method) to insure an accurate accounting. Coal companies often prevented the checkers from taking their stations at the colliery because of conflicts with the companies' DOCKING BOSSES or WEIGHMEN. See PIECE PAY METHOD, WEIGHT PAYMENT METHOD.

Shutdown—Another term for STRIKE and WORK SUSPENSION, whereby workers close a mine to express GRIEVANCES and demand redress from employers. See WILDCAT STRIKE.

Sliding Scale—A method of paying MINEWORKERS that tied wages to the wholesale price of coal. The first step was to establish a base price for coal and a schedule for wages. As coal prices rose so did wages, but if the price of coal dropped, wages did too, but not below a guaranteed minimum. WYOMING ANTHRACITE REGION or NORTHERN COAL FIELD workers were not paid according to the sliding scale, but by the PIECE (CAR OF COAL) or in some instances by weight (the long ton of 2,240 lbs.). The sliding scale was the main demand of anthracite's first successful labor union, the WORKINGMEN'S BENEVOLENT ASSOCIATION, during a failed strike in 1869. It was finally attained as part of the 1902 strike settlement but lasted for only a short time.

Slope—An inclined-tunnel mine entrance driven from the surface into a VEIN OF COAL, or from one vein to another within a mine, and laid with rail tracks used to transport COAL CARS, men, and equipment into and out of the mine.

Socialism, Socialist(s)—Conceived during the nineteenth century as an alternative to CAPITALISM, it is an economic system based on collective ownership of the means of production. Marxist socialism involved working-class ownership of the means of production via the state, collectivized farms, and state management of the economy. Other socialist thinkers and activists, such as Eugene Debs, Robert Owen, and Norman Thomas, had alternative understandings of attaining and sustaining collective ownership of the means of production and land, and of managing the economy. Variants include STATE SOCIALISM, public ownership, and cooperatives. See COMMUNISM, IWW, SOVVERSIVI, SYNDICALISM.

Soft Coal—Another term for BITUMINOUS.

Southern Anthracite Field—The largest and southernmost of the four ANTHRACITE COAL FIELDS, it was the first to produce coal in large quantities in the mid-nineteenth century. It ranges for approximately 55 miles in a southwesterly direction beginning near Jim Thorpe (Mauch Chunk) in Schuylkill County, and ending in two long tails in Dauphin County. It is headquartered at Pottsville and other settlements included Coaldale, Lansford, Lykens, Minersville, Nesquehoning,Tamaqua, Tower City. It is coterminous with the SCHUYLKILL ANTHRACITE REGION. The field had numerous small operators but was eventually dominated by the Philadelphia & Reading Coal and Iron Co. See EASTERN-MIDDLE ANTHRACITE FIELD, NORTHERN ANTHRACITE FIELD, WESTERN-MIDDLE ANTHRACITE FIELD.

Southern Coal Field—See SOUTHERN ANTHRACITE FIELD.

Sovversivi—An Italian word for subversive, it referred to various types of domestic RADICALS, including SOCIALISTS, COMMUNISTS, SYNDICALISTS, and ANARCHISTS. The *sovversivi* became a transnational group when many emigrated to other countries and resumed militant activities overseas. The NORTHERN ANTHRACITE FIELD had *sovversivi* such as Dr. Alberico Molinari, a physician and a Socialist who emigrated from the northern Italian city of Cremona to Scranton in 1903, where he continued as a physician and a Socialist organizer. The Sicilian *sovversivi* likely participated in *FASCI* organizations and *JACQUERIE* uprisings.

Speed-up—A management-induced increase in the pace and output of work in order to secure greater productivity and profits.

Special Contract—A form of legally binding INDIVIDUAL CONTRACT between a MINER and a coal company for specific tasks such as ROBBING PILLARS, quarrying ROCK TUNNELS, or doing DEVELOPMENT WORK. Special contracts formed the basis of the SUBCONTRACTING SYSTEM.

Split Labor Market Hypothesis—The proposition that immigrant and other marginalized workers tend to divide a labor market into higher and lower sectors because, unlike established employees, the newcomers are willing to accept lower wages and inferior working conditions. By extension, such workers are expected to look less favorably upon unions and collective bargaining because they fear both management and unemployment. Research studies on ANTHRACITE workers and their labor movements of the late nineteenth and early twentieth centuries do not support the hypothesis. See STRIKES OF 1887–1888.

Sprag—A thick wooden peg approximately 16 inches long, pointed on either side, used by MINEWOKERS to stop a COAL CAR within a mine in lieu of breaks. The worker thrust one or more sprags between the spokes of a mine car's wheels. Injuries were common, including loss of fingers.

Squire—Another term for a Justice of the Peace, in Pennsylvania it is a judicial position with limited municipal powers to hear cases involving civil controversies, conserving the peace, minor criminal complaints, and to committing offenders to jail.

Stablemen—MINEWORKERS charged with taking care of mules used to transport loaded CARS OF COAL from the COAL FACE to the main SHAFT or, in later mining, to the CONVEYOR from where the coal was sent to the BREAKER for processing.

State Socialism—A variant of SOCIALISM whereby a state or central government owns the means of production and manages the economy. See CAPITALISM, SOCIALISM, SYNDICALISM.

Stint—A worker-supported production limit reinforced by the WORK CULTURE and often by a labor union. Workers set stints below their actual production capabilities so as to control both the pace of work and the final output. Stints were attacked by managers in coal and other industries for their inefficiency and became part of the labor-management conflict over WORKPLACE CONTROL. ANTHRACITE coal companies used SUBCONTRACTING and LEASING to, among other things, circumvent and ultimately destroy the workers' stints.

Stoker—Developed by the ANTHRACITE LABORATORIES, it was an anthracite-burning furnace that added the convenience of an automatic feeder. It was intended to bolster anthracite's main market, home heating, but failed to stem the changeover to natural gas and heating oil.

Stope—An excavation in the form of steps made by mining ore from steeply inclined or vertical veins.

Strike(s)—A work stoppage called by MINEWORKERS and/or their union based on GRIEVANCES related to wages, working conditions, work rules, safety, discrimination, or some other issues. Also referred to as a WALKOUT or a WORK SUSPENSION. See BOYCOTT, SHUTDOWN, WILDCAT STRIKE.

Strikebreaker—A person who crosses a PICKET LINE to work during a STRIKE. See SCAB, BLACKLEG.

Strike of 1875—Known as The Long Strike, it began in the SOUTHERN COAL FIELD with a LOCKOUT initiated by the Philadelphia & Reading Coal Company and its president, Franklin B. Gowen. The company sought to reduce workers' wages and destroy the ANTHRACITE INDUSTRY's first successful union, the WORKINGMEN'S BENEVOLENT ASSOCIATION. When Gowen and other OPERATORS in the field cut wages, the union called a strike that lasted six months. Workers from the southern and MIDDLE COAL FIELDS supported the action, but those in the NORTHERN COAL FIELD did not. The strike failed and the companies broke the union. The workers took a 20 percent wage reduction as a consequence.

Strikes of 1887–1888—The STRIKES began on September 10, 1887, under the sanction of two unions—the KNIGHTS OF LABOR and the MINERS AND LABORERS' AMALGAMATED ASSOCIATION—that cooperated in demanding higher wages and other terms from management. The first strike occurred when certain coal companies refused the demand to increase wages 15 percent. The strike weakened when the NORTHERN COAL FIELD workers refused to support the demand and when the SOUTHERN COAL FIELD workers accepted the Philadelphia & Reading's lower wage increase. The strike continued in the MIDDLE COAL FIELDS with certain OPERATORS using SCABS, some of whom were Italian immigrants, to break the strike. However, much to the operators' surprise, most recent Italian, Slavic, Hungarian, Lithuanian, and other immigrant workers supported the strike. The Philadelphia & Reading workers again suspended work, on January 3, 1888, when the company rescinded its wage increase. Violence flared during January and February, 1888, as the strike continued in the SOUTHERN and MIDDLE COAL FIELDS. The companies refused to budge and finally broke the strike when the workers could no longer hold out. The unions relented on March 4. The failed strikes of 1887–1888 ended the influence of both the Knights and the Amalgamated organizations in the ANTHRACITE COAL FIELDS.

Strike of 1900—The ANTHRACITE MINEWORKERS' first successful GENERAL STRIKE, it was conducted by the UMWA, which began to organize the ANTHRACITE COAL FIELDS in 1894. It was led by union president John Mitchell, and continued from September 17 to October 24, 1900. The terms of the strike settlement included an average ten percent wage increase for the ANTHRACITE INDUSTRY'S 146,000 MINEWORKERS. The elimination of the INDIVIDUAL CONTRACT, which was a main issue for workers at THE ERIE COAL COMPANIES and at some other companies, was among the union's demands but the issue was dropped during negotiations. Despite the settlement, many problems remained unresolved, and the OPERATORS refused to recognize the legitimacy of the UMWA as the workers' bargaining agent.

Strike of 1902—An historic WALKOUT that sought to rectify the unresolved GRIEVANCES —including elimination of the INDIVIDUAL CONTRACT—remaining from the STRIKE OF 1900. The vast majority of ANTHRACITE's 150,000 workers honored the 164-day action, which continued from May 12 to October 23, 1902. The episode represented the first time in American labor history when the U.S. government intervened in a major industrial WORK SUSPENSION and did not take the side of management. UMWA President John Mitchell led the workers. The SHUTDOWN ended when the COAL OPERATORS and union representatives agreed to arbitration conducted by a seven-member ANTHRACITE STRIKE COMMISSION appointed by President Theodore Roosevelt, who took an even-handed approach in the matter. Famed attorney Clarence Darrow served as the UMWA's legal counsel during the Strike Commission's hearings. The union also received support from Rev. John J. Curran of Wilkes-Barre, "the labor priest," and other clergy and community leaders. The coal corporations were represented by various company presidents and lawyers. The commission's final ruling included an average ten percent pay increase for workers as well as the formation of the ANTHRACITE BOARD OF CONCILIATION to negotiate future labor disputes. Despite the settlement, the coal operators would not recognized the UMWA as the workers' bargaining agent and continued to disregard the union's legitimacy for another two decades.

Strike of 1910—A WILDCAT STRIKE at THE ERIE COAL COMPANY's collieries, involving 12,000 workers and lasting over three weeks, from May 18 to June 8, 1910. The action was called by an early DISSIDENT group that protested issues such as excessive dockage, wages, and work-rule decisions. The company asked DISTRICT 1 President Benjamin McEnaney to intervene and settle the strike even though the large majority of Erie's 12,000 workers were not UMWA members. The strikers refused to negotiate through the union, as they had lost faith in its effectiveness, and said they would negotiate only with company General Manager Captain William A. May. The STRIKE was finally settled through the mediation of the Italian Consul, Chevalier Fortunato Tiscar, who was based in Scranton. He was asked to intercede by union, company, and public officials because about a quarter of Erie's employees were Italian immigrants (the largest single ethnic group), and they were among the most militant.

Strike of 1916—A WORK SUSPENSION conducted under the auspices of the INDUSTRIAL WORKERS OF THE WORLD during the fall of 1916, it concentrated on the COLLIERIES of THE ERIE COAL COMPANIES and the Delaware & Hudson Coal Company. Worker dissatisfaction with TENANCY in the form of SUBCONTRACTING was the main cause. The action proved unsuccessful in all of its intentions and marked the end of the IWW's involvement in the ANTHRACITE COAL FIELDS.

Strike of 1920—A WORK SUSPENSION at THE ERIE COAL COMPANIES, it was led by Alex Campbell and Rinaldo Cappellini and lasted seven weeks. TENANCY in the form of the SUBCONTRACTING SYSTEM was the main contentious issue. The outcome provided a victory for the workers. Erie ceased issuing SUBCONTRACTS (at least for a time) and virtually all of the companies' 12,000 workers joined the UMWA. Up to this time, PaCC and HC&I had been the only two major non-union companies in the ANTHRACITE INDUSTRY.

Strike of 1922—A GENERAL STRIKE sanctioned by the UMWA, it continued between April 1 and September 2 for a total of 163 days. In a separate action, workers in the BITUMINOUS industry also walked out on April 1 marking the first time HARD COAL and SOFT COAL employees staged a simultaneous WORK STOPPAGE; the strikes called out a total of 600,000 mineworkers. The ANTHRACITE strike was fought not merely over the usual issues such as wages (management demanded a 2.5 percent wage cut despite healthy profits, while the workers wanted a 20 percent increase), but the companies seemed intent on destroying the UMWA in line with the corporate anti-unionism of the 1920s. For example, in addition to the wage decrease, company executives insisted on annual arbitration between labor and management, and refused to consider the CHECK-OFF, which was another prime union demand. With both sides holding unmovable positions, President Warren G. Harding intervened but could not negotiate a settlement. He then proposed a return to work under the existing contract followed by arbitration, a proposal the UMWA rejected. Congress then passed the Cummins-Winslow Act of 1922, which led to a temporary resolution. Among other provisions, the legislation created the United States Coal Commission to investigate both the anthracite and bituminous industries. The anthracite parties agreed to extend the existing contract for one year until the Commission completed its study. A new round of labor-management negotiations began on April 1, 1923. The strike of 1922 reduced demand in hard coal's main market, home heating, as consumers began considering other fuels.

Strike of 1925–26—A GENERAL STRIKE sanctioned by the UMWA, it lasted from September 1, 1925, to February 12, 1926, and stands as ANTHRACITE's longest SHUTDOWN, 170 days. Workers demanded a 10 percent pay hike as well as the CHECK-OFF, while management called for a wage decrease and would not consider the check-off. Because Republican President Calvin Coolidge at first adopted a "hands-off" policy, Pennsylvania Governor Gifford Pinchot, a progressive Republican, invited both labor and management representatives to a negotiating conference in Harrisburg; however, only the labor delegates attended. Pinchot criticized the operators for their aloofness and responded with a proposal to end the strike. His plan included modified check-off and arbitration plans, no wage increases, and no coal price increase for five years. The union accepted the offer but the operators declined, stating that only "the market" should determine prices. They, furthermore, rejected the additional requirement that they "open the books" to the state regulators who would be charged with insuring compliance. The governor condemned the coal executives as self-interested monopolists, while they viewed him as an anti-business, pro-labor regulator. Pinchot persevered and sought stronger authority over the coal companies through legislation that would have permitted a state government takeover of the mines if and when conditions warranted; however, the proposed laws failed because of the companies' pressures on state legislators. President Coolidge finally took some initiative but his tenders received no support from either side. Finally, with a fuel shortage looming and markets continuing to dwindle, UMWA President John L. Lewis met with the operators' chief negotiators and they hammered out a truce that extended the existing contract for five years, avoided the subject of the check-off, and stipulated that arbitration could occur only if both side consented. The internecine conflict cost the industry thousands of customers and marked the beginning of a steep decline in the demand for hard coal.

Strike of 1928—A STRIKE that began in January 1928 at the No. 6 SHAFT at PaCC's No. 6 Colliery, it was called because the workers protested the SUBCONTRACTING SYSTEM, which THE ERIE COAL COMPANIES' management was intent on expanding. The strike was led by a militant group under the direction of Alex Campbell and others who moved to depose the incumbent officers of L.U. 1704 and elect an INSURGENT slate. The strike spread to the other three No. 6 Colliery mines and involved over 1,700 workers, although the other 11 Erie collieries declined to go along. The strike quickly turned violent. It led to the murder of three L.U. 1704 leaders by organized crime assassins as well as the shooting death by a mineworker (the charge was involuntary manslaughter) of a former president of L.U. 1703. Pittston fell into a state of panic and siege. Strikers and managers reached a temporary accord in April and extended it to June, but the WORK SUSPENSION recommenced in July. Meanwhile, insurgents initiated SHUTDOWNS and RUMP elections at other Erie collieries during the spring, summer, and fall of 1928, in one of the most tumultuous years any northern field company had seen.

Strike of 1932—A GENERAL STRIKE called by INSURGENTS under the leadership of Thomas J. Maloney Sr., it began on March 11, and ended on April 1. The main issues were the UMWA's inability to uphold established work rules and wage rates, an end to the INDIVIDUAL CONTRACT and the SUBCONTRACTING SYSTEM, and attainment of an EQUALIZATION plan. Early reports indicated that up to one-half of the NORTHERN ANTHRACITE FIELD's workers stayed out. However, support began to wane due to UMWA threats and workers began returning to their posts. Insurgent leaders negotiated an end to the strike. The coal companies refused to take back many strikers even though rehiring was part of the strike settlement.

Strike of 1935—Initiated by the UAMP on February 4, 1935, the STRIKE was directed mainly against the Glen Alden Coal Company and the Hudson Coal Company, which had the largest number of unemployed MINEWORKERS. It was led by UAMP President Thomas J. Maloney, and other union officers and it precipitated considerable violence and even death. Characterized by the introduction of STRIKEBREAKERS by the company, the main issues were unemployment, COLLIERY closings, and the refusal of the company to recognize the UAMP. Seventy percent of Glen Alden's employees honored the WALKOUT. Luzerne County Judge W.A. Valentine ruled the action illegal and issued a court injunction. The INSURGENTS ignored the ruling and, on March 16, 1935, police arrested and jailed Maloney, vice president Henry Schuster, and 27 other UAMP members for contempt of court. Maloney and his colleagues were detained in the county jail for over a month until April 18, 1935. A peace proposal by Governor George H. Earle brought a pause in the conflict, but when the OPERATORS rejected Earle's plan, the disorder recommenced. The insurgents and companies finally reached an uneasy armistice in June 1935. The strike marked the last major labor action by the UAMP, which disbanded a few months later.

Subcontract—An INDIVIDUAL CONTRACT between a company and a certified CONTRACT MINER, it provided the right to remove coal from a section of a mine or to perform specific tasks such as digging a ROCK TUNNEL or a HEADING, or conducting DEVELOPMENT WORK. See SUBCONTRACTING SYSTEM, SUBCONTRACTOR.

Subcontracting System—A form of TENANCY whereby a MINERAL RIGHTS-controlling company granted an INDIVIDUAL or a SPECIAL CONTRACT to a certified CONTRACT MINER, or an independent agent who hired a certified miner. The SUBCONTRACTS were issued to remove VIRGIN COAL from certain veins, but most commonly for SECOND or THIRD MININGS involving ROBBING THE PILLARS. Subcontracts were also granted for DEVELOPMENT WORK such as driving GANGWAYS, HEADINGS, and ROCK TUNNELS (see PETTY OPERATOR). Subcontracting first became an issue at PaCC, HC&I, and other companies in the 1890s and, after a hiatus between 1902 and 1913, continued as a source of workers' GRIEVANCES for decades. It figured prominently in numerous STRIKES and DUAL UNION MOVEMENTS. The UMWA regularly called for its elimination in labor-management contract negotiations, although with little effect. The structure persisted into the 1950s but was largely replaced by the LEASING SYSTEM after 1935.

Subcontractor—The term referred to a certified CONTRACT MINER—or an independent, non-mine-working agent who hired a certified miner—who secured an INDIVIDUAL CONTRACT or a SUBCONTRACT from a coal company to remove coal in certain sections of a mine, sometimes for VIRGIN COAL but most commonly for a SECOND or THIRD MINING involving ROBBING THE PILLARS. The contract could also involve the completion of specific tasks such as conducting DEVELOPMENT WORK, and driving GANGWAYS, HEADINGS, and ROCK TUNNELS. See PETTY BOSS, SPECIAL CONTRACT, INDIVIDUAL CONTRACT, TENANCY.

Subsidence—A term referring to CAVE-INS on the surface, it was usually caused by the removal of underground support PILLARS during SECOND and THIRD MININGS conducted by TENANTS. Buildings, streets, and gas and water lines were often damaged and persons were injured or even killed by the cave-ins. In the PENN COAL CASE (1922), the U.S. Supreme Court protected the coal companies from any liability associated with subsidence. See DAVIS ACT, FOWLER ACT, KOHLER ACT, ROOM-AND-PILLAR MINING.

Sulfur Mines, Mining, Miners, Mineworkers, Workers—Sulfur was mined in Sicily from ancient times until the early twentieth century. The island had the world's leading sulfur industry during the 1700s and 1800s, where it was centered in the West Central area around 14 mining towns, among them Cianciana, Montedoro, San Cataldo, and Serradifalco. The settlements experienced widespread labor RADICALISM in the late nineteenth century because of the exploitation and injustices that permeated the industry. Many mines were owned by the landed class or the Catholic Church and were organized around SUBCONTRACTORS and a SUBCONTRACTING SYSTEM, many of whom were corrupt and/or had ORGANIZED CRIME affiliations. The mines were known as some of the most onerous and inhumane work places in Europe. Largely as a response to the abuse and corruption, sulfur workers formed SOCIALIST and other RADICAL movements during the last quarter of the nineteenth century that called regular STRIKES. Numerous sulfur miners emigrated from Sicily to the NORTHERN ANTHRACITE FIELD in the late nineteenth and early twentieth centuries and found work at THE ERIE COAL COMPANIES. They continued with militant actions in the IWW, UMWA, NMU and UAMP, and many were prominent among the INSURGENTS in the northern anthracite field. See *JACQUERIE*, *FASCI*.

Suspension, Work Suspension—Another term for STRIKE or WALKOUT.

Syndicalism, Syndicalist(s)—An alternative to CAPITALISM and STATE SOCIALISM, syndicalism proposes an economic system based on a federation of industrial or trade unions that control the means of production through numerous collectively owned organizations that function in a non-competitive manner through union democracy. It represents a form of decentralized socialist economic management. See ANARCHISM.

Taylorism—The philosophy and practice of business management promulgated by Frederick W. Taylor, who put forth a general criticism of the American workplace and worker in *The Principles of Scientific Management* (1911) and other publications. To achieve maximum economic efficiency and, hence, profitability for owners and shareholders, Taylor advocated complete WORKPLACE CONTROL by management and the use of "scientific" methods of organizing production. Workers in several industries fought the idea and it led to considerable labor-management conflict.

Tenancy, Tenancy System—Term for a system of mining where MINERAL RIGHTS-controlling corporations signed agreements with INDEPENDENT COAL COMPANIES (TENANTS) that allowed the mining of coal from a particular VEIN within a mine, or taking control of an entire COLLIERY. The NORTHERN ANTHRACITE FIELD had two types of TENANTS: SUBCONTRACTORS and LEASEHOLDERS. The legally-binding agreements led to the

SUBCONTRACTING SYSTEM and the LEASING SYSTEM, which were pioneered and propagated by PaCC and HC&I. Some of the tenants had ORGANIZED CRIME affiliations.

Tenant(s)—General term for SUBCONTRACTOR or LEASEHOLDER, i.e., a person or a company that secured a legal contract allowing MINERAL RIGHTS access to unmined coal. See SUBCONTRACTING SYSTEM, TENANCY.

Third Mining—The removal of all the remaining coal in the PILLARS and other areas following a SECOND MINING. An often dangerous undertaking, it was also referred to as "retreat mining" because workers started in the innermost sections of a vein and worked toward (or retreated to) the SHAFT or SLOPE, the ROOF usually caving in as they progressed. See FIRST MINING, ROOM AND PILLAR METHOD.

Timber(s)—The term for the underground wooden supports put up by a MINER and his LABORER(s) to protect from ROOF falls and CAVE-INS within a CHAMBER or other part of a mine.

Tonnage—Wage payments to MINEWORKERS based upon the tons of coal produced. See CAR OF COAL, WEIGHT PAYMENT METHOD.

Topping—A reference to RUN-OF-MINE COAL piled above the sides of a CAR OF COAL. Companies demanded that MINEWORKERS load the cars well above the sides otherwise pay would be DOCKED.

Tri-District Conventions—Periodic meetings of UMWA ANTHRACITE Districts 1, 7, and 9, the conventions were attended by union officers as well as representatives of the RANK-AND-FILE. The UMWA published the deliberations as convention *Proceedings*.

Underworld—A term referring to ORGANIZED CRIME and criminal activity, which has connotations of operating "below" or "underneath" the established institutional and normative order. However, researchers have often found close cooperation between the underworld and the UPPERWORLD of established institutions. Indeed, the underworld could not exist as it does without cooperation from, and collusion with, individuals and businesses from the upperworld.

United Anthracite Miners of Pennsylvania (UAMP)—A labor union established on August 7, 1933, it represented the culmination of the INSURGENT MOVEMENT in DISTRICT 1 that began among MINEWORKERS at THE ERIE COAL COMPANIES in the 1910s. It became the alternative to the UMWA, which was viewed as remote and unconcerned with the plight of the ANTHRACITE worker. Led by Rinaldo F. Cappellini and its first president, Thomas J. Maloney, Sr., the organization called a series of STRIKES between 1933 and 1935, which ended in failure. The DUAL UNION could not fight the combined resources of the coal companies, the courts, the police, and John L. Lewis and the UMWA, and even the Franklin D. Roosevelt Administration, which saw it as a drag on New Deal recovery programs. The UAMP disbanded in late 1935. Maloney was killed by a package bomb on April 10, 1936, almost surely because of his involvement with UAMP. A series of legal and family problems removed Cappellini from a leadership position. Despite its failures, the UAMP stands as one of the most prominent dual unions in anthracite history.

United Mine Workers of America (UMWA)—Founded by BITUMINOUS workers in 1890 in Columbus, Ohio, it became the largest MINEWORKERS' union and one of the largest labor organizations in the U.S. The UMWA began organizing the three ANTHRACITE fields in 1894 and strengthened its position with the STRIKES of 1900 and 1902. However, RANK-AND-FILE dissatisfaction with its BUSINESS UNIONISM and conservative leadership continued over the next three decades among numerous anthracite workers, including those at PaCC and HC&I. Large numbers of ERIE COAL COMPANY men had left the union by 1904, choosing to fight their own battles through WILDCAT STRIKES, and they remained unaffiliated until the STRIKE OF 1920, when nearly 12,000 joined local unions, thus becoming the last workforce employed by the major OPERATORS to join the UMWA. Numerous GRIEVANCES against the companies and the UMWA persisted through the 1920s with SUBCONTRACTING being the main one. The dissatisfaction culminated in a major insurgency in 1928 that spread to the whole of DISTRICT 1 in the early 1930s, and led to the formation of the UAMP in 1933. Much of the displeasure was directed at John L. Lewis who was first elected UMWA president in 1920. See MINERS AND LABORS' AMALGAMATED ASSOCIATION, IWW, KNIGHTS OF LABOR, PROGRESSIVE MINERS UNION, UAMP, WORKINGMEN'S BENEVOLENT ASSOCIAITON.

UMWA Districts—The UMWA created three districts within ANTHRACITE: No. 1 in the NORTHERN COAL FIELD, No. 7 in the EASTERN-MIDDLE AND WESTERN-MIDDLE COAL FIELDS, and No. 9 in the SOUTHERN COAL FIELD. They convened periodic TRI-DISTRICT CONVENTIONS where the three districts met together to consider matters such as a labor-management contracts. The districts also met in separate conventions. Each district had officers and an executive board and contributed a member to the International UMWA board.

Upperworld—A term referring to the established institutional and normative order in any community or society. It includes corporations, banks and other financial concerns, small businesses, political institutions at various levels, the judiciary, law enforcement, government agencies, unions, universities and colleges, schools, religious associations, various professions, etc. It also includes the leaders of these organizations. Researchers have found close cooperation between the upperworld and the UNDERWORLD of ORGANIZED CRIME. Indeed, the underworld could not exist as it does without cooperation from, and collusion with, elements within the upperworld.

Vein(s) of Coal—Strata of coal of varying thicknesses with rock above and below. The number of veins in a mine varied throughout northeastern Pennsylvania. Mines in the Pittston area of Luzerne County typically had six to eight veins, which were relatively close to the surface, as compared to Nanticoke, which had 23 veins. In Lackawanna County, Old Forge had 19 veins, Scranton 17 veins, and Olyphant 12 veins. Thickness could vary from several inches to over 20 feet in the NORTHERN ANTHRACITE FIELD. See SEAMS OF COAL.

Virgin Coal—Coal taken in a FIRST MINING. See ROOM-AND-PILLAR METHOD, SECOND MINING, THIRD MINING.

Volpe Coal Company—An ANTHRACITE enterprise owned by alleged crime boss Santo Volpe from the 1930s to the 1950s. He secured his first SUBCONTRACT with PaCC in 1913. Volpe continued gaining subcontracts with THE ERIE COAL COMPANIES and then became a major LEASEHOLDER who garnered 56 PaCC or TPC leases between 1934 and 1948 through this and other coal companies. He grew into one of the ANTHRACITE INDUSTRY's larger producers.

Wagner Act (1935)—Formally known as the National Industrial Relations Act, it was sponsored by New York Senator Robert F. Wagner and gave employees of private corporations the right to form labor unions, engage in collective bargaining, call STRIKES, and otherwise engage in concerted action. The Act created the National Labor Relations Board (NLRB), appointed by the president of the U.S. to enforce the law's provisions. UMWA President John L. Lewis was a powerful member of the NLRB in the late 1930s.

Walkout—Another term for STRIKE. See BOYCOTT, SHUTDOWN, WORK SUSPENSION.

Washery—A COLLIERY building where impurities are removed from coal by washing.

Waste—A colloquial term for the rock, slag, and dirt other unusable materials that are the by-product of coal mining, and for which the MINERS and LABORERS were not paid, or were DOCKED.

Weighman—In mines that paid by the WEIGHT PAYMENT METHOD, coal companies employed a person with this title to examine the cars of RUN-OF-MINE COAL coming out of a mine for insufficient weight, and "dock" or reduce a miner's wages accordingly. Workers regularly charged that the weighman reduced wages unfairly. The final ruling of the ANTHRACITE COAL STRIKE COMMISSION in 1903 gave members in LOCAL UNIONS the option to elect and pay a CHECK-WEIGHMAN to work alongside the weighman and, at least in principle, ensure against unfair weighing and excessive DOCKAGE. In practice, coal company management often prevented the check-weighmen from assuming their duties by denying them access to the COLLIERY. See DOCKING BOSS and CHECK- DOCKING BOSS.

Weight Payment Method—Wage payments to MINEWORKERS based on the tonnage produced. See CHECK-WEIGHMAN, PIECE PAYMENT METHOD, WEIGHMAN.

Western-Middle Anthracite Field—One of four ANTHRACITE COAL FIELDS, it is part of the LEHIGH ANTHRACITE REGION. It runs in an east and west direction for about 33 miles and includes parts of Northumberland, Schuylkill, and Columbia counties. Major cities include Ashland, Centralia, Mahanoy City, Shenandoah, Mount Carmel, and Shamokin. The Philadelphia and Reading Coal & Iron Company and the Susquehanna Coal Co. were the most prominent operators. See EASTERN-MIDDLE ANTHRACITE FIELD, NORTHERN ANTHRACITE FIELD, SOUTHERN ANTHRACITE FIELD.

Western-Middle Coal Field—See WESTERN-MIDDLE ANTHRACITE FIELD.

White Hand Society—A voluntary organization formed in northeastern Pennsylvania during the early twentieth century by law-abiding sons of Italy who were determined to oppose the BLACK HAND gang and its criminal members. At the same time, the White Hand Society wanted to counteract the stereotypes cast on the entire Italian community as a result of the menacing illegal behaviors of the Black Hand.

Wildcat Strike—An unsanctioned WALKOUT or WORK SUSPENSION called by workers at a particular industrial facility in violation of a collectively bargained agreement and/or a law. The ANTHRACITE INDUSTRY's labor-management contracts beginning in the 1920s forbade wildcats strikes and stated that sanctioned STRIKES could be called only by the UMWA. The origin of the term is unclear.

Wobbly(ies)—An alternative name for members of the INDUSTRIAL WORKERS OF THE WORLD, a radical union founded in Chicago in 1905.

Work Culture—A term referring to the norms, values, practices, and traditions surrounding and supporting the conduct of work in an employment setting, it is shaped by both labor and management and has much to do with LABOR-MANAGEMENT RELATIONS and WORKPLACE CONTROL. See STINT, TAYLORISM.

Work Suspension—See STRIKE.

Workingman's Benevolent Association (WBA)—Formed on March 17, 1869 by MINEWORKERS in the SCHUYLKILL ANTHRACITE REGION, the WBA became the first union to organize workers in all four ANTHRACITE COAL FIELDS. Led by President John Siney, an Irish immigrant, the union called one of the first successful industry-wide STRIKES in May, 1869. The manner in which the strike was broken in late August, 1869, particularly by Welsh workers from the Hyde Park section of Scranton, may have figured into arson at the Avondale mine in Plymouth Township, Luzerne County, that led to 110 deaths. The failure of the Long STRIKE OF 1875 precipitated the union's defeat.

Workplace Control—A term referring to the concerns surrounding who will exercise more power over the conduct of work—labor or management? Industrial engineer Frederick W. Taylor argued that the conflicts between the parties over work organization, work rules, technological implementation and other matters must be settled in favor of management. TAYLORISM advocated complete workplace control by management. TENANCY became an instrument at THE ERIE COAL COMPANIES and at other companies to enhance managerial control and thereby end the so-called MINER'S FREEDOM, enhance "discipline" and control, boost productivity, and increase profits.

Wyoming Anthracite Region—One of three ANTHRACITE REGIONS fashioned by the COAL OPERATORS for marketing purposes. It encompassed the NORTHERN COAL FIELD, which extended from Forest City in Lackawanna County to the northeast, to Mocanaqua in Luzerne County to the southwest, as well as small areas of Wayne and Susquehanna counties to the northeast. Main cities included Carbondale, Nanticoke, Pittston, Scranton, and Wilkes-Barre. Major producers were D&H, THE ERIE COAL COMPANIES, Glen Alden, LVCC, and Susquehanna Collieries Company. See LEHIGH ANTHRACITE REGION, SCHUYLKILL ANTHRACITE REGION

Wyoming Valley—Part of the Wilkes-Barre/Scranton Metropolitan Statistical Area (population 563,631 in 2010) in northeastern Pennsylvania, it is situated in Luzerne County. It extends from Duryea, Exeter, and West Pittston on the north, to Shickshinny on the south for a length of about 43 miles. The valley extends to the northeast and southwest for a distance of about seven miles, bounded by two Appalachian mountain chains. The Susquehanna River bisects the valley, whose largest city is Wilkes-Barre (population 41,498 in 2010). The area contained one of the richest ANTHRACITE deposits in the world. Coal mining in the valley ended in the 1970s.

Yardage—Wage payments to MINEWORKERS for removing rock or WASTE in a CHAMBER during the mining of coal.

Yellow Dog Contract—A type of INDIVIDUAL CONTRACT between a company and a worker whereby, as a condition of employment, the worker must agree not to join a labor union. Federal legislation outlawed such contracts in the 1930s.

* The glossary drew upon mining terms contained in the Hudson Coal Company's *The Story of Anthracite* published by the company in 1932; and *A Dictionary of Mining, Mineral, and Related Terms*, compiled and edited by Paul W. Thrush and the staff of the U.S. Bureau of Mines, Department of Interior, USGPO, 1968. The authors would like to express special thanks to Thomas P. Supey Jr., a former anthracite mineworker and current superintendent of the Lackawanna Coal Mine tour in Scranton, for assistance with certain definitions.

APPENDIX II

Short Biographies

This appendix provides biographical sketches of most persons discussed in the volume. Birth and death dates have been included in those cases where they were known. A person whose name appears in capital letters has his or her own entry.

Adams, James D.—A miner and subcontractor at HC&I's Butler Colliery, he was one of over 30 subcontractors at the Erie Coal Companies who resigned from their positions following a major strike in 1920, after which the companies returned the mines (for a time) to the traditional work organization of one miner harvesting coal in a single chamber with one or two laborers.

Adonizio, Charles—A miner and subcontractor at HC&I's Butler Colliery, he was one of over 30 subcontractors at the Erie Coal Companies who resigned from their positions following a major strike in 1920, after which the companies returned the mines (for a time) to the traditional work organization of one miner harvesting coal in a single chamber with one or two laborers.

Adonizio, Joseph—A miner and subcontractor at HC&I's Butler Colliery, he was one of over 30 subcontractors at the Erie Coal Companies who resigned from their positions following a major strike in 1920, after which the companies returned the mines (for a time) to the traditional work organization of one miner harvesting coal in a single chamber with one or two laborers. He became a partner in the Berge-Rose Coal Company, formed on September 26, 1931, with WILLIAM H. BERGE, PASQUALE ADONIZIO, and JOHN KEHOE.

Adonizio, Pasquale (aka Tony Rose)—A miner and subcontractor at HC&I's Butler Colliery, he was one of over 30 subcontractors at the Erie Coal Companies who resigned from their positions following a major strike in 1920, after which the companies returned the mines (for a time) to the traditional work organization of one miner in a single chamber with one or two laborers. He went on to acquire 49 subcontracts and leases from the Pennsylvania Coal Company (PaCC) and the Hillside Coal & Iron Company (HC&I) between 1921 and 1933. He served as a partner in the Berge-Rose Coal Company, formed on September 26, 1931, with WILLIAM H. BERGE, JOSEPH ADONIZIO, and JOHN KEHOE. He also leased the Beaver Meadow and Derringer collieries in Hazleton until 1940.

Aellio, Dominick—A mineworker at PaCC's No. 6 Colliery, he and FRANK FORDUTTO were violent intruders to the home of SAMUEL ALFILI Sr., in Pittston in March 1928. He was shot and killed by 18-year-old SAMUEL ALFILI Jr., who defended his mineworker father from a gun assault by the intruders. The senior Alfili had fought with the assailants over the subcontracting system at the No. 6 Colliery, where they all worked.

Agati, Frank—A resident of West Wyoming and a mineworker at PaCC's No. 6 Colliery, he was elected president of L.U. 1703 in the mid-1920s. In 1925, he was appointed to the District 1 scale committee by the executive board. In 1926, he left mine working when District 1 President RINALDO F. CAPPELLINI appointed him as an organizer and a personal bodyguard. He was shot dead on February 16, 1928 by SAMUEL BONITA in a gun battle at the District 1 headquarters in the Miner's National Bank building in downtown Wilkes-Barre. Bonita, as well as fellow insurgents STEVEN MENDOLA and ADAM MOLESKI were convicted of voluntary manslaughter and sentenced to prison despite their arguing self-defense because Agati also fired his gun. The father of seven children, Agati was alleged to have had organized-crime connections and was also reported to have visited the murderers of THOMAS LILLIS the day before his (Lillis's) assassination on January 18, 1928. Following a wake attended by hundreds of mourners, Agati's body was taken to St. Rocco's Catholic Church in Pittston, and then to the cemetery, in an enormous funeral procession. See GUY AGATI.

Agati, Guy—He married SANTO VOLPE's daughter EUPHAMIA ("Fanny") in 1939, more than a decade after the murder of his father, FRANK AGATI, which occurred during the labor conflict at PaCC's No. 6 Colliery in 1928.

Alaimo, Dominick—A mineworker for the Knox Coal Company, he was elected as a grievance committee member in L.U. 8005 at PaCC's Ewen Colliery during the 1950s. A reputed "soldier" in the Russell Bufalino organized crime family, he attended the Apalachin crime-syndicate meeting in 1957 with three other gang members from northeastern Pennsylvania and was allegedly one of those who escaped immediate arrest, although he was later captured. He also owned and operated a dress manufacturing shop in Pittston in the 1950s. In the investigations that followed the Knox Mine Disaster of January 22, 1959, he was charged with and pled guilty to accepting bribes from management to keep "labor peace." He was sentenced to two years in prison. See ANTHONY ARGO, CHARLES PIASECKI.

Alba, Charles—In a rump election on January 11, 1928, insurgents deposed him as vice president of L.U. 1703 at PaCC's No. 6 Colliery, along with the other incumbent officers including the president, PATSY PAGLIOCA, and the other vice president, JOSEPH SHIMOKONIS. See also JOSEPH MOLESKI, PHILIP JULIANI.

Alba, Ross—After he was fired from his job at PaCC during the insurgent-sponsored wildcat strike of March 1932, the company refused to give him his job back even though rehiring strikers was part of the strike settlement. See JOHN ASAKAVAGE, FRANK GALLO, CATAL GIORDANO, GEORGE KIETYANNA.

Alfili, Samuel Jr.—An 18-year-old from Pittston who defended his mineworker father, SAMUEL ALFILI Sr., in March 1928, from a gun assault by two intruders to their home. The teenager used a shotgun to kill DOMINICK AELLIO, age 33, and seriously wound FRANK FORDUTTO. Both victims were mineworkers at the No. 6 Colliery where they had fought with the senior Alfili over the subcontracting system.

Alfili, Samuel Sr.—A mineworker at PaCC's No. 6 Colliery, he suffered a pistol wound in March, 1928 during an assault by two intruders to his Pittston home, DOMINICK AELLIO and FRANK FORDUTTO. His son, SAMUEL ALFILI Jr., shot and killed Aellio while seriously wounding Fordutto. The three mineworkers had fought over the subcontracting system at the No. 6 colliery, where they all worked.

Antonello, James—He was one of four L.U. 1703 grievance committee members at PaCC's No. 6 Colliery who were deposed by insurgents in a rump election on January 11, 1928. The others were E. AQUILINA, SAM LATORE and SAM SEIT.

Aquilina, E.—He was one of four L.U. 1703 grievance committee members at PaCC's No. 6 Colliery who were deposed by insurgents in a rump election on January 11, 1928. The others were JAMES ANTONELLO, SAM LATORE and SAM SEIT.

Argo, Anthony—A committeeman at the Ewen Colliery's L.U. 8005 and a mineworker at the Knox Coal Company, he pleaded guilty to labor law violations following the Knox Mine Disaster of January 22, 1959, and was given a suspended sentence. See DOMINICK ALAIMO, ANTHONY PIASECKI.

Armat, James—Insurgents elected him as a committeeman in L.U. 1703 at PaCC's No. 6 Colliery on January 25, 1928. See SAMUEL BONITA, JOSEPH VICTOR, PETER REILLY.

Asakavage, John—After he was fired from his job at PaCC during the insurgent-sponsored wildcat strike of March, 1932, the company refused to give him his job back even though rehiring strikers was part of the strike settlement. See ROSS ALBA, FRANK GALLO, CATAL GIORDANO, GEORGE KIETYANNA.

Attardo, Anthony—A mineworker at PaCC's Ewen Colliery, he was elected as one of seven delegates from L.U. 1487 to attend the insurgent (rump) convention of August 1933.

Attardo, Charles—A vigorous subcontracting and organized-crime critic from Pittston as well as an avid Industrial Workers of the World (IWW) supporter, he was murdered in 1914, allegedly by the region's organized-crime element. Alleged mobster CHARLES CONSAGRA was charged with the murder but was released for lack of evidence.

Austin, William—A District 1 mineworker, delegates elected him as one of four United Anthracite Miners of Pennsylvania (UAMP) state board members at the insurgent (rump) convention of August 7, 1933, the other three being JOSEPH GALLIG, STANLEY GORGAS and WILLIAM SAXTON.

Babolo, Rev. John—A clergyman from Pittston, he served as a member of the committee formed by Italian Consul CHEVALIER FORTUNATO TISCAR to settle the wildcat strike at the Erie Coal Companies in 1910.

Baer, George F. (1842–1914)—In 1901, financier J.P. MORGAN made him president of the Philadelphia & Reading Railroad. He spoke for the company during the Anthracite Strike Commission hearings of 1902–03, during which time he authored the infamous "Divine Right of God" letter, which attempted to justify the dominance of capital over labor as a consequence of divine selection, or God's will.

Baker, G.F.—He was a member of the PaCC board of directors in 1910.

Barbara, Joseph—Born in Montedoro, Sicily, he established residence in Buffalo, New York, where he became involved in organized crime. He developed ties with mobsters in northeastern Pennsylvania and allegedly succeeded JOHN SCIANDRA as top boss in 1950, a position he held until 1959. He brought RUSSELL BUFALINO back to the Wyoming Valley from Buffalo and made him underboss in the crime family. Barbara owned a company identified by the Pennsylvania Crime Commission as having lent money to Bufalino as well as LOUIS CONSAGRA, ANGELO POLIZZI, and SANTO VOLPE. He was the convener of the infamous Apalachin crime-syndicate meeting at his estate in upstate New York in 1957. See DOMINICK ALAIMO.

Barral, John—A member of L.U. 699 from Edwardsville, in late 1928 he was one of 45 insurgents expelled from the United Mine Workers of America (UMWA) by JOHN BOYLAN and the District 1 executive board for dual unionism in making plans to establish the Anthracite Mine Workers Union (AMWU).

Barrett, James R.—He was a member of the Communist Party of the United States of America (CPUSA), which had about 50 members across the four anthracite fields at the end of the 1920s.

Bauman, Andrew—The members of L.U. 1495 at PaCC's No. 9 Colliery elected him as a delegate to the insurgent (rump) District 1 convention of August 1933.

Bauman, Charles—The members of L.U. 1495 at PaCC's No. 9 Colliery elected him as a delegate to the insurgent (rump) District 1 convention of August 1933.

Belfonte, Nick—He served as a member of the grievance committee of L.U. 1703 at PaCC's No. 6 Colliery in 1928.

Bellas, Joseph—A member of L.U. 699 from Edwardsville, in late 1928 he was one of 45 insurgents expelled from the UMWA by JOHN BOYLAN and the District 1 executive board for dual unionism in making plans to establish the AMWU.

Bellefield, John—He was elected to the District 1 inspection board at the insurgent (rump) convention of May 21–June 1, 1928. Later in 1928, he was one of 45 insurgents expelled from the UMWA by JOHN BOYLAN and the District 1 executive board for dual unionism in making plans to establish the AMWU.

Benefati, Nick—A mineworker at PaCC's No. 6 Colliery, the members of L.U. 1703 elected him as a delegate to the insurgent (rump) District 1 convention of August 1933.

Benjamin, Atty. Frank—As the assistant district attorney of Lackawanna County, he joined with legal counsel Atty. JOHN R. EDWARDS to start deportation proceedings against non-citizens during the IWW-sponsored strike of 1916 at the Erie Coal Companies and the Hudson Coal Company.

Berge, William H.—He was a partner in the Berge-Rose Coal Company, formed on September 26, 1931, with PASQUALE and JOSEPH ADONIZIO, and JOHN KEHOE. The firm leased coal properties mainly from the LVCC.

Birbeck, Charles—During the labor uprising at the No. 6 Colliery and at other Erie Coal Company operations in 1928, he was transferred as a foreman in the Thomas Shaft at HC&I's Butler Colliery to the same position at the Fernwood Slope at the same colliery.

Bird, Morgan—In partnership with his brother, SAMUEL BIRD, he operated the Bird Mining Company in the 1940s and 1950s. They leased mineral rights from the Glen Alden Coal Company, the Kingston Coal Company, and others to take coal from deep mines and strip mines in Plymouth and Wyoming boroughs and Hanover Township.

Bird, Samuel—See MORGAN BIRD.

Biscontini, Albert—A partner in the Newport Excavating Company in Luzerne County, he was indicted with former Governor JOHN S. FINE and DONALD P. MORGAN for personal and corporate tax evasion. All of the defendants were acquitted after lengthy trials and arduous jury deliberations. Two other partners, his father LAWRENCE BISCONTINI and SANTO VOLPE, had passed away by the time the indictments were issued. The company began deep mining and strip mining coal using leases in 1950.

Biscontini, Lawrence—He was a partner in the Newport Excavating Company in Luzerne County with former Governor JOHN S. FINE, his son ALBERT BISCONTINI, DONALD P. MORGAN, and SANTO VOLPE. The company began deep mining and strip mining coal using leases in 1950. He and Volpe had passed away by the time the other partners were indicted for personal and corporate tax evasion.

Boland, John J.—One of PaCC's earliest subcontractors following the strike of 1902, he secured an individual contract in 1909 for a second mining in Dunmore and continued taking such agreements until 1930.

Bonita, Frank—An anthracite mineworker from Pittston and a PaCC employee, he was murdered by unknown assailants one week before the Save-the-Union meeting in Pittsburgh in September 1928.

Bonita, Samuel—An ardent opponent of the subcontracting system, insurgents elected him as president of L.U. 1703 at PaCC's No. 6 Colliery in a rump election on January 11, 1928, after they had unseated the incumbent officers, including President (and subcontractor) PATSY PAGLIOCA. The UMWA negated the results but allowed a second ballot, held on January 25, 1928. Bonita was again chosen as president, with JOSEPH VICTOR as vice president and PETER REILLY as recording secretary. On February 16, 1928, Bonita traveled to the District 1 headquarters in the Miner's National Bank building in downtown Wilkes-Barre for negotiations with union officials regarding the ongoing strike at the No. 6 Colliery. The discussions turned violent and gunshots were fired. Bonita was convicted of voluntary manslaughter in the shooting death of former L.U. 1703 president and District 1 organizer FRANK AGATI. During the trial, Bonita admitted to shooting Agati but argued for self-defense because Agati also fired shots. The jury's initial verdict of "guilty of involuntary manslaughter" was rejected by Luzerne County Judge WILLIAM S. McLEAN because it was not among the charges in the indictment. In reconsidering the case, the jury ruled that Bonita was guilty of "voluntary manslaughter" and added a caveat: "with a plea for mercy." McLean rejected the mercy appeal and sentenced Bonita to the maximum six to twelve years in prison. L.U. 1703 grievance committee members STEVEN MENDOLA and ADAM MOLESKI were charged and convicted as accomplices in separate trials and sentenced by Judge McLean to four to eight years in prison.

Borsellino, Paolo (1940–1992)—A prominent Italian prosecuting magistrate who spent much of his professional life pursuing the Mafia, he was murdered in Palermo, Sicily, by a car bomb less than two months after his fellow Mafia-fighting magistrate GIOVANNI FALCONE had been assassinated. Borsellino delivered the eulogy at Falcone's funeral. The Palermo-area airport is named in their honor.

Boylan, John J.—A miner employed by the Penn Anthracite Collieries Company in Scranton, he was a member of the District 1 executive board during the labor wars at the Erie Coal Companies in 1928. He maintained private communications with UMWA President JOHN L. LEWIS detailing the unrest. He was Lewis's handpicked choice to replace RINALDO F. CAPPELLINI after the latter's resignation as District 1 president in July 1928. Boylan held the presidency until 1935 when he resigned following the UAMP uprising and received an appointment as secretary of the Anthracite Board of Conciliation (ABC).

Branco, Vito—A prominent Italian citizen from Pittston, he served as a member of the committee formed by Italian Consul CHEVALIER FORUNATO TISCAR to settle the wildcat strike at the Erie Coal Companies in 1910.

Brandeis, Justice Louis (1856–1941)—He was the U.S. Supreme Court justice who wrote the minority opinion in the Penn Coal Case in 1922.

Brennan, Martin F.—He was president of UMWA District 7 in 1930.

Brennan, William J.—A PaCC mineworker from North Scranton, he was elected president of District 1 in 1921 on an anti-subcontracting platform that promised other reforms. ENOCH WILLIAMS was elected as vice president. Brennan appointed RINALDO F. CAPPELLINI as a district organizer in 1920 but soon fired him following an altercation. Cappellini ran against Brennan for the presidency in 1923 and won. In 1926, Brennan unsuccessfully challenged JOHN L. LEWIS and his administration by running for the secretary-treasurer of the International UMWA on the Save-the-Union ticket with presidential candidate JOHN BROPHY of central Pennsylvania's soft-coal fields. Lewis, who often "red-baited" opponents, incorrectly labeled Brennan and Brophy as communist sympathizers. Save-the-Union supporters challenged the election results, arguing that large-scale voter fraud deprived Brophy and Brennan of victory. Brennan was banned from the UMWA for a period of time for dual unionism. He remained active in District 1 affairs into the late 1920s and early 1930s. He was elected as the executive committee chairman of the District 1 insurgent group in charge of planning the special (rump) convention May 21–June 1, 1928.

Brislin, Dennis—An International Board member of District 1 and a loyal supporter of the RINALDO F. CAPPELLINI administration, the insurgents "impeached" him at their special (rump) convention of May 21–June 1, 1928, when they elected a slate of alternative officers.

Brophy, John (1883–1963)—President of UMWA District 2, in the bituminous fields in central Pennsylvania, he ran as the Save-the-Union candidate in 1926 in an unsuccessful challenge to JOHN L. LEWIS for the UMWA presidency. WILLIAM J. BRENNAN, former District 1 president and a PaCC mineworker ran as his vice presidential companion. Brophy was charged with dual unionism and banned from the UMWA for a time. In 1933, he made peace with Lewis and rejoined the organization. In 1935, Lewis made him the first director of the Congress of Industrial Organizations (CIO).

Brown, Hugh V.—He was president of UMWA District 7 in 1928.

Brown, Mayor P.R.—He was mayor of Pittston during the labor wars at the Erie Coal Companies during the early 1920s. He blamed District 1 President RINALDO F. CAPPELLINI for the labor-management turmoil and banned him from the city in order to prevent his attendance at union meetings. On local union meeting nights, police stopped cars entering to search for Cappellini.

Brown, Superintendent—As superintendent at PaCC's No. 6 Colliery, his refusal to negotiate with the newly elected insurgent officers at L.U. 1703 in August 1928, precipitated a two-week strike.

Brownell, G.F.—He was a member of the PaCC board of directors in 1910.

Bryden, Andrew—A prominent Scottish immigrant who lived in Pittston, he served as a colliery superintendent at PaCC operations during the mid-nineteenth century.

Brydon, John—In May 1928, during the rancorous labor conflicts at the Erie Coal Companies, he succeeded JOSEPH P. JENNINGS as general superintendent of all PaCC and HC&I operations.

Bufalino, Rosario (Russell) (1903–1994)—Born in Montedoro, Sicily, in 1903, his family bought him to the U.S. when he was an infant. They returned with him to Italy for a time, and again emigrated to the U.S. when he was ten years old. He began working as a laborer and as a mechanic at PaCC's No. 6 Colliery in the 1920s. He moved to Buffalo, NY, where he participated in various rackets such as bootlegging alcohol, trucking, and labor racketeering. He also ran an automobile repair shop called the Centennial Garage in the 1930s. He returned to northeastern Pennsylvania as the alleged underboss to former Buffalo resident, JOSEPH BARBARA, who allegedly became the area's top boss in 1950, succeeding JOHN SCIANDRA. Bufalino allegedly succeeded Barbara as head of the northeastern Pennsylvania crime organization in 1959 and ruled until 1989. In the late 1950s, the McClellan Committee hearings in the U.S. Senate (headed by Sen. JOHN L. McCLELLAN) identified Bufalino as the head of the area's crime family and described him as "one of the most ruthless and powerful leaders of the Mafia in the United States." Because of his connections with the Teamsters Union, Bufalino became a prime suspect in the 1976 disappearance of ex-Teamsters President JAMES R. HOFFA. In 1978, he was convicted of extortion and sentenced to four years in prison. He was later sentenced to ten years' imprisonment for attempted murder. He resisted efforts by the federal government to have him deported, claiming that he was born in the U.S. WILLIAM D'ELIA allegedly succeeded him as top boss in 1989.

Buffalino, Charles C.—He was a partner in a new Volpe Coal Company, incorporated on November 23, 1937, with CHARLES J. BUFFALINO and SANTO VOLPE. He was Santo Volpe's son-in-law. The firm became a major lessee of the Erie Coal Companies.

Buffalino, Charles J.—He was a partner in a new Volpe Coal Company, incorporated on November 23, 1937, with CHARLES C. BUFFALINO and SANTO VOLPE. The firm became a major lessee of the Erie Coal Companies.

Buss, Sheriff George—As Luzerne County sheriff during the IWW-sponsored wildcat strike at the Erie Coal Companies and the Hudson Coal Company in 1916, he joined with District Attorney FRANK A. SLATTERY in starting deportation proceedings against numerous Wobblies.

Butler, William—A third generation coal operator in the Hazleton area, he sold his company in 1964 partly because of the growing influence of organized crime.

Buzzarelli, Feeney—Insurgents elected him as president of L.U. 1487 at PaCC's Ewen Colliery during the early 1930s. He assisted in directing a gathering of 4,000 striking insurgents at the Pittston Armory in March, 1932. See LEWIS CASTERLINE, JOHN SHIPLEY, STANLEY OCHESKI.

Calamara, Charles—A native of Montedoro, Sicily, he was a PaCC miner and a strong supporter of the insurgent movement. He was murdered on January 4, 1931, in Pittston. His close friend CHARLES (SAM) LICATA suffered a similar fate near the same spot in Pittston on March 2, 1931.

Callaway, M.P.—He was appointed to the Erie Coal Companies' board of directors in 1928 following the resignation of the previous board and the company president, C.S. GOLDSBOROUGH. The shake-up followed the VAN SWERINGEN brothers' appointment of MICHAEL GALLAGHER as board president in 1926 and as company president in 1928. See D.W. COOKE, H.A. KNAPP, C.R. NASH, H.T. PETERS, D.E. POMEROY, A.W. PUTNAM, G.T. SLADE.

Calvini, Arturo—An IWW organizer during the IWW-sponsored strike of 1916 at the Erie Coal Companies and the Hudson Coal Company.

Campamassi, Angelo—Newspapers called him "the ringleader" of 300 strikers at PaCC's Central Colliery in Duryea during the wildcat strike of 1910. He was arrested and jailed.

Campbell, Alexander (Alex) (1882–1928)—A PaCC mineworker since the late 1880s, he was elected check-docking boss on several occasions; however, the company usually prevented him from assuming the position. He became an insurgent leader in 1920 and was elected president of L.U. 1703 at PaCC's No. 6 Colliery. He led the company-wide anti-subcontracting strike of 1920 with organizer RINALDO F. CAPPELLINI and local vice president JOSEPH YANNIS. A resident of Railroad Street in Pittston with his wife and seven children, the 46-year-old labor leader was murdered on February 28, 1928, with PETER REILLY during the labor wars at No. 6 Colliery. Following a funeral service attended by thousands, he was buried in the Pittston cemetery. Police charged RALPH MELISSARI of New Jersey and PETER DeLUCCA and VINCENZO DAMIANO of New York City with the homicides. A fourth person described as "an Italian, about 35 years old" and listed as "John Doe" was also indicted but the charges were dropped. Acquitted of Campbell's murder but convicted of Reilly's, Melissari was sentenced to life in prison. DeLucca was also convicted and given a life sentence. Hired gunmen murdered Damiano before his trial began.

Campbell, Andrew—A member of L.U. 699 from Edwardsville, he was one of 45 insurgents expelled from the UMWA in late 1928 by JOHN BOYLAN and the District 1 executive board for dual unionism in making plans to establish the AMWU.

Campbell, Mrs. Anna Jenkins—The spouse of ALEX CAMPBELL and mother of seven children, she supported her husband's union cause following his murder by writing a letter to the editor urging workers to attend the District 1 insurgent (rump) convention of May 21–June 1, 1928. She also wrote a letter to the insurgents gathered at the rump convention, which was read by EDWARD McCRONE.

Cappellini, Marie (1909–1997)—Born in the Polish provinces of Austria-Hungary as Marie Evan, she emigrated to the U.S. with her parents in the early twentieth century. Her father was a mineworker and the family settled in Larksville, PA. She married RINALDO F. CAPPELLINI in 1922 and supported his labor activities in the 1920s and early 1930s, but divorced him on January 26, 1933, for unclear reasons. They had two sons during the 1920s, Robert and Gifford. They reunited sometime after 1935 when Rinaldo had completed his prison term. She and the children joined him in his travels after he made peace with UMWA President JOHN L. LEWIS and became a labor organizer for the CIO in the latter 1930s. She was the proprietor of the Villa Cappellini restaurant in downtown Wilkes-Barre from 1929 to 1937.

Cappellini, Rinaldo F. (1895–1966)—Born in Nocera Umbra in the central Italian province of Perugia, the charismatic Cappellini emigrated to the U.S. with his mother and two siblings in 1902. Shortly after the family settled in Pittston, he found work as a breaker boy at PaCC. Within a few years he moved underground as a mule driver. At age 15, a trip of loaded coal cars ran over his right arm and left hand. Physicians amputated the arm and the top parts of three fingers were lost. A financial settlement brought nearly $10,000 and the promise of a "job for life." In 1920, he was hired as an organizer at PaCC's No. 6 Colliery at the suggestion of fellow mineworker and union supporter ALEX CAMPBELL. Cappellini and Campbell led a 12-week strike in 1920 that ended the Erie Coal Companies' subcontracting system (only temporarily), and brought 12,000 PaCC and HC&I workers into the UMWA. He was elected president of L.U. 1703 at the No. 6 Colliery was also elected chairman of the General Grievance Committee that covered all PaCC operations. He married the former Marie Evan in 1922. He defeated WILLIAM J. BRENNAN for the presidency of District 1 in 1923 and was reelected in 1925 and 1927. He resigned the position on July 10, 1928, during the labor wars that, while having begun at the Erie Coal Companies, had spread throughout District 1. He was succeeded to the top post by JOHN BOYLAN, the handpicked choice of UMWA President JOHN L. LEWIS. A reinvigorated Cappellini then ran against Boylan in June 1929 at the head of an insurgent ticket that included vice-presidential candidate THOMAS LAVALLE. The challengers lost the election amid cries of voter fraud. After taking a break from labor activities, he reemerged in the early 1930s as a leader, with THOMAS J. MALONEY Sr., of a District-wide insurgent movement that led to the formation of the UAMP in 1933. He was chosen as state chairman of the UAMP at the inaugural convention but left the scene following a series of personal and legal problems. Charged and convicted of arson to his own home, he was sentenced, on February 2, 1931, to serve two to four years in Philadelphia's Eastern State Penitentiary. He denied the charges and he and his allies, including wife MARIE CAPPELLINI, pleaded with Governor GIFFORD PINCHOT to issue a pardon, which he did on March 25, 1932. For unclear reasons, Marie Cappellini divorced her husband on January 26, 1933, while he was incarcerated. Upon his release from prison, Cappellini returned to organizing for the UAMP. He was again arrested on January 14, 1934, for a disturbance at Marie's home, and the parole violation sent him to the Graterford State Penitentiary where he served the remaining 13 months of the original sentence. He and Marie reunited soon after his second prison release. He made peace with John L. Lewis, who appointed him as an organizer for the CIO in the late 1930s. He died at his home at Harvey's Lake, PA, at age 71.

Carcivenek, Frank—The members of L.U. 1495 at PaCC's No. 9 Colliery elected him as a delegate to the insurgent (rump) District 1 convention of August 1933.

Cardone, Michael—The members of L.U. 1495 at PaCC's No. 9 Colliery elected him as a delegate to the insurgent (rump) District 1 convention of August 1933.

Cardoni, Frank—A PaCC mineworker, he served as president of L.U. 1581 at No. 14 Colliery and as chairman of the General Grievance Committee at PaCC until he was deposed by insurgents in 1928. He supported RINALDO F. CAPPELLINI and the District 1 administration during the labor wars of the period; Cappellini appointed him as a district organizer that same year. In the 1950s he served as a member of the District 1 executive board under District 1 President AUGUST J. LIPPI. In December 1960, he was indicted for tax evasion with Lippi, LEONARD STATKIEWITZ, and ROBERT L. DOUGHTERY for income received as kickbacks from Dougherty's Avon and Peeley mining companies; all of the defendants were acquitted in May, 1963, although the companies were found guilty of tax evasion.

Carey, Vincent—Delegates elected him as one of three UAMP auditors at the District 1 insurgent (rump) convention of August 1933, the other two being STANLEY OCHESKI and JOSEPH SCHUSTER.

Cassley, Jacob—The members of L.U. 1495 at PaCC's No. 9 Colliery elected him as a delegate to the insurgent (rump) District 1 convention of August 1933.

Casterline, Lewis (1903–1995)—Born as Angelo Castellani of northern Italian immigrant parents, he began as a miner and eventually became a PaCC superintendent. A resident of Wyoming, PA, he maintained a strong concern for the insurgent movement at PaCC and in District 1. He was present at many important meetings, including the founding of the UAMP, and he worked closely with THOMAS J. MALONEY Sr. and RINALDO F. CAPPELLINI in fostering the organization. In cooperation with FEENEY BUZZARELLI, president of L.U. 1487 at PaCC's Ewen Colliery, and others, he assisted in directing a gathering of 4,000 striking insurgents at the Pittston Armory in March 1932. Archival evidence suggests that he was an informant for the federal government during the labor wars of the 1930s. See STANLEY OCHESKI, JOHN SHIPLEY.

Cavanaugh, Daniel—A mineworker at PaCC's Ewen Colliery, he was elected president of L.U. 1487 in June, 1928 by an insurgent revolt that deposed the incumbent officers, including President GEORGE MOLESKI.

Cavanaugh, John—He was elected to the executive committee of the District 1 insurgent group in charge of planning the special (rump) convention of May 21–June 1, 1928.

Chamberlain, Alex—A mining engineer and second-generation coal operator, he witnessed the "rip and tear" mining practices of the tenants at his father's Jermyn Green Coal Company in the 1940s.

Chaplain, Edwin—An assistant foreman at PaCC's No. 1 Colliery in Dunmore, he was shot at by picketers but missed when he attempted to cross a picket line and enter the pit during the wildcat strike of 1910.

Chavez, Klemento—He was a 42-year-old striking miner in Walsenburg, Colorado, whom the police shot and killed in January 1928 during a wildcat strike. Police claimed that he had fired on them from a meeting hall used by the IWW.

Chiavinko, Frank—He was elected vice president of L.U. 1495 at PaCC's No. 9 Colliery when insurgents overthrew the "Orlando regime" in April 1928, headed by President ERNEST ORLANDO, who supported the Cappellini administration. See JACOB COLLIER.

Collier, Jacob—He was elected president of L.U. 1495 at PaCC's No. 9 Colliery when insurgents overthrew the "Orlando regime" in April 1928, headed by President ERNEST ORLANDO, who supported the Cappellini administration. Cappellini and the UMWA did not recognize the legitimacy of the election. Collier and his fellow officers promised to rectify the workers' grievances against subcontracting and other matters. Upon assuming office, he took control of the local union's charter, seal, checkbook, and equipment, prompting Orlando to take him to court for larceny. See FRANK CHIAVINKO.

Cometa, Thomas—Union members retained him as recording secretary of the Ewen Colliery's L.U. 1487 in June 1928.

Connelly, Daniel H.—He was the chief state mine inspector in the Pittston area at the time of the Knox Mine Disaster of January 22, 1959. In the early 1960s, he became president of PaCC and served as the company's last chief executive.

Connolly, Harry J.—He was the president and chairman of the PaCC and TPC's board of directors in the 1940s.

Consagra, Charles—Born in Montedoro, Sicily, he was an alleged leader of the Black Hand gang in Pittston at the turn of the century and an organized crime associate of SANTO VOLPE. In 1907, he was indicted, with STEFANO LaTORRE, SALVATORE "SAM" LUCCHINO, Volpe, and seven others for conspiracy to extort protection money from mineworkers. He was among the nine convicted and sentenced to a year in solitary confinement in the Luzerne

County jail. He was charged with attempted murder in the gun assault and wounding of Pittston detective Lucchino in 1911, and was arrested for Lucchino's murder on July 21, 1920. The charges were dropped in both cases for lack of evidence. He secured a mining subcontract at PaCC with Volpe in 1913, which began the modern era of tenancy at the Erie Coal Companies. He was implicated but not charged in hiring the two gunmen—PETER ERICO and ANTONIO PUNTARIO—who shot Lucchino. He was murdered gangland-style in Buffalo, New York, in 1922.

Cooke, D.W.—He was appointed to the Erie Coal Companies' board of directors in 1928 following the resignation of the previous board and the company president, C.S. GOLDSBOROUGH. The shake-up followed the VAN SWERINGEN brothers' appointment of MICHAEL GALLAGHER as board president in 1926 and as company president in 1928. See M.P. CALLAWAY, H.A KNAPP, C.R. NASH, H.T. PETERS, D.E. POMEROY, A.W. PUTNAM, G.T. SLADE.

Cooney, William—He served as chairman of the Underwood Colliery grievance committee in 1928 when PaCC announced that it would remove the subcontractors from the mines and return the chambers to regular contract miners and their laborers. See RIGO MADTUCCI.

Correale, Fred and Palmer—Lessees with the Glen Alden Corporation, their alleged organized crime dealings were detailed in the Pennsylvania Crime Commission's 1980 *Report*.

Costa, Joe (1900–1999)—A Portuguese immigrant whose birth name was Costanzio Lopes, he was a certified miner from Swoyersville, PA. He was elected as one of two financial secretaries by insurgents in L.U. 1703 at the No. 6 Colliery on January 25, 1928, the other one being ALFRED MOLESKI. Because of his ties to organized crime at this time, he may have been working as a labor infiltrator. He worked as a contract miner and as a subcontractor for several companies including PaCC in the 1920s and 1930s, Penn Anthracite Collieries Company in the 1930s, and the Pagnotti Company's Harry E Colliery during the 1940s. His job with an organized-crime-affiliated subcontractor at PaCC's No. 14 Colliery in the mid-1930s involved roaming the mines "stealing" cars of coal by erasing the code markings of the proper mining crew and substituting his subcontractor's markings. One of his tasks while working at the Harry E was to distract and entertain state mine inspectors so they would ignore mining-law violations. He briefly served as a minor operative under Swoyersville's alleged organized-crime boss (who was also the Republican Party head), LORENZO FERRARO, in the early 1920s. He ran his own alcohol bootlegging and distribution business during the Prohibition Era, where he developed business relationships with LOUIS PAGNOTTI Sr. and JOHN KEHOE. He remembered with anger a Scottish superintendent at Penn Anthracite to whom he had to pay a kickback for a subcontract in the late 1930s. He was the first author's uncle through marriage.

Costanza, James—As a young boy, he witnessed the ALEX CAMPBELL and PETER REILLY murders on February 28, 1928, which occurred across the street from his home on Railroad Street in Pittston.

Curran, Rev. Monsignor John J. (1859–1936)—Born in Hawley, PA, in 1859, he worked eight years as a breaker boy and mule driver, most of the time for PaCC. His mining days convinced him that workers had a just cause in unionization. He was ordained a Catholic priest in 1887 by Bishop Gerald P. O'Hara of the Diocese of Scranton. He served as the founding pastor of Holy Savior parish in Wilkes-Barre, the church hall often being used for labor meetings and rallies. His support for the mineworkers brought him into the anthracite strike of 1902, where he worked on a settlement with President THEODORE ROOSEVELT and UMWA President JOHN MITCHELL, both of whom he befriended. Roosevelt often called upon Curran for advice in labor disputes. Mitchell's conversion to Catholicism in 1908 was influenced by his wife and his cleric friend. Curran presided at Mitchell's Requiem Mass and burial service in Scranton in 1919. Curran's life was threatened several times because of his labor involvements. He suffered a heart attack on Good Friday, April 10, 1936, when the rectory at St. Mary's

Church in Wilkes-Barre (where he had assumed pastoral duties in 1918) was firebombed. That same day, a cigar-box bomb was delivered to his close associate and UAMP leader THOMAS J. MALONEY Sr., maiming and eventually killing Maloney and his son, THOMAS J. MALONEY Jr. Curran never fully recovered from the episodes and died six months later at the age of 77. He was buried in St. Mary's Cemetery in Hanover Township, Luzerne County. He was also an active participant in the Temperance Movement.

Damiano, Vincenzo—A hired assassin from New York City, he was charged for the murders of ALEX CAMPBELL and PETER REILLY, with co-conspirators PETER DeLUCCA and RALPH MELISSARI. He was murdered before his trial began.

Darrow, Atty. Clarence (1857–1938)—A nationally renowned attorney, he served as the UMWA's chief counsel during the Anthracite Strike Commission hearings of 1902–03.

Davis, Oliver L.—Chief inspector at the Pennsylvania Anthracite Collieries Company in the 1930s, when the firm expanded subcontracting and leasing. Company records indicate that he knew about and tolerated wage-rate violations by the company's subcontractors and leaseholders. See CHARLES DORRANCE, JOHN H. HARVEY.

Davis, Thomas—A federal mediator from Wilkes-Barre, he made an unsuccessful attempt to intervene in the wildcat strike at PaCC's No. 6 Colliery in April, 1928.

Davis, William—A mineworker at the Glen Alden Coal Company, he served as chairman of the insurgent committee that organized the first special (rump) convention of District 1 of May 21–June 1, 1928.

DeGerolamo, George—A mineworker for PaCC and TPC and a local union official, he experienced the labor wars of the early 1930s, which he documented in an unpublished autobiography (see References).

Delaney, Ray—He was elected as an International Board member at the District 1 insurgent (rump) convention of May 21–June 1, 1928. Later in 1928, he was one of 45 insurgents expelled from the UMWA by JOHN BOYLAN and the District 1 executive board for dual unionism in making plans to establish the AMWU.

D'Elia, William—He allegedly became the top boss of northeastern Pennsylvania's organized crime family in 1994. He succeeded Edward Sciandra, who allegedly took control in 1989 when the sitting boss, RUSSELL BUFALINO, was sent to prison. In 2006, D'Elia was indicted on federal charges of laundering drug money. He was later indicted for the attempted murder of a witness and was served with new charges of money laundering. He plea-bargained with prosecutors to lessen his prison sentence and, once in jail, provided evidence in another case. He will be eligible for parole in 2014.

Delmont, James—The members of L.U. 1495 at PaCC's No. 9 Colliery elected him as a delegate to the insurgent (rump) District 1 convention of August 1933.

DeLucca, Peter—A resident of New Jersey, he was charged for the murders of ALEX CAMPBELL and PETER REILLY, with co-conspirators VINCENZO DAMIANO and RALPH MELISSARI. He was acquitted of Campbell's murder but convicted, with Melissari, of Reilly's murder and sentenced to life in prison.

Dembicki, Stanley—He was one of three IWW members from Scranton, and one of 95 Wobblies nationwide, who were arrested and detained by the federal government during a series of strikes in 1917. IWW members ALBERT B. PRASHNER and SALVATORE ZUMPANO of Scranton were also held.

Dessoye, Joseph—The members of L.U. 1495 at PaCC's No. 9 Colliery elected him as one of two assistant secretaries in April 1928, the other one being JOHN MARKOWSKI. See JACOB COLLIER.

Donovan, Mary (1886–1973)—Political activist, labor organizer, and spouse of POWERS HAPGOOD, whom she married in 1927, she wrote about anthracite issues including the labor wars at the Erie Coal Companies in 1928. Like her husband, she was a member of the Socialist Party where she served as the corresponding secretary for the NICOLA SACCO and BARTOLOMEO VANZETTI defense team; she delivered a eulogy following their executions in 1927. She was arrested and jailed with her husband along with local activist CHARLES (SAM) LICATA for inciting a riot in Pittston in 1928 by leading a parade through the city after MAYOR WILLIAM H. GILLESPIE refused to grant a permit; the charges were dropped.

Dorrance, Charles—A descendant of one the Wyoming Valley's most prestigious families, he was president of the Penn Anthracite Collieries Company in the 1930s when the company expanded its subcontracting and leasing systems. Company records indicate that he knew about and tolerated wage-rate violations by the company's subcontractors and leaseholders. He also served on the boards of directors of the Wyoming National Bank of Wilkes-Barre and other companies. See OLIVER L. DAVIS, JOHN H. HARVEY.

Dorsey, George—A mineworker from Scranton, District 1 members elected him to the executive board in 1927. A loyal member of the RINALDO F. CAPPELLINI administration, the insurgents "impeached" him at the special (rump) convention of May 21–June 1, 1928, when they elected a slate of alternative officers. The UMWA ignored the convention and the election.

Dougher, Joseph—A union activist from Archbald, Lackawanna County, and an officer in L.U. 1782, he was elected as an inspection-board member from the first inspection district at the District 1 insurgent (rump) convention of May 21–June 1, 1928. He was listed as a member of a "small cadre" of communists from the northern field who attended the Save-the-Union meeting in Pittsburgh in September 1928. Also in 1928, he was one of 45 insurgents expelled from the UMWA after being charged by JOHN BOYLAN and the District 1 executive board with dual unionism for making plans to establish the AMWU.

Dougherty, Robert L.—A mineworker from Hazleton, in 1924, at age 25, he became the youngest mine foreman in Pennsylvania. He worked as the superintendent at the Exeter Colliery in the late 1940s and then served as the general manager of the Knox Coal Company in 1950, in which he became a partner and an officer in the mid-1950s with LOUIS FABRIZIO, Mrs. JOSEPHINE (JOHN) SCIANDRA, and AUGUST J. LIPPI. After defeating charges of conspiracy, mining-law violations, bribery, and manslaughter, he pled guilty to personal tax evasion associated with his income from the Knox Coal Company. He received a sentence of one year and one day plus a $10,000 fine. Dougherty was released after serving four and one-half months in the Federal Correctional Institution at Danbury, CT, part of it in the company of Fabrizio. In another case, in May 1963, he was acquitted with AUGUST J. LIPPI, LEONARD STATKIEWITZ, and FRANK CARDONI on charges of personal income-tax evasion in association with two other coal companies that he (Dougherty) owned, the Peeley and the Avon, although the companies were found guilty of corporate tax evasion.

Dulkas, Adam—Insurgent members of L.U. 1495 at PaCC's No. 9 Colliery elected him as treasurer in 1928.

Durkin, James—As a lessee of the Glen Alden Coal Company and owner of the Blue Coal Corporation and Pocono Downs race track (where he allegedly had business dealings with investor JAMES R. HOFFA), his alleged crime dealings were detailed in the Pennsylvania Crime Commission's 1980 *Report*.

Dziengelewski, Stanley—A member of the CPUSA, he served as chairman of the Save-the-Union Committee in the Anthracite Tri-District area. He represented District 1 at the committee's convention in Pittsburgh in April 1928. See GEORGE PAPCUN.

Edmonson, Robert E.—Publisher of the Nazi journal, *American Vigilante Bulletin*, he was originally based in New York City, but in 1939, his Edmondson Economic Service moved to the remote Luzerne County community of Stoddartsville, from where he continued publishing. His move to northeastern Pennsylvania was likely influenced by his friend and fellow bigot, Ku Klux Klansman PAUL M. WINTER.

Edwards, Atty. John R.—Legal counsel to Lackawanna County Sheriff BENJAMIN S. PHILLIPS in the 1910s, he joined with Assistant District Attorney FRANK BENJAMIN in starting deportation proceedings against non-citizens during the IWW-sponsored strike of 1916 at the Erie Coal Companies, the Hudson Coal Company, and the Jermyn Coal Company.

Enderline, Larry—Delegates elected him as secretary-treasurer of the UAMP at the District 1 (rump) convention in August 1933.

Enfileman, Samuel—The members of L.U. 1495 at PaCC's No. 9 Colliery elected him as a delegate to the insurgent (rump) District 1 convention of August 1933.

Erico, Peter—A hired assassin of Pittston detective SALVATORE "SAM" LUCCHINO, the Trenton, NJ, resident was convicted of first degree homicide and sentenced to death. ANTONIO PUNTARIO was also convicted in the murder. The death penalty was carried out at Rockview State Penitentiary in Centre County, PA, on September 25, 1922, as Erico became the second person (after Puntario) to die by electrocution for a crime committed in Luzerne County.

Ettor, Joseph J. (1885–1948)—A New York City-born Italian-American, he was an anarchist, union organizer, and leader in the IWW who was fluent in Italian and English. He mobilized support for the IWW in the northern anthracite field during 1908–1909. He established a local union at PaCC's Old Forge Colliery, which remained among the most militant leading up to the strike of 1916. See LUIGI GALLEANI, CARLO TRESCA.

Eustice, Joseph—A mineworker at PaCC's No. 6 Colliery, he was elected as a delegate from L.U. 1703 to the insurgent (rump) District 1 convention of August 1933.

Ewen, John—He served as president of PaCC from 1851 to 1877 and was a member of the board of directors from 1851 to 1876. PaCC's Ewen Colliery in Jenkins Township, PA—where the original breaker, built in 1886, fell to fire in 1914, and was replaced by a steel-and-concrete structure in June, 1915—was named in his honor.

Fabrizio, Louis—He took his first TPC lease on June 22, 1933, with partner Atty. FRANK J. FLANNERY and they went on to secure four other leases, the last one expiring on June 11, 1940. He became a partner with JOHN SCIANDRA in the Knox Coal Company after having secured a lease for sections of the Pittston and Marcy veins at the Schooley Shaft in 1943. AUGUST J. LIPPI and ROBERT L. DOUGHERTY joined the partnership in 1950. Fabrizio served as the president of the Knox Coal Company from its founding in 1943 until it went out of business in 1960. He and his partners obtained 16 additional PaCC leases and lease adjustments between 1943 and 1958, as Knox became one of the Erie Coal Companies' larger tenants. One lease obtained in 1943 allowed for mining in sections of the Pittston and Marcy Veins at the Schooley Shaft. In 1954, they secured a lease to take virgin coal in the Pittston and Marcy Veins at PaCC's Ewen Colliery. They called the mine the River Slope because the entrance was adjacent to the Susquehanna River. The mine became the site of the infamous Knox Mine Disaster of January 22, 1959. After defeating charges of conspiracy, mining-law violations, bribery, and manslaughter, he pleaded guilty in two separate trials dealing with personal and corporate tax evasion associated with his income from the Knox Coal Company. He was sentenced to 181 days in Danbury (CT) Federal Correctional Institution on January 24, 1964, where he served nearly five months, part of it in the company of Knox partner, ROBERT L. DOUGHERTY. See Atty. FRANK J. FLANNERY.

Facciponte, Sam—A mineworker at PaCC's Ewen Colliery, he was elected as one of seven delegates from L.U. 1487 to attend the insurgent (rump) convention of August 1933.

Fahy, John—A UMWA organizer in the Hazleton area in the 1890s, he was later elected president of District 7.

Falcone, Giovanni (1939–1992)—An Italian prosecuting magistrate born in Palermo, Sicily, he spent most of his professional life in the Palace of Justice trying to break the Mafia's stranglehold on Sicilian life. After a distinguished career, culminating in the famous Maxi Trial in which he indicted numerous mob members and obtained their convictions, he was assassinated on the motorway near the town of Capaci. The Palermo-area airport is named in his honor and that of his murdered colleague and fellow magistrate, PAOLO BORSELLINO.

Falkowski, Ed—He was a member of a small group of communist mineworkers on District 1 who also joined the Save-the-Union Committee in the 1920s.

Falls, William H.—He was a member of the PaCC board of directors between 1851 and 1853.

Falzone, Anthony—A mineworker at PaCC's Ewen Colliery, he was elected one of seven delegates from L.U. 1487 to attend the insurgent (rump) convention of August 1933.

Feraini, Joseph—A prominent Italian-American citizen from Pittston, he served as a member of the committee formed by the Italian consul, CHEVALIER FORTUNATO TISCAR, to settle the wildcat strike at the Erie Coal Companies in 1910.

Ferraro, Lorenzo—The alleged local crime boss in Swoyersville from the 1920s to the 1950s, he was also the head of the borough's Republican Party, which dominated politics for decades until the election of 1932, when a significant realignment occurred and the Democrats gained (and continue to hold) a council majority. He was also prominent in the Luzerne County Republican Party, including the period when JOHN S. FINE dominated the local GOP. Ferraro owned and operated a trucking business, a construction company, and a grocery store. See ANGELO SIRACUSE, CHARLES SIRACUSE.

Ferretti, Joseph—A miner and subcontractor at HC&I's Butler Colliery, he was one of over 30 subcontractors at the Erie Coal Companies who resigned from their positions following a major strike in 1920, after which the companies returned the mines (for a time) to the traditional work organization of one miner harvesting coal in a single chamber with one or two laborers.

Ferry, Neal J.—He served as labor's representative to the Anthracite Coal Commission in 1920, appointed by President Woodrow Wilson. He convinced fellow commissioners to allow subcontracts only in isolated cases but the ruling was open to interpretation and carried no enforcement authority, so the practice continued. He also served as the chairman of a special committee appointed in 1925 by UMWA President JOHN L. LEWIS to investigate the labor turmoil at the Erie Coal Companies.

Figlock, Anthony—A District 1 executive board member, he witnessed the shooting of FRANK AGATI on February 16, 1928, at UMWA headquarters in downtown Wilkes-Barre, and was himself nearly hit by bullets. The trauma caused serious distress, which, according to a court ruling related to his wife's pension, contributed to his early death some months later. See SAMUEL BONITA, STEVEN MENDOLA, ADAM MOLESKI.

Fine, Governor John S. (1893–1978)—He served as Luzerne County Republican Party Chairman 1922–1923 and continued to control the party after he was appointed as judge to the county's Court of Common Pleas by Governor GIFFORD PINCHOT in 1927. He was elected governor of Pennsylvania for one term, 1951–1955. After leaving the governor's office, he became a partner with ALBERT BISCONTINI, LAWRENCE BISCONTINI, DONALD P. MORGAN, and SANTO VOLPE in the Newport Excavating Company, which was a deep-mining and strip-mining venture in Luzerne County that led to his indictment for corporate and personal income-tax evasion. Albert Biscontini and Donald P. Morgan were also indicted (Volpe and Lawrence Biscontini had passed away by this time). The defendants were found "not guilty" after a lengthy trial and arduous jury deliberations.

Flannery, Atty. Frank J.—He obtained five PaCC leases between 1933 and 1940 in partnership with LOUIS FABRIZIO.

Flynn, Patrick—Insurgent members of L.U. 1495 at PaCC's No. 9 Colliery elected him as the recording secretary in 1928.

Flynn, Peter—Insurgent members of L.U. 1495 at PaCC's No. 9 Colliery elected him as one of three auditors in 1928, the other two being ARTHUR RENFER and FRANK TROSS.

Flynn, William J.—A U.S. Secret Service agent, he was in charge of the Lupo-Morello organized-crime case in 1909 in New York City. In pursuing the gang, he supervised SALVATORE "SAM" LUCCHINO of Pittston, who went undercover to bring gang members to justice.

Fordutto, Frank—A mineworker at PaCC's No. 6 Colliery, he and DOMINICK AELLIO were violent intruders to the home of SAMUEL ALFILI Sr. in Pittston in March, 1928. He was shot and wounded by 18-year-old SAMUEL ALFILI Jr., who defended his mineworker father from a gun assault by the intruders. The senior Alfili had fought with the assailants over the subcontracting system at the No. 6 Colliery, where they all worked.

Fredericks, C.H.—He became the secretary-treasurer of PaCC following MICHAEL GALLAGHER's appointment as chairman of the board of directors in 1926.

Fugmann, Michael—An immigrant mineworker born in Germany, a UAMP board member, and an associate of UAMP President THOMAS J. MALONEY Sr., he was charged with murder in the death of Maloney. He denied the charges but was convicted on circumstantial evidence and sentenced to death. After an appeal to the state Supreme Court and a brief reprieve from Governor George Earl for a psychiatric examination, Fugmann died in the electric chair on July 17, 1938.

Fulmer, Jule Ann—She was a two and one-half-year-old girl from Pittston who perished in 1943 when the sidewalk caved in beneath her feet due to mine subsidence. She was walking on Mill Street a short distance behind her aunt when she plummeted into the mine. Searchers recovered her body a few days later. Her death gained wide attention and highlighted the dangers associated with the practice of "robbing the pillars" of coal within a mine.

Galick, John—Insurgents deposed him as the president of L.U. 265 at HC&I's Butler Colliery in February 1928.

Gallagher, James—He was elected to the executive committee of the District 1 insurgent group in charge of planning the special (rump) convention May 21–June 1, 1928.

Gallagher, Michael (1870–1957)—He was selected by the VAN SWERINGEN brothers to become the chairman of the PaCC board of directors in 1926 and president of PaCC and The Pittston Company (TPC) in 1929. He maintained headquarters in New York City and Cleveland, Ohio. He had formerly been a business partner with RICHARD F. GRANT of Cleveland and had served for 18 years as president of the M.A. Hanna Company's bituminous operations. He was instrumental in expanding tenancy systems at the Erie Coal Companies. See C.S. GOLDSBOROUGH, FREDERICK D. UNDERWOOD.

Galleani, Luigi (1861–1931)—A multi-lingual Italian-born anarchist and intellectual, as a member of the IWW he worked to unionize PaCC and HC&I workers in a drive that culminated in the strike of 1916. He edited the principal Italian anarchist newspaper, *Cronaca Sovversiva,* from 1903 until 1918, when it was shut down by the U.S. government. See JOSEPH ETTOR, CARLO TRESCA.

Gallig, Joseph—A District 1 mineworker, delegates elected him as one of four UAMP state board members at the insurgent (rump) convention of August 7, 1933, the other three being WILLIAM AUSTIN, STANLEY GORGAS and WILLIAM SAXTON.

Gallo, Frank—After he was fired from his job at PaCC during the insurgent-sponsored wildcat strike of March 1932, the company refused to give him his job back even though rehiring strikers was part of the strike settlement. See ROSS ALBA, JOHN ASAKAVAGE, CATAL GIORDANO, GEORGE KIETYANNA.

Gavonis, Edward—A member of L.U. 699 from Edwardsville, in late 1928 he was one of 45 insurgents expelled from the UMWA by JOHN BOYLAN and the District 1 executive board for dual unionism in making plans to establish the AMWU.

Gelso, Charles—An alleged member of the area's organized-crime gang, he and his sons SAMUEL and PHILIP GELSO secured several PaCC and TPC leases between the 1930s and 1950s. They established the No. 14 Coal Company and leased the No. 14 Colliery in Port Blanchard, Jenkins Township, which they operated from 1949 until 1961.

Gelso, Philip—An alleged member of the area's organized-crime gang, he secured several PaCC and TPC leases with his father CHARLES GELSO and brother SAMUEL GELSO. They established the No. 14 Coal Company in Port Blanchard, Jenkins Township, and leased the No. 14 Colliery, which they operated from 1949 to 1961. He was convicted of wage violations with his brother in the investigations that followed the Knox Mine Disaster of January 22, 1959. He was fined and given a suspended sentence. Charles Gelso was deceased when the indictments were issued.

Gelso, Samuel—An alleged member of the area's organized-crime gang, he secured several PaCC and TPC leases with his father CHARLES GELSO and brother PHILIP GELSO. They established the No. 14 Coal Company in Port Blanchard, Jenkins Township, and lease the No. 14 Colliery, which they operated from 1949 to 1961. He was convicted of wage violations with his brother in the investigations that followed the Knox Mine Disaster of January 22, 1959. He was fined and given a suspended sentence. Charles Gelso was deceased when the indictments were issued.

Giarrtano, Calogero (1876–1963)—He was born in Serradifalco, Sicily, in 1876 where he began work in the sulfur mines at age eight. He emigrated to the U.S. and eventually moved to Pittston where he worked as an anthracite miner. He died of black lung disease in 1963.

Gilbert, Harry—He was one of three heading men who worked as subcontractors for the Hudson Coal Company in Scranton around the turn of the century, the other two being MICHAEL McHALE and HENRY RICHARDS. They were denied empty coal cars and their tools were destroyed by militant workers when they defied the local union's order to surrender their subcontracts and uphold the production stint.

Gildea, Squire (first name unknown)—He was the Squire (Justice of the Peace) of Archbald, PA, in 1896 during the Battle of Archbald, where Italian immigrant laborers staged a wildcat strike over wages. The owners of the Forest Mining Company solicited him to break the strike, which he accomplished with the assistance of hired deputies from the Barrington & McSweeney detective agency.

Gillespie, William H.—Mayor of Pittston from 1926 to 1930, he tried to calm the citizenry and facilitate an end to the bloodshed at the No. 6 Colliery during the winter of 1928. He had run for office hoping to end the mining-related violence in Pittston, the slogan for his campaign being "Gillespie or Guns."

Giordano, Catal—After he was fired from his job at PaCC during the insurgent-sponsored wildcat strike of March 1932, the company refused to give him his job back even though rehiring strikers was part of the strike settlement. See ROSS ALBA, JOHN ASAKAVAGE, FRANK GALLO, GEORGE KIETYANNA.

Gislon, Rev. William—A clergyman from Pittston, he served as a member of the committee formed by the Italian consul, CHEVALIER FORTUNATO TISCAR, to settle the wildcat strike at the Eire Coal Companies in 1910.

Gleason, James—He was an elected member of the District 1 executive board during the 1920s until his death in 1927. GEORGE DORSEY took his place in the biennial election in 1927.

Glover, Superintendent—He was superintendent of PaCC's No. 6 Colliery during the labor wars of 1928.

Golden, Christopher—As president of UMWA District 9, he attended a special meeting of high-ranking union officials and No. 6 Colliery employees in mid-March 1928 that dealt with the subcontracting system and the No. 6 Colliery strike. He wanted to learn more about the structure and operation of the system and why it had been causing such chaos at the Erie Coal Companies. He reported no successful examples of the subcontracting system in his district. No. 6 Colliery employees FRANK McGARRY, JAMES LAMARCA, and others explained that violent organized criminals were the prime movers behind the system at Erie operations, a problem not found in District 9. See ANDREW MATTEY.

Goldsborough, C.S.—President of the Erie Coal Companies' board of directors, he resigned in 1928 and was replaced by MICHAEL GALLAGHER. See M.P. CALLAWAY, H.A KNAPP, C.R. NASH, H.T. PETERS, D.E. POMEROY, A.W. PUTNAM, G.T. SLADE.

Goodlavage, Michael R.—A member of L.U. 699 from Edwardsville, in late 1928 he was one of 45 insurgents expelled from the UMWA by JOHN BOYLAN and the District 1 executive board for dual unionism in making plans to establish the AMWU.

Gorgas, Stanley—A District 1 mineworker, delegates elected him as one of four UAMP state board members at the insurgent (rump) convention of August 7, 1933, the other three being WILLIAM AUSTIN, JOSEPH GALLIG, and WILLIAM SAXTON.

Gould, Jay (1836–1892)—He won the battle to control the Erie Railroad in the 1860s and served as company president between 1868 and 1872. He eventually relinquished control following a scandal that damaged his reputation. During his tenure as president, he purchased the Hillside Coal & Iron Company and made it an Erie subsidiary.

Graber, Joseph—A Wobbly leader in northeastern Pennsylvania arrested for espionage as a German spy during World War I, he was convicted and sentenced, on August 30, 1918, to serve five years at the federal penitentiary in Leavenworth, Kansas.

Graham, William—The last president of the Penn Anthracite Collieries Company in Scranton, he served into the late 1990s.

Grant, Richard F.—As president of the Lehigh Valley Coal Company (LVCC), he was one of six coal company representatives during the labor-management contract negotiations of 1930.

Gray, Judge George (1840-1925)—A member of the U.S. Circuit Court, he was appointed by President THEODORE ROOSEVELT to serve as chairman of the Anthracite Coal Strike Commission in 1902.

Great, Ed—A member of L.U. 699 from Edwardsville, in late 1928 he was one of 45 insurgents expelled from the UMWA by JOHN BOYLAN and the District 1 executive board for dual unionism in making plans to establish the AMWU.

Great, Walter—A member of L.U. 699 from Edwardsville, in late 1928 he was one of 45 insurgents expelled from the UMWA by JOHN BOYLAN and the District 1 executive board for dual unionism in making plans to establish the AMWU.

Great, Zigmund—A member of L.U. 699 from Edwardsville, in late 1928 he was one of 45 insurgents expelled from the UMWA by JOHN BOYLAN and the District 1 executive board for dual unionism in making plans to establish the AMWU.

Grecio, Samuel—On February 19, 1928, "Big Sam" Grecio became the fourth target for murder during the labor wars at PaCC's No. 6 Colliery. A 35-year-old member of L.U. 1703's grievance committee and a devotee of the insurgent movement, he was waylaid as he and his wife walked along a Pittston street at 9:00 p.m. Two culprits lurched from the dark and grabbed Grecio while a third fired three rounds to his head. The attack purportedly occurred in retaliation for FRANK AGATI's demise at the hands of L.U.1703 President SAMUEL BONITA. Grecio miraculously survived the assault, although he lost sight in both eyes and incurred other disabilities. As he drifted in and out of consciousness, he summoned fellow insurgent ALEX CAMPBELL to his hospital bed. Grecio warned the No. 6 Colliery labor leader that he would likely become the next target, which proved correct when assassins executed Campbell, along with insurgent PETER REILLY, ten days later.

Greco, Charles—Union members retained him as the sentinel of the Ewen Colliery's L.U. 1487 in June 1928.

Green, E.M.—Owner of Jermyn-Green Coal Company with WILLIAM JERMYN, he became one of the Erie Coal Companies' most prominent lessees.

Griffin, Thomas—Insurgents elected him as a trustee of L.U. 1703 at the No. 6 Colliery in 1928.

Griffith, William R.—He served as PaCC president from 1848 to 1850 and was also a member of the PaCC board of directors from 1851 to 1853.

Groves, Robert (1895–1973)—Born in Logan Lea, West Calder, Scotland, he began working in the coal mines in Shotts, Scotland, at age 12. After serving in the British Army in World War I, he was married in 1921 and emigrated to the bituminous regions of Pennsylvania and West Virginia in 1922. In 1926, he moved to Scranton and later to Pittston where he resided until 1959, before living out his retirement years in West Pittston in 1959. He took employment with subcontractor PASQUALE ADONIZIO in 1926 at the Schooley Colliery in Exeter, and began working for PaCC at the No. 7 Shaft of the Ewen Colliery in 1931. He also worked for the lease-holding Saporito Coal Company from 1938 to 1943, and then for the Knox Coal Company from 1943 to 1959. Between 1926 and 1943, he held positions as laborer, miner, charge man, fire boss, assistant foreman, and foreman. Between August 1947 and February 1959 he held the position of inside superintendent for the Knox Coal Company's River Slope mine. Following the Knox Mine Disaster of January 22, 1959, he was among the company officials indicted for manslaughter. In a trial at the Luzerne County Courthouse in 1960, he and assistant foreman William Receski were found not guilty. He was the father-in-law of WILLIAM A. HASTIE Sr.

Guizor, Barney—When insurgents at HC&I's Butler Colliery decided, in late February 1928, to "clean house" and elect new officers of L.U. 265, they chose him as president of the local. Guizor had been a former official of L.U. 1703 at PaCC's No. 6 Colliery before securing a job at the Butler Colliery. See also JOHN PALUSKI, ALBERT STRUCKE, MICHAEL TELLO, JOHN GALICK.

Hanahoe, Matthew—In a move to retake L.U. 1703 at PaCC's No. 6 Colliery, insurgents called an unofficial meeting of the local on January 11, 1920, and deposed him as one of two trustees, the other being SYLVESTER MORRIS. See PATSY PAGLIOCA.

Hapgood, Powers (1899–1949)—A political activist and labor organizer born to an upper-class family, he began mining coal following graduation from Harvard University in 1921. He wrote about mining and worked as an organizer for the UMWA after 1922 but was banished from the organization following the campaign by JOHN BROPHY for the UMWA presidency in 1926. Like his spouse, MARY DONOVAN, whom he married in 1927, he was active in the Socialist Party. He was arrested and jailed with his spouse and local labor insurgent, CHARLES (SAM) LICATA, for inciting a riot in Pittston in early March 1928 by leading a parade through the city after MAYOR WILLIAM H. GILLESPIE refused to grant a permit; however, the charges were eventually dropped. The event was part of a Save-the-Union Committee effort to raise funds for the defense of SAMUEL BONITA, STEVEN MENDOLA, and ADAM MOLESKI. Hapgood was said to have been employed in a Wilkes-Barre coal mine at the time. He organized the defense team for the trials of Bonita, Mendola, and Moleski. He reconciled with Lewis in 1935 and worked for the CIO as an organizer and in other capacities until his death.

Harris, Walter—An insurgent from the Parsons section of Wilkes-Barre, he was elected secretary-treasurer of District 1 at the insurgent (rump) convention of May 21–June 1, 1928, for which JOHN BOYLAN and the District 1 executive board expelled him from the UMWA.

Hart, Daniel L. (1866–1933)—A popular three-term mayor of Wilkes-Barre from 1919 until his death while serving in office. He was also a playwright.

Hartneady, Michael—He served as president of UMWA District 7 from 1929 to 1935.

Harvemeyer, William—He was a member of the PaCC board of directors between 1851 and 1853.

Harvey, John H.—He was general superintendent of the Penn Anthracite Collieries Company in the 1930s as the firm expanded its subcontracting and leasing systems. Company records indicate that he knew about and tolerated wage-rate violations by the company's subcontractors and leaseholders. See OLIVER DAVIS, CHARLES DORRANCE.

Hastie, William A. Sr.—An historian of the anthracite coal industry, he worked as a laborer for the Knox Coal Company during the 1950s and was off-shift when the Knox Mine Disaster occurred on January 22, 1959. He assisted in the rescue of survivors. He was the son-in-law of Knox superintendent, ROBERT GROVES.

Hawley, Irad—He was a member of the PaCC board of directors between 1851 and 1853.

Hawley, Ira—He served as PaCC president between 1850 and 1851.

Hennig, Walter S.—He was a sergeant in charge of the Pennsylvania State Police during a wildcat strike incident on May 24, 1910, at PaCC's Ewen Colliery.

Hermann, Steve—An insurgent mineworker, he ran as for vice president in the District 1 election of May, 1931 on a ticket that included insurgent THOMAS J. MALONEY Sr. as president. The challengers lost to incumbent JOHN BOYLAN and his slate of candidates.

Hermansen, John—He was elected as an International Board member during the District 1 insurgent (rump) convention of May 21–June 1, 1928. Later in 1928, he was one of 45 insurgents expelled from the UMWA by JOHN BOYLAN and the District 1 executive board for dual unionism in making plans to establish the AMWU.

Hidock, John—Insurgent union members retained him as secretary-treasurer of the sinking fund of the Ewen Colliery's L.U. 1487 in June, 1928.

Hine, Lewis Wickes (1874–1940)—An American social-documentary photographer born in Oshkosh, WI, he became a photographer for the National Child Labor Committee in 1908. His images were influential in the group's campaign against child labor. He took numerous iconic photographs of breaker boys and other mineworkers at PaCC and other anthracite collieries in the 1910s.

Hoban, Rev. (and Bishop) Michael J. (1853–1926)—A Catholic clergyman, he was appointed bishop of the Diocese of Scranton in 1899. He supported POPE LEO XIII's encyclical *Rerum Novarum* (1891), which focused on workers' rights and responsibilities in the industrial economy. He allowed Msgr. JOHN J. CURRAN to participate as a workers' advocate in the labor-management negotiations during the Strike of 1902.

Hobbs, Gerald—Insurgent union members elected him as one of four trustees of L.U. 1495 at PaCC's No. 9 Colliery in 1928, the other three being STANLEY MOSKAITIS, EDWARD RICHARDS, FRANK WEITZ.

Hoffa, James R. (1913–1975)—A former national president of the Teamsters union, he disappeared in 1975 and was never found. He was an alleged investor in the Blue Coal Corporation with JAMES DURKIN and others. Stories have circulated throughout northeastern Pennsylvania about his remains being buried in abandoned coal workings, and about the role of alleged crime boss RUSSELL BUFALINO in his disappearance.

Hogan, Edward—He was elected as an executive committee member and as an inspection-board member at the insurgent (rump) District 1 convention of May 21–June 1, 1928. Later in 1928, he was one of 45 insurgents expelled from the UMWA by JOHN BOYLAN and the District 1 executive board for dual unionism in making plans to establish the AMWU.

Holland, Richard—A miner at HC&I during the early 1900s, he was chastised by Erie general superintendent VICTOR L. PETERSON during the Anthracite Strike Commission hearings in 1902–03 for having reduced from ten to six the number of coal cars he and his crew produced each shift since the strike of 1900.

Holmes, Justice Oliver Wendell Jr. (1841–1935)—A U.S. Supreme Court justice who wrote the majority opinion in the Penn Coal Case of 1922, which protected the coal companies from liability for damage caused by subsidence from mining.

Hopkins, John—He was indicted and pled guilty to dynamiting the McLane Silk Company factory in Scranton with his brother, JULIAN HOPKINS, in March 1928 as such violence spread beyond the mining industry.

Hopkins, Julian—He was indicted and pled guilty to dynamiting the McLane Silk Company factory in Scranton with his brother JOHN HOPKINS in March 1928 as such violence spread beyond the mining industry.

Howat, Alex (1876–1945)—A bituminous mineworker and labor leader born in Scotland, in 1906 he was elected president of UMWA District No. 14 in Kansas and served until 1914. After being cleared of false criminal charges, he was reelected in 1916 and remained in office until 1921. In the early 1920s, he was arrested and jailed for calling a wildcat strike in violation of the Kansas no-strike law. In 1936, he backed the Save-the-Union ticket of JOHN BROPHY and WILLIAM J. BRENNAN in their challenge of JOHN L. LEWIS for control of the International UMWA. He was banned from the UMWA in 1930 for rebellious activities.

Hoy, George A.—He was a member of the PaCC board of directors between 1851 and 1853.

Inglis, Major William W.—After serving as general superintendent of the Erie Coal Companies in the 1910s, he became general manager of the DL&W coal division in 1916, and became vice president and then president of the Glen Alden Coal Company in the early 1920s. He was one of six coal-company representatives during the labor-management contract negotiations of 1930.

Irvin, Charles—An immigrant from Shotts, Scotland, during the 1930s, he worked as a miner for PaCC into the 1940s and for the Knox Coal Company thereafter.

Isaacs, George—Insurgents elected him as District 1 vice-president during the insurgent (rump) convention of May 21–June 1, 1928. See FRANK McGARRY.

James, Gov. Arthur (1883–1973)—Born in Plymouth, PA, he was the Luzerne County district attorney from 1920 to 1926. He prosecuted and secured the convictions of ANTONIO PUNTARIO and PETER ERICO, two hired assassins from Trenton, NJ, for the murder of Pittston detective SALVATORE "SAM" LUCCHINO. He also served as governor of Pennsylvania from 1939 to 1943. He worked as a breaker boy and mule driver in his youth.

Jamieson, James—An immigrant from Shotts, Scotland, he worked as a miner for PaCC during the 1930s and 1940s and then as a miner and assistant foreman for the Knox Coal Company. He survived the Knox Mine Disaster of January 22, 1959, and twice re-entered the River Slope mine in search of missing workers.

Jennings, Joseph P.—An employee of PaCC and HC&I for 35 years, he started as a slate picker in 1893 and joined the engineering corps in 1899. He enrolled at Lafayette College, Easton, in 1904 and graduated from the Pardee-Scientific Department. He was appointed superintendent of the Old Forge, Central, and Consolidated collieries and then became the division superintendent. He was later appointed general superintendent of all PaCC & HC&I operations. As the person in charge of expanding tenancy, in 1913 he offered subcontracts to mineworkers FRANK McHALE and THOMAS MITCHELL at PaCC's No. 6 Colliery. When they declined, he offered the agreements to SANTO VOLPE and CHARLES CONSAGRA, who accepted, thus beginning the "modern" subcontracting era at the Erie Coal Companies. He resigned as general superintendent in May 1928 during the rancorous labor wars at the Erie Coal Companies and was replaced by JOHN BRYDON.

Jermyn, Mayor Edmund B.—During his first term as mayor of Scranton between 1914 and 1918, he formed the Cave Mine Bureau that sought a remedy for the city's constant and dangerous subsidence problems. He served a second term in office between 1926 and 1930.

Jermyn, William S.—Owner of Jermyn-Green Coal Company with E.M. GREEN, he became one of the Erie Coal Companies' most prominent lessees.

Johnson, Emily—A resident of West Pittston, she served as the personal secretary of Luzerne County Judge Henry A. Fuller for 35 years. She was a charitable person who befriended many Italian immigrant families in Pittston, assisting them with various needs and problems, and thereby gaining their trust and confidence. She learned about many aspects of their lives, including their difficulties dealing with organized crime. She discussed her many experiences, including the assassination of SALVATORE "SAM" LUCCHINO, in several conversations with WILLIAM A. HASTIE Sr.

Johnson, James C.—He was transferred as superintendent of PaCC's No. 6 Colliery to the same position at the Ewen Colliery in mid-April 1928, during the intense conflict at the No. 6 and other PaCC and HC&I collieries.

Jones, David—A mineworker at the Glen Alden Coal Company, he served as vice chairman of the insurgent committee that organized the first special (rump) convention of District 1 of May 21–June 1, 1928.

Jones, James—An international UMWA organizer who worked in the northern coal field, he sent regular private communications to JOHN L. LEWIS regarding strikes, rallies, insurgent leaders, District 1 officers, and other "inside" information.

Jones, Judge Benjamin R.—Governor Edwin S. Stuart appointed him to the Luzerne County Court of Common Pleas in 1910 after he had served two terms as Luzerne County district attorney. During the anthracite labor wars of the 1920s, he released radical activists and spouses POWERS HAPGOOD and MARY DONOVAN from jail for lack of evidence after they had been held without bail from March 5 until April 12, 1916. He released local activist CHARLES (SAM) LICATA much earlier in the same incident. During a UAMP strike in January 1934, he issued a restraining order against the upstart union. He was the target of one of six package bombs mailed on April 10, 1936, one of which killed UAMP leader THOMAS J. MALONEY Sr. and his son; however, the bomb sent to Jones was intercepted by authorities.

Jones, William—Delegates elected him as one of six members of the Resolutions Committee at the District 1 insurgent (rump) convention of May 21–June 1, 1928.

Jordan, Rev. R.D.—A Catholic priest, he publicly backed the workers and legitimized their cause during the wildcat strike of 1920 at the Erie Coal Companies.

Joyce, James A.—Pittston businessman and insurgent advocate, he worked to broker a settlement during the strike at the Erie Coal Companies in 1920. When he was near compromise, his home and business were bombed, allegedly by dissident mineworkers who wanted to maintain the strike.

Juliani, Philip—In a rump election on January 11, 1928, insurgents deposed him as one of two financial secretaries of L.U. 1703 at PaCC's No. 6 Colliery, along with the other incumbent officers including the president, PATSY PAGLIOCA. See also CHARLES ALBA, JOSEPH SHIMOKONIS.

Kaschinsky, Charles—He was elected to the executive committee of the District 1 insurgent group in charge of planning the special (rump) convention May 21–June 1, 1928.

Kearney, James—Insurgents elected him as secretary and as a member of the grievance committee of L.U. 1703 at PaCC's No. 6 Colliery in February 1928. See FRANK LICATA, FRANK McGARRY, JOSEPH VICTOR.

Kehoe, John—He was a powerful and cantankerous coal operator, brewery owner, alcohol bootlegger, and founder and editor of the *Sunday Dispatch* in Pittston. As the political boss of Pittston and a leader in the county Republican Party, he became a close ally of Governor JOHN S. FINE. He formed a partnership with PASQUALE and JOSEPH ADONIZIO and WILLIAM H. BERGE in the Berge-Rose Coal Company, established on September 26, 1931. He also formed Kehoe-Berge Coal Company in partnership with William H. Berge, which secured dozens of leases and contracts from major coal companies between 1932 and 1960. The firm stood as TPC's fourth-largest leaseholder in 1941 with access to 1,935,237 tons of unmined coal. It employed thousands of workers and operated numerous collieries, deep mines, and strip mines in Luzerne and Lackawanna counties, including the Seneca, Broadwell, William A., O&W, Kehoe-Berge, Hallstead, Hunter, Red Ash, Mosca, Stevens, Exeter, Fox Hill, Hunter, No. 10, Cold Creek. The firm also operated the Last Chance breaker in Avoca where it processed coal until building the William A. breaker in Duryea. Its pillar robbing caused widespread subsidence in Duryea. Following the insurgent strike of 1932, the company was charged with discrimination in rehiring UAMP strikers. Kehoe was indicted and convicted of income-tax evasion and, in a case that went to the U.S. Supreme Court in 1925, he was required to pay $316,000. He eventually became a pauper although he continued to reside in a mansion along the Susquehanna River in Harding, PA. He died in 1961.

Kennedy, Thomas (1887–1963)—Born in Lansford, PA, and known as "The Little Giant of Anthracite," he was elected president of District 7 in 1910 and secretary-treasurer of the UMWA in 1928. He suspended union work for a time when he was elected lieutenant governor of Pennsylvania as a Democrat in 1934. After serving one term, he ran for governor in 1938 but did not obtain sufficient party backing. He was elected vice president of the UMWA in 1947 and assumed the presidency upon JOHN L LEWIS's retirement in 1960. He held position until failing health caused him to retire in 1962.

Kietyanna, George—After he was fired from his job at PaCC during the insurgent-sponsored wildcat strike of March 1932, the company refused to give him his job back even though rehiring strikers was part of the strike settlement. See ROSS ALBA, JOHN ASAKAVAGE, FRANK GALLO, CATAL GIORDANO.

King, Alfred—He was transferred as foreman of the No. 7 Shaft of PaCC's Ewen Colliery to the same position at the Thomas Shaft of HC&I's Butler Colliery in mid-April, 1928, during the intense conflict at the No. 6 and other PaCC and HC&I collieries.

Kirby, Alan P.—Born in Wilkes-Barre on July 31, 1892, he was the son of F.M. Kirby whose fortune came from his "dime store" partnership with F.W. Woolworth. In August 1937, Alan Kirby attempted to garner the lease on PaCC's No. 6 Colliery in partnership with SANTO VOLPE and two other men; Volpe turned down their offer and took the lease with other partners in his Volpe Coal Company. Also in 1937, Kirby and other investors gained an interest in the Alleghany Corporation, formerly held by the VAN SWERINGEN brothers before their bankruptcy. He became president of Alleghany and successfully revived the company. Alleghany again rose to prominence on Wall Street through a series of takeovers, including the purchase of over one million shares in the New York Central Railroad by 1953. When the railroad suffered financial problems, Alleghany sustained major losses. Its precarious fiscal situation forced Alleghany to sell TPC in 1954. TPC became an independent corporation with 30 subsidiaries, including the Brinks Armored Car division and several coal companies in the bituminous fields. Kirby held the top position at Alleghany until 1961 and served as chairman of the board and chief executive officer from 1963 to 1967, and as chairman emeritus from 1967 to 1973.

Kmetz, John—A mineworker from Nanticoke, he was elected to the District 1 executive board in 1928 as a supporter of the Rinaldo Cappellini administration.

Knapp, H.A.—He was appointed to the Erie Coal Companies' board of directors in 1928 following the resignation of the previous board and the company president, C.S. GOLDSBOROUGH. The shake-up followed the VAN SWERINGEN brothers' appointment of MICHAEL GALLAGHER as board president in 1926 and as company president in 1928. See M.P. CALLAWAY, C.R. NASH, H.T. PETERS, D.E. POMEROY, A.W. PUTNAM, G.T. SLADE.

Kosik, Michael J.—He started working as a breaker boy in PaCC's No. 9 Colliery at age 10 in 1910 and later worked as a miner at HC&I's Butler Colliery. He was elected vice president of District 1 in 1925 and won reelection in 1927, 1929, 1931, and 1933. He was appointed president of District 1 in 1935 upon the resignation of JOHN BOYLAN, and won presidential elections over the next 16 years. He left office in July 1951 to accept an appointment as secretary of the ABC, a post he held until his retirement on January 1, 1968. AUGUST J. LIPPI followed him as president of District 1.

Kosloski, Michael—The members of L.U. 1495 at PaCC's No. 9 Colliery elected him as a delegate to the insurgent (rump) District 1 convention of August 1933.

Koval, Charles—Members of L.U. 1495 at PaCC's No. 9 Colliery elected him as financial secretary in 1928.

Kuma, Frank—A mineworker at PaCC's Ewen Colliery, he was elected as one of seven delegates from L.U. 1487 to attend the insurgent (rump) convention of August 1933.

Kuratz, Michael—He was elected to the executive committee of the District 1 insurgent group in charge of planning the special (rump) convention May 21–June 1, 1928.

Lamarca, James—A mineworker at PaCC's No. 6 Colliery, members of L.U. 1703 elected him as check-docking boss in April 1928, but the company would not allow him to assume the position. After some delay the company consented, resulting in a temporary truce between labor and management regarding the protracted strike at the colliery.

Langan, Ambrose—He served as mayor of Pittston from 1930 to 1938, during many of the labor wars at the Erie Coal Companies.

Laquasto, Sam—Delegates elected him as one of six members of the Resolutions Committee at the District 1 insurgent (rump) convention of May 21–June 1, 1928.

Latona, Samuel—He was elected financial secretary of the Ewen Colliery L.U. 1487 in June, 1928.

Latore, Sam—He was one of four L.U. 1703 grievance committee members at PaCC's No. 6 Colliery who were deposed by insurgents in a rump election on January 11, 1928. The others were E. AQUILINA, JAMES ANTONELLO, SAM SEIT.

LaTorre, Stefano ("Steve") (1886–1984)—A sulfur miner born in Montedoro, Sicily, he emigrated to Pittston in 1903. He was the reputed co-founder, with his brother-in-law SANTO VOLPE, of the second-oldest Sicilian crime organization in the U.S., which formed in Pittston in 1908. A coal miner at PaCC, he garnered 21 tenancy agreements between 1931 and 1936 and became a partner in the Saporito Coal Company with JOHN SCIANDRA and CARLO SAPORITO. In 1943, he and Saporito were allegedly forced out of a lease that PaCC turned over to the newly established Knox Coal Company and its owners Sciandra and LOUIS FABRIZIO.

Lavelle, Thomas—A mineworker and UMWA member, in June 1929 he ran for vice president of District 1 on the insurgent ticket with presidential candidate RINALDO F. CAPPELLINI. The challengers lost the election to JOHN BOYLAN and other incumbents amid cries of voting fraud.

Law, William (1824–1889)—An immigrant from Wanlockhead, Scotland, where he was a lead miner, he settled in Providence (later part of Scranton) in 1850. He moved to Pittston where he became a colliery superintendent for PaCC and a prominent citizen.

Lazar, Peter—A miner at the Harry E Colliery in Swoyersville, he survived an explosion in 1941 that took the life of his laborer, PAUL WOLENSKY.

Lelack, Michael—A mineworker at PaCC's Ewen Colliery, he was retained as treasurer of the Ewen L.U. 1487 in June 1928.

Lenardo, Frank—A mineworker at PaCC's No. 6 Colliery, he was elected as a trustee of L.U. 1703 by insurgents in 1928.

Lewis, John L. (1880–1969)—One of the most prominent twentieth-century labor leaders, he was born on February 12, 1880, to Welsh immigrant parents in Cleveland, Iowa. He began working in the bituminous mines at age 17. He was the president of the UMWA from 1919 to 1960, becoming more autocratic over time. His inability to understand the labor grievances associated with tenancy in the northern anthracite field figured prominently in the labor wars of the 1920s and 1930s. By the time of his departure from the UMWA, the organization had become very corrupt within District 1 and elsewhere. He was crucial to establishing the CIO in 1938. President Lyndon B. Johnson awarded him the Presidential Medal of Freedom in 1964.

Lewis, T.L.—He was the president of the UMWA from 1908 to 1910.

Lewis, Thomas—A 43-year-old mineworker from Pittston and active supporter of the insurgent cause, he was struck in the back with six bullets on the day after THOMAS LILLIS's murder in 1928. Passersby found his body lying along the Erie Railroad tracks. Police pursued two suspects who fled by rail. No arrests were made in the case. Lewis was an outspoken critic of subcontracting at PaCC's No. 6 Colliery where he had served as president of L.U. 1703 in the early 1920s.

Licata, Charles (Sam)—He was a member of a small group of radicals within District 1 who actively supported the Save-the-Union movement in 1928. Insurgents elected him as the acting recording secretary of L.U. 1703, to replace the murdered PETER REILLY. In March 1928, Pittston police arrested and jailed Licata, as well as POWERS HAPGOOD and MARY DONOVAN, for "inciting a riot" by conducting a march through Pittston without a permit; the charges quickly dropped. As he walked home on March 2, 1931, the 38 year-old Pittstonian was assassinated by three men who stepped from an automobile and fired six bullets into his body. The assailants were not found. Licata was a friend of insurgent CHARLES CALAMARA who had suffered a similar death near the same spot in Pittston on January 4, 1931.

Licata, Frank—A mineworker at PaCC's No. 6 Colliery, he served as a member of the colliery's grievance committee during the insurgent strikes of 1928. In February 1928, insurgents elected him treasurer of L.U. 1703. See JAMES KEARNEY, FRANK McGARRY, JOSEPH VICTOR.

Licata, Lee—The members of L.U. 1495 at PaCC's No. 9 Colliery elected him as a delegate to the insurgent (rump) District 1 convention of August 1933.

Lillis, Thomas (1884–1928)—On January 18, 1928, the 43-year-old miner and union activist became the first of four persons murdered in the so-called Feud at PaCC's No. 6 Colliery in

Pittston. Born in 1884 of Irish parents, he had served as a treasurer for L.U. 1703 in the early 1920s and followed ALEX CAMPBELL as the No. 6 check-weighman in 1925. A bachelor, he lived with his aged mother, brother, and sister in the Browntown section of Pittston Township.

Lindy, Clarence—Delegates elected him as one of six members of the Resolutions Committee at the District 1 insurgent (rump) convention of May 21–June 1, 1928.

Lippi, August J. (1900–1970)—A mineworker from Exeter, he joined L.U. 1192 in Wyoming, PA, in July, 1914, at age 14. In 1922, he secured work at PaCC's No. 14 Colliery where he was elected secretary of L.U. 1581. In 1925, President RINALDO F. CAPPELLINI appointed him to an administrative position in District 1, and in 1927 he was elected to the District executive board. He supervised a second election at L.U. 1703 at PaCC's No. 6 Colliery on January 25, 1928, when the insurgents again ousted the incumbents and took control of the local. A loyal member of the Cappellini administration, District 1 insurgents "impeached" him and other incumbent officials at their special (rump) convention of May 21–June 1 1928; the UMWA did not recognize the legitimacy of the convention or the ballot. He was elected secretary-treasurer of the District in 1941. The executive board, with JOHN L. LEWIS's approval, appointed him as acting District 1 president in July, 1951, following the appointment of the incumbent, MICHAEL J. KOSIK, as secretary of the ABC. Lippi was first elected president in 1952 and continued winning elections until 1964—even after he had been charged and convicted of felonies related to his corrupt practices. In July 1946 he became vice president and board member of the First National Bank of Exeter and rose to the presidency in January 1958 upon the death of bank president Major William A. Clark. Lippi was also elected chairman of the bank's board. In 1964, he was convicted of misappropriating bank funds, a decision affirmed on appeal and denied a hearing by the U.S. Supreme Court. In another business deal, he was a secret (and illegal) partner in the Knox Coal Company beginning in 1950 while he was also president of District 1, in violation of the federal Taft-Hartley law. Following the Knox Mine Disaster of January 22, 1959, he was indicted for manslaughter, conspiracy to violate the mining laws, conspiracy to hide ownership in the Knox Coal Company, personal income-tax evasion, corporate income-tax evasion, and bribery. His convictions for manslaughter and conspiracy were overturned on appeal. He was convicted on April 13, 1961, for evading $80,445 of income taxes at the Knox Coal Co. A separate court case against him and fellow UMWA officials FRANK CARDONI and LEONARD STATKIEWITZ for income-tax evasion in association with ROBERT L. DOUGHERTY's Avon and Peeley Coal Companies ended in acquittal. The Federal Bureau of Investigation twice caught Lippi attempting to exit the country: in February 1961 (airline ticket to Brazil) and December 1962 (airline ticket to Santiago, Chile). He entered the Lewisburg, PA, federal penitentiary in November 1965 and, after being denied parole in September 1967, April 1968, and July 1968, he was finally released in March 1968. With debts of at least $327,337, he declared bankruptcy shortly thereafter. He allegedly facilitated the activities of organized crime within District 1 and within the northern anthracite industry.

Loquasto, Sam—A mineworker and insurgent activist at PaCC's Ewen Colliery, insurgents elected him as temporary chairman of L.U. 1487 in July 1933, and also as a delegate to the District 1 insurgent (rump) convention of August 1933. At the convention, he was chosen as one of six members of the UAMP Resolutions Committee.

Lucchino, Pasqualina ("Ester")—Born in Montedoro, Sicily, she was the mother of SALVATORE "SAM" LUCCHINO and PIETRO LUCCHINO, and the spouse of TRANQUILO LUCCHINO. She was convicted of larceny and receiving stolen goods in 1911.

Lucchino, Pietro ("Peter")—Born in Montedoro, Sicily, he was the son of PASQUALINA LUCCHINO and TRANQUILO LUCCHINO and the brother of SALVATORE "SAM" LUCCHINO. An early member of the area's Black Hand gang, in 1907 he was one of nine persons convicted of conspiracy to extort protection money from mineworkers, which resulted in a sentence of one year in solitary confinement in the Luzerne County jail.

Lucchino, Rose—She was the wife of STEFANO LaTORRE, the alleged co-founder of the region's organized-crime gang and a relative of SALVATORE "SAM" LUCCHINO and his family.

Lucchino, Salvatore "Sam" (1884-1920)—A Pittston city detective, he was murdered on July 21, 1920, at age 33, by ANTONIO PUNTARIO and PETER ERICO, two hired assassins from Trenton, NJ. Lucchino had been conducting investigations into the subcontract mining system at PaCC and HC&I and his murder was undoubtedly ordered by organized criminals, with CHARLES CONSAGRA being the main suspect. An immigrant from Montedoro, Sicily, he joined the Pittston area's Black Hand gang as a youth along with his brother, PIETRO LUCCHINO. In 1907, he was one of nine persons convicted of conspiracy to extort protection money from mineworkers, and spent one year in solitary confinement in the Luzerne County jail. After his release, he began working as an undercover agent for the U.S. Secret Service where he participated in the LUPO-MORELLO counterfeiting case in 1909 in New York City. He served as a main witness in the trial that sent gang leaders GIUSEPPE MORELLO and IGNAZIO SAIETTA to long sentences in the federal penitentiary in Atlanta. He worked under the supervision of Secret Service agent WILLIAM J. FLYNN in pursuing the case. He left the Secret Service and secured a position on the Pittston police force. He survived two assassination attempts and received numerous death threats while on the police force. He was the son of PASQUALINA and TRANQUILO LUCCHINO.

Lucchino, Mrs. Salvatore ("Nellie")—The spouse of murder victim SALVATORE "SAM" LUCCHINO, she served as a key witness in the trial in 1920 that led to the convictions of hired gunmen PETER ERICO and ANTONIO PUNTARIO. She was the recipient of several death threats following her husband's murder. See ORRA R. WHITE.

Lucchino, Tranquilo ("Joseph")—An immigrant from Montedoro, Sicily, he was the father of SALVATORE "SAM" and PIETRO LUCCHINO and the spouse of PASQUALINA LUCCHINO. He had Black Hand connections and likely committed suicide on August 10, 1911, allegedly in despair over his son's work for the Secret Service and the Pittston police, although there was some suspicion that he might have been murdered because of his son's law enforcement activities.

Loomis, Superintendent (first name unknown)—He served as superintendent of DL&W's Woodward Colliery in Edwardsville, PA, and negotiated with District 1 President THOMAS D. NICHOLLS over the use of the subcontracting system in 1901.

MacVeagh, Esq., Wayne (1833–1917)—A former attorney general under President James A. Garfield, he was the chief counsel for the Erie Coal Companies. As such, he led the questioning of witnesses (including JOHN MITCHELL) on the companies' behalf during the Anthracite Strike Commission hearings of 1902–03.

Madtucci, Rigo—He directed the subcontracting system at PaCC's Underwood Colliery in 1928 when the company agreed to remove (for a time) all subcontractors from this particular operation and return the mines to the traditional work organization of one miner harvesting coal in a single chamber with one or two laborers.

Maloney, A.J.—As president of the Philadelphia and Reading Coal & Iron Company, he was one of six coal-company representatives during the labor-management contract negotiations of 1930.

Maloney, Thomas J. Jr. (1932–1936)—On April 11, 1936, at age 4, he died from injuries caused by a cigar-box bomb mailed to his home in Wilkes-Barre Township. His father, UAMP President THOMAS J. MALONEY Sr., also died from the blast. He was buried alongside his father at St. Mary's Cemetery, Hanover Township, PA.

Maloney, Thomas J. Sr. (1892–1936)—An outside laborer at the Glen Alden Coal Company's Stanton Colliery in Wilkes-Barre, he was elected to the colliery grievance committee and was also elected president of L.U. 466. He emerged as a District 1 insurgent leader during the UMWA Tri-District convention in August 1930 during heated debates over the recently negotiated labor-management contract. He ran unsuccessfully against JOHN BOYLAN for the District 1 presidency in May 1931. He became the leader of the insurgent movement, which called a wildcat strike in March 1932. He worked to found the UAMP with RINALDO F. CAPPELLINI and others. He was elected president of the UAMP at its inaugural convention on August 7, 1933, and he held the post until the union disbanded in late 1935. In 1932, the UMWA banished him and 34 other UAMP members for 15 years for dual unionism. He relied on his friend and confessor, Msgr. JOHN J. CURRAN, during the insurgent campaigns during the first half of 1930s. When Maloney and 28 other UAMP members defied a court order to end a strike in March 1935, Luzerne County Judge W.A. VALENTINE sent them to jail for several weeks. Maloney was mortally wounded on April 10, 1936, upon opening a cigar-box bomb, mailed to his home in the Georgetown section of Wilkes-Barre Township. He lingered for a week and died on April 16, 1936, at age 43. His son, THOMAS J. MALONEY Jr., age 4, was standing next to him when the blast occurred and died the next day. Father and son were buried side-by-side at St. Mary's Cemetery, Hanover Township, PA. MICHAEL FUGMANN, a former UAMP board member and Maloney friend, was charged and convicted of first degree murder punishable by death. Maloney's daughter MARGARET, age 16, who was also present when the bomb exploded, suffered severe injuries but survived after a long convalescence. Maloney's wife was in another room and suffered no blast injuries.

Mancini, Anthony—A mineworker at PaCC's Ewen Colliery, he was elected president of the Colliery's L.U. 1487 on June 5, 1928, by an insurgent revolt that deposed incumbent President GEORGE MOLESKI.

Manos, Joseph—As an organizer for District 1 during a wildcat strike in 1910, he complained that the Erie Coal Companies' management regularly failed to keep their promises to address the workers' grievances over docking, wage rates, and other matters.

Markowski, John—Insurgents elected him as one of two assistant secretaries of L.U. 1495 at PaCC's No. 9 Colliery in April 1928, the other being JOSEPH DESSOYE. See JACOB COLLIER.

Martin, James—The son of an English miner from Schuylkill County, he began work as a breaker boy and then became a mule driver at a D&H mine when his family moved to Plains Township, Luzerne County. After becoming a miner, the six-foot-three-inch, 200-pound Martin worked his way up to colliery foreman. He became the Republican leader of the town and was elected Luzerne County sheriff in 1895. He led a corps of deputies who fired upon marching mineworkers at Lattimer in 1897, killing 19 and wounding 38 men, most of whom were of Slavic heritages. As a test case, Martin and 73 deputies were charged and tried for the murder of one victim, Mike Cheslak. After five weeks of testimony from nearly 200 witnesses, the jury entered a verdict of "not guilty" in what many thought was a biased and unfair decision.

Martin, William—Delegates elected the Jermyn resident as one of four UAMP board members at the District 1 insurgent (rump) convention of August 1933.

Mattey, Andrew—As president of District 7 between April 1926 and January 1929, he attended a special meeting of high-ranking UMWA officials and No. 6 Colliery employees in mid-March 1928, dealing with the subcontracting system and the No. 6 Colliery strike. He said that certain companies in his district tried to institute subcontracting but the workers walked out and would not allow it. He asked why that was not the case at the Erie collieries in District 1. No. 6 Colliery workers FRANK McGARRY, JAMES LAMARCA and others informed him that what distinguished District 7 from District 1 was the presence of violent organized criminals as subcontractors in District 1 and at the Erie collieries. See CHRISTOPHER GOLDEN.

May, Captain William A.—Appointed as the first general manager of the Erie Coal Companies in the early twentieth century, he had previously been an official at HC&I. He testified before the

Anthracite Strike Commission that miners consistently produced coal with up to 21 percent waste. At first, he refused to negotiate an end to the wildcat strike at the Erie companies in 1910, but soon agreed to join the negotiations started by the Italian consul, CHEVALIER FORTUNATO TISCAR. Workers perennially complained that May had not lived up to his promises to address their grievances. He became president of the Erie companies later in the decade and moved to expand the subcontracting system. He was president during the company-wide strike of 1920, led by ALEX CAMPBELL and RINALDO F. CAPPELLINI, which temporarily ended the subcontracting system and brought 12,000 company mineworkers into the UMWA.

McClellan, Sen. John L. (1896–1977)—He was chairman of the U.S. Senate Select Committee on Improper Activities in Labor and Management established by the upper chamber on January 30, 1957. The committee investigated the alleged organized-crime dealings of SANTO VOLPE, among many other alleged Mafia members.

McCrone, Edward—A resident of North Scranton, he served as secretary of the insurgent committee that organized the first insurgent special (rump) convention of District 1 on May 21–June 1, 1928. At the start of the convention, he read the letter from Mrs. ANNA JENKINS CAMPBELL (widow of ALEX CAMPBELL) that amounted to a call to action by the insurgent delegates. He was a member of L.U. 151 in Scranton and served as a delegate to the Reconvened Tri-District Convention of August 1930, where he was one of many who spoke against the proposed labor-management agreement negotiated by President JOHN L. LEWIS. See THOMAS J. MALONEY Sr.

McDade, James B.—A mining entrepreneur from Scranton, he was the principal owner of the Heidelberg Coal Company in Avoca, among other coal companies.

McEnaney, Benjamin—President of UMWA District 1 between 1909 and 1911, at management's request in 1910 he worked to settle the wildcat strike at the Erie Coal Companies even though the great majority of the firm's 12,000 employees did not belong to the union.

McGarry, Frank—An insurgent mineworker from Pittston, he was dismissed by PaCC management for labor activism but an appeal to the ABC led to his reinstatement, although the company would not allow him to resume his former position. He withdrew from union matters for a time but reemerged and was elected president of L.U. 1703 at PaCC's No. 6 Colliery following the arrest and imprisonment of incumbent President SAMUEL BONITA on manslaughter charges in February 1928. He was also elected as the No. 6 Colliery's check-weighman, a position held by ALEX CAMPBELL until his murder in February 1928. McGarry was elected to the executive committee of the District 1 insurgent group in charge of planning the special (rump) convention for May 21–June 1, 1928. Delegates elected him as District 1 president at the convention after they had "impeached" RINALDO F. CAPPELLINI and his administration. The UMWA national office nullified the election but McGarry and the other officers continued in their dual-union positions at UAMP headquarters in Scranton. Later in 1928, he was one of 45 UAMP members expelled from the UMWA by JOHN BOYLAN and the District 1 executive board for dual unionism. For reasons that remain unclear, he did not actively participate in the labor wars after 1929 when THOMAS J. MALONEY Sr. and Rinaldo F. Cappellini assumed the leadership positions. See GEORGE ISAACS, JAMES KEARNEY, FRANK LICATA, JOSEPH VICTOR.

McHale, Frank—In 1913, PaCC Superintendent JOSEPH P. JENNINGS offered him and fellow miner THOMAS MITCHELL a subcontract at PaCC's No. 6 Colliery, which they turned down out of fear of resentment and possible violence from fellow workers. Jennings then offered the subcontract to alleged gangsters SANTO VOLPE and CHARLES CONSAGRA, who took it, thus beginning the modern subcontracting system at the Erie Coal Companies.

McHale, Michael—He was one of three heading men who worked as subcontractors for the Hudson Coal Company in Scranton around the turn of the century, the other two being HARRY

GILBERT and HENRY RICHARDS. They were denied empty coal cars and their tools were destroyed by militant mineworkers when they defied the local union's order to surrender their subcontracts and uphold the production stint.

McLean, Judge William S.—Luzerne County judge who sentenced SAMUEL BONITA, STEVEN MENDOLA, and ADAM MOLESKI to maximum prison sentences for voluntary manslaughter in the shooting death of District 1 organizer FRANK AGATI on February 16, 1928. McClean also served as chairman of the Industrial Relations Board of the Wilkes-Barre/Wyoming Valley Chamber of Commerce.

McMillan, James L.—A mineworker and immigrant from Scotland, he lived in Pittston and became a high-ranking executive at PaCC.

Melissari, Ralph—A hired assassin from New Jersey, he was charged for the murders of ALEX CAMPBELL and PETER REILLY with co-conspirators PETER DELUCCA and VINCENZO DAMIANO of New York City. He was acquitted of Campbell's murder but convicted, with DeLucca, of Reilly's murder and was sentenced to life in prison.

Memolo, Martin—As the burgess (mayor) of Old Forge, PA, he and his family received threats, presumably sent by IWW backers, when he refused to order State Police out of town during the IWW-sponsored strike of 1916 at the Erie Coal Companies and the Hudson Coal Company. For safety considerations, his family moved to an undisclosed location in Scranton.

Mendola, Steven—A mineworker at PaCC's No. 6 Colliery, he was elected by insurgent L.U. 1703 members as a grievance committee member in a rump election on January 11, 1928, that deposed the incumbent officers, including the president, PATSY PAGLIOCA. On February 16, 1928, Mendola traveled with fellow local union officials SAMUEL BONITA and ADAM MOLESKI to the District 1 headquarters in the Miner's National Bank building in downtown Wilkes-Barre for negotiations with District officials regarding the ongoing strike at the No. 6 Colliery. The discussions turned violent and gunshots were fired. FRANK AGATI, a former L.U. 1703 president and current District 1 organizer, was shot and died. Bonita, the sitting L.U. 1703 president, was convicted of voluntary manslaughter as the principal in the shooting and given the maximum sentence of six to twelve years in prison by Luzerne County Judge WILLIAM S. McLEAN. Mendola and Moleski were charged as accomplices and also convicted of voluntary manslaughter in separate trials. Judge McLean gave them each the maximum sentence of four to eight years in prison.

Menzies, William—He was the inventor of the Cone Hydro Separator in 1915, which was used in breakers such as the Huber in Ashley, PA, to clean coal using an advanced spinning-washing technology.

Meroski, Edward—A mineworker at PaCC's Ewen Colliery, he was elected as one of seven delegates from L.U. 1487 to attend the insurgent (rump) convention of August 1933.

Mikulski, John—A mineworker from Plymouth Township, he worked as a motor runner and local union officer at the Avondale Colliery when it was owned by ROBERT L. DOUGHERTY's Avon Coal Company in the late 1950s. He was a strong critic of the individual contract and the subcontracting system.

Miller, Elliott—He was elected to the executive committee of the District 1 insurgent group in charge of planning the special (rump) convention in Scranton on May 21–June 1, 1928.

Milner, E. Stewart—A PaCC employee since 1926, he retired as the last company vice president in 1978, but remained a consultant and keeper of the company's papers and maps at the former corporate headquarters in Dunmore, near Scranton. He donated most of the records to the Pennsylvania State Archives in Harrisburg and to the Anthracite Heritage Museum in Scranton, but he either gave or sold a collection of several hundred mine and property maps to a local businessman with alleged criminal connections.

Milton, Benjamin—He was transferred as superintendent of PaCC's Ewen Colliery to the same position at No. 6 Colliery in mid-April 1928, during the intense conflict at the No. 6 and other Eire Coal Companies collieries.

Minch, Adam—He was elected to the executive committee of the District 1 insurgent group in charge of planning the special (rump) convention for May 21–June 1, 1928.

Minerich, Tony—During a walkout at PaCC's No. 6 Colliery in November 1928, the communist-led National Miners Union (NMU) sent him in to take control of the strike. The workers paid no attention to Minerich or his union and did not complain when the State Police jailed him on a spurious charge of planting dynamite.

Minturn, Robert B.—He was a member of the PaCC board of directors between 1851 and 1853.

Mitchell, John (1870–1919)—Born in Braidwood, Illinois, to a Scottish-immigrant father and an English-immigrant mother, he was orphaned at an early age and became a mineworker. He worked his way up the UMWA ranks until he became the fifth president of the UMWA, serving from 1898 to 1907. He came to be called the "Patron Saint of Anthracite" after leading historic strikes of 1900 and 1902, which brought important gains to the mineworkers. He worked to settle the 1902 strike with President THEODORE ROOSEVELT and Rev. JOHN J. CURRAN. Beginning in 1902, and continuing for decades, anthracite mineworkers celebrated October 29 (the day the 1902 strike ended) as "Johnny Mitchell Day." Among his most important strategies was the successful unification of workers from more than two dozen ethnic groups on behalf of the UMWA. A convert to Catholicism in 1907, he died in New York City on September 9, 1919. Curran presided at his Requiem Mass in the Diocese of Scranton's main cathedral, after which Mitchell was buried in the cathedral cemetery. In June 1924, the men of anthracite assembled a great parade in Mitchell's honor and unveiled a monument to his memory on the main square in downtown Scranton. Pennsylvania Governor GIFFORD PINCHOT, UMWA President JOHN L. LEWIS, and American Federation of Labor Secretary FRANK MORRISON attended the dedication. Ten thousand mineworkers marched as 100,000 citizens looked on. The inscription on the monument reads: "Champion of Labor, Defender of Human Rights." An image of the monument can be seen as the frontispiece to Chapter One in this volume.

Mitchell, Thomas—In 1913, PaCC Superintendent JOSEPH P. JENNINGS offered Thomas and fellow miner FRANK McHALE a subcontract at PaCC's No. 6 Colliery, which they turned down out of fear of resentment from fellow workers. Jennings then offered the subcontract to alleged gangsters SANTO VOLPE and CHARLES CONSAGRA who took it, thus beginning the modern subcontracting system at the Erie Coal Companies.

Moffat, Robert—He and his brothers formed the Moffat Coal Company of Taylor, PA, in 1937. The company originally worked under leases from major coal companies, including Glen Alden and Hudson. In July 1953, the Moffat operation purchased all of Glen Alden's mines in the Taylor and Scranton areas for eight million dollars. The company became the largest coal producer in Lackawanna County, with eight deep mines and five strip mines. It was out of business by 1968. The company's Continental mine is now the site of the Lackawanna Coal Mine Tour in Scranton's McDade Park.

Moleska, James—Delegates elected him as one of six members of the Resolutions Committee for the District 1 insurgent (rump) convention of May 21–June 1, 1928.

Moleski, Adam—A mineworker at PaCC's No. 6 Colliery, he was elected by insurgent members of L.U. 1703 as a member of the grievance committee in the rump election of January 11, 1928, that deposed the incumbent officers, including the president, PATSY PAGLIOCA. On February 16, 1928, Moleski traveled with fellow local union officials SAMUEL BONITA and STEVEN MENDOLA to the District 1 headquarters in the Miner's National Bank building in downtown Wilkes-Barre for negotiations with District officials regarding the ongoing strike at the No. 6 Colliery. The discussions turned violent and gunshots were fired. FRANK AGATI, a former L.U. 1703 president and current District 1 organizer, was shot and died. Bonita, the sitting L.U. 1703 president, was convicted of voluntary manslaughter as the principal in the shooting and given the maximum sentence of six to twelve years in prison by Luzerne County Judge WILLIAM S. McLEAN. Moleski and Mendola were charged as accomplices and also convicted of voluntary manslaughter in separate trials. Judge McLean gave them each the maximum sentence of four to eight years in prison.

Moleski, Alfred—Insurgent members of L.U. 1703 at PaCC's No. 6 Colliery elected him as one of two financial secretaries on January 11, 1928, the other one being JOE COSTA.

Moleski, George—He was unseated as president of L.U. 1487 at PaCC's Ewen Colliery by insurgents on June 5, 1928, and replaced by insurgent ANTHONY MANCINI.

Moleski, Joseph—In a rump election on January 11, 1928, insurgent members of L.U. 1703 at PaCC's No. 6 Colliery deposed him as one of two financial secretaries, along with other incumbent officers including the president, PATSY PAGLIOCA. The other financial secretary at the time was PHILIP JULIANI. See also JOSEPH SHIMOKONIS, CHARLES ALBA.

Morello, Giuseppe (1867–1930)—Also known as "The Grey Fox," he was the brother-in-law of IGNAZIO SAIETTA and co-leader of the Lupo-Morello gang of counterfeiters and murderers active in New York City around the turn of the century. With the assistance of undercover Secret Service agent SALVATORE "SAM" LUCCHINO of Pittston, he was convicted of counterfeiting and sentenced to 30 years in the Atlanta federal penitentiary. See WILLIAM J. FLYNN.

Morgan, Donald P.—Brother-in-law and tax consultant to former Governor JOHN S. FINE, he became a partner in the Newport Excavating Company in Luzerne County with Fine, LAWRENCE BISCONTINI, ALBERT BISCONTINI, and SANTO VOLPE. The company began deep mining and strip mining coal using leases in 1950. He was indicted with Fine and Albert Biscontini for personal and corporate income-tax evasion. All three defendants were acquitted after lengthy trials and arduous jury deliberations. Lawrence Biscontini and Volpe had passed away by the time the indictments were issued.

Morgan, J. Pierpont (1837–1913)—One of the most powerful financiers and monopolists in U.S. history, his business empire extended to banking, railroads, telephones, electronics, steel, manufacturing, and other sectors. From his base in New York City, he controlled the anthracite industry during the early twentieth century through his hold on the coal-carrying railroads, which, in turn, owned or controlled the major coal companies. He was also a philanthropist, yachtsman, and one of the most influential members of the Episcopal Church.

Morgan, Willard—An insurgent from Dickson City, he was elected as vice president of the UAMP in August 1933. He was one of 35 UAMP members charged with dual unionism by JOHN BOYLAN and the District 1 executive board and expelled from the UMWA for 15 years. See THOMAS J. MALONEY Sr., HENRY SCHUSTER.

Morris, A.K.—He served as vice president of the Erie Coal Companies in the 1920s and 1930s and participated in all of the period's labor-management negotiations.

Morris, Sylvester—In a move to retake L.U. 1703 at PaCC's No. 6 Colliery, insurgent members called an unofficial meeting on January 11, 1920, and deposed Morris as one of two trustees, the other one being MATTHEW HANAHOE. See PATSY PAGLIOCA.

Moskaitis, Stanley—Insurgent union members elected him as one of four trustees of L.U. 1495 at PaCC's No. 9 Colliery in 1928, the others being GERALD HOBBS, EDWARD RICHARDS, FRANK WEITZ.

Muir, Thomas—The foreman at PaCC's No. 14 Colliery, he was accused of wounding an Italian worker during the wildcat strike of 1910.

Mullarkey, Joseph—Insurgent members elected him as a trustee of L.U. 1703 at PaCC's No. 6 Colliery in 1928.

Mundernar, Michael—He was serving as president of L.U. 1495 at PaCC's No. 9 Colliery when he submitted his resignation because of infighting, but the members convinced him to stay and preside over the election of 14 delegates to the insurgent (rump) convention of August 1933.

Murnin, William—A union activist who participated in the District 1 insurgent (rump) convention of May 21–June 1, 1928, in late 1928 he was one of 45 insurgents expelled from the UMWA by JOHN BOYLAN and the District 1 executive board for dual unionism in making plans to establish the AMWU.

Musto, James A. (1899–1971)—An insurgent mineworker at PaCC's No. 6 Colliery in the 1920s, he was blacklisted and pushed out of the industry at which point he became a barber and grocer. He was elected to the Pennsylvania General Assembly in 1948 and served until his death. He was appointed to the Joint Legislative Committee that investigated the Knox Mine Disaster of January 22, 1959, which conducted hearings into the disaster as well as the subcontracting and leasing systems.

Naples, Mike—A miner and subcontractor at HC&I's Butler Colliery, he was one of over 30 subcontractors at the Erie Coal Companies who resigned from their positions following a major strike in 1920, after which the companies returned the mines (for a time) to the traditional work organization of one miner harvesting coal in a single chamber with one or two laborers.

Nash, C.R.—He was appointed to the Erie Coal Companies' board of directors in 1928 following the resignation of the previous board and the company president, C.S. GOLDSBOROUGH. The shake-up followed the VAN SWERINGEN brothers' appointment of MICHAEL GALLAGHER as board president in 1926 and as company president in 1928. See M.P. CALLAWAY, H.A KNAPP, H.T. PETERS, D.E. POMEROY, A.W. PUTNAM, G.T. SLADE.

Neill, Charles F.—As the ABC umpire, he delivered a blow to the insurgent strike at the No. 6 Colliery in March 1928 when he ruled in favor of the Hudson Coal Company's right to issue subcontracts for machine mining. Neill concluded that he had no authority to disallow the agreements without violating the 1916 labor-management agreement.

Nelson, Clarence—A mineworker at PaCC's Ewen Colliery, he was elected by members as one of seven delegates from L.U. 1487 to attend the insurgent (rump) convention of August 1933.

Nelson, Steve—The CPUSA assigned him to organize anthracite mineworkers, especially the unemployed, during the 1930s.

Nicholls, Thomas D.—A miner born in Wilkes-Barre, he was elected president of UMWA District 1 in 1899 and continued holding the office until 1909, when he resigned for health-related reasons.

Ocheski, Stanley—As president of L.U. 1703 at PaCC's No. 6 Colliery, he assisted in directing a gathering of 4,000 striking insurgents at the Pittston Armory in March 1932. Delegates elected him as one of three UAMP auditors at the District 1 insurgent (rump) convention of August 1933, the other two being JOSEPH SCHUSTER and VINCENT CAREY. See FEENEY BUZZARELLI, LEWIS CASTERLINE, JOHN SHIPLEY.

Offendafer, Officer Jaspar—A policeman during the wildcat strike of 1910 at the Erie Coal Company, he sustained injury when a striker, PIETRO ZURLA, beat him about the head after he had fallen from his horse.

Olyphant, Robert M.—President of the Delaware & Hudson Company between 1884 and 1903, he participated in the Anthracite Strike Commission hearings of 1902–03.

Orlando, Ernest—In April 1928, insurgents deposed him as president of L.U. 1495 at PaCC's No. 9 Colliery and selected JACOB COLLIER as the new president. When Collier confiscated the local's charter, seal, checkbook, and equipment, Orlando took him to court for larceny.

Orlando, Albert—The members of L.U. 1495 at PaCC's No. 9 Colliery elected him as a delegate to the insurgent (rump) District 1 convention of August 1933.

Orr, John—He was elected to the executive committee of the District 1 insurgent group in charge of planning the special (rump) convention of May 21–June 1, 1928.

Paglioca, Patsy—A subcontractor at PaCC's No. 6 Colliery during the 1920s, he was deposed as president of L.U. 1703 on January 28, 1928, in a rump election conducted by insurgents. When the UMWA negated the results and agreed to supervise a second election, he again lost to SAMUEL BONITA. As confirmation of his outcast status, in April 1928 he was barred from attending a special meeting of the local but he was allowed to sit in the rear of the hall with other onlookers.

Pagnotti, Louis Sr. (1894–1966)—In a rags-to-riches story, his parents emigrated from northern Italy to Old Forge, where he was born Luigi Pagnotto. Orphaned at age six, he returned to Italy to live with family and then came back to Old Forge where he was raised by other relatives. He began working as a door tender for the Jermyn Coal Company at age 10 and was promoted to mule driver one year later. Within a few years, he became a laborer. He began subcontracting in the 1920s and, in partnership with his nephew JAMES TEDESCO, formed a coal company and became one of the area's largest independent producers in the 1930s, having secured leases from PaCC and LVCC. He was known to have little tolerance for active unionists and, to bolster productivity and output, Pagnotti Company mines used among the industry's largest coal cars, five tons each. He and his nephew began purchasing collieries in the 1940s and 1950s, especially from LVCC. Operations included the Harry E, Sullivan Trail, and Franklin. Pagnotti Enterprises formed in 1942 and grew into a diversified company that continues to mine coal in the middle field. He served as burgess (mayor) of Old Forge between 1929 and 1933. Although not listed by state or federal investigators as a member of the region's organized-crime group, the FBI and the Pennsylvania Crime Commission concluded that he had business dealings with crime-family members. He was also alleged to have engaged in alcohol bootlegging in the 1920s. He was an avid sportsman and boxing promoter and well known throughout the northern field.

Paluski, John—Insurgents elected him as vice president of L.U. 265 at HC&I's Butler Colliery in late February 1928 in an election that brought President BARNEY GUZIOR to office. See JOHN GALICK, ALBERT STRUCKE, JOHN TELLO.

Panzitta, David—He worked for his father's subcontracting and lease-holding firm, the Panzitta Coal Company, in the 1940s and 1950s. He recognized the corruption problem within the anthracite tenancy systems.

Papcun, George—A member of the CPUSA, he served as secretary of the Save-the-Union Committee in the Anthracite Tri-District area. In late 1928, he was one of 45 insurgents expelled from the UMWA by JOHN BOYLAN and the District 1 executive board for dual unionism in making plans to establish the AMWU. Due to his communist ties, he was banned from, and escorted out of, the District 1 Convention of May 21–June 1, 1928. See STANLEY DZIENGELEWSKI.

Parish, Daniel—He was a member of the PaCC board of directors between 1851 and 1853.

Pazello, Americo—A mineworker at PaCC's Ewen Colliery, he was elected as one of seven delegates from L.U. 1487 to attend the insurgent (rump) convention of August 1933.

Peters, H.T.—He was appointed to the Erie Coal Companies' board of directors in 1928 following the resignation of the previous board and the company president, C.S. GOLDSBOROUGH.

The shake-up followed the VAN SWERINGEN brothers' appointment of MICHAEL GALLAGHER as board president in 1926 and as company president in 1928. See M.P. CALLAWAY, D.W. COOKE, H.A KNAPP, C.R. NASH, D.E. POMEROY, A.W. PUTNAM, G.T. SLADE.

Peterson, Victor L.—He served as general superintendent of coal operations at HC&I during the early twentieth century.

Petrini, Bart—Delegates elected the Old Forge resident as one of four UAMP board members at the District 1 insurgent (rump) convention of August 1933.

Petrowski, Andrew—A member of L.U. 699 from Edwardsville, in late 1928 he was one of 45 insurgents expelled from the UMWA by JOHN BOYLAN and the District 1 executive board for dual unionism in making plans to establish the AMWU.

Phillips, Sheriff Benjamin S.—He directed the Lackawanna County police force that joined with State Police Lieutenant PRICE and his officers in a raid on an IWW meeting in Old Forge in 1916. They arrested 267 Jermyn Coal Company employees who, according to authorities, had illegally assembled. See Atty. JOHN R. EDWARDS, Atty. FRANK BENJAMIN.

Piasecki, Charles—President of the Ewen Colliery's L.U. 8005 in the 1950s and a mineworker for the Knox Coal Company, he pled guilty of income tax evasion, conspiracy, accepting bribes from management to keep "labor peace" during investigations following the Knox Mine Disaster of January 22, 1959. He was given a two-year prison sentence.

Piccillo, Joseph—As a young boy, he witnessed the ALEX CAMPBELL and PETER REILLY shootings on February 28, 1928, which took place across the street from his home on Railroad Street in Pittston.

Pierson, Charles T.—He was president of PaCC from 1838 to 1848.

Pinchot, Gov. Gifford (1865–1946)—A Progressive Republican who served as Secretary of the Department of Interior under President THEODORE ROOSEVELT, he was elected governor of Pennsylvania in 1924 and again in 1930.

Platt, Isaac L.—He was a member of the PaCC board of directors between 1851 and 1853.

Policca, Tony—A participant in the IWW-sponsored strike in1916 at HC&I Butler Colliery, he was sent to the hospital with lacerations to the head caused by a policeman's billy club.

Polizzi, Angelo—He was an alleged criminal associate of SANTO VOLPE.

Pollard, Richard—He was transferred as foreman of the Fernwood Slope of HC&I's Butler Colliery to the same position at the No. 7 Shaft of PaCC's Ewen Colliery in mid-April 1928, during the intense conflict at the No. 6 and other PaCC and HC&I collieries.

Pomeroy, D.E.—He was appointed to the Erie Coal Companies' board of directors in 1928 following the resignation of the previous board and the company president, C.S. GOLDSBOROUGH. The shake-up followed the VAN SWERINGEN brothers' appointment of MICHAEL GALLAGHER as board president in 1926 and as company president in 1928. See M.P. CALLAWAY, D.E. COOKE, H.A KNAPP, C.R. NASH, H.T. PETERS, A.W. PUTNAM, G.T. SLADE.

Pope Leo XIII (1810–1903)—Born as Vincenzo Gioacchino Raffaele Luigi Pecci, he issued an encyclical (a teaching "letter") termed *Rerum Novarum* (Of New Things) on May 15, 1891. The full title of the document included "Rights and Duties of Capital and Labour," with subtitle of "On the Conditions of Labour." A path-breaking treatise, the encyclical focused on the worker in an industrial society and emphasized economic fairness in labor-management relations as well as the importance of labor unions in securing workers' rights.

Powderly, Terrence V. (1849–1924)—He served as Grand Master of the Knights of Labor from 1879 to 1893, and as mayor of Scranton from 1878 to 1884. He spoke about the *Padrone System* before the Immigration Investigation Commission in 1895 and acknowledged its existence in the mining regions. He served as U.S. Commissioner General of Immigration from 1897 to 1902 and as the Chief Information Officer for the U.S. Bureau of Immigration from 1907 to 1921.

Prashner, Albert B.—He was one of three IWW members from Scranton detained during a series of national strikes in 1917, which resulted in the detention of 95 Wobblies by the federal government. The two other local IWW members held were STANLEY DEMBICKI and SALVATORE ZUMPANO.

Price, Lieutenant (first name unknown)—He led a State Police force that joined with Lackawanna County Sheriff BENJAMIN S. PHILLIPS and his deputies in a raid on an IWW meeting in Old Forge in 1916. They arrested 267 Jermyn Coal Company employees who, according to authorities, had illegally assembled.

Puma, Joe—A young Pittstonian who worked as a nipper at PaCC's No. 6 Shaft, No. 6 Colliery, in 1911, he was the subject of a photograph by LEWIS WICKES HINE.

Puntario, Antonio—A hired assassin of Pittston detective SALVATORE "SAM" LUCCHINO, the Trenton, NJ, resident was convicted of first degree homicide and sentenced to death. PETER ERICO was also convicted in the murder. The judgment was carried out at Rockview State Penitentiary in Centre County, PA, on September 25, 1922, when Puntario became the first person to die by electrocution for a crime committed in Luzerne County.

Putnam, A.W.—He was appointed to the Erie Coal Companies' board of directors in 1928 following the resignation of the previous board and the company president, C.S. GOLDSBOROUGH. The shake-up followed the VAN SWERINGEN brothers' appointment of MICHAEL GALLAGHER as board president in 1926 and as company president in 1928. See M.P. CALLAWAY, D.E. COOKE, H.A KNAPP, C.R. NASH, D.E. POMEROY, H.T. PETERS, G.T. SLADE.

Reath, N.B.—He was a member of the PaCC board of directors in 1910.

Reilly, Peter (1900–1928)—A 28-year-old Lithuanian-born mineworker whose birth surname was Saudargis, he was murdered with fellow insurgent ALEX CAMPBELL on February 28, 1928, while driving in his car on Railroad St., Pittston. The assassins used 12-gauge Remington automatic shotguns to strike the victims with some 18 rounds each. Following a funeral service attended by thousands, he was buried in St. Casimir's Lithuanian Church cemetery in Pittston. Police arrested and charged RALPH MELISSARI of New Jersey and PETER DELUCA and VINCENZO DAMIANO of New York City with the homicides. A fourth person listed as "John Doe" was also indicted but the charges were dropped; he was described as "an Italian, about 35 years old." Acquitted of Campbell's murder but convicted of Reilly's, Melissari was sentenced to life in prison. DeLucca was also convicted and given a life sentence. Hired gunmen murdered Damiano before his trial began. Shortly before his death, insurgents had elected Reilly as recording secretary of L.U. 1703 at PaCC's No. 6 Colliery. He was first elected as secretary of the local following the strike of 1920 and he served in that capacity until 1925 when he lost to a candidate opposed to the insurgents.

Renfer, Arthur—Members of L.U. 1495 at PaCC's No. 9 Colliery elected him as one of three recording secretaries in 1928, the other two being PETER FLYNN and FRANK TROSS.

Ricci, Anthony—He attended national Save-the-Union Committee meeting in Pittsburgh in September 1928. He was also a supporter of the NMU, which failed to organize anthracite mineworkers in the late 1920s.

Richards, Edward—Insurgent union members elected him as one of four trustees of L.U. 1495 at PaCC's No. 9 Colliery in 1928, the other three being STANLEY MOSKAITIS, GERALD HOBBS, and FRANK WEITZ.

Richards, Henry—He was one of three heading men who worked as subcontractors for the Hudson Coal Company in Scranton around the turn of the century, the other two being MICHAEL McHALE and HARRY GILBERT. They were denied empty coal cars and their tools were destroyed by militant mineworkers when they defied the local union's order to surrender their subcontracts and uphold the production stint.

Richardson, G.A.—He was a member of the PaCC board of directors in 1910.

Ridgeley, Thomas—He was transferred as superintendent of PaCC's No. 9 Colliery to the same position at the Central Colliery in mid-April 1928, during the intense conflict at the No. 6 and other PaCC collieries.

Rinker, Rev. Richard A.—He was the pastor of the First Presbyterian Church, Pittston, who directed the service at ALEX CAMPBELL's funeral in March 1928.

Rinn, John—He worked as a UMWA organizer in the southern field in the 1890s.

Roberts, Rev. Peter (1859–1932)—A Congregationalist minister and sociologist, he wrote two books on the anthracite industry and region in the early twentieth century (see References), and testified before the Anthracite Strike Commission of 1902. He tended to portray Italians, Slavs, and other Eastern and Southern Europeans as racially and culturally inferior, but he acknowledged their vital role in promoting unionism in the late 1890s and early twentieth century.

Robertson, Margaret Campbell—The daughter of murder victim ALEX CAMPBELL, she was interviewed by the press immediately following his assassination. In oral history conversation, she stated her belief that SANTO VOLPE ordered the death of her father and PETER REILLY on February 28, 1928, because of their insurgent activities at PaCC's No 6 Colliery.

Rodda, Sheriff Frederick—As Luzerne County sheriff, he led deputies against 300 Erie Coal Company strikers at PaCC's Central Colliery in Duryea during the wildcat strike of 1910.

Roosevelt, President Franklin D. (1882–1945)—The thirty-second president of the U.S., he was elected to four terms in office (1933–1945). He died early in his fourth term and was succeeded by Vice President Harry Truman. He termed his domestic policies The New Deal, which included major reforms and programs intended to address the effects of the Great Depression that commenced in 1929, when Herbert Hoover served as president. During Roosevelt's second term, UAMP president THOMAS J. MALONEY Sr. and insurgent advisor Msgr. JOHN J. CURRAN traveled to the nation's capital seeking the assistance of the president and his administration in settling the labor wars in the northern anthracite field. They argued that the recently passed National Industrial Recovery Act (1933), a centerpiece of the New Deal, guaranteed the right of workers to select any union they wanted. They also spoke to Senator ROBERT F. WAGNER, author of the National Labor Relations Act (1935), who agreed to conduct an investigation. Overall, the insurgents did not secure the type of aid and treatment from Washington D.C. that they had hoped for, probably because of the influence of JOHN L. LEWIS, who was a presidential appointment to the National Labor Board.

Roosevelt, President Theodore (1858–1919)—The twenty-sixth president of the U.S., he served in office from 1901 to 1909. He intervened in the anthracite strike of 1902 and appointed a seven-member Anthracite Strike Commission to issue a final settlement. He became a personal friend of Rev. JOHN J. CURRAN of Wilkes-Barre, who assisted in the strike negotiations. He visited northeastern Pennsylvania on a number of occasions following the strike, often in the company of Curran.

Ross, Angelo—A breaker boy at PaCC's No. 14 Colliery, he was the subject of a 1911 photograph by LEWIS WICKES HINE.

Ruck, Rob—Was involved with the CPUSA, which had about 50 members across the four anthracite fields at the end of the 1920s.

Rudis, John—Insurgents elected him as the sentinel of L.U. 1495 at PaCC's No. 9 Colliery in 1928.

Sacco, Ferdinando Nicola (1891–1927)—He was an Italian immigrant anarchist who, with BARTOLOMEO VANZETTI, was convicted of murdering two men in 1920 during a South Braintree, Massachusetts, robbery. Following two controversial nationally publicized trials followed by two appeals to the State Supreme Judicial Court, the pair were executed on August 23, 1927. Mineworkers in UMWA District 1 established a Sacco-Vanzetti Committee in 1927, which included representatives from 40 UMWA local unions as well as several Italian-American organizations. The media presentation of the case fostered a radical and subversive stereotype of Italians and other immigrants. The debate over their guilt or innocence has continued through the years. Although he did not issue a pardon, Governor Michael Dukakis officially declared August 23, 1977, as Nicola Sacco and Bartolomeo Vanzetti Memorial Day, on the 50th anniversary of their executions.

Saietta, Ignazio (1877–1947)—Known as "Lupo the Wolf," he was the brother-in-law of GIUSEPPE MORELLO and co-leader of the Lupo-Morello gang of counterfeiters and murderers active in New York City during the first decade of the twentieth century. With the assistance of undercover Secret Service agent SALVATORE "SAM" LUCCHINO of Pittston, Saietta and other gang members were convicted of counterfeiting and sentenced to 30 years in the Atlanta federal penitentiary. See WILLIAM J. FLYNN.

Saietta, Sam—A mineworker at PaCC's No. 6 Colliery, he was elected as a delegate from L.U. 1703 to the insurgent (rump) District 1 convention of August 1933.

Sapolis, Thomas—The members of L.U. 1495 at PaCC's No. 9 Colliery elected him as a delegate to the insurgent (rump) District 1 convention of August 1933.

Saporito, Carlo—A mineworker from Pittston, he was a partner in the PaCC lessee firm, the Saporito Coal Company, formed March 21, 1939, along with JOHN SCIANDRA and STEFANO LaTORRE. In 1943, he and LaTorre were allegedly forced out of a lease that PaCC turned over to the newly established Knox Coal Company and its owners, Sciandra and LOUIS FABRIZIO.

Savage, Joseph—On January 25, 1928, insurgents chose him as one of two financial secretaries of L.U. 1703 at PaCC's No. 6 Colliery, in a second election that again deposed the incumbent officers. JOE COSTA was the other financial secretary. See SAMUEL BONITA, PATSY PAGLIOCA.

Saxton, William—A District 1 mineworker, delegates elected him as one of four UAMP state board members at the insurgent (rump) convention of August 7, 1933, the other three being WILLIAM AUSTIN, JOSEPH GALLIG, and STANLEY GORGAS.

Scalzo, Vincent—A mineworker at PaCC's Ewen Colliery and an insurgent activist, he was elected temporary secretary of L.U. 1487 by insurgents in 1933.

Scovich, Stanley—A member of L.U. 699 in Edwardsville, in late 1928 he was one of 45 insurgents expelled from the UMWA by JOHN BOYLAN and the District 1 executive board for dual unionism in making plans to establish the AMWU.

Schuster, Henry—A mineworker from Scranton, he was elected as secretary-treasurer of the UAMP in August 1933. He was one of 35 UAMP members charged with dual unionism by JOHN BOYLAN and the District 1 executive board and expelled from the UMWA for 15 years. See THOMAS J. MALONEY Sr., WILLARD MORGAN.

Schuster, Joseph—Delegates elected him as one of three UAMP auditors at the District 1 insurgent (rump) convention of August 1933, the other two being STANLEY OCHESKI and VINCENT CAREY.

Sciandra, John (1899–1949)—Born in Montedoro, Sicily, he emigrated to the U.S. in 1908, he became a mineworker at PaCC. On March 21, 1939, he established the Saporito Coal Company with partners CARLO SAPORITO and STEFANO LaTORRE. The firm secured seven leases from TPC and PaCC between March 21, 1939, and January 27, 1943. He formed the Knox Coal Company in 1943 with partner LOUIS FABRIZIO and they secured sixteen PaCC leases and lease adjustments between 1943 and 1958. One lease obtained in 1943 allowed for mining in sections of the Pittston and Marcy Veins at the Schooley Shaft. In 1954, they secured a lease to take virgin coal in the Pittston and Marcy Veins at PaCC's Ewen Colliery. The called the mine the River Slope because the entrance was adjacent to the Susquehanna River. The mine became the site of the infamous Knox Mine Disaster of January 22, 1959. Sciandra allegedly followed SANTO VOLPE as the region's organized-crime boss in 1933 and remained in power until his death from natural causes.

Sciandra, Josephine (Mrs. John)—The wife of JOHN SCIANDRA, she was a co-owner of the Knox Coal Company following her husband's death in 1949, and, as such, she was indicted with ROBERT L. DOUGHERTY, LOUIS FABRIZIO, and AUGUST J. LIPPI and pled guilty to personal and corporate income-tax evasion in the investigations that followed the Knox Mine Disaster of January 22, 1959. Unlike her colleagues, she received a suspended sentence, probably because she served as a "front" for the company's ownership by other members of her family.

Scovitch, Stanley—A member of L.U. 699 from Edwardsville, in late 1928 he was one of 45 insurgents expelled from the UMWA in 1928 by JOHN BOYLAN and the District 1 executive board for dual unionism in making plans to establish the AMWU.

Seit, Sam—He was one of four L.U. 1703 grievance-committee members at PaCC's No. 6 Colliery who were deposed by insurgents on January 11, 1928, in a rump election. The others were E. AQUILINA, JAMES ANTONELLO, and SAM LATORE.

Serafini, Anthony—A mineworker at LVCC's Exeter Colliery, insurgents elected the Old Forge resident as elected temporary president of L.U. 1084. He was also elected as a delegate to the insurgent (rump) convention of August 1933, where he was chosen as one of four UAMP board members.

Shiffko, Frank—He was present in the room during the February 16, 1928, shooting death of UMWA District 1 official FRANK AGATI, and was nearly struck by bullets. See SAMUEL BONITA, ANTHONY FIGLOCK, STEVEN MENDOLA, ADAM MOLESKI.

Shimokonis, Joseph—In a move to retake L.U. 1703 at PaCC's No. 6 Colliery, insurgents called an unofficial meeting on January 11, 1928, and deposed him as vice president, along with the other incumbent officers, including President PATSY PAGLIOCA, and the other vice president, CHARLES ALBA. See JOSEPH MOLESKI, PHILIP JULIANI.

Shipley, John—In cooperation with LEWIS CASTERLINE of L.U. 1581 at PaCC's No. 14 Colliery, he assisted in directing a gathering of 4,000 striking insurgents at the Pittston Armory in March 1932. See FEENEY BUZZARELLI, STANLEY OCHESKI.

Shoemaker, T.S.—He served as the chief mining engineer for the Penn Anthracite Collieries Company of Scranton during the 1930s as the firm expanded its subcontracting and leasing systems.

Shovlin, Joseph—Delegates elected him as one of six members of the UAMP Board of Directors and also as a member of the Resolutions Committee, at the District 1 insurgent (rump) convention of August 1933.

Sinclair, William (1911–1959)—Born in Bellshill, Bothwell, Scotland, his mother, Mary H. Todd Sinclair, was a first cousin of Knox Coal Company Superintendent ROBERT GROVES. Sinclair worked in the coal mines at Shotts, Scotland, before emigrating to Pittston in 1927, where he joined relatives. He worked as a miner and laborer for PaCC into the 1940s and then as a laborer for the Knox Coal Company at the River Slope mine. He was one of 12 victims of the Knox Mine Disaster of January 22, 1959.

Siracuse, Angelo J. (1912–1995)—An American-born mineworker whose parents emigrated from the sulfur-mining town of Cianciana, Sicily, he worked as a pump runner at the Harry E Colliery in Swoyersville, where his father and brothers were also employed. He participated in the political realignment of Swoyersville in 1932–33, when the town's predominantly Slavic and Italian voters rebelled against political boss LORENZO FERRARO and the Republicans who had controlled politics, and elected a Democratic majority to the borough council. He was the first author's maternal uncle. See CHARLES SIRACUSE, CALOGERO "CARL" SIRACUSE, FRANK SIRACUSE, JOHN SIRACUSE.

Siracuse, Charles (1903–1933)—Born in Cianciana, Sicily, he emigrated to the U.S. with his mother and siblings in 1907. He became a miner at the Harry E Colliery in Swoyersville and rose to the presidency of the Colliery's L.U. 452 in the late 1920s, where he served until his premature death from cancer. He supported RINALDO F. CAPPELLINI's rise to the District 1 presidency in the early 1920s and backed the insurgent movement during the second half of the 1920s and into the early 1930s. In 1932, he was elected as a member of the borough's first Democratic Party majority as part of the borough's political realignment away from political boss LORENZO FERRARO and the previously dominant Republicans. He was the first author's maternal uncle. See ANGELO SIRACUSE, CALOGERO "CARL" SIRACUSE, FRANK SIRACUSE, JOHN SIRACUSE.

Siracuse, Calogero (1875–1933)—A sulfur miner from Cianciana, Sicily, he emigrated to the U.S. in 1904 and spent his working life as a coal miner at the Harry E Colliery in Swoyersville. He was the first author's maternal grandfather. See ANGELO SIRACUSE, CHARLES SIRACUSE, FRANK SIRACUSE, JOHN SIRACUSE.

Siracuse, Frank (1899–1976)—A Sicilian immigrant, he became a mineworkers at the Harry E Colliery and other companies. He was the first author's maternal uncle. See ANGELO SIRACUSE, CHARLES SIRACUSE, CALOGERO "CARL" SIRACUSE, JOHN SIRACUSE.

Siracuse, John (1905–1987)—A mineworker at the Harry E Colliery in Swoyersville for a brief period, he became the co-owner of a coal-trucking business with his brothers. He was the first author's maternal uncle. See ANGELO SIRACUSE, CHARLES SIRACUSE, CALOGERO "CARL" SIRACUSE, FRANK SIRACUSE.

Slade, G.T.—He was appointed to the Erie Coal Companies' board of directors in 1928 following the resignation of the previous board and the company president, C.S. GOLDSBOROUGH. The shake-up followed the VAN SWERINGEN brothers' appointment of MICHAEL GALLAGHER as board president in 1926 and as company president in 1928. See D.W. COOKE, M.P. CALLAWAY, H.A KNAPP, C.R. NASH, H.T. PETERS, D.E. POMEROY, A.W. PUTNAM.

Slattery, Attorney Frank A.—As Luzerne County district attorney during the IWW-sponsored wildcat strike at the Erie Coal Companies and the Hudson Coal Company in 1916, he joined with Sheriff GEORGE BUSS in starting deportation proceedings against numerous Wobblies.

Smiles, Norman—He was transferred as foreman at the Sax Drift of PaCC's No. 14 Colliery to the same position at the No. 10 Shaft of the No. 9 Colliery in mid-April 1928, during the intense conflict at the No. 6 and other PaCC collieries.

Smith, George—The members of L.U. 1495 at PaCC's No. 9 Colliery elected him as a delegate to the insurgent (rump) District 1 convention of August 1933.

Smith, John B.—He was the general superintendent of PaCC's Gravity Railroad between 1850 and 1886. In 1886, he was elected president of PaCC's Erie & Wyoming Valley Railroad Company.

Sobeck, Frank—His family's home was one of many wrecked by dynamite in Pittston during the labor wars of 1928.

Sobers, Frank—A union activist in L.U. 699 in Edwardsville, he served as an inspection board member of the fourth inspection district and as an executive committee member of the group that organized the first special (rump) convention of District 1 on May 21–June 1, 1928.

Sobey, John—During labor unrest at D&H mines in 1901, union-supporting saboteurs wrecked his home with dynamite when he refused to comply with the established production stint of two cars per man per shift.

Spachia, Samuel—A miner at PaCC who served as a member of the General Grievance Committee and vice president of the Ewen Colliery's L.U. 1487, he was killed by assassins on January 6, 1925, in Pittston, at age 31. The murderers were never found.

Spalding, Bishop John Lancaster (1840–1916)—The Roman Catholic bishop of Peoria, Illinois, he was appointed by President THEODORE ROOSEVELT as one of seven members of the Anthracite Strike Commission in 1902.

Spernatz, Andrew—Insurgents deposed him as treasurer of L.U. 1703 at PaCC's No. 6 Colliery on January 11, 1928, as they moved to take control of the local. See PATSY PAGIOCA.

Sperrazza, Giuseppe—A 53 year-old mining subcontractor from Pittston Township, he was found dead in his automobile in a remote section of a neighboring town on May 26, 1934. He had been shot nine times by unknown assailants.

Spudis, Joseph—A mineworker at PaCC's No. 6 Colliery, he was elected as a delegate from L.U. 1703 to the insurgent (rump) District 1 convention of August 1933.

Statkiewicz, Leonard—A District 1 official and business partner of AUGUST J. LIPPI, he was indicted and acquitted with Lippi of corporate tax evasion and payroll conspiracy in the investigations following the Knox Mine Disaster of January 22, 1959.

Steele, Charles—He was a member of the PaCC board of directors in 1910.

Stella, Pacifico (Joe) (1923–2007)—A resident of Pittston Township, he worked underground for PaCC prior to serving in World War II. After the war, he secured a position as a surveyor for the company. He surveyed the Knox Coal Company's River Slope mine some months prior to the Knox Mine Disaster of January 22, 1959, and found illegal mining, which he reported to superiors, who did not act to curtail the mining. He was in the River Slope mine on the day of the disaster and, using survey maps, he led six workers to the Eagle Air Shaft through which a total of 33 men were rescued. He often spoke about tenancy's harmful effects on mine safety.

Stout, John—He was a member of the PaCC board of directors between 1851 and 1853.

Strucke, Albert—Fellow insurgents elected him as treasurer of L.U. 265 at HC&I's Butler Colliery in February 1928, when the incumbent officers, including President JOHN GALICK, were deposed. See BARNEY GUZIOR, JOHN PALUSKI, JOHN TELLO.

Stuart, J.C.—He was a member of the PaCC board of directors in 1910.

Suender, E.H.—As president of the Madeira Hill Coal Company, he was one of six coal company representatives during the labor-management contract negotiations of 1930.

Taylor, Moses—A coal and railroad baron, he served as a member of the PaCC board of directors between 1851 and 1853.

Taylor, Russell J.—Chief of police in Pittston, he pledged to protect all workers, including those that wanted to cross picket lines, during the insurgent strike of March 1932.

Tedesco, James (1910–2008)—He was the nephew and business partner of LOUIS PAGNOTTI Sr. in coal mining and other operations. They began with leases from PaCC and LVCC and grew to become one of the area's largest coal producers. Tedesco was a founding partner in

Pagnotti Enterprises, which formed in 1942. He served as a member of Anthracite's Committee of Twelve in the early 1960s, which was an appointed body consisting of labor and management representatives. The FBI and the Pennsylvania Crime Commission associated both Pagnotti and Tedesco with JOSEPH BARBARA Sr., the alleged head of the region's organized-crime family between 1950 and 1959. Tedesco was convicted, with SANTO VOLPE and others, of price fixing in 1944, and again with other persons and companies in 1977. He served as burgess (mayor) of Old Forge from 1937 to 1945, and was president of the Old Forge Bank for many years.

Tello, Michael—Fellow insurgents elected him as secretary of L.U. 265 at HC&I's Butler Colliery in February 1928, when the incumbent officers, including President JOHN GALICK, were deposed. See BARNEY GUZIOR, JOHN PALUSKI, ALBERT STRUCKE.

Thomas, Eben B.—He was the chairman of the PaCC board of directors and president of the Erie Railroad between 1894 and 1901.

Thomas, Thomas R.—A foreman at the D&H, he testified before the Anthracite Coal Strike Commission in 1902 regarding his aversion toward the UMWA, stating that it had ruined discipline and undermined his control over the workers.

Tiscar, Chevalier Fortunato (1857–1945)—He emigrated from Catania, Sicily, in 1891 and served as an Honorary Italian Consul from 1896 to 1929, with his home and office in Scranton. A distinguished citizen of the city, he was made a Knight of the Order of the Crown of Italy in 1900, and at the international exposition at Milan in 1906, he was awarded a grand silver medal and a diploma of honor. Also in 1906, he was appointed a Chevalier of the Order of the Crown of Italy by King Victor Emanuel. His responsibilities as a diplomatic envoy covered 20 counties in northeastern PA. Because of the large number of Italian (particularly Sicilian) immigrants involved, he figured prominently in negotiating the wildcat strike of 1910 at the Erie Coal Companies. He also became involved in legal proceedings related to Italian immigrants during the IWW-sponsored strike in 1916. His residence was severely damaged by dynamite on November 11, 1931, allegedly by anti-Fascist elements who protested the visit of the foreign minister of Italy to the U.S. Although asleep at the time of the blast, he and his wife escaped injury. He died at age 87 in an accidental fall and was buried in St. Catherine's Cemetery in Moscow, Lackawanna County.

Tomchak, Frank—A union activist and member of the UAMP from Plymouth, PA, he was an avid supporter of THOMAS J. MALONEY Sr. and the insurgent movement during the early 1930s. He served as a delegate of L.U. 466 at the Glen Alden Coal Company's Stanton Colliery in Wilkes-Barre, which was also Maloney's home local.

Torchia, Atty. Samuel H.—He served as legal counsel to the District 1 insurgent (rump) convention of May 21–June 1, 1928.

Torma, John—A mineworker from Nanticoke, delegates elected him as one of four UAMP board members at the District 1 insurgent (rump) convention of August 1933. Because of his ardent dual unionism and generally militant disposition, in 1933 the UMWA banished him from the organization for 99 years.

Tracy, Judge William—A member of the Pennsylvania Bureau of Mediation and Arbitration, he was invited to initiate a series of meetings between the Erie Coal Companies' president, WILLIAM A. MAY, Pittston MAYOR WALSH, workers' representatives, and local businessmen in an effort to end the wildcat strike of 1920 at the Erie Coal Companies. He failed to settle the dispute.

Tresca, Carlo (1879–1943)—An Italian-born anarchist and IWW organizer, he was active in organizing workers in various industries, including bituminous miners in western Pennsylvania. See JOSEPH ETTOR, LUIGI GALLEANI.

Trettis, Vincent—A member of L.U. 699 from Edwardsville, in late 1928 he was one of 45 insurgents expelled from the UMWA by JOHN BOYLAN and the District 1 executive board for dual unionism in making plans to establish the AMWU.

Tross, Frank—Insurgent members of L.U. 1495 at PaCC's No. 9 Colliery elected him as one of three recording secretaries in 1928, the other two being PETER FLYNN and ARTHUR RENFER.

Truesdale William. H. (1851–1935)—He became president of the DL&W Railroad in 1899 and helped form the coal division's arguments during the hearings of the Anthracite Strike Commission in 1902–03.

Try, Anthony—The members of L.U. 1495 at PaCC's No. 9 Colliery elected him as a delegate to the insurgent (rump) District 1 convention of August 1933.

Try, James—The members of L.U. 1495 at PaCC's No. 9 Colliery elected him as a delegate to the insurgent (rump) District 1 convention of August 1933.

Tutsellino, Billy—Known as "Billy the Bird," he was allegedly SANTO VOLPE's strong man and "enforcer," charged with roaming the mines to insure discipline and take care of "trouble makers."

Uncavage, George—A member of L.U. 699 from Edwardsville, in late 1928 he was one of 45 insurgents expelled from the UMWA by JOHN BOYLAN and the District 1 executive board for dual unionism in making plans to establish the AMWU.

Underwood, Frederick D.—The first president of the Erie Coal Companies, he held the office from 1901 to 1926. He was also a member of the board of directors. He was dismissed by the new owners of the Erie Railroad, the VAN SWERINGEN brothers, and replaced by MICHAEL GALLAGHER.

Valentine, Judge W.A.—A Luzerne County judge who sent UAMP President THOMAS J. MALONEY Sr., Vice president WILLARD MORGAN, Secretary-treasurer HENRY SCHUSTER, and 26 other UAMP members to jail when they defied a court order to end a strike in March 1935. A year earlier, in late March 1934, a bomb shattered a car owned by Valentine but driven by his daughter, Mary. She escaped harm by exiting the vehicle minutes before the device exploded; however, a passing newsboy sustained injuries. Wilkes-Barre printer and insurgent sympathizer Emerson Jennings along with an accomplice named Charles Harris were convicted of the bombing, which occurred during a UAMP strike at the Glen Alden Coal Company collieries in 1935.

Van Sweringen Brothers, Oris Paxton (1879–1936) and Mantis James (1881–1935)—As railroad entrepreneurs whose empire encompassed ten percent of the nation's rail industry at its peak in the 1920s, the Cleveland, Ohio, natives gained control of the Erie Railroad, including subsidiaries PaCC and HC&I, in 1924. In 1926 they hired MICHAEL GALLAGHER to oversee coal operations. They had decided to move their anthracite subsidiaries beyond subcontracting into the leasing business. In 1929 they established a new subsidiary within PaCC called The Pittston Company whose main purpose was to issue coal leases in the northern anthracite field. The brothers' company purchases were highly leveraged with the support of the Morgan bank in New York City and also banks in Cleveland, and their empire collapsed with the stock market crash of 1929. See C.S. GOLDSBOROUGH.

Vanzetti, Bartolomeo (1888–1927)—An Italian immigrant anarchist who, with FERDINANDO NICOLA SACCO, was convicted of murdering two men in 1920 during a South Braintree, Massachusetts, robbery. Following two controversial, nationally publicized trials and two appeals to the State Supreme Judicial Court, the pair was executed on August 23, 1927. The media presentation of the case fostered a radical and subversive stereotype of Italians and other immigrants. The debate over their guilt or innocence has continued through the years.

Mineworkers in UMWA District 1 established a Sacco-Vanzetti Committee in 1927, which included representatives from 40 UMWA local unions as well as several Italian-American organizations. Although he did not issue a pardon, Governor Michael Dukakis officially declared August 23, 1977, as Nicola Sacco and Bartolomeo Vanzetti Memorial Day, on the 50th anniversary of their executions.

Verdine, Charles—Fellow insurgents of L.U. 1703 at PaCC's No. 6 Colliery elected him as a grievance committee member during the insurgent takeover of the local in January 1928.

Victor, Joseph—Insurgents elected him as vice president of L.U. 1703 at PaCC's No. 6 Colliery in February 1928. He was one of five L.U. 1703 members who were also avid Save-the-Union Committee members during the latter half of the 1920s. See SAMUEL BONITA, JAMES KEARNEY, FRANK LICATA, FRANK McGARRY, PETER REILLY.

Volpe, Euphamia—One of SANTO VOLPE's daughters, she married FRANK AGATI's son, Guy.

Volpe, Charles—A mineworker from Pittston and a distant relative of SANTO VOLPE, he worked for different coal companies, including some owned by LOUIS PAGNOTTI, during the 1940s and 1950s. Lax safety measures became a constant source of concern to him and, after two close brushes with serious injury, he decided to quit mining in the 1950s.

Volpe, Santo (1880–1958)—A sulfur miner born in Montedoro, Sicily, he emigrated to the U.S. in 1904. In 1908, with his brother-in-law STEFANO LaTORRE, he allegedly co-founded the second Sicilian organized-crime "family" in the U.S., in Pittston, as an outgrowth of the area's already active Black Hand gang. He allegedly served as the region's top crime boss from 1908 to 1933, when he was succeeded by JOHN SCIANDRA. He became a subcontractor for PaCC with CHARLES CONSAGRA in 1913, marking the beginning of the Erie Coal Companies' renewed tenancy program. He went on to secure dozens of tenancy agreements with the Erie Coal Companies, including 56 leases between 1934 and 1948. Only LOUIS PAGNOTTI had secured a greater number. He was a partner in the new Volpe Coal Company, incorporated on November 23, 1937, with CHARLES C. BUFFALINO and CHARLES J. BUFFALINO. Among the firm's major leases were the Forest City Colliery and the No. 6 Colliery in 1937, and the Butler Colliery in 1938. When PaCC and TPC conducted an inventory of coal reserves held by 13 leaseholders in 1941, Volpe controlled 14,480,780 tons, which was more than any other tenant and second only to the 25,839,600 tons held by the parent company, PaCC, at the Ewen Colliery. He established other partnerships such as the Volpe Mining Company, the Volpe-Adonizio Company, and the Daley & Volpe Coal Company. By 1940, Volpe's coal interests had become a major independent mining operation. He exerted considerable influence over Erie's subcontracting and leasing systems, as well as District 1 union affairs. Because of his participation in PaCC's subcontracting system, he opposed RINALDO F. CAPPELLINI, ALEX CAMPBELL, and other insurgents in the anti-subcontracting strike of 1920 and other strikes throughout the 1920s and early 1930s. He also opposed the UAMP movement in the early 1930s. He was reputed to have ordered the assassinations of No. 6 Colliery insurgent leaders Alex Campbell and PETER REILLY in 1928, although he was never indicted and the two hired gunman from New York, RALPH MELISSARI and PETER DELUCA, who were convicted in Reilly's death, never revealed who had hired them. In the 1930s, he received a gubernatorial appointment to the Pennsylvania Coal Commission, an agency intended to stabilize the market by allocating production quotas to each company. In the 1950s, he became a business partner with former Pennsylvania Governor JOHN S. FINE and others in the mining operations of the Newport Excavating Company in Luzerne County, a venture for which Fine and the surviving partners were indicted but not convicted for tax evasion (Volpe had passed away by the time the indictments were issued). He enjoyed a friendship with UMWA President JOHN L. LEWIS, who backed him in securing a lease in the middle coal field in the 1950s. Lewis and other dignitaries attended Volpe's viewing upon his death in West Pittston in 1958.

Vrataric, Frank—A mineworker and member of the CPUSA, he supported the Save-the-Union Committee in the Anthracite Tri-District during the latter part of the 1920s. See STANLEY DZIENGELEWSKI, GEORGE PAPCUN.

Wagner, Senator Robert F. (1877–1943)—A U.S. Senator from New York who, with members of the National Labor Board, asked the ABC to investigate the northern coal field strikes of the early 1930s. He had been conducting talks with Msgr. JOHN J. CURRAN regarding the field's labor unrest. Wagner was criticized by Curran, THOMAS J. MALONEY Sr., and the rank-and-file insurgents for not doing enough to address the workers' grievances, especially their right to join whatever union they wished. Many were insurgents who wanted to leave the UMAW and join the UAMP, a prospect that Wagner evidently did not support.

Wallo, Mike—A member of L.U. 699 from Edwardsville, in late 1928 he was one of 45 insurgents expelled from the UMWA by JOHN BOYLAN and the District 1 executive board for dual unionism in making plans to establish the AMWU.

Walsh, Mayor (first name unknown)—As mayor of Pittston, he partook in a series of meetings with Judge WILLIAM TRACY of the Pennsylvania Bureau of Mediation and Arbitration, Erie Coal Companies President WILLIAM A. MAY, workers' representatives, and local businessmen in an effort to settle a general strike at the Erie Coal Companies in 1920, led by ALEX CAMPBELL and RINALDO F. CAPPELLINI.

Walter, W. Alfred—At the behest of J. PIERPONT MORGAN, he was installed as president of the Lehigh Valley Railroad in 1897.

Warren, Major Everett—He represented PaCC and HC&I during the Anthracite Strike Commission hearings of 1902–03.

Warriner, J.B.—As president of the Lehigh Navigation Coal Company (a subsidiary of the Lehigh Coal & Navigation Company), he was one of six coal-company representatives during the labor-management contract negotiations of 1930.

Warriner, Samuel D.—As president of Lehigh Coal & Navigation Company, he was one of six coal-company representatives during the labor-management contract negotiations of 1930.

Wassinci, Enrico—A miner and subcontractor at HC&I's Butler Colliery, he was one of over 30 subcontractors at the Erie Coal Companies who resigned from their positions following a major strike in 1920, after which the companies returned the mines (for a time) to the traditional work organization of one miner harvesting coal in a single chamber with one or two laborers.

Weitz, Frank—Insurgent union members elected him as one of four trustees of L.U. 1495 at PaCC's No. 9 Colliery in 1928, the others being GERALD HOBBS, STANLEY MOSKAITIS, and EDWARD RICHARDS.

White, John B.—He was elected UMWA president and served from 1911 to 1918.

White, Orra R.—He joined Mrs. SALVATORE ("NELLIE") LUCCHINO as a key witness in SALVATORE "SAM" LUCCHINO's murder trial in 1920, which led to the convictions of hired gunmen PETER ERICO and ANTONIO PUNTARIO.

Whitey, John—He was elected as an inspection-board member of the second inspection district of District 1 at the insurgent (rump) convention of May 21–June 1, 1928.

Williams, Enoch—He was elected secretary-treasurer of District 1 in 1921 on a reform platform headed by WILLIAM J. BRENNAN, who was elected president. The pair was among the officers defeated in the 1923 District 1 election by an insurgent ticket headed by RINALDO F. CAPPELLINI as president and GEORGE ISSACS as vice president. Williams was later elected secretary of District 1, a post he held for several years during which time he remained loyal to JOHN L. LEWIS and the established District 1 administration. He often sent private letters to Lewis informing him about various problems, personalities, and issues within the district.

Williams, Jack ("Jock") (1896–1960)—An immigrant from Shotts, Scotland, during the 1920s, he was employed as a miner for PaCC into the 1940s, and then as an assistant foreman for the Knox Coal Company. On the day of the Knox Mine Disaster, January 22, 1959, he was working at the River Slope mine and investigating cracking sounds in a particular chamber when the Susquehanna River broke into the workings. He managed to escape by running up the slope with two other men. He later described the underground scene as akin to "Niagara Falls."

Williams, Richard—He was transferred as foreman at the No. 10 Shaft of PaCC's No. 9 Colliery to the same position at the Sax Drift of the No. 14 Colliery, in mid-April 1928, during the intense conflict at the No. 6 and other Erie collieries.

Wilson, Andrew—An immigrant from Shotts, Scotland, during the 1920s, he began as a miner for PaCC and was promoted to the person in charge of safety for the entire company. He worked at PaCC into the 1940s and then became a state mine inspector where he stayed into the 1960s.

Winter, Paul M.—Born in Sunbury, PA, in August 1895, he had moved with his parents to Wilkes-Barre prior to the U.S. Census of 1910. Before World War I he worked as a surveyor and then an insurance agent. He served in World War I and afterwards joined the American Legion at Wilkes-Barre, representing the post at statewide meetings in the early 1920s. He was listed in public records as a reporter, investigator, and insurance agent in the 1920s and 1930s. By 1920 he had organized a Ku Klux Klan klavern in Queens County, NY, and by 1930, as a resident of Hempstead, NY, he became a leader in the KKK. He maintained ties to Wilkes-Barre while living in New York, and in the early 1940s, he moved to Shavertown in Luzerne County where he headed the Wilkes-Barre Klavern No. 311. He was also the chief actor in the anti-semitic Civil Intelligence Bureau in the 1940s. His colleague and fellow bigot ROBERT E. EDMONDSON probably moved to Luzerne County because of his relationship with Winter.

Wolansky, Fr. John (1857–1926)—The first Ukrainian Greek Catholic priest to serve in the U.S., he was sent by church authorities in Ukraine to minister to immigrant anthracite mineworkers and their families. He arrived in Pennsylvania in 1884, whereupon he formed a parish in Shenandoah, Schuylkill County. He pioneered the establishment of the Ukrainian Greek Catholic Church in the U.S. As the most prominent anthracite-region Ukrainian, he also supported labor unions and headed a Knights of Labor-sponsored assembly at Shenandoah in the late 1880s. He was a distant relative of the first author.

Wolensky, Nicholas (1917–1971)—A laborer at the Maltby Colliery in Swoyersville before World War II, he worked as a breakerman at the Harry E Colliery in Swoyersville from 1946 until the facility closed in the early 1960s. He was the first author's father.

Wolensky, Oney (1885–1958)—A 1912 immigrant from Ukraine, he worked as a miner at the Harry E Colliery in Swoyersville. He was the first author's grandfather.

Wolensky, Paul (1920–1941)—A laborer at the Harry E Colliery in Swoyersville, he died in an explosion in 1941. His supervising miner, PETER LAZAR, was injured but survived. He was the first author's first cousin once removed.

Wynne, John—During the labor uprising at PaCC's No. 6 Colliery and other Erie operations in 1928, he was transferred as superintendent of PaCC's No. 9 Colliery to the same position at the Central Colliery.

Yannis, Joseph—Interim District 1 president from 1918 to 1919, he was elected vice president of L.U. 1703 at PaCC's No. 6 Colliery in 1920, with ALEX CAMPBELL as president. He remained an active insurgent during the 1920s.

Zaldokis, Mike—A union activist in District 1, he attended the national Save-the-Union Committee meeting in Pittsburgh in September 1928, and also supported the NMU, which failed to organize anthracite mineworkers in the late 1920s.

Zarek, Alex—He was an anthracite region member of the CPUSA and union activist during the 1920s.

Zarn, William—A PaCC miner and subcontractor who, in 1901, lost money on a subcontract because UMWA members prevented the driver boys from delivering empty coal cars to his workplace, after which he was forced to surrender the agreement.

Zeto, William—A watchman at PaCC's Ewen Colliery, he was accosted and shot by strikers allegedly because he remained on the job during the wildcat strike of 1910. He survived the assault.

Zumpano, Salvatore—A mineworker from Scranton, he was one of three IWW members detained during a series of strikes in 1917, which resulted in the detention of 95 Wobblies nationwide by the federal government. Two other northeastern Pennsylvania IWW members were also held: STANLEY DEMBICKI and ALBERT B. PRASHNER.

Zurla, Pietro—A participant in the wildcat strike of 1910 at the Eric Coal Companies, he was arrested for beating officer JASPAR OFFENDAFER after the officer had fallen from his horse.

INDEX

D

Q

R

S

T

U

V

W

About the Authors

Robert P. Wolensky is a Professor Emeritus of Sociology and Acting Co-director of the Center for the Small City at the University of Wisconsin-Stevens Point. He also serves as Adjunct Professor of Sociology and History at King's College in Wilkes-Barre, PA. A native of the Wyoming Valley of northeastern Pennsylvania, he has authored or co-authored books on the Tropical Storm Agnes flood of 1972 (1995), the Knox Mine Disaster of 1959 (1999 and 2005), the ladies' garment workers' industry in Eastern Pennsylvania between 1944 and 2000 (2002), and the Avondale Mine Disaster of 1869 (2008). He is also the director of the Northeastern Pennsylvania Oral and Life History Project. He has been appointed as a fellow or visiting professor at the University of Wisconsin-Madison, the University of Wisconsin-Milwaukee, the London School of Economics & Political Science, and Wilkes University. He earned an undergraduate degree from Villanova University and a doctoral degree from Penn State University.

William A. Hastie Sr. was born in West Pittston, PA, on May 28, 1919, to a Welsh-American mother and Scottish-American father, the second of six children. He graduated from West Pittston High School in 1936. He enlisted in the U.S. Army in 1938 and served two tours of duty, which included defense of the outer perimeter of the Panama Canal at Puerto Rico, and the North African, Sicilian, and Italian Campaigns, including at Anzio beachhead, during World War II. After the war, he returned to West Pittston and, from 1952 to 1959, he was employed by the Knox Coal Company as a laborer at the main slope in Exeter and later at Port Griffith. During the Knox Mine Disaster in January 1959, he was an off-shift worker and first responder in rescue efforts, after which he worked to construct the coffer dam as part of the effort to secure the disaster site. From 1983 into the early 1990s, he was employed by the Wyoming Historical & Geological Society (now the Luzerne County Historical Society) in Wilkes-Barre, PA, where he was appointed Acting Curator and held other positions. While his formal education ended in 1936, he has continued with self-education and research throughout his adult life. He maintains a substantial archive of records, pamphlets, articles, newspaper clippings, photographs, and other items relevant to anthracite, ethnic, and regional history. He has been highly sought as a public speaker since 1960 concerning anthracite mining practices and history, Welsh and Welsh-American history and culture, and municipal and regional history. He has also been interviewed extensively on audiotape, film, and video for documentaries, research monographs, school projects, and news reports.